Scientific and Technical Challenges in the Well Drilling Process 2014

Advanced Engineering Problems in Drilling

IOP Conference Series: Earth and Environmental Science
Volume 24

Tomsk, Russia
24 - 27 November 2014

ISBN: 978-1-63439-965-4
ISSN: 1755-1307

Printed from e-media with permission by:

Curran Associates, Inc.
57 Morehouse Lane
Red Hook, NY 12571

Some format issues inherent in the e-media version may also appear in this print version.

Printed by Curran Associates, Inc. (2015)

For permission requests, please contact the Institute of Physics
at the address below.

Institute of Physics
Dirac House, Temple Back
Bristol BS1 6BE UK

Phone: 44 1 17 929 7481
Fax: 44 1 17 920 0979

techtracking@iop.org

Additional copies of this publication are available from:

Curran Associates, Inc.
57 Morehouse Lane
Red Hook, NY 12571 USA
Phone: 845-758-0400
Fax: 845-758-2634
Email: curran@proceedings.com
Web: www.proceedings.com

TABLE OF CONTENTS

PETROLEUM POTENTIAL OF LOWER-JURASSIC DEPOSITS IN NUROLSK MEGADEPRESSION 1
G. Lobova, E. Osipova, V. Isaev, D. Terre

RARE-EARTH ELEMENT DISTRIBUTION PATTERNS IN METASOMATITES OF EASTERN KAZAKHSTAN GOLD-ORE DEPOSITS 7
A. Yu

ECOLOGICAL RISK ANALYSIS AS A KEY FACTOR IN ENVIRONMENTAL SAFETY SYSTEM DEVELOPMENT IN THE ARCTIC REGION OF THE RUSSIAN FEDERATION 12
Y. Bolsunovskaya, L. Bolsunovskaya

CLASSIFICATION OF HYDRAULIC BOREHOLE MINING TECHNOLOGICAL PROCESSES DURING PAY ZONE DEVELOPMENT 17
I. Bondarchuk, I. Shenderova

IMPRESSION BLOCK WITH ORIENTATOR 21
V. Brilin, O. Ulyanova

WELL EXTENSION IN FRACTURED AND BROKEN ROCKS BY WIRE-LINE TOOL (SSK-59) 26
V. Brilin, O. Ulyanova

EXPLORATION DRILLING FOR PRE-MINING GAS DRAINAGE IN COAL MINES 31
E. Shubina, V. Brylin, V. Lukyanov, T. Korotchenko

THE EFFECT OF LIGHTWEIGHT AGENTS ON THE DENSITY OF CEMENT SLURRY APPLIED DURING OIL AND GAS WELL DRILLING 38
A. Bubnov, V. Khorev, I. Boyko

INFLUENCE OF CHEMICAL REAGENTS AND ADDITIVES ON THE RHEOLOGICAL PROPERTIES OF LIGHTWEIGHT CEMENT SLURRY WITH EXFOLIATED VERMICULITE 43
A. Bubnov, I. Boyko, V. Khorev

STRENGTH PROPERTIES OF CEMENT SLURRIES WITH LIGHTWEIGHTS APPLIED IN OIL AND GAS WELLS 49
A. Bubnov, I. Boyko, V. Khorev

METHODOLOGICAL APPROACHES IN ESTIMATING ANOMALOUS GEOCHEMICAL FIELD STRUCTURE 54
R. Gavrilov, M. Rudmin

REGIONAL STRATIGRAPHY AND ITS DEPENDENCY ON TECTONIC MOVEMENTS (CASE STUDY: UPPER CRETACEOUS AND PALEOGENE STAGES IN WESTERN SIBERIA) 60
V. Podobina

LITHO-GEOPHYSICAL STRUCTURE OF PALEOZOIC-MESOZOIC CONTACT ZONES IN NORTH-OSTANINSK OIL FIELD (TOMSK OBLAST) 65
A. Ezhova, V. Merkulov, V. Chekanstev, R. Abramova

STRATIGRAPHY AND CORRELATION OF LOWER-MIDDLE JURASSIC SEDIMENTS IN SE WEST-SIBERIAN PETROLEUM-BEARING PROVINCE 73
A. Ezhova, A. Ostankov, E. Panova, R. Abramova

PELLET IMPACT DRILLING OPERATIONAL PARAMETERS: EXPERIMENTAL RESEARCH 81
A. Kovalyov, S. Ryabchikov, Y. Isaev, F. Aliev, M. Gorbenko, A. Baranova

DESIGNING THE EJECTOR PELLET IMPACT DRILL BIT FOR HARD AND TOUGH ROCK DRILLING 89
A. Kovalyov, S. Ryabchikov, Y. Isaev, F. Aliev, M. Gorbenko, A. Strelnikova

PRODUCTION OF PURE RAW QUARTZ BY ENRICHING QUARTZITES 95
M. Korovkin, L. Ananyeva, V. Ginsar, O. Razva

LITHO-MINERALOGICAL CHARACTERISTICS AND FACIES FORMATION CONDITIONS OF OIL-RESERVOIR ROCKS $J1^1$ AND $J1^2$ IN KAZAN OILFIELD (TOMSK OBLAST) 100
L. Krasnoshchekova, D. Cherdansteva, R. Abramova

RELIABILITY ASSESSMENT OF TUNNELING FLOW CHARTS 106
N. Antropova, V. Krets, V. Luk'yanov, A. Baranova

FUNCTIONAL-ANALYTICAL CAPABILITIES OF GIS TECHNOLOGY IN THE STUDY OF WATER USE RISKS 109
O. Nevidimova, E. Yankovich, K. Yankovich

TEXTURAL AND STRUCTURAL FEATURES, COMPOSITION AND FORMATION CONDITIONS OF ARENACEOUS ROCKS IN PK_1 HORIZON, POKURSK SUITE IN SOUTH-EASTERN PUR-TAZOVSK AREA (YAMALO-NENETS AUTONOMOUS TERRITORY) 116
N. Nedolivko, T. Perevertailo, A. Barkalova, T. Dolgaya

FOREST FIRE SITUATION ANALYSIS OVER FOREST RESERVE LAND IN TOMSK PETROLEUM PROVINCE 122
O. Pasko, A. Baranova

VASYUGAN HORIZON STRUCTURE FEATURES WITHIN JUNCTION ZONE OF UST-TYM DEPRESSION AND PARABEL MEGASWELL (TOMSK OBLAST) 127
T. Perevertailo, N. Nedolivko, T. Dolgaya

METALLOGENIC PROBLEMS OF HYDROTHERMAL GOLD DEPOSIT FORMATION: FACTS AND ARGUMENTS 133
I. Kucherenko, Z. Yuxuan

NOBLE METALS IN ROCKS AND ORES OF MAYSKO–LEBED ORE FIELD (MOUNTAIN SHORIYA) 141
A. Pshenichkin, T. Timkin, Y. Oskina

PROBLEM ANALYSIS OF GEOTECHNICAL WELL DRILLING IN COMPLEX ENVIRONMENT 146
A. Kasenov, M. Biletskiy, B. Ratov, T. Korotchenko

EVALUATION OF PLAIN RIVER CHANNEL DEFORMATION IN THE ABSENCE OF OBSERVATION DATA 152
O. Savichev, M. Reshetko, I. Matveenko, Y. Ivanova

FACIES AND SEDIMENTATION MODEL OF IRON-ORE SEQUENCE IN BAKCHAR DEPOSIT 158
M. Rudmin, A. Mazurov, A. Ruban

LIABILITY OF THE SUPERVISOR UNDER PETROLEUM DRILLING CONTRACT 165
A. Sizov, K. Tretyakov, G. Boyarko, I. Shenderova

MODELING OF TUNNELING-INDUCED GROUND SURFACE MOVEMENT 170
L. Strokova

LITHO-FACIES FACTORS OF LATE JURASSIC PRODUCTIVE SEDIMENTS IN MYLDZHINO GAS-CONDENSATE FIELD (TOMSK OBLAST) 176
T. Ten, E. Panova, R. Abramova

DESIGNING ECONOMIC AND LEGAL MECHANISM OF LAND MANAGEMENT IN OIL AND GAS COMPANIES 182
M. Tsibulnikova, O. Pogharnitskaya, A. Strelnikova

FIRST NATIVE SILICA FINDINGS IN BISMUTH FROM GARNET SKARNS OF RIBNY LOG - 2 GOLD ORE TARGET IN TOPOLNINSK ORE DEPOSIT (GORNY ALTAI) 187
T. Cherkasova, T. Timkin, O. Savinova

TIME CYCLE CALCULATION PROCEDURE FOR THE SPECIAL CREW DURING THE MINING MOBILE MACHINE COMPLEX OPERATION 192
V. Shmurygin, V. Lukyanov, A. Maslovsky

UNDERSTANDING THIN BEDS WITHIN THE BAZHENOV FORMATION USING SEISMIC MODELING 195
A. Bashirov, V. Belozerov, N. Bukhanov, A. Evdokimov

PETROLEUM GENERATION POTENTIAL OF BITUMINOUS MUDSTONES IN TOMSK REGION BAZHEN SUITE (WESTERN SIBERIA) 202
V. Belozerov, V. Baranov, A. Dmitriev

DEVELOPMENT PROSPECTS FOR BAZHENOV FORMATION BITUMINOUS CLAYSTONES IN THE SOUTHEAST OF THE WEST SIBERIAN PLATE (TOMSK REGION) 208
V. Belozerov, V. Baranov, A. Dmitriev

OPTIMIZATION OF GAS CONDENSATE FIELD A DEVELOPMENT ON THE BASIS OF "RESERVOIR – GATHERING FACILITIES SYSTEM" INTEGRATED MODEL 215
E. Demidova, O. Maksyutina

REGULARITIES OF CHANGES IN FLUID COMPOSITION AND PROPERTIES IN VANKOR FIELD POOLS: FROM LIGHT TO HEAVY OIL 229
I. Goncharov, N. Oblasov

MDBT ESTIMATION RATIO FOR TRANSFORMATION ORGANIC MATTER RATIO IN BAZHENOV FORMATION OF WESTERN SIBERIA (TOMSK OBLAST, RUSSIA) 239
I. Goncharov, V. Samoilenko, N. Oblasov, S. Fadeeva

SUBMARINE FAN RESERVOIR ARCHITECTURE AND HETEROGENEITY INFLUENCE ON HARD-TO-RECOVER RESERVES.ACHIMOV FM. 251
A. Kondratyev, V. Rukavishnikov, L. Shakirzyanov, K. Maksyutin

EXISTING APPROACHES TO TIGHT ROCK LABORATORY PETROPHYSICS: A CRITICAL REVIEW 264
D. Konoshonkin, S. Parnachev
OIL EXTRACTION OPTIMIZATION CASE: HORIZONTAL PERMEABILITY HETEROGENEITY EVALUATION AND APPLICATION 275
M. Korovin, V. Merkulov, S. Gojkovic
PERMEABILITY INHOMOGENEITY ACCOUNTING IN TERMS OF EFFICIENT DEVELOPMENT STRATEGY 283
M. Korovin, V. Merkulov, S. Gojkovic
APPLICATION OF MISCIBLE DISPLACEMENT FOR FIELD MTX LOW PERMEABILITY FORMATIONS 291
V. Kuntsevich, S. Slivkin, V. Belozerov
GEOLOGICAL REASONS CAUSING RAPID RATE DECLINE. A CASE STUDY OF FIELD T, RUSSIA. 306
S. Kuzmin, V. Rukavishnikov, V. Sukovatyi
LITHOLOGY EVALUATION OF A BAZHENOV FORMATION RESERVOIR USING SEISMIC AND WELL LOG ANALYSIS 320
N. Lomakina, V. Belozerov, N. Bukhanov
INTEGRATED APPROACH TO GAS ACCUMULATION IDENTIFICATION IN FIELD M 330
K. Malyshevskaya, V. Rukavishnikov, B. Belozerov, A. Podnebesnikh
DEPOSITIONAL FEATURES OF THE MIDDLE JURASSIC FORMATION OF FIELD N AND THEIR INFLUENCE ON OPTIMAL DRILLING SCHEDULE 343
D. Mishina, V. Rukavishnikov, B. Belozerov, A. Bochkov
BAZHEN FM MATURED RESERVOIR EVALUATION (WEST SIBERIA, RUSSIA) 357
S. Parnachev, A. Skripkin, V. Baranov, S. Zakharov
Author Index

The Conference "Advanced Engineering Problems in Drilling" was devoted to the 60th anniversary of the Drilling Department, Institute of Natural Resources. Today this Department is the "descendant" of two existing departments - Mining Exploration Technology and Oil and Gas Drilling. It should be mentioned that this remarkable date is associated with the first graduation class of mining engineers in "Mining Exploration Technologies", as well as the 30th anniversary of the Oil and Gas Well Drilling Department.

Anniversary is an excellent occasion to remember one's historical past. At the beginning of the last century within the Tomsk Technological Institute n.a. Emperor Nikolai II the Mining Department was established which soon embraced the Obruchev-Usov Mining-Geological School. This School became the parent of mining-geological education in the Asian region of Russia, as well as the successor of mining-geological science. It was and is today one of the leading schools in the spheres of mineral resources exploration, surveying and mining.

1927 is the year of the establishment of the Department of Technology in Mineral Exploration. SibGeokom (Western-Siberia branch of the Geological Committee) under the supervision of M.A. Usov obtained the first Krelis rotary boring drill. Prior to that only the Keystone cable drilling rig was used in exploration. It was I.A. Molchanov who was responsible for the development and implementation of new technology in the field of exploration. In the yard of SibGeokom (now it is Building № 6, Usov St.) the first drilling rig was mounted. This was the beginning of the first training courses for Krelis drilling foremen under the supervision of I.A. Molchanov. In 1931 I.A. Molchanov headed the Department of Exploration which was located in Building № 6. In the outside territory of this building a drilling site was launched, including Keystone cable drilling rig, CAM-500 drilling rig and others. In the Building itself, i.e. in one study room (now № 107), the floor was lowered to 2m and a drilling rig was mounted where students could obtain practical skills in drilling.

The Department of Exploration became the foundation of the future department of Mining Exploration Technologies. However, the Department of Exploration, headed by A.A. Belitshky from 1944, furthered its work in the research sphere of drilling (including such leading specialists as P.F. Palyanov, V.I. Molchanov, I.S. Mitushkin, V.M. Matrosov, V.P. Krendelev) and in 1949 a new speciality was introduced "Technologies in Mineral Exploration."

In 1952 the graduate of Moscow Geological Institute (now Geological Institute, Russian Academy of Science) PhD. S.S. Sulakshin began working in this Department, and in 1954 headed the newly established Department of Technologies in Mineral Exploration and was its continuous Head for more than 32 years and during the last 60 years has been a close associate of this department. Due to his brilliant supervision, this Department flourished and during the last 20 years has been one of the top departments in Russian affliated departments.

In 1962 within the framework of this Department a new speciality was introduced- Oil and Gas Well Drilling. 125 full-time and 50 part-time students were enrolled in the two above-mentioned specialities. As a result, there was a necessity to open a new independent department which was in 1984 (October 1).The Department of Oil and Gas Well Drilling was located in Building № 8 and then in Building № 15. The Department staff included graduates of the Department of Technologies in Mineral Exploration and was headed by Yu. L. Boyarko, one of the first graduates of this Department.

Time passed by and life made its own adjustments which influenced the further existence of these two departments. Due to the decrease of exploration and development drilling scope the student enrollment in the two above-mentioned specialties also decreased many-fold. As a result the

IOP Conf. Series: Earth and Environmental Science **24** (2015) 011001 doi:10.1088/1755-1315/24/1/011001

two departments - Technologies in Mineral Exploration and Technologies in Mineral Exploration were merged into one department. In 2003 the newly merged Department of Drilling was established within the Institute of Petroleum Engineering, now the Institute of Natural Resources and is located in Building № 6 where "it began its life." During these 60 years more than 3 thousand specialists have graduated the Department of Drilling, many whom are highly-qualified and dedicated professionals.

There is no doubt that this Conference involved comprehensive advanced engineering problems in drilling and issues on relevant personnel training. It is extremely important to understand how the 60-year progress and contribution in the field of drilling has left its trace in the history of this Department; and, that, now, it is necessary to move further and seek new and new horizons in drilling.

IOP Conf. Series: Earth and Environmental Science **24** (2015) 011002 doi:10.1088/1755-1315/24/1/011002

Peer review statement

All papers published in this volume of *IOP Conference Series: Earth and Environmental Science* have been peer reviewed through processes administered by the proceedings Editors. Reviews were conducted by expert referees to the professional and scientific standards expected of a proceedings journal published by IOP Publishing.

IOP Conf. Series: Earth and Environmental Science **24** (2015) 012001 doi:10.1088/1755-1315/24/1/012001

Petroleum potential of Lower-Jurassic deposits in Nurolsk megadepression

Lobova G[1], Osipova E[2], Isaev V[3] and Terre D[4]

[1 2 3] Department of Geophysics, Institute of Natural Resources, National Research Tomsk Polytechnic University
[4] Department of Foreign Languages, Institute of Natural Resources, National Research Tomsk Polytechnic University

E-mail: [2]osipovaen@list.ru

Abstract. Based on reservoir temperature measurement data from twenty-three reference well cross-sections and paleotemperature modeling, the thermal history of Lower-Jurassic Togur source rock within Nurolsk megadepression and its framing structures have been reconstructed. Plotted maps showed the density distribution of initial accumulated oil resources for Lower J_{15}-J_{16} formations. Based on performed zoning of reservoirs the areas for priority hydrocarbon prospecting were proposed.

1. Introduction

The developed deposits in the Nurolsk-Koltogorsk area, mainly referring to Upper-Jurassic oil-gas complex (OGC) within which the anticline structures, have been completely explored. Therefore, prospecting and exploration of hydrocarbon deposits in complicated traps being confined to Lower-Jurassic oil-gas complex is of top- priority objective in geological survey. Based on geo-temperature regime reconstruction of Lower-Jurassic Togur source rock, the intensive petroleum charge systems were plotted, as well as the zoning of studied area according to the density of possible resources for Lower J_{15}-J_{16} formations was performed.

2. Potential petroleum characteristics

Nurolsk megadepression is a negative I order structure. According to the reflection horizon F_2 of the basement top, its contour curve is -3240 m. and its amplitude is 400 m. within an area of 11800 km^2. The main hydrocarbon source for Cretaceous and Upper-Jurassic oil-gas complex embraces disseminated organic matter (DOM) of Bazhenov sapropelite -type suite *(J_3vv)* [1].

The major oil generating formation was defined as Togur source rock suite (J_1t_1) for Middle-, Lower-Jurassic and Paleozoic oil and gas bearing complexes. Its location in Nurolsk megadepression is restricted to lowered relief zones, bay-like pinching out towards positive framing structures and crystalline basement protrusions (figure 1a). Generation potential is determined by high content of humus- sapropelite organic matter (up to 10%) and catagenetic DOM maturation from Mesocatagenesis$_1^2$-Mesocatagenesis$_2$. Hydrocarbon deposits of Lower-Jurassic oil - gas complex are associated with structural-lithological and fault traps. This OGC embraces oil-gas bearing complex $J_{17\text{-}16}$ (J_1h-p) and J_{15} (J_1t_2-J_2a_1) layers of Urmansk and Salatsk suites, respectively.

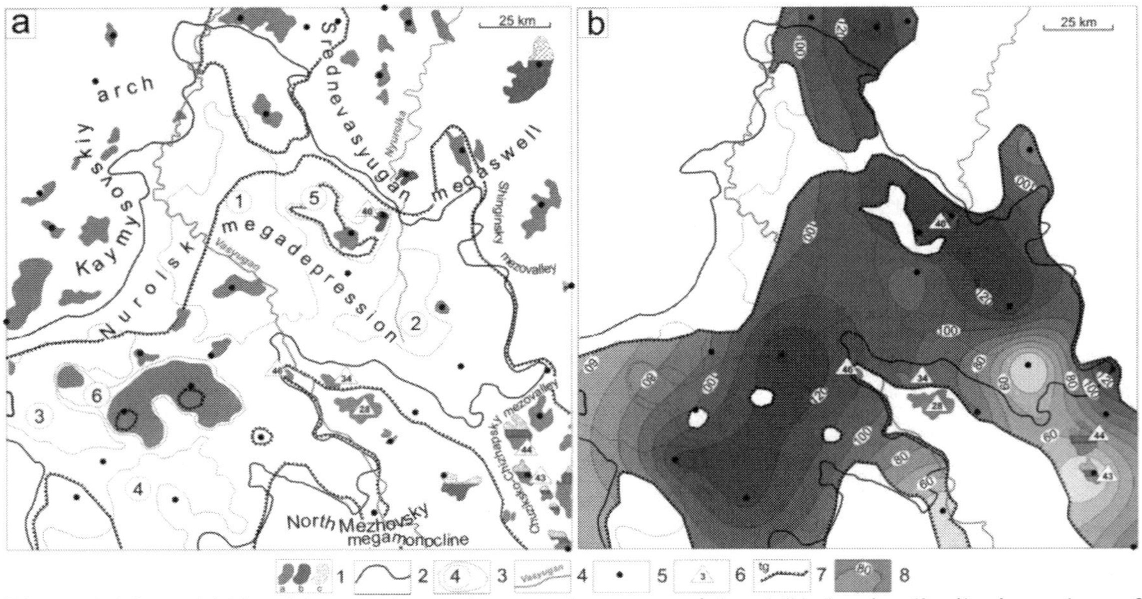

Figure 1. Maps: (a) Nurolsk megadepression petroleum potential and (b) density distribution values of generated Togur oil resources (b). 1- fields: a- oil, b- condensate, c- gas; 2 – Nurolsk megadepression borderlines; 3 – III order structure and its corresponding numbers: 1 – Kulan-Igaysk depression, 2 - Tamradsk depression, 3 – Osevoy syncline, 4- Tamyansky syncline, 5 –Festival bar, 6 – Igolsko-Talovoye dome-shaped uplift, 4 – river system; 5 – well where paleotemperature modeling was performed; 6 – field arbitrary number; 7 – Togur suite borderlines; 8 – density value isolines of generated Togur oil (c.u.). Map (b) shows field deposits in Lower-Jurassic OGC: 28 – Juzhno-Mayskoye; 34 – Mayskoye; 40 – Severno-Festivalnoye; 43 – Archinskoye ; 44 – Urmanskoye; 46 – Sredne-Mayskoye

Late Pliensbach – Early-Toarcian J_{16} formation is characterized by more transgressive lateral bedding than that of the basal J_{17} formation. The deposit formation is confined to the channels of paleorivers (figure 2a). The areas for priority exploration [4] are those zones of river flow discharge from the of eroded protrusion basement paleo-tops formed from felsic intrusive rocks. These deposits are associated with commercial oil deposits in Mayskoye field and gas-condensate deposits in Severno-Festivalnoye field.

Sand J_{15} formation has a more significant lateral bedding (figure 2b). Pinching out towards Pre-Jurassic basement paleo-protrusions, it often interbeds with the overlying layer forming a single deposit ($J_{14\text{-}15}$), similar to that of the one in Srednemayskoye field. Nowadays, three oil and two oil-gas-condensate fields have been discovered, including those deposits in Salatsk suite.

3. Research methods

The reconstruction of the thermal history of the Togur suite deposits was performed by paleotemperature computer modeling [5, 6]. Twenty-three reference wells located within Togur deposits (figure 1b) were selected for paleotemperature modeling. The paleo-temperature calculation technique and schematic map of distribution computational values of mantle basement heat flow density are given in [1].

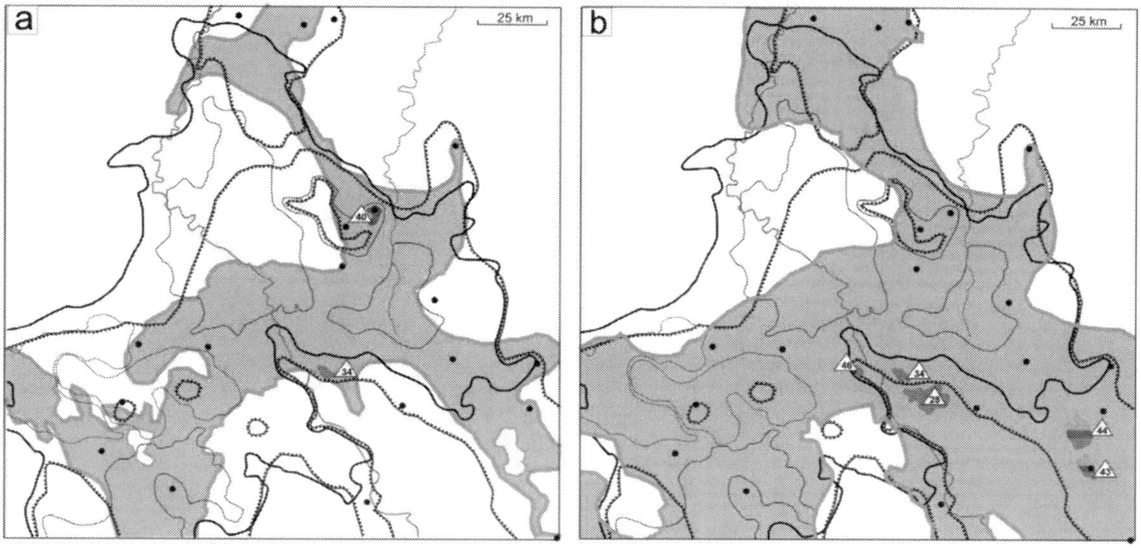

Figure 2. Schematic maps of J_{16} (a) and J_{15} (b) distribution (stroke and fill) of Lower-Jurassic oil-gas complex of Nurolsk megadepression (based on [7]). The figure shows Mayskoye field, Severo-Festivalnoye field with the J_{16} deposits (a); Yuzhno-Mayskoye field, Mayskoye field, Archinskoye field, Urmanskoye field, Srednemayskoye field with J_{15} deposits (b). The remaining symbols are indicated in figure 1.

The calculated paleotemperatures in Togur suite sediments (humus disseminated organic matter) at set geological time intervals and temperature gradation catagenesis zones of intensive generation and hydrocarbon migration [8] extrapolate the fact that the source rock migrates into the major oil kitchen zone where intensive generation of Togur oils (from 95 ^{0}C) begins.
According to calculated paleotemperatures the schematic maps involving 22 key geological time intervals from origin to full maturation of each suite formation were plotted. Figure 3 shows the schematic maps of hydrocarbon charge location for six geological time intervals - from origin to full maturation today.

To estimate the relative density distribution magnitude of generated Togur oil, the conditional integral index *R* was calculated by the formula [9]:

$$R = \sum_{i=1}^{20} (U_i t_i \cdot 10^{-2}),$$

where, U_i – calculated temperature of oil generation, °C; t_i – duration of oil generation, millions of years; number of time intervals i = 1, ..., 20 defined by the number of suites during their formation stages. Schematic map (figure 1b) plotted on the basis of R value interpolation.

The applied approach in estimating hydrocarbon resources cumulatively determines the geotemperature dynamics of source deposit in major oil generation zone [10]. Resource density evaluation is performed using conventional units, which, in its turn, is appropriate for subsequent areal zoning.

4. Results and discussion

The conditions for Togur oils major in oil generation zone occurred 91.6 Ma during the Alb-Senoman stage. Sources of oil generation are located both in the central part of Nurolsk megadepression and on the north-eastern depression flank (figure.3a).

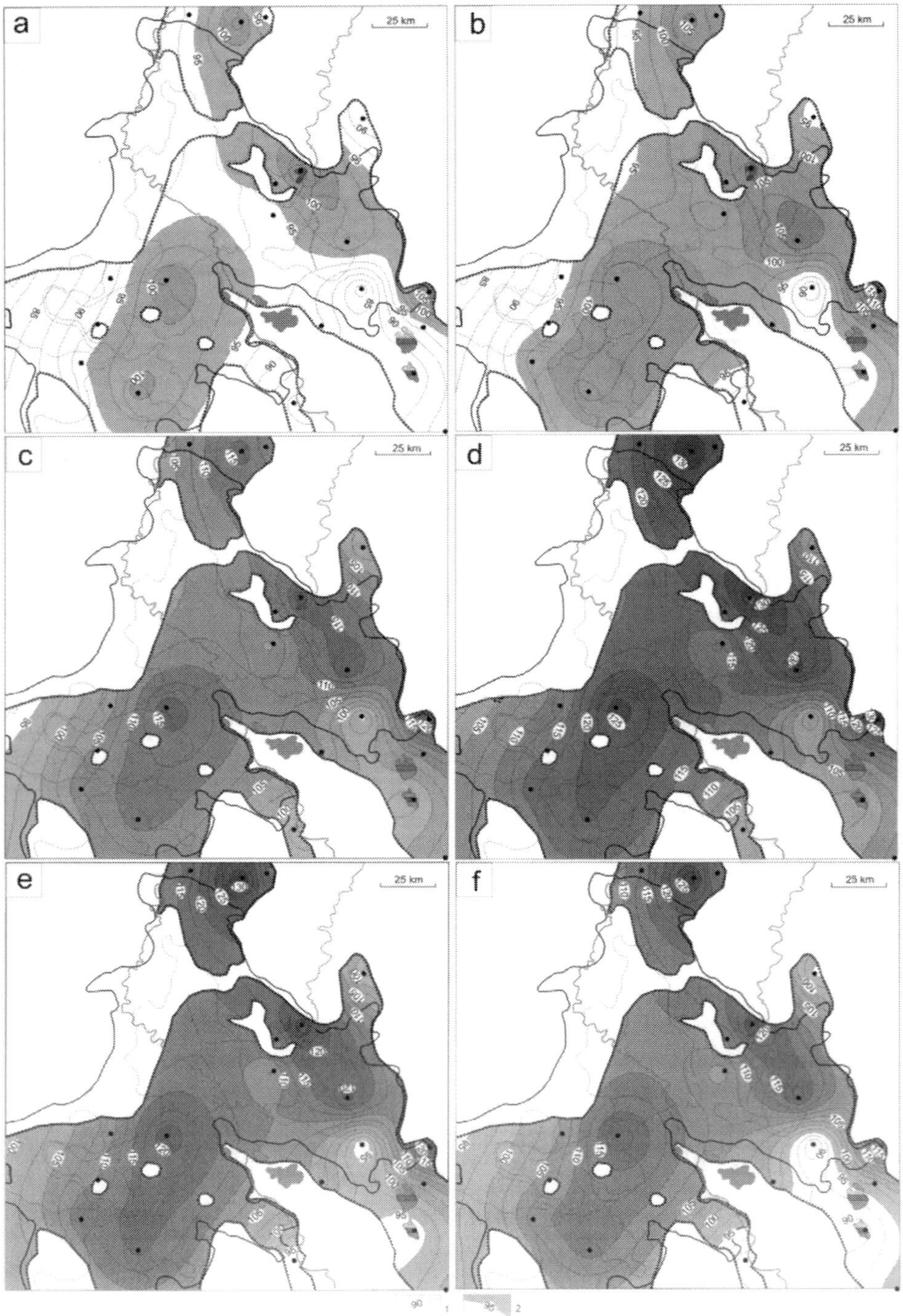

Figure 3. Schematic maps of computed geotemperature distribution and source location of intensive generation of Togur oils: (a) 91.6 Ma ; (b) 86.5 Ma ; (c) 61.7 Ma ; (d) 37.6 Ma ; (e) 24 Ma ; (f) present day section . 1 - isotherm, ^{0}C; 2 - source contour. The map depicts fields with Lower-Jurassic OGC deposits. The remaining symbols are indicated in figure 1.

From 86.5 Ma (formation period of Ipatovsk suite) the oil generation source territory increases embracing the entire central Nurolsk megadepression. Maximum heating of these Togur sediments reaches 115^0 C (figure3b). During the formation of the Gankinsk suite, i.e. 61.7 Ma ago, oil generation expands practically throughout the entire area of Togur sediments, except a small area in south-east of the studied area (figure 3d). These sedimentary formations are heated up to their maximum temperature- 37.6 Ma ago- and the major oil generation zone embraces the entire Togur suite territory. Maximum paleotemperatures are above 130^0 C (figure 3d). At the beginning of the Neogene paleotemperatures decrease, whereas, in the south and south-east, small areas "break out" of the major oil generation zone (figure 3e). The sedimentary formation cooling is governed by climatic changes in the Oligocene period and is still taking place nowadays. Maximum Togur suite temperature decreased to 115-120^0 C. The major oil generation zone in the south-east has also "shrunk" due to the lack of relevant conditions (figure 3f).

Based on the areal distribution of J_{16} and J_{15} layers, layer zoning pattern was plotted in accordance to the resource density of initial accumulated Togur oils (figure 4). J_{16} layer boundaries do not extend outside the contour Togur oil distribution zone (figure 4a). Severo-Festivalnoye field- gas-condensate deposit, and Mayskoe field -oil deposit in one and the same formation are located in the areas with high integral index *R*.

J_{15} areole distribution overlaps the boundaries of Togur deposits (figure 4b). The major oil deposits are on the northern Lavrovsk mezo-flank. Mayskoye field is also located there. Srednemayskoe field is at the interface of two areas of the highest and lowest *R* values, while there are no Togur oil source rocks in this zone. Formation of oil deposit in J_{15} layer within Uzhno-Mayskaya is possible due to the lateral migration of hydrocarbons from the suite source rock distribution zone.

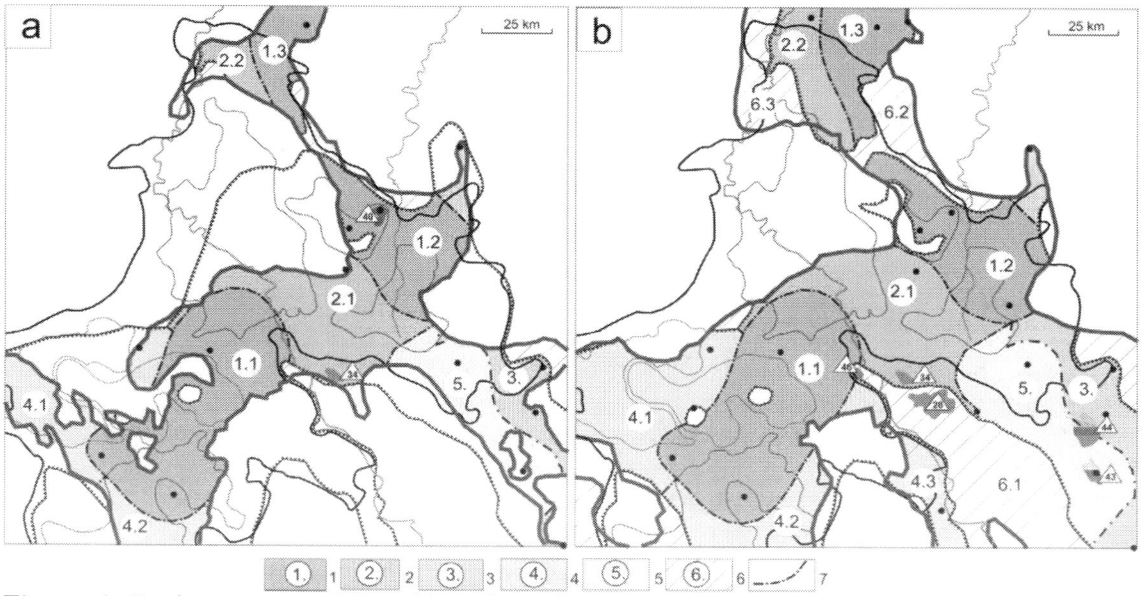

Figure 4. Zoning pattern of J_{16} (a) and J_{15} (b) of Lower-Jurassic oil-gas complex in Nurolsk megadepression according to resource density of initial accumulated Togur oils. 1-6 –areas (rating number; range of resource density values, c.u.): 1. – 110....130, 2. – 90 ......110, 3. – 70...110, 4. – 50...110, 5. – 40...90, 6. – less than 40; 7- area boundaries. The remaining symbols are indicated in figure 1

5. Conclusion

The potentially productive Lower-Jurassic deposit areas have been identified in central Nurolsk megadepression, on the northern Tamradsk depression flank and in the eastern North Nurolsk megadepression barrier. Coincidence of zones of maximum computational values of initial accumulated Togur oil resource density for J_{16} and J_{15} layers considerably increase the significance of these potential productive areas for exploration, as well as geological and economical interest. Such areas could be prospective targets in hydrocarbon prospecting in Lower-Jurassic oil-gas complex.

References

[1] Osipova E N, Prakoyo F S and Kudryashova L K 2014 Petroleum potential of the Neocomian deposit of Nyurolsky megadepression. XVIII International Scientific Symposium in Honour of Academician M. A. Usov: Problems of Geology and Subsurface Development 7–11 April, Tomsk, Russia: IOP Conference Series: Earth and Environmental Science. URL: http://iopscience.iop.org/1755-1315/21/1

[2] Isaev V I and Fomin A N 2006 Areas of generation of Bazhenov and Togur types of oil in southern part of Nyurolskoy megadepression *Russian Geology and Geophysics* **47 (6)** 734–745

[3] Fomin A N 2011 *Katagenez organicheskogo veschestva i neftegazonost mezozoyskih i paleozoyskih otlozheniy Zapadno-Sibirskogo megabasseyna* (Novosibirsk: INGG SO RAN) 331p

[4] Gurari F G and Ekhanin A E 1987 Zakonomernosti razmescheniya zalezhey v nizhneyurskikh otlozheniyakh Zapadno-Sibirskoy plity *Geologiya i geofizika* **10** 19-26.

[5] Isaev V I, Gulenok R U, Veselov O V, Byichkov A V and Soloveychik Yu G 2002 Kompyuternaya tehnologiya kompleksnoy otsenki neftegazovogo potentsiala osadochnyih basseynov *Geologiya nefti i gaza* **6** 48-54

[6] Gulenok R Yu, Isaev V I, Kosygin V Yu, Lobova G A and Starostenko V I 2011 Estimation of the Oil-and-Gas Potential of Sedimentary Depression in the Far East and West Siberia Based on Gravimetry and Geothermy Data *Russian Journal of Pacific Geology* **5** (4) 273–287

[7] Lifanov V A, Nassonova N V and Lapina L V 2011 The geological structure of the basal layers of Ju10–11 in the western part of Tomsk region *Geology, geophysics and development of oil and gas fields* **12** 4–11

[8] Burstein L M, Zhidkova L V, Kontorovich A E and Melenevskii V N 1997 Model katageneza organicheskogo veshestva (na primere bazhenovskoy svity) *Geologija i Geofizika* **6** 1070–1078

[9] Lobova G A, Popov S A and Fomin A N 2013 Lokalizatsiya prognoznaykh resursov nefti yursko-melovykh neftegazovykh kompleksov Ust-Tymskoy megavpadiny *Neftianoe Khoziaystvo* **2** 36-40

[10] Tissot B P 2003 Preliminary data on the mechanisms and kinetics of the formation of petroleum in sediments. Computer simulation of a reaction flowsheet *Oil & Gas Science and Technology-Rev. IFP* **58** (2) 183-202

IOP Conf. Series: Earth and Environmental Science **24** (2015) 012002 doi:10.1088/1755-1315/24/1/012002

Rare-earth element distribution patterns in metasomatites of Eastern Kazakhstan gold -ore deposits

Yu. Ananyev
Department of Mineral Exploration, Institute of Natural Resources,
National Research Tomsk Polytechnic University, Tomsk, Russia

E-mail: AnanyevYS@ignd.tpu.ru

Abstract. The article describes the distribution patterns of rare-earth elements in metasomatites of Eastern Kazakhstan gold-ore deposits. The results demonstrated that the formation of all metasomatites is associated in the rebalancing process of rare-earth elements. The investigation established the vertical differentiation of rare-earth elements in beresites accompanying gold-ore mineralization of various structural and morphological types. Three types of distribution of rare-earth elements in wallrock beresites have been identified.

1. Introduction

Conventionally, the distribution characteristics of rare-earth elements have been extensively discussed and are used as geochemical indicators of substance source and geodynamic genesis conditions of magmatic systems. However, the information on the behavior of rare-earth elements within ore-metasomatic formations of gold -ore deposits in black-shale thicknesses is considerably limited.

Most Eastern Kazakhstan gold-ore deposits are concentrated within Eastern Kalbin metallogenic zone of Zaysan fold system. Within this zone, 18 gold-ore deposits were revealed in 2nd-order cross structures. Ore deposits are united into three structural and morphological groups: 1) ore quartz-veined and stockwork types in terrigenous, volcanogenic and carbonate thicknesses and gabbro-plagiogranite thicknesses; 2) mineralized sulphide zones in Carbonaceous terrigenous thicknesses; 3) combined type, including quartz veins, stockworks, gold-pyrite-arsenopyrite deposits in various terrigenous and volcanogenic thicknesses and intrusions. It is assumed that various structure and morphological mineralization types are located in extensive unified complex ore-metosomatic layered horizons [1]. Furthermore, the bottom section of these layered horizons predominantly embraces impregnated mineralization within black-shale thicknesses and intrusion massifs; in the middle section the gold mineralization is primarily associated with vein-disseminated morphological type occurring in black-shale thicknesses and mottled-structured dikes; in the top section the gold mineralization is primarily associated with the veined morphological type. The following metasomatic process types have been detected in the ore fields: albite-amphibolitic, quartz-feldspathic, propylite and listvenite-beresite [2].

The REE distribution located in Carboniferous black-shale rock thicknesses was investigated within the metasomatites of Baladzhal, Kyzyl and Espinsk gold-ore areas. The Baladzhal ore area embraces impregnated, vein-impregnated and vein mineralization types not only in the mafic intrusions but also in the exocontact of terrigenous thicknesses, whereas major gold reserves are concentrated within impregnated ores of beresitized gabbroids. In the Kyzyl ore area, the major gold reserves are concentrated in Bakyrchik deposit that is confined to superimposed near NW fracture systems having formed the Kyzyl thrust [3]. The main deposit ore bodies are comprised of veinlet-impregnated gold-pyrite-arsenopyrite associations within mylonitized and beresitized Carboniferous terrigenous rocks.

The Espe ore area is characterized by the development of predominate quartz-veined deposits within Carboniferous molassoid formation. Such gold-bearing quartz veins are in the cross-cutting position with host rocks and demonstrate variable thickness and accompanying beresitization.

2. Investigation methods

Analytical analysis employed the following methods: mass spectrometry with inductively coupled plasma (conducted in the Chemical Analysis Center, Tomsk) and instrumental neutron activation analysis (conducted in the Nuclear Geochemical Laboratory, Department of Geoecology and Geochemistry, Tomsk Polytechnic University). Graphic representation of the results involved normalization chondrite [4] and unaltered rocks.

3. Results and discussion

All mineragenic types of Baladzhal deposit metasomatic formations are characterized by the predominance of light lanthanides overlying heavy ones (figure 1). Integrally, the distribution trends of rare-earth elements in metasomatites reflect the distribution patterns of elements in replacement gabbro. In this case, the maximal total concentration of rare-earth elements (REE) contains quartz-feldspathic metasomatites (from 221 to 242 g/t) and minimal concentration of beresites (from 88 to 78 g/t). The cerium anomalies in metasomatites are not pronounced, preassuming the involvement of magmatogene fluids during their formation. The regular REE concentration reduction level was determined as from alkali quartz-feldspar to late listvenite-beresites, which, in its turn, indicated a well-defined flow of REE during the hydrothermal process.

The REE distribution was investigated in the top section of the 9th ore body of the Bakyrchik deposit located in Kyzyl ore area. The analytical samples were selected from underground and surface mine openings, as well as core samples from bore holes. The vertical sampling range was 120 meters and ore body a-dipping more than 250 meters. Samples of unaltered hosted silty - sandstones were selected at 8 to 10km from the ore area in blocks uncontaminated during the hydrothermal processes.

The data analysis demonstrated that the REE concentration in beresites varies from 70 to 122 g/t, whereas, the maximum enriched concentration was determined in beresites of the upper horizons. Light lanthanides dominated over heavy and intermediate lanthanides, while intermediate lanthanides dominated over the heavy ones within all horizons. Europium anomaly within all horizons is slightly negative, and its magnitude increases along the ore body uprising angle. This indicates the fact that the relative oxidation of metalliferous fluid is due to its interaction with host rock pore solutions.

IOP Conf. Series: Earth and Environmental Science **24** (2015) 012002 doi:10.1088/1755-1315/24/1/012002

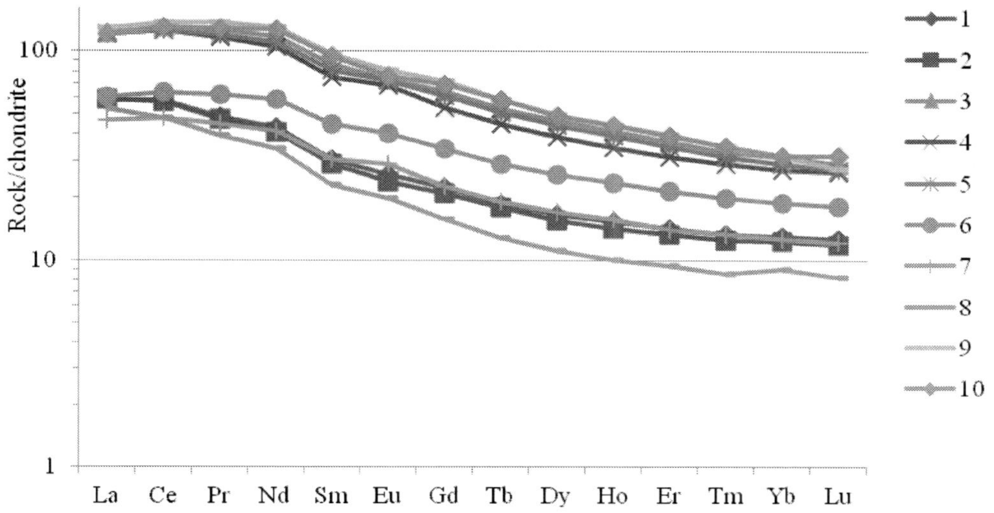

Figure 1. REE distribution in apogabbro metasomatites: 1 and 2 albite-amphibolitic metasomatites; 3 – 5 apogabbro quartz-feldspathic metasomatites; 6 apogabbro propylites; 7 and 8 apogabbro beresites; 9 and 10 unweathered gabbro.

The relation of REE content in beresites to that of unaltered rocks allows identifying the areas of relative element input / loss. For example, the REE normalization to unaltered silty sandstone revealed (figure 2) the relative loss of all REE at horizon +254 meters, partial loss at horizon +288 meters and fixation of all elements at horizon +405 meters. Such a partial REE differentiation in an ore body may be governed by the gravitational factor, while insignificant concentration variations of lanthanides within different horizons indicate a rather significant process uprising, which, in its turn, formed the deposit ore bodies.

The lateral and vertical REE distribution in beresites was investigated within the Espinsk ore area. The investigation of lateral REE distribution patterns in Severvo beresite veins (horizon +60 m) showed that the minimum element concentration was detected in fore-set of the metasomatic horizon, while maximum concentration was found in the back of the horizon, which proves the fact that REE input into beresites during the metasomatic process. Moreover, there is a regular concentration exchange of light, intermediate and heavy lanthanides, i.e. maximum concentrations are identified in the back zone, and the minimum concentrations in the fore-set zone. The ratio of light, intermediate and heavy lanthanides confirms the REE differentiation in the metasomatic horizon: fraction of maximal light lanthanides in the fore-set horizon, while heavy lanthanides - in the back. Such a REE distribution could be the result of only a potassium metasomatism [5].

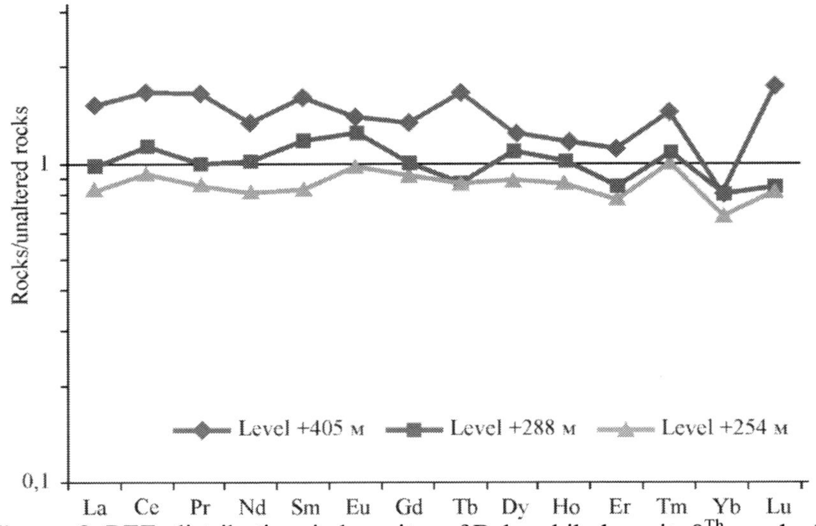

Figure 2. REE distribution in beresites of Bakyrchik deposit- 9^{Th} ore body

The vertical REE distribution was investigated in beresites accompanying the Severo vein in three horizons: +60, +180 and +240 m. The investigation showed that the total REE concentration in beresites increased along the vein uprising. Furthermore, the concentration of light lanthanides in beresites uprises, while the content of intermediate and heavy elements slightly reduces. The ratio of total concentrations of light, intermediate and heavy rare-earth elements in beresites uprises in the veins, notably, while the most contrasting is the ratio of light and heavy lanthanides. The normalization of lanthanide concentrations to unaltered rocks showed that the lanthanides are fixated in all horizons.

The beresites in all studied ore areas have very close distribution parameters, which is probably conditioned by a uniform origin of fluids (figure 3)

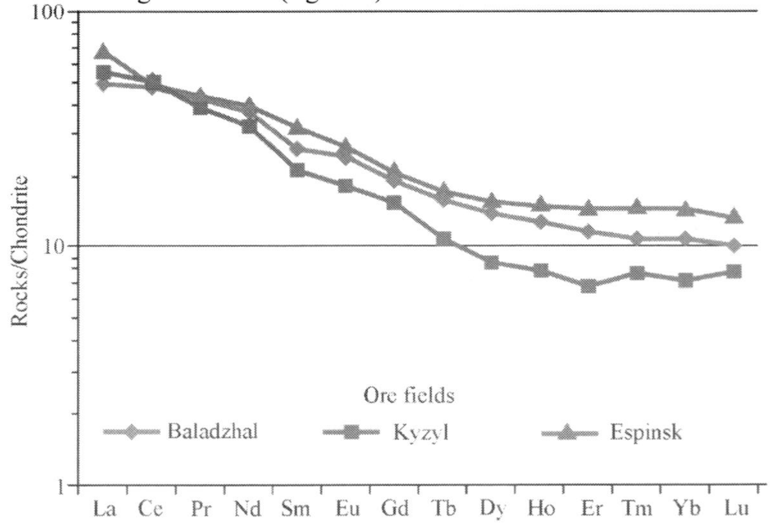

Figure 3. Distribution curves of average REE concentration in beresites of Baladzhal, Bakyrchik and Espinsk ore areas

The distribution pattern of noncoherent elements could be the uniform feature of hydrothermal fluid origin. Thus, the distribution pattern of Y, Zr, La, Sm and Nb in beresites within the studied targets is indicative of their uniform origin. The distribution of K and Rb could be correlated to the main magmatic trend [6], which with [7] could be regarded as indirect evidence of the fluid origin magmagene of formed the beresites with associated gold mineralization.

Conclusions

1. All metasomatite types in Eastern Kazakhstan gold-ore areas were formed as a result of the redistribution of rare-earth element concentrations.
2. All metasomatite types are characterized by a uniform feature of REE distribution, i.e. the predominance of light lanthanides over intermediate and heavy lanthanides.
3. Near-vein lateral zoning is governed by the maximum fraction of light lanthanides in the fore-set of the metasomatic horizon, while heavy lanthanides are vividly observed in the back sections of the metasomatic horizon.
4. There is an axial distribution zoning for rare-earth elements within these deposits, i.e. maximal fraction of light lanthanides within the upper sections of ore bodies, and heavy lanthanides - in the bottom sections.
5. There are three types of lanthanide distribution in the beresites of the Eastern Kazakhstan gold -ore areas: 1) loss of all lanthanides at formation stage of impregnated ores; 2) redistribution in black-shale thicknesses at formation stage of vein-disseminated ores; 3) fixation in black-shale thicknesses at formation stage of vein bodies.
6. Beresites of investigated gold-ore areas have a uniform origin, whereas it is the magmagene source that formed these fluids.

Acknowledgement

The investigations were supported by National Research Tomsk Polytechnic University, project No. 114 - 2014.

References

[1] Korobeynikov A and Maslennikov V 1994 Regularities of formation and location of noble metal deposits in black-shale rock masses of Northeastern Kazakhstan, Tomsk TSU p. 337

[2] Ananyev Yu 2010 *Exploration and protection of mineral resources* **11** 3

[3] Antonov Yu 2010 *Geology and protection of mineral resources* **35** (2) 15

[4] Evensen N, Hamilton P and O'Nions R 1978 Geochim. et Cosmochim. Acta **42** (8) 1199

[5] Balashov Yu 1976 Geochemistry of rare-earth elements Moscow Science p 267

[6] Shaw D 1968 Geochim. et Cosmochim. Acta **32** (6) 573

[7] Bortnikov N, Gamyanin G, Vikenteva O, Prokofev V, Alpatov V and Bakharev *A Geology of ore fields* **49** (2) 99

IOP Conf. Series: Earth and Environmental Science **24** (2015) 012003 doi:10.1088/1755-1315/24/1/012003

Ecological risk analysis as a key factor in environmental safety system development in the Arctic region of the Russian Federation

Y A Bolsunovskaya[1] and L M Bolsunovskaya[2]
[1,2]Tomsk Polytechnic University, Tomsk, Russia

E-mail: [1]ju_al@inbox.ru, [2]bolsunovskl@tpu.ru

Abstract
Due to specific natural and climatic conditions combined with human intervention, the Arctic is regarded as a highly sensitive region to any environmental pressures. Arctic projects require continuous environmental monitoring. This poses for the government of the Russian Federation (RF) a tremendous task concerning the formation and implementation of sustainable nature management policy within the international framework. The current article examines the basic constraints to the effective ecological safety system implementation in the Arctic region of the RF. The ecological risks and their effects which influence the sustainable development of the region were analyzed. The model of complex environmental safety system was proposed.

1. Introduction

The Arctic region is characterized by rather vulnerable environmental conditions. Specific natural and climatic settings combined with intensive industrial activity have a negative impact on the Arctic ecosystems. This can cause significant environmental damage in a global context. The current climatic changes transfer the geopolitical setting of the Arctic. Decrease in sea ice coverage provides new transportation routes and access to resources, which in its turn can cause extra environmental and social threats. Therefore, the Arctic environmental conditions should be regarded as a complex system of interrelated factors. On the one hand, this region is exposed to the extensive global development; on the other hand, the changes in the Arctic region itself cause the response of the whole community.

The Arctic safety system is a key factor of sustainable development of the region. To address this issue effectively, it is essential to advance the cooperation between countries, transnational companies and research institutes. Arctic Monitoring and Assessment Programme (AMAP) [1], Protection of the Arctic Marine Environment (PAME) [2] and the Institution Building for Northern Russian Indigenous Peoples' Project (North Development) are some examples of such cooperative work.

The global character of the Arctic environmental safety system proves the urgent need for the analysis of ecological risks, exposure and effects, their relationship with each other, as well as the necessity for the revision of the environmental law.

2. Materials and methods

The current study involves the following stages: identification of the basic constraints to the effective ecological safety system implementation, analysis of the Arctic ecological risks, and elaboration of the complex environmental safety system for the Arctic region of the RF. The study is based on a large bulk of scientific articles, nominative documents, environmental government programmes and reports presented at international non-governmental conferences.

2.1. Basic constraints to the effective ecological safety system implementation in the Arctic region of the RF

The commercial development of the Arctic is restrained by a number of specific features of the region itself: difficult operating environment and climate, substandard infrastructure and high operating costs [3]. In due course, these regional features would have even greater effect on the hydrocarbon extraction, as well as the Arctic environmental conditions. Therefore, they should be considered while taking any operating and political decisions.

Every country develops natural resources in accordance with internal legal regulations, ecological monitoring and relevant international agreements. Despite the special status of the region and government intervention, the Arctic projects are more highly dependent on the economic viability than environmental concerns. For example, the present legal framework that governs petroleum resource production in many countries stimulates petroleum companies to monitor the compliance with the safety and environmental regulations by themselves, with economic viability being one of the basic motivating factors. It means that it is rather profitable for petroleum companies to invest money in safety and security arrangements as environmental costs may significantly exceed the amount of money spent on risk minimization.

Unlike other regions, the Arctic is characterized by a rather sensitive environment, which in its turn requires the RF and other Arctic states to comply with certain environmental standards, as well as to apply eco-friendly production technologies. Like other countries, Russia does not possess a complex of state-of-the-art eco-friendly production technologies that guarantee environmental preservation of the Arctic region. It highlights the urgent need to facilitate international cooperation in environmental safety system development.

Political risks and political background in general are also considered as a constraint to the effective ecological safety system development [4, 5]. The current political processes interfere with the reinforcement of international cooperation. A ban on the supply of state-of-the-art equipment and technology required for the development of Arctic Shelf to Russia, which is likely to cause significant environmental effects, is a vivid example of such processes. Thus, the absence of the required production technologies and the environmental negligence would definitely lead to negative effects within the global scale.

2.2. Ecological risk analysis in Russian Arctic

Risk analysis is one of the key issues to be addressed in elaborating effective Arctic development policy including environmental safety system.

Environmental risks relating to the Arctic region of the RF can be divided into three groups: natural-ecological, techno-ecological and socio-ecological (table 1).

Table 1. Environmental risk classification within the Arctic region of the RF

	Risk type	Description	Effects
Natural-ecological	Methane emission	Methane emissions from underwater wells significantly speed up the process of global warming, which in its turn leads to the release of new amount of methane. Methane emission volume can be compared with that of global ocean [6].	Sharp increase in atmospheric methane concentration, which leads to unpredictable climatic changes.
	Contamination by resistant organic matter	The process of global warming and permafrost degradation may lead to the release of toxic substances from chemical and nuclear disposal sites [7, 8].	Sharp increase in air pollution concentration, population health risk.
	Increase in ground ice thaw	Thermokarst development, ground distortion, thermal erosion.	Infrastructure deformation, reduction of building

	depth		operation term, high flood risk.
Techno-ecological	Accumulated ecological damage	The effects of anthropogenic activities within the sites of mining, heavy and processing industries, military-industrial complex.	Constant environmental threats in the Arctic region.
	Constant technogenic burden	• During geophysical survey: hydraulic impact, noise exposure. • During drilling: discharge of brine water, toxic liquid and solid wastes into the sea, emergency oil spills, atmospheric emissions, seismic hazard factor increase. • During hydrocarbon transportation: emergency or intentional discharge of petroleum products (transportation mode - tanker); excavation and trenching in pipeline construction activities, post trenching and backfilling, rock dumping (marine pipeline engineering).	Permanent adverse impact on the Arctic ecosystems, extinction of living organisms, fresh water contamination, human losses.
Socio-ecological	Damages to the traditional lifestyle of aboriginal population	Constraints to the on-time adaptation of aboriginal communities to the ongoing changes.	Threat to traditional aboriginal population lifestyle.
	Health impact	Specific climatic conditions.	Polar stress syndrome.
		Integrated or isolated impact of natural-climatic and techno-ecological risks.	Increase in the frequency of reproductive losses.

The above-mentioned risks were outlined and classified into three main groups with due regard to the influencing factors, as well as natural and climatic conditions of the region. The global character of climate change in the Arctic region primarily manifested in the decrease in extent and thickness of Arctic sea ice, as well as anthropogenic impact, cause a number of interrelated effects which could influence different spheres of regional development. These effects should be analyzed both separately and in an integrated manner.

3. Results and discussion

The analysis has revealed that the Arctic region is subjected to various large-scale ecological risks which can have both long-term and short-term effects. Besides, combined with other impacts, these effects could significantly increase the threat level and impede necessary response actions. Thus, the Arctic environmental safety system is comprised of a complex of factors which should be considered not only in various arctic projects implementation but also for sustainable development of the Arctic region itself.

As the RF has extensive territorial claims in the Arctic region and rather high population density in comparison with other Arctic countries, the environmental safety system is of great importance. The RF places emphasis on Arctic environment protection and minimization of the adverse effects of oil and gas activity in the region, especially with regard to continuing this work. The diversity and global character of the ecological risks reveal their strong interrelation with economic and social spheres of the Arctic region development, which in its turn proves the urgent need for the development of such an environmental safety system that would consider the specific features of the Arctic region and contribute to its sustainable development within the global scale. This fact is proved by the following initiatives: "Declaration on the protection of the environment and conservation of biological diversity during the exploration and development of mineral resources of the Arctic continental shelf in the Russian Federation" (Rosneft in cooperation with business partners: ExxonMobil, Statoil, Eni);

establishment of the committee for health, safety and environment for exploration works in the Kara sea (Rosneft in cooperation with ExxonMobil); establishment of the Russian Center for Development of the Arctic on Bely island (initiated by authorities of the Yamal-Nenets Autonomous District); Strategic Action Programme for Protection of the Russian Arctic Environment [9] and others. However, despite the above-mentioned initiatives, it is only the beginning of a rather long process. Insufficiency and fragmentation of the research data concerning the effects of ecological risks, as well as analyzed constraints prevent from minimizing adverse impact on the Arctic environment.

In this context, it is required to incorporate environmental issues into all spheres of the Arctic region development programme forming a complex environmental safety system (figure 1). Ecological risk analysis carried out at different levels should be a key factor of such an environmental safety system, which facilitates information exchange at the national and international levels. To increase the efficiency of the proposed system, the following actions should be taken:

- creation of research database relating to potential risks, exposure and effects;
- formulation of environmental standards which would clearly regulate the activities to be performed in the Arctic region;
- active cooperation of the government and companies in the Arctic environment protection;
- facilitation of international cooperation in environmental safety system development including the development of uniform legal regulation criteria for the Arctic environmental protection.

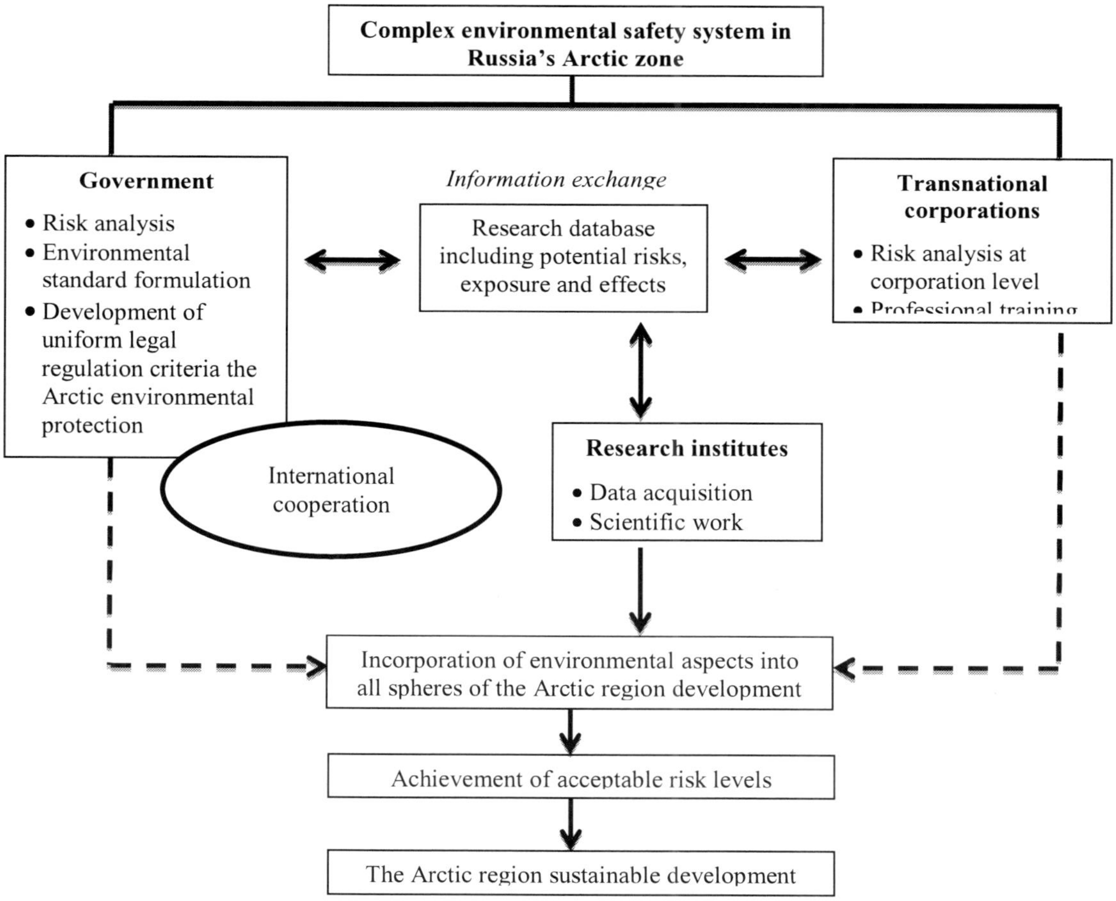

Figure 1. A complex environmental safety system in Russia's Arctic zone

In the context of reinforced international cooperation, it is necessary to consider political instability which may cause the situation to escalate and lead to the tension between economic growth and environmental protection. Therefore, it is of great importance to draw the right lessons from the worldwide experience [10, 11] and find such solutions that would be effective not only in Russian Arctic territory but within the whole Arctic region.

4. Conclusion

Based on the scale of the Arctic environmental problems revealed due to the ecological risk analysis, it is possible to conclude that the contemporary environmental safety policy can hardly guarantee sustainable development of the RF Arctic region. The absence of systemic approach and complex ecological risk analysis impedes the development of strategies to minimize the negative impact on the environment and reduce the dependence of the Arctic projects on their economic viability.

Despite increasing number of environmental protection projects, the scale of the development and production activities that is planned in the Arctic region of the RF also suggests a turn towards ecological issues. To address this issue, the complex environmental safety system which considers the negative effects of the risks not only at the stage of strategy development but also during project implementation has been developed. The risk analysis, being an integral component of the proposed system, is of great importance as it enables to identify the definite hazard sources and predict their potential damage in order to provide the sustainable development the Arctic region within the acceptable risk levels.

References

[1] Arctic Pollution 2011 Arctic Monitoring and Assessment Programme (Oslo) p 50
[2] The Arctic Council Regional Programme of Action for the Protection of the Arctic Marine Environment from Land-based Activities (RPA) 2009 p 15
[3] Johnston P 2012 Arctic Energy Resources: Security and Environmental implications *Journal of Strategic Security* vol 5 No. **3** 13-32
[4] Bolsunovskaya Y A, Boyarko G Yu and Bolsunovskaya L M 2014 Political risks of hydrocarbon deposit development in the Arctic seas of the Russian Federation *IOP Conference Series: Earth and Environmental Science* vol 21 1-6
[5] Sentsov A, Bolsunovskaya Y and Bolsunovskaya L 2014 Effective Planning of the Future of the Arctic *IOP Conference Series: Earth and Environmental Science* vol 21 1-6
[6] Shakhova N, Semiletov I, Salyuk A, Yusupov V, Kosmach D and Gustafsson O 2010 Extensive Methane Venting to the Atmosphere from Sediments of the East Siberian Arctic Shelf *Science* vol 327 1246-1250
[7] Johannessen O M, Volkov V A, Pettersson L H, Maderich V S, Zheleznyak M J, Gao Y, Bobylev L P, Stepanov A V, Neelov I A, Tishkov V P and Nielsen S P 2010 *Radioactivity and Pollution in the Nordic Seas and Arctic region. Observations, Modeling, and Simulations* (UK: Praxis Publishing Ltd.) p 408
[8] Whiteman G, Hope C and Wadhams P 2013 Vast costs of Arctic change *Nature* vol 499 (7459) 401-403
[9] Strategic Action Programme for Protection of the Russian Arctic Environment 2009 Ministry for Economic Development of the Russian Federation on a basis of the materials of the UNEP/GEF Project «Russian Federation – Support to the National Programme of Action for the Protection of the Arctic Marine Environment» p 26
[10] Arctic Resource Development. Risks and Responsible Management 2012 A Joint Report from FNI and DNV p 36
[11] Environmental management in Arctic oil and gas operations: Good practice guide 2013 International Association of Oil and Gas Producers p 120

IOP Conf. Series: Earth and Environmental Science **24** (2015) 012004 doi:10.1088/1755-1315/24/1/012004

Classification of hydraulic borehole mining technological processes during pay zone development

I B Bondarchuk[1] and I V Shenderova[2]
[12]National Research Tomsk Polytechnic University Tomsk, Russia

E-mail: [1] bond_ib@mail.ru, [2]inna-shenderova@yandex.ru

Abstract. Relevance of the work is defined by the need of solid mineral deposits development by hydraulic borehole mining. The main advantage of the method is that the extraction of minerals could be carried out in difficult geological conditions, excluding tunneling of mine workings and quarries construction. The article presents a generalized and systematic classification of hydraulic borehole mining technological processes during pay zones development. According to the classification three groups of technological processes were defined: main, auxiliary and hydraulic borehole cutting head monitoring. The main technological processes are: rocks fracturing, suction and lifting of the slurry to the surface, delivery of the slurry to the slurry pump. Auxiliary processes include: cleaning of intake ports of slurry retrieval device, drilling of pilot hole and maintenance of mining chambers roof sustainability. To hydraulic borehole cutting head monitoring processes refer: operation modes control, tripping operation and rotation.

1. Introduction

The whole complex of hydraulic borehole mining (HBM) could be divided into four technological processes [1]: well drilling (construction); pay zone development (minerals extraction); slurry processing; pre- and post-works – producing of working agents, facilities construction, assembling-disassembling, mine reclamation and etc.

One of the most time-consuming HBM process is cutting of particles from the formation to create slurry of liquid and rock material and delivery of the slurry to the surface. For more deep and systematic view, a classification of HBM technological processes during pay zone development was created (figure1), at that, 3 groups of technological processes are defined: main, auxiliary and hydraulic borehole cutting head monitoring.

2. Materials and Methods

Main technological processes are realized throughout the whole process of minerals extraction. To these processes refer: rock fracturing by a jet of pressurized water (cutting of minerals from the formation), suction and lifting of the slurry to the surface, delivery of the slurry to the slurry pump.

Rock fracturing during HBM process could be realized by the following methods: hydrodynamic (water jet, hydropercussion and depression), mechanic and their combinations [2].

Suction and lifting of the slurry to the surface could be done through: discharging process, displacement process, mechanical process and their combinations. At that, discharging could be created by airlifts, hydraulic elevators (jet pumps) and vortex unit. [2].

Delivery of the slurry to the slurry pump could be realized by means of directed or free flow.

The following could be referred to the directed flow:

- discharging, created by airlifts, jet pumps and vortex device;
- displacement, created by back pressure in extraction chamber and working agents;
- water jet, created through application of pressurized water jet produced by special water jet pumps[3].

Free flow of slurry is provided by specially formed bottom of the cone chamber. The cone form of the chamber could be created due to the angle of natural repose or due to the application of water jets with specially placed jet nozzles. [4].

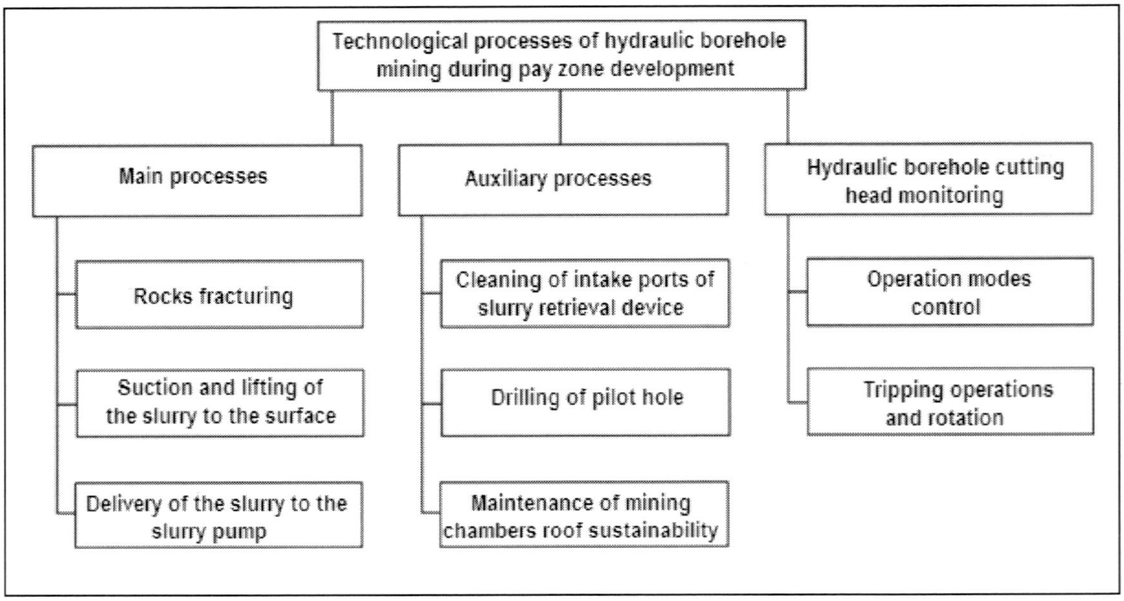

Figure 1. Classification of hydraulic borehole mining technological processes during pay zone development

Auxiliary technological processes are realized in case of necessity during the process of mineral extraction, namely: cleaning of intake ports of slurry retrieval device, drilling of pilot hole and maintenance of mining chambers roof sustainability.

Irregularity of rock fracture, heterogeneity of formation, roof caving and insufficient slurry dilution during HBM lead to constant clogging (blocking) of slurry retrieval device intake ports. The following methods could be applied for cleaning of slurry retrieval device intake ports:

- hydrodynamic, which in its turn is divided to:
  - water jet - realized with the help of special gumming nozzles, forming jets directed to the intake [5];
  - water impulse - created by hydraulic bumps in bottom-hole zone, which in its turn are produced by airlifts, hydraulic elevators, hydrodynamic generators, explosions, airguns, acoustic guns as well as by raising/lowering and rotating of hydraulic cutting head.
- mechanical, realized by shells, blades, pins and other destructive element which are the parts of hydraulic cutting heads [6];
- pneumatic, realized via injection of compressed air into the intake port through special downhole pneumatic units (gas accumulating high pressure caps) [7]
- mixed method.

Drilling of pilot hole could be fulfilled by hydraulic cutting head without replacing of drilling assembly, which allows increasing of working productivity due to quick transformation from the one process to another. This technological process could be realized according to three technological schemes:

- drilling of pilot hole during extraction [6, 8, 9];
- drilling of pilot hole and further extraction [10, 11];
- drilling of pilot hole with casing and further extraction [10, 11].

Mining chambers originate during pay zone development by HBM method. At that, if the roof is made from a weak rock it may collapse. Following methods could be used to provide mining chamber roof sustainability:

- construction of artificial roof (mine working support) [12, 13];
- applying of special systems for pay zone development, namely: trapping [14]; successive working out of mine sectors [15], cut-and-fill mining [15], creating of back pressure in chamber by working agents (water, compressed air) [6, 16]; creating of chamber with the diameter not exceeding the dome of natural collapse [12];
- mixed method.

Technological processes of hydraulic borehole cutting head monitoring are the processes for cutting head operating during minerals extraction, namely: operating mode control, tripping operations and rotation.

Operating mode control of hydraulic borehole cutting heads could provide three modes of operation: combined operation of jet drilling and slurry retrieval units, jet drilling unit autonomous operation, slurry retrieval unit autonomous operation. The following technologies could be used to control the operating mode: valves, open-close channels for supplying of working agents [10]; separate channels to supply working agents for jet drilling and slurry retrieval unit [11], turn on and turn off of working agents supply [9].

Joining of hydraulic borehole cutting heads during HBM process could result in difficulty of tripping operations and rotation. It is explained by the fact that hydraulic borehole cutting heads have special double pipe column which require nonstandard casing connection. This results in inability of rotary drilling. Besides, there is a small distance between the cutting head and the casing string. When the cutting head is rotated from the surface it increases the risk of cutting head locking and unscrewing of casing pipe bottom. The mentioned problems could be solved by:

- reducing of pipe number in the construction of hydraulic borehole cutting heads [10, 17];
- applying of downhole hydraulic engines based on turbodrill or screw motor construction [8], or Segner turbine [17].

3. Conclusion

The presented classification, organize and generalize the information concerning technological processes for pay zone development by HBM method. The classification data will significantly simplify the method of choosing and improving of solid minerals deposits development by hydraulic borehole mining technology.

References

[1] V Z Arens, N I Babichev, A D Bashkatov, O M Gridin, A S Khrulev and G K. Khcheyan 2011*Skvazhinnaya gidrodobycha poleznykh iskopayemykh* (Moscow: Gornaya kniga) p 295
[2] I B Bondarchuk and S Ya Ryabchikov Modern concepts of technological processes in development of productive formations by hydraulic borehole mining *Bulletin of the Tomsk Polytechnic University* Vol **1** 325
[3] Borehole water jet assembly: patent 374453 USSR № 1386206/22-3
[4] Borehole water jet unit: patent 1317131 USSR № 3991025
[5] Borehole water jet unit: patent 1320419 USSR № 4009262
[6] Hydraulic borehole mining unit: patent 1370244 USSR № 4088854; filed 07.07.1986; published 30.01.1988
[7] Downhole pneumatic units: patent 1273565 USSR № 3941320; filed 30.07.1985; published

30.11.1986
[8] Hydraulic borehole mining unit: patent 1265341 USSR № 3884993; filed 18.04.1985; published 23.10.1986
[9] Hydraulic borehole mining unit: patent 1229346 CCCP. № 3672807; filed 14.12.1983; published 07.05.1986
[10] Hydraulic borehole cutting head: patent 2302526 Russian Federation № 2006102464/03
[11] Hydraulic borehole cutting head: patent 2301337 Russian Federation № 2006102408/03
[12] Wellfield construction: patent 1278446 USSR № 3936045
[13] Mine working support construction during hydraulic borehole mining: patent 2101507 Russian Federation № 96120980/03;
[14] Hydraulic borehole mining: patent 611001 USSR № 2180093
[15] Hydraulic borehole mining: patent 2086768 Russian Federation № 95105942/03
[16] Mineral extraction from solid rock formations: patent 2447287 Russian Federation № 2010135859/03;
[17] Hydraulic borehole cutting head: patent 89610 Russian Federation № 2009125785/22

Impression block with orientator

V I Brilin[1] **and O S Ulyanova**[2]

[1]Department of Well Drilling, Institute of Natural Resources, National Research Tomsk Polytechnic University Tomsk, Russia
[2]Department of Foreign Languages, Institute of Natural Resources, National Research Tomsk Polytechnic University Tomsk, Russia

Abstract. Tool review, namely the impression block, applied to check the shape and size of the top of fish as well as to determine the appropriate tool for fishing operation was realized. For multiple application and obtaining of the impress depth of 3 cm and more, the standard volumetric impression blocks with fix rods are used. However, the registered impress of fish is not oriented in space and the rods during fishing are in the extended position. This leads to rods deformation and sinking due to accidental impacts of impression block over the borehole irregularity and finally results in faulty detection of the top end of fishing object in hole. The impression blocks with copy rods and fixed magnetic needle allow estimating the object configuration and fix the position of magnetic needle determining the position of the top end of object in hole. However, the magnetic needle fixation is realized in staged and the rods are in extended position during fishing operations as well as it is in standard design. The most efficient tool is the impression block with copy rods which directs the examined object in the borehole during readings of magnetic needles data from azimuth plate and averaging of readings. This significantly increases the accuracy of fishing toll direction. The rods during fishing are located in the body and extended only when they reach the top of fishing object.

1. Introduction

A wide range of different impression blocks are used to determine dimension, configuration, condition (of drilling bit, cutter, etc) and location of the top end of fishing object in the borehole.

The lead impression blocks or alloy impression blocks (the materials consist of aluminium 90 – 98.5% and antimony 1.5 -2% and others) are applied to get impress of the frontal surface of the objects which are inside the borehole. Good-quality impresses are taken by the impression blocks made of material with high plastic properties of the following content: 45-50% of aluminium powder, 18-27% of ethoxyline resin, 15-20% of castor oil and 10-12% of stearin [1].

To get a frontal impress it is necessary to design reinforcing wires which will be 2-3 mm lower than the height of the impression block or mastic foundation before filling it up with lead or other binding materials over the working surface. After that the impression block is filled with the material which allows getting the impress. The flat impression block should not have borders outstanding the body otherwise it could run out of operation causing more serious accidents. The applied impression blocks of this type are usually of single use and allow getting only shallow non-orientable impress. At that if a broken tool has sharp edges, some parts of the impression blocks could be left in a well.

2. Methods

The standard volumetric impression blocks (SVB) with the diameter from 112 to 370 mm are applied for multiple use and taking the impress of 3 cm in depth and more [1]. SVB particularities lie in the

IOP Conf. Series: Earth and Environmental Science **24** (2015) 012005 doi:10.1088/1755-1315/24/1/012005

possibilities of multiple applications and getting the impress with the help of the extended rods (figure 1). Before running into the well, the impression block is joined to a drill string, at that it is necessary to check the absence of rod displacement.

It is necessary to run the impression block accurately into the well especially after running the casing shoe. Upon reaching the place of impress taking, the load of 20-50 kN is created on a drill string. After reaching the fishing object which is inside the well, the rods begin to run up, then fix and copy the shape of the upper part of fishing object. The disadvantage of the impression block application is that the rods are in extended position during run. This could lead to deformation and loosing of rod due to accidental impacts of impression block over the borehole irregularity, which could result in faulty detection of the top end of a fish object. Besides, the impress is not oriented inside the well.

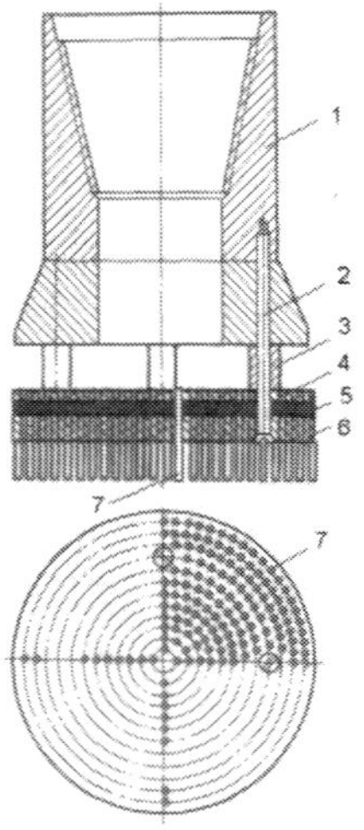

Figure 1. Standard volumetric block

1 – body, 2 –screw, 3 – bushing; 4 – fixed table; 5 –elastic packing; 6 – guide plate; 7 – rods

There is an impression block which has a body with watercourse and a connection thread for the drill string (figure 2) [2]. The lower part of the body has a flange which is connected with a fixed plate having a hollow rod. The plate has an elastic packing and a disk with equally located and matched holes in which copy rods are installed capable to run up. Extended copy rods are retained by elastic packing (with the diameter of holes less than that of the copy rod) before running into a well.

The hollow rod has a magnetic needle, with the holders, floating in liquid, placed in a leakproof annular camera limited by: disk 5, flange 4 and a transparent body side part.

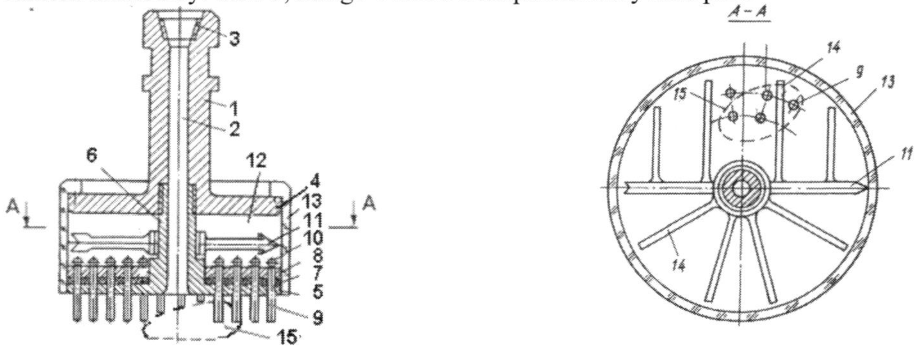

Figure 2. Impression volumetric block

1 –body, 2 – watercourse, 3 – connecting thread, 4 – flange, 5 – plate, 6 – hollow rod, 7 – elastic packing, 8 –

IOP Conf. Series: Earth and Environmental Science **24** (2015) 012005 doi:10.1088/1755-1315/24/1/012005

disk, 9 – copy rods, 10 – ends of copy rods, 11 –magnetic needle, 12 – annular camera, 13 – transparent side part of the body, 14 – holder, 15 – failure object.

During failure investigation the impression blocks with preliminarily extended copy rods run down the borehole and fluid is started to be pumped through a drill string to wash the cutting away cleaning the fishing object. Then the impression block is lowered to the fishing object and copy rods run up copying the object configuration by their ends, at that the upper ends with heads fix the position of magnetic needle. After that, the impression block is lifted from the well and the magnetic needle is matched with the direction of magnetic meridian in order to fix the position and shape of fishing object.

The disadvantages of the equipment lies in inaccuracy of fishing objects azimuthal location assessment, as during running up of copy rod, displacement of the magnetic needle (fixed by the rod) could occur due to the fact that rods fix the magnetic needle not gradually, but in stages according to the rare placement of rods.

Besides, the rods during well run are in the expended position which leads to deformation or loosing of rod in the well.

3. Results

The offered impression block (figure 3) is intended to increase reliability and efficiency of fishing operations by obtaining of volumetric impress and azimuthal location detection of fishing object inside a well, at that protecting copy rods from deformation during well run [3].

The impression block consists of upper 1 and lower 2 parts of a split body which has a spring hollow bushing 3 inside to forming together with pusher-disk 4 a mechanism to run in/out of operation the copy rods 5.

The main parts of copy tool – copy rods 5 installed in the holes of backing disk 6 and run through elastic packing 7 and guide disk 8, which together with packing 7 and base disk 6 is attached to the upper part of body 1 by tightening screws 9 and bushing 10. Disks 4, 6, 8 and packing 7 are made with a central hole through which runs a washing tube 11, connected with pusher disk 4 and rod 3.

The lower part of body 2 has a row of cameras 12 (not less than 3) filled with liquid where cylindrical fitting elements 13 with limb (outer ring) 14 in their upper ends are installed. Upper 15 and lower 16 axes for magnetic needles 17 fixation are installed in the central part of the fitting elements. Relatively to magnetic needles 17, in the lower part of body 2 with an exit to the lower part of cameras 12, holes 18 are located under needles 17 and thus attached by spring connecting rods 19 above pusher disk 4. To the upper end of connecting rods 19, elastic retainers 20 are fixed, which have the central hole intended for lower axis 16 of the magnetic needle. The upper part of cameras 12 has a reamer 21 with plug 22 and is covered with cap 23 made from the transparent material. Control sphere 24 is essential for impression block operating. Hole 25 is intended for liquid release from the tool during its run to the surface after the investigation.

Before running into a well, rods 5 are installed in initial position along the lower border of guide disk 8, at that disk 4 blocks the magnetic needles by retainers 20 through spring connecting rods 19. Cameras 12 and reamer 21 are filled with transparent liquid. The impression block is attached to a drill string and is lowered to the place of impress taking. When the impression block runs into the well, well washing is started to remove the cutting in order to clean the fishing object. Then the control sphere 24 is placed into the tool. It delivers the impression block under its own weight and flow of liquid. Then under the liquid pressure, the rod 3 runs down compressing the spring and moving pusher disk 4 into the lower position. Disk 4 moves rods 5 to lower operating position and simultaneously spreads the magnetic needles during rods 19 moving to the lower position under the strings influence. The tool is left steady (for some period of time) without pressure of liquid in order to still magnetic needles. Then fluid injection is stopped and the system (connecting rod 3 and disk 4) returns to initial position under the string influence. Disk 4 moves connecting rods 19 up fixing the magnetic needle position. After that the impression block covers the fishing object and is loaded with the drill string weight.

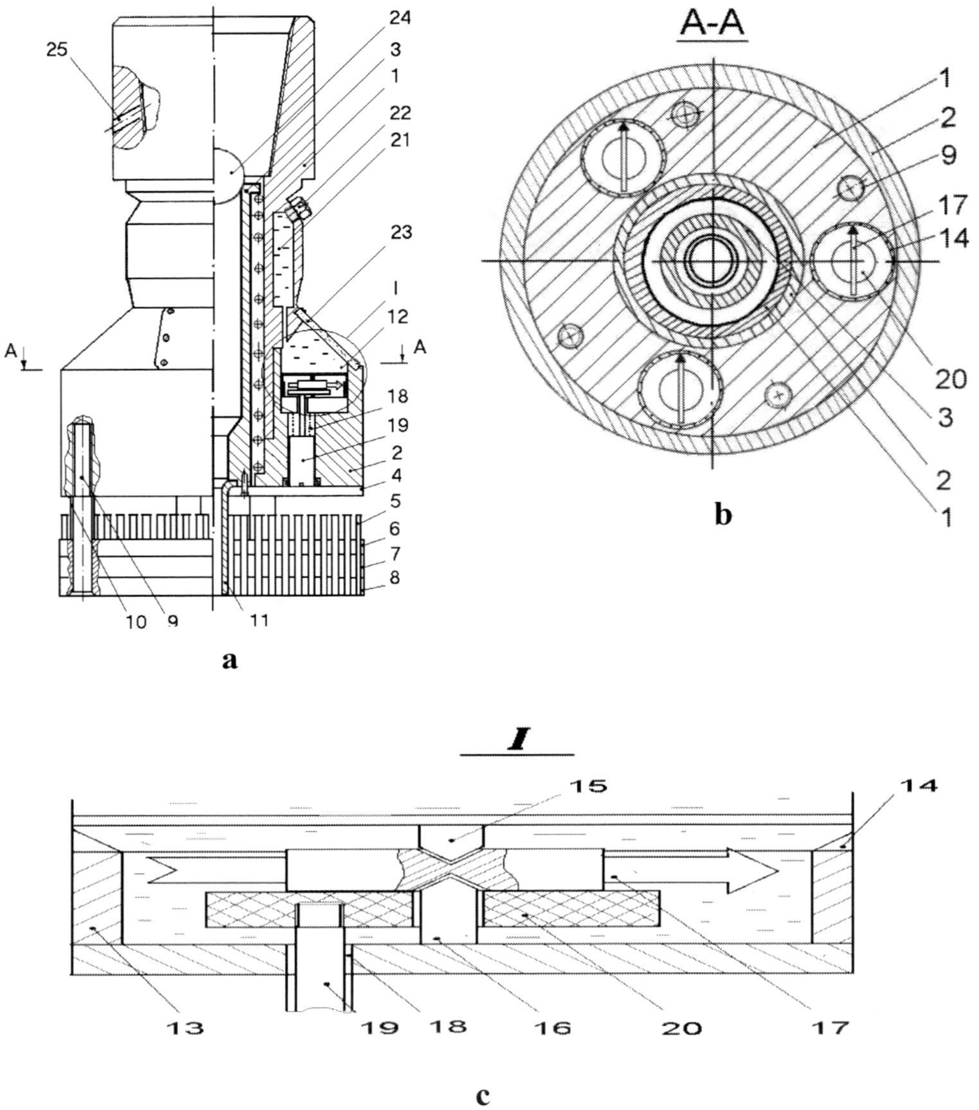

Figure 3. The impression block with orientator

a – layout of the impression block (profile plan), b – the impression block (cross sectional drawing with cameras and magnetic needle), c – camera with magnetic needle (cross-section) 1 – upper part of the split body, 2 – lower part of the split body, 3 – spring hollow bushing, 4– pusher, 5 – copy rods, 6 – holes of backing disk, 7 – elastic packing, 8 – guide disk, 9 – tightening screws, 10 – bushing, 11 – washing tube, 12 – camera, 13 – fitting element, 14 – limb (outer ring), 15– upper semiaxis, 16 – lower semiaxis, 17 – magnetic needle, 18 – hole, 19 – rod, 20 – retainer, 21 – reamer, 22 –plug, 23 – cap, 24 – control sphere, 25 – hole.

Rods 5 during interaction with the fishing object move copying its configuration. When rods 5 run up to the surface, they keep their new configuration due to flexible elastic packing 7 in the holes of which they are fixed. The guiding of the fishing object inside the well is realizes after matching of the magnetic needles with magnetic meridian direction during readings of magnetic needles data from azimuth plate and averaging of readings to increase the guide accuracy. For further tool application it is necessary to press the tool over the horizontal surface for putting rods 5 to the lower part of guide disk 8.

Conclusion

The present impression block allows:

- defining precisely the azimuthal projection of fishing object inside the well;
- avoiding copy rods deformation in the process of running inside the well as well as deformation and displacement of the magnetic needles during their fixation in definite position.

All the above mentioned provides the reliable impression block operation and high efficiency of well investigation.

References

[1] Pustovojtenko I P 1988 *Preduprezhdenie i Likvidacija Avarij v Burenii.* – 3-rd ed. (M.: Nedra) 279
[2] Patent № 1506099, USSR, MPK E 21V 47/09, Impression block/ Inventors Piterskij V M, Goj V L, Panin N M, et al. Applied Sept. 29, 1987; Patented Sept. 07, 1989. Bul.№ 33.
[3] Patent № 33779 Russia, 7MPK E21V 47/09, Impression block / Inventors Brylin V I and Rybakov Y A, Applied June 30, 2003; Patented Nov. 10, 2003. Bul. № 31

IOP Conf. Series: Earth and Environmental Science **24** (2015) 012006 doi:10.1088/1755-1315/24/1/012006

Well extension in fractured and broken rocks by wire-line tool (SSK-59)

V. I. Brilin[1] and O. S. Ulyanova[2]
[1]Department of Well Drilling, Institute of Natural Resources, National Research Tomsk Polytechnic University Tomsk, Russia
[2]Department of Foreign Languages, Institute of Natural Resources, National Research Tomsk Polytechnic University Tomsk, Russia

Abstract. Conditions of increasing the drilling rate per run, which influence wire-line tool application efficiency, are analyzed. Decreasing of drilling rate is usually caused by alarm initiation which indicates core blocking. Producing of longitudinal and lateral vibrations during drilling process and initiation of fluid upward flow inside the inner core pipe, significantly increases the drilling rate. For initiating an upward flow inside the inner core pipe it is suggested to install the bottom hole pump inside the wireline inner core barrel to be applied as a mechanical motor of downhole vibrator which starts to operate only in a case of core blocking. It was found out that the bottom hole pump provided the necessary speed of upward flow (0.3 – 0.5 m/s) at the rotation speed of more than 150 rpm. Field test was carried out to drill siltstone and sandstone formations of 9 -10 drilling category and their alternation with the formation dip angle of more than 70° (rough and disintegrated rocks). The upward flow inside the inner core pipe SSK-59 allowed increasing the drilling rate by 10% as well as to increase core recovery ratio. Besides, it was noted that a large amount of the cuttings (sized 8 – 10 mm) were accumulating inside the inner core pipe during downhole pump application, that in its turn, proved the initiation of upward flow inside the inner core pipe.

1. Introduction

Drilling rate is a key factor defining the efficiency of wire-line tool application. 15-30% of total drilling operation time is spent on auxiliary operations connected with core pipe tripping depending on geological conditions [4].

Besides, during wire-line tool drilling the useful length of the inner core pipe is not fully put in operation. As a rule, a lot of drill operations stop because of core blocking inside the inner core pipe. Depending on drilling conditions the admission coefficient of the inner core pipe (average drilling ratio to useful length of the wireline core barrel) in rough and disintegrated rock formations can decrease up to 30% and less.

At the same time a number of the complete runs (core recovery ratio - 100%) are not more than 25-50%, at that 10-20% of runs have a length of less than 1 meter. These numbers show that the increasing of drilling rate per run is one of the essential factors which prove the efficiency of wire-line tool application. In most cases the reason for decreasing of run length is connected with alarm initiation indicating core blocking inside the inner core pipe. Usually core blocking occurs during fractured and disintegrated rocks drilling due to getting of small rock pieces between the core plug and pipe or at core cleavage under sharp angle to borehole axis and splitting of core.

2. Methods

The following factors influence the increasing of drilling rate per run and length of run: operating parameters (rotation rate, weight on the bit), stability of outer core pipe (cementing while drilling), quality of inner core pipe inner surface (surface condition and hardness), influence of longitudinal and lateral vibrations on inner core pipe during drilling process (or occasionally at the beginning of core blocking), initiation of upward fluid flow inside the inner core pipe, etc.

Relative influence of weight on the bit and centrifugal force, caused by tool rotation, can lead to instability of outer core pipe and its bending inside the borehole [6]. The bending of core pipe while drilling, negatively influence the core keeping conditions. Due to the small distance between outer and inner pipes (δ = 1.7 mm) an inner core pipe touches the walls of a core pipe and bends. This leads to its rotation and vibration. These processes result in core disintegration and blocking.

The investigations have shown [2], that in a range of applied weights (from 600 to 2000 daN) and rotation rate (from 200 to 1200 rpm), the rotary rate influence on the stability of core retrieving barrel is greater than the influence of weight on the bit. Only one core retrieving barrel (among all SSK-59 assembly) with the length of 2.5 m saves the stability during the whole range of researches.

Increasing of core pipe stability is either possible by the application of centralizer with hard facing, installed in such a way so that the length of separate parts of outer core pipe will not exceed the collapsing length or by the application of equipment with higher hardness resource as the outer core pipe [9]. Rigid structure allows applying rational operating parameters. This results in reaching of maximum mechanical drilling speed and increasing of core recovery ratio. Increasing of mechanical speed reduces core keeping period inside the core pipe as well as core disintegration and blocking.

It has been noted [4], that increasing of mechanical speed from 2 to 8 m/h leads to the increasing of inner core pipe filling per run from 31 to 76%.

The possibility of core blocking inside the inner core pipe depends on the quality of its inner surface (surface condition and hardness) i.e. from core moving resistance. The less is friction between the core and the pipe, the less is the possibility of core blocking.

Produced inner core pipes for wire-line tool SSK-59 have inner surface with average roughness height of 0.4-8 um. During exploitation the quality of inner surface worsens: appearing of scratches, increasing of bending; all these reduce drilling rate per run.

Russian Federation Research Institute of Exploration Technique (VITR) [3] has researched the influence of post operating inner core pipe application on the drilling rate. Two inner core pipes – new one and with the run rate of 400 – 500 m were compared. It was proved that the application of new inner core pipe with undisturbed inner surface leads to the increase of drilling rate per run by 1.5 times in comparison with post operating inner core pipe.

Consequently, the increasing of inner core pipe inner surface hardness and its condition, positively influence the drilling rate. For that purpose, it was advised to treat the inner surface of inner core pipe with the steel rod pulled by drawing bench.

After treating the quality of inner surface improved: roughness grade raised to 10 (average roughness height 0.10 – 0.12 um). Due to cold work hardening of inner surface, hardness increased by

10%. Drilling rate per run increased in average from 3.05 to 4.55m with the application of treated pipes during the process of weak and rough formations drilling.

Influence of longitudinal and lateral vibrations on inner core pipe significantly raise the drilling rate. Small core particles which usually block core plugs or large pieces could be considered as friable material. If they are treated by vibration, friction force between core and the inner surface of inner core pipe will decrease. This will provide free passing of friable material through core and increase drilling rate due to eliminating of core blocking. At that, core recovery ratio reduces due to its disintegration during inner core pipe vibration [7].

3. Results and Discussion

To raise the length of the run and to get high core recovery ratio, a core wire-line tool (SSK-59EB) was invented [5]. It allows initiating of upward flow inside the inner core pipe with a help of ejector through lateral vibrations. However vibrations destruct the core and reduce its recovery. At the same time according to the research, at the minimum charge of fluids Q = 15 and 19 l/min, flow resistance between a core barrel and inner core pipe exceeds the permissible rate for ejector operating. As a result, ejector "turns up side down" and the fluid starts to flow into the inner core pipe. This leads to the decrease of core recovery or core blocking.

For creating an upward flow inside the inner core pipe it is suggested to install the bottom hole pump inside the wireline wireline inner core barrel to be applied as a mechanical motor of downhole vibrator which starts to operate only in a case of core blocking [8].

Wire-line tool (fig.1) assembly includes drill string 1 with outer pipe 2 and bit 3. Inside of the outer pipe 2, a wireline inner core barrel 6 with locking coupling and mechanic vibrator in the form of upper and lower coupling 8 and 9 are assembled to stop ring 4.

Upper coupling 8 is fixed to locking coupling 5, the lower coupling 9 is spring-loaded by spring pins 10 and 11 through lower coupling housing 12 interacting with them through bearings 13 and 14. Lower coupling 9 is attached to the wireline inner core barrel 6 through housing 15, at that housing 15 is a housing of the piston pump having valves 16

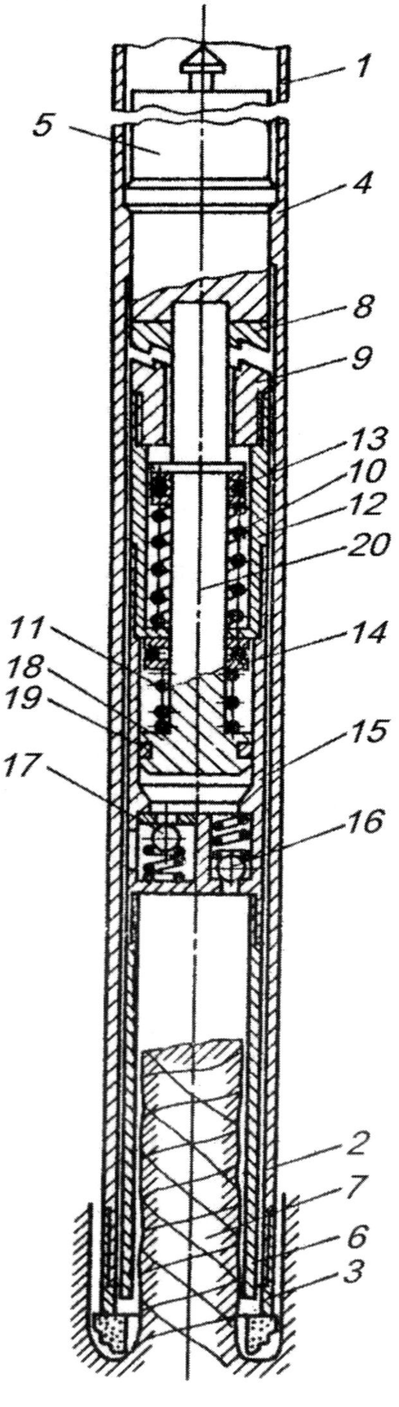

Figure 1. Wire-line tool assembly
1 – drill string; 2 – outer pipe, 3 – bit, 4 – stop ring, 5 – locking coupling, 6 – wireline inner core barrel, 7 –core, 8 – upper coupling, 9 –lower coupling, 10, 11 – spring pins, 12 – lower coupling housing, 13, 14 –bearings, 15 – piston pump housing, 16 – suction valve, 17 – discharge valve; 18 – pump piston, 19 – pump cup, 20 – rod

and 17. Inside the housing, a pump piston 18 with pump cup 19 is installed which is joint to locking coupling by rod 20.

After running of wire-line tool to the bottom hole, the fluid is injected through drill string 1 and casing between outer pipe 2 and wireline inner core barrel to the bottom hole and then through casing to the surface. After that, the drill rod starts to rotate and weight on the bit is created. This results in realizing of core drilling operations. Core 7 gets into inner core pipe 6. During rough and disintegrated rocks drilling, core blocking may occur due to the accumulation of core particles inside the inner core pipe 6. However, bit 3 together with core pipe 2 runs deeper destroying the rock, at that the inner core pipe 6 does not rotate. The locking coupling 5 is fixed inside the core pipe 2 and limit the wireline inner core barrel run to the surface. Spring 10 is compressed and lower coupling 9 contacts with upper coupling 8 rotating with core pipe 2.

At joining of rotating and non-rotating couplings, slipping or jumping may occur. This results in longitudinal vibrations of the wireline inner core barrel joint to mechanical vibrator through coupling 9. Due to these longitudinal vibrations of inner core pipe, the cuttings are shaken inside the wireline inner core barrel partially eliminating the core blocking. Besides, relative motion of piston pump housing 15 and pump piston 18 creates discharging in piston part of the pump (during run down of the pump 15). As a result fluid from inner core pipe 6 flows through suction valve 16 into the underside part of piston pump. When piston pump 15 runs up, the fluid is dispersed through discharge valve 17 from underside part of piston pump to casing. This relative motion of piston pump 15 provides the fluid upward flow inside the inner core pipe 6.

The purpose of laboratory testing was to estimate the basic parameters - pump inlet capability and speed of upward flow inside the inner core pipe at various pump response speed, which is defined by wire-line toolrotation speed at operating vibrator. The results are given in table 1.

Table 1. Results of pump parameters tests considering wire-line toolrotation rate.

Parameters	RPM		
	155	390	680
Fluid flow rate inside core pipe, l/min.	2.3	4.6	8.9
Upward fluid flow speed between inner walls of inner core pipe and core (nominal core diameter - 35 mm), m/sec.	0.34	0.68	1.31

According to table 1, pump with mechanical vibrator has provides the required speed of upward flow (0.3 – 0.5 m/sec.) at wire-line tool rotation rate of $\geq$ 150 rpm.

4. Conclusion

Laboratory tests have proved the efficiency of wire-line tool with piston pump and mechanical vibrator as a motor.

After laboratory tests, the piston pump (PVN-01) was produced for field testing carried out in Ognevskaya party by Ust-Kamenogorskaya geological survey expedition. The purpose of field testing was evaluation of piston pump (PVN-01) performance reliability, detection of weak details, comparative analysis of SSK-59 and PVN-01 operating parameters.

Drilling was conducted in siltstone of 9th drilling category, sandstone (10th category) and their alternations. Formation dip angle was 70°. Rock formations were from fractured (5-15 pieces per core meter) to rough and disintegrated (more than 15-70 pieces per meter).

Drilling was realized out by wire-line tools (ZIF-650M, SSK-59) with drill bits (K-01-2).

A bottom hole vibration pump was installed between inner core pipes and mechanical vibrator of the wireline inner core barrel. Core pipe was joined by special connector with centralizer.

Drilling modes were kept within the limits applied during drill operations without piston pump (PVN-01) operation: weight on the bit 460-1000 daN, rotation rate - 800 rpm, fluid flow rate – 10 l/min, fluid pressure rate 1.0 – 1.6 MPa. Water was used as fluid.

Table 2. Results of comparative analysis

Drilling interval, m	Type of tool	Drilling depth, m	Average length of the run, m	Average core recovery, %
135,5-166 203-207,2	Wire-line tool (SSK-59) with a vibrator	34.7	2.07	84
166-203	Wire-line tool (SSK-59) with pump (PVN-01)	37	2.3	87

According to table 2 the installation of pump (PVN-01) which initiates the upward flow inside the inner core pipe of SSK-59 allows increasing the drilling rate by 10% at relative core recovery rate increasing. Besides, it was noted that a large amount of the cuttings sized up to 8-10 mm had been accumulated in the inner core pipe at PVN-01application during drilling operations. When drilling operations were conducted by wire-line tool SSK-59 without PVN-01pump, the cuttings were not accumulating. That was proved due to creation of upward flow inside the inner core pipe by the bottom hole pump.

References

[1] V S Arbit 1974 *Issledovanija i Razrabotka Metodiki Rascheta Jezhektornyh Burovyh Ustrojstv*: *PhD thesis* (Tomsk) p 204

[2] V V Bol'shakov, N A Litvin and V I Lisjanskij 1982 Burenie s Primeneniem SSK-59 *Razvedka i ohrana nedr* **9** 54 – 57

[3] I M Ginzburg 1978 Vlijanie Kachestva Vnutrennej Poverhnosti Kernopriemnoj Truby na Velichinu Prohodki za Cikl pri Burenii SSK-59 *ONTI VITR* **122** 35 – 37

[4] I M Ginzburg 1978 Vlijanie Parametrov Rezhima Burenija na Velichinu Prohodki za Cikl Snarjadami so Sjemnymi Kernopriemnikami *ONTI VITR* **122** 33 – 36

[5] V P Onishhin, G A Blinov V G Vartykjan and D N Plavskij 1990 *Metody, Tehnologija i Organizacija Burovyh Rabot s Ispol'zovaniem S'emnogo Instrumenta* (Leningrad: Nedra) p 268

[6] N N Miheev 1968 O Dline Dvojnyh Kolonkovyh Trub dlja Almaznogo Burenija *ONTI VITR* **104** 34 – 37

[7] V A Orehov 1974 Tehnika i tehnologija razvedochnogo burenija (Alma-Ata: ONTI KazIMS) **2** 61 – 63

[8] V I Brylin and S V Pikunov Coring tool, patent № 1466306, Russia, MKI E 21 **V25/02** 1996.

[9] *Tehnologija Burenija Kompleksom SSK-59 s Primeneniem Zhestkoj Komponovki ZhK-840-7* (Ust'-Kamenogorsk) 1983, p 4

[10] A I Urzhumov 1986 Mehanicheskij Vibrator dlja SSK *Razvedka i ohrana nedr* **4** 33 – 35

Exploration drilling for pre-mining gas drainage in coal mines

E A Shubina[1], V I Brylin[2], V G Lukyanov[3] and T V Korotchenko[4]
[1]National Research Tomsk Polytechnic University, 30 Lenina Ave., Post index – 634050, Tomsk, Russia;
e-mail: Lena_shubina@mail.ru
[2] National Research Tomsk Polytechnic University, 30 Lenina Ave., Post index – 634050, Tomsk, Russia;
e-mail: brylinvi@mail.ru
[3] National Research Tomsk Polytechnic University, 30 Lenina Ave., Post index – 634050, Tomsk, Russia;
e-mail: ktxng@tpu.ru
[4] National Research Tomsk Polytechnic University, 30 Lenina Ave., Post index – 634050, Tomsk, Russia;
e-mail: tatyana1003@mail.ru

Abstract. High natural gas content in coal seams and low gas drainage efficiency are the basic issues to be addressed in order to ensure coal mining safety. A great number of wells being drilled within various gas drainage techniques significantly increase the costs of coal mining and do not reduce the gas content levels within the coal beds up to the required parameters in a short period of time. The integrated approach toward exploration well spacing applied at the stage of project development could make it possible to consider coal seam data to provide more effective gas drainage not only ahead of mining but also during further gas content reduction and commercial production of methane. The comparative analysis of a closely spaced grid of exploration program compiled in accordance with the recommendations on applying mineral reserves classification and inferred resources of coal and shale coal deposits and currently effective stimulation radius proves the necessity and possibility to consider exploration well data for gas drainage. Pre-mining gas drainage could ensure the safety of mining operations.

1. Introduction

Due to the fast growth of coal mining industry, there is an increase in mining depth which in its turn means gas content increase, as well. This fact significantly influences the safety of mining operations, volumes of coal output and associated gas production, as well as the extent of environmental pollution.

Only 20 % of total coal reserves in Kuzbass can be produced by the most cost-effective method, i.e. open-pit mining. Therefore, the rest coal reserves can only be extracted by underground mining. However, increase in mining depth and gas content inevitably results in coal production costs.

Due to high methane level, such techniques as in-seam gas drainage (shortly before mining) and gas drainage while mining are widely applied in coal mines. Usually, to reduce gas content in a coal mine, it is required to drill from 10 to 150 m and stop all mining operations within the proven territory.

IOP Conf. Series: Earth and Environmental Science **24** (2015) 012007 doi:10.1088/1755-1315/24/1/012007

2. Materials and methods

In practice, a combination of various methods is used to degasify coal seams characterized by high gas content as much as possible before they are mined, and to decrease the amount of emissions during mining.

Techniques that reduce coal seam gas content before mining include: barrier degassing and in-seam drainage, shortly before mining. Pre-mining drainage that is of rare use in Russia is also classified as a technique applied to reduce coal seam gas content.

Due to low efficiency factor of in-seam drainage (shortly before mining – 0.2 – 1.4 %), the basic portion of methane is released during gas drainage while mining, particularly, by means of ventilation systems (10-40%) and powerful fans (60-90%) placed at the coal face.

Therefore, all the actions taken to reduce coal seam gas content result in significant production costs and do not allow increasing coal production rate. Challenges in methane extraction from coal seams are caused by low permeability of reservoir coal seam, with methane being occluded. To increase permeability of reservoir coal seam, a great number of stimulation technologies have been developed.

Pre-mining drainage is carried out by drilling vertical, directional and horizontal wells from the surface. Most wells do not release gas until the permeability of the reservoir coal seams is enhanced by stimulation treatment aimed at creating various fractures for further methane extraction from coal seam before mining operations themselves.

Depending on coal geology of a mine and stimulation technology to be used, gas can be released within a radius of 200-300 m from the vertical well.

Based on the results of hydraulic fracturing implementation in the Karaganda and Donetsk coal basins, the parameters of this technology are defined.

Pre-mining drainage involving reservoir coal seam fracturing is more suitable for use in coal seam confined by impervious rocks of not lower than mean stability [1-5], with gas content being 8 m^3/tonne when reported on a dry, ash-free basis.

The efficiency of drainage wells depends on various factors; therefore, methane production period is more than 3 years.

Based on the research carried out by Inter-industry scientific center VNIMI, it has been revealed that the most effective way to prevent coal mine burst in deep mining is to apply stimulation technologies which are widely used in petroleum industry as they may significantly increase the effective drainage radius.

One of the solutions to control strain-stress characteristics of large coal seams is to apply acoustic simulation of production wells which has been already tested in coal deposits and gas condensate fields of the Russian Federation. This method makes it possible to control strain-stress characteristics of a coal mass by means of the wells drilled directly into the coal seam from the mine workings, or drilled down from the surface. Local acoustic excitation is one of the types of acoustic simulation method.

When applying the method of acoustic excitation, it is possible to influence strain-stress characteristics of the coal seam as the system "well-formation"/well-rock stratum" is considered resonance.

The above-mentioned methods are unique as they allow carrying out pre-mining gas drainage and decreasing high strain energy up to safety parameters over great areas – up to 200 thousand m^2 (R~500 m) when the wells are drilled from the surface; for in-mine boreholes, the effective area is up to 30 thousand m^2 (R~200 m).

In order to control mechanical properties of rock effectively [6], the above-mentioned methods can be applied in the following way: with the amplitude being more than critical compressing strain or elastic strain, elastic waves are excited at a set distance from the well or acoustic waves are excited within elastic strain values.

70 -90 % of total methane volume is extracted by means of modern pre-mining gas drainage techniques based on drilling wells from the surface. There some cases when methane is produced in coal deposit, with gas content being 3-4 m^3/tonne when reported on a dry, ash-free basis [7].

In worldwide mining practice, there is a special term "methane control" which implies that 75% of the total volume of methane is extracted by means of pre-mining gas drainage, 15% - by the drainage shortly before mining, 10% - be ventilation systems during mining itself.

Unfortunately, the implementation of pre-mining gas drainage technique is a rather complicated process as it faces such challenges as high costs of well drilling and extended period of drilling and degassing works (minimum - 3 years).

The number of wells, its depth and spacing depend on many factors and these parameters including the direction of drilling are defined based on the specific features of a given mining site.

Unlike horizontal drilling, vertical well drilling is not so expensive. In most cases, vertical well drilling does not require any extra facilities. An alternative method of pre-mining drainage is to drill directional boreholes, i.e. deliberate inclination of the bore hole from the vertical, and horizontal wells within the coal seams characterized by great thickness or complicated geology when it is too expensive to drill vertical wells. A single horizontal well can replace the need to drill several vertical wells; however, horizontal well drilling is rather cost-intensive. Besides, it horizontal drilling is impossible without extra equipment and trained specialists. Well spacing is basically defined by time, i.e. the less field development period is, the more wells are required [7].

In some cases, a combined well spacing pattern including vertical, horizontal and directional wells is applied. All these types of wells have their own advantages and limitations. Since for exploration purposes vertical wells are basically drilled and only in some cases directional drilling is applied, the main focus of this article is the application of vertical wells in coal gas drainage.

In order to extract as much methane as possible from coal seam and surrounding strata, it is proposed to apply pre-mining gas drainage technique based on drilling exploration wells planned to be drilled at the exploration stage of coal mine development project.

3. Results and discussion

Gas content data obtained at different stages of prospecting and exploration present reasonable interpretations of the geology of the area, which is enough to decide upon whether to continue exploring the area or carry out the pre-mining gas drainage.

Based on the experience gained while exploring the mines within Kuznetsk coal basin, it is possible to predict the thickness of gas weathering zone and further increase in gas content with increasing depth.

On the average, the thickness of gas weathering zone in Kuzbass region varies from 200 m to 100 m, which corresponds to the horizons from +200 to ± 0 m absolute. Today, mining is basically carried out within the horizon ± 0 m absolute and lower.

The dependence of gas content on the horizon is given in table 1.

As indicated in table 1, coal seam gas content within the mines currently developed by coal-producing companies is 10-30 m^3/tonne when reported on a dry, ash-free basis. It increases with increasing depth, which in its turn will result in significant gas drainage costs. A more vivid example of gas content increase with increasing depth is illustrated in figure 1.

Table 1. Gas content data in Kuzbass basin

Horizon, m abs.	Gas content, m^3/t on a dry, ash-free basis	Mining depth, m
+ 200	0-9	100-200 (gas weathering zone)
+100	10-15	200-300
±0	13-22	300-400
-300	22-30	600-900

IOP Conf. Series: Earth and Environmental Science **24** (2015) 012007 doi:10.1088/1755-1315/24/1/012007

The application of exploration drilling for pre-mining drainage will certainly allow coal mine operators to manage the time more effectively which is usually spent on developing mine plan, evaluating mine plan stages, constructing the main facilities within the mine workings and conducting pre-mining gas drainage.

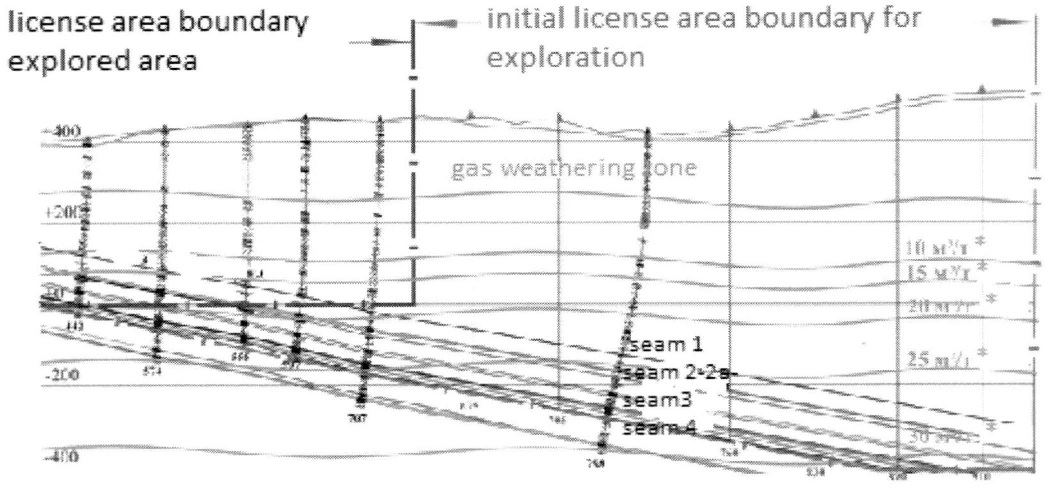

*- when reported on a dry, ash-free basis

Figure1. Coal seam gas content cross-section

In accordance with the Decree of the Government of the Russian Federation № 315 of April 25, 2011, it is required to degasify coal seams, with gas content being more than 13 m^3/tonne when reported on a dry, ash-free basis.

In coal producing countries, to ensure the safety of mining operations this value is decreased up to 9 m^3/tonne when reported on a dry, ash-free basis.

Due to devastating mine explosions in mine workings, the proposal to degasify coal seams, with gas content being 9 m^3/tonne when reported on a dry, ash-free basis is currently under consideration in the Government of the Russian Federation. Therefore, it can be assumed that pre-mining gas drainage of coal seams with gas content being 13-30 m^3/tonne when reported on a dry, ash-free basis will become an obligatory stage. Besides, depending on the relative methane-bearing capacity, mines are divided into categories. Thus, in accordance with the item № 267 "Coal mine safety rules", a mine is classified as "very gassy mine" when relative methane-bearing capacity is 15 m^3/tonne or more when reported on a dry, ash-free basis [8].

The analysis of relative methane-bearing capacity of Kuzbass coal mines (lower than horizon. ± 0 m absolute) has revealed that most mines can be classified as "very gassy mines" as relative methane-bearing capacity is 30-60 m^3/tonne and more when reported on a dry, ash-free basis.

Table 2. Dependence of exploration well spacing on pre-mining gas drainage area

Continuity of seam	Gas drainage area	Well spacing based on reserves category, m					
		A		B		C_1	
		between exploratory lines	between wells	between exploratory lines	between wells	between exploratory lines	between wells
		600–800	200–400	800–1200	400–600	до 2000	до 1000
Continuous seam	Stimulation radius 200 m	45-78%		20-45 %		13-28 %	
	Stimulation radius 300 m	72-100 %		35-75 %			
Relatively continuous seam		300–400	150–250	400–600	200–300	до 1000	до 500
	Stimulation radius 200 m	100 %		75-100%		25-55%	
	Stimulation radius 300 m	100 %		100%			

The above-mentioned data prove the relevance problem in question. i.e. high coal seam gas content. It is also proved by the fact that relative methane-bearing capacity is 2-4 times higher than the value set to such a category of mines as "very gassy mine". Therefore, it is proposed to use exploration wells not only for information gathering but also for pre-mining gas drainage, which should be considered even at the stage of mine planning.

The application of well spacing pattern in accordance with "Recommended practice for application of the classification of reserve and possible resources of mineral deposits" [9] and the stage of mine-site exploration allows further pre-mining gas drainage by means of the exploration wells planned to be drilled at the exploration stage of mine development. The well data should be preserved for the duration of compiling geological record including coal seam methane reserves confirmation.

Table 2 illustrates the analysis results of gas drainage area when stimulation radius is 200 - 300 m and exploration wells are spaced with the above mentioned recommendations.

For example, based on the exploration well spacing pattern requirements, i.e. 200 m between wells and 400 m between exploratory lines, it is possible to degasify 100% of mine site (figure 2) and decrease gas content up to 13 m^3/tonne when reported on a dry, ash-free basis, with the fracturing radius being 200 m.

The proposed well spacing pattern can also be applied to decrease the gas content during mining, which could prevent a sudden emanation of methane from the roof and floor during the pillar recovery process.

Therefore, a closely spaced well pattern applied at the exploration stage will provide more accurate reserves estimates, which in its turn will ensure 100% gas drainage and gas content reduction up to 70-90% [7]. This will ensure the safety of mining operations and reduce the urgency of gas drainage while mining.

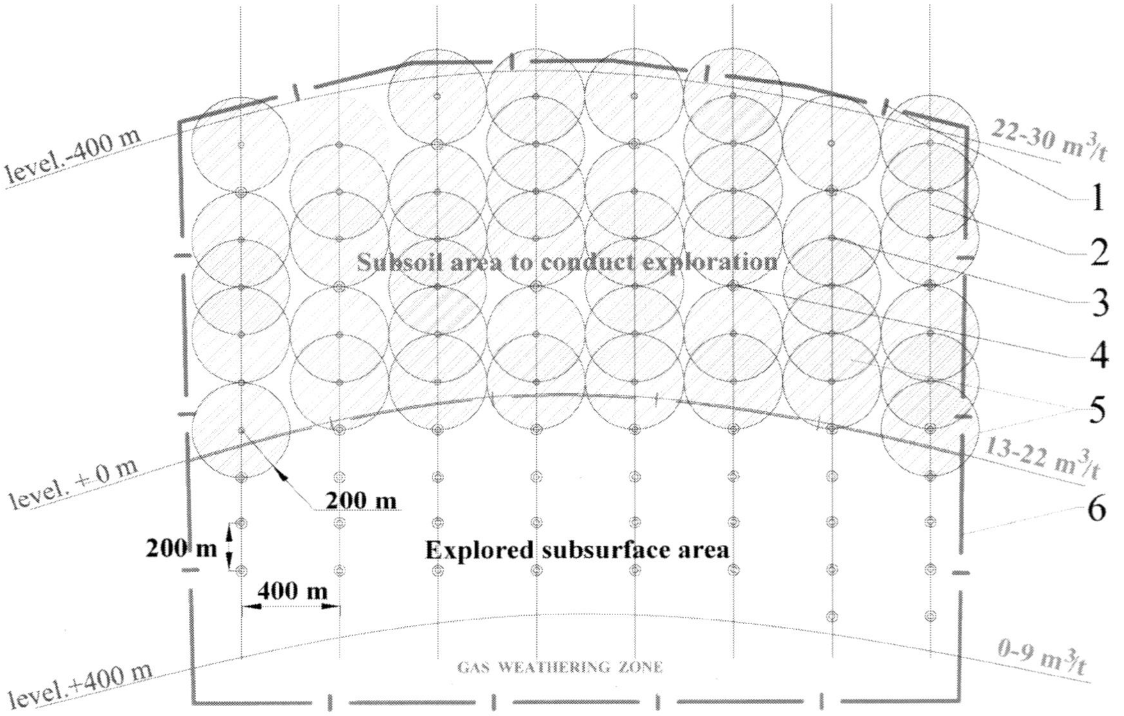

Figure 2. Well spacing pattern: 1 – initial mining license area; 2 – cross-section; 3-planned borehole; 4 – drilled borehole; 5 – stimulation zone; 6 – license area boundary

The optimal well spacing pattern determined at the exploration stage, as well as a properly defined number of exploratory wells which can be further applied for pre-mining gas drainage make it possible to use effectively mine planning period so that gas content is reduced by the commencement of mining operations. Besides, while building coal mining facilities at the mine site, it is possible to degasify the territory carrying out pre-mining gas drainage of mine workings.

The volume of coal seam methane extracted by means of pre-mining drainage makes it possible to use it in various industries. In many drainage projects worldwide, the drained gas is basically used as an energy source for the mine itself or local region. Depending on produced gas quality and volumes, it can be commercially used as well.

4. Conclusion

The application of exploration drilling for pre-mining gas drainage can be effective not only in terms of gas content reduction but also because of the following reasons:

- reduction in drilling costs when it comes to barren spots and gas weathering zone (100-200 m) drilling that accounts for 50-70% of total exploration drilling costs;
- possibility to degasify surrounding strata from which methane also emanates as the overlying strata subside and the underlying strata have;
- possible application of exploration well to degasify the roof of pillar-caving;
- elimination of expensive machine downtime while mining;
- reduction in the amount of methane emissions contributing to greenhouse effect, which has been increased 8 times over the last decade because of higher mining productivity in Kuzbass basin;
- mining industry cost-efficiency increase due to the use of methane for industrial purposes at the mine site.

References

[1] Puchkov L A, Slastunov S V and Fedunec B I 2004 *Perspektivy dobychi metana v Pechorskom ugol'nom bassejne* (Moscow: Izdatel'stvo MGGU) p 557
[2] Puchkov L A, Slastunov S V and Prezent G M 2002 *Mining Informational and Analytical Bulletin* **6** 6-10
[3] Puchkov L A, Slastunov S V and Kolikov K S 2009 Problems of realization of the concept of methane safetyon mines of Russia *Ugol* **1** 28–30
[4] Malyshev Ju N, Hudin Ju L, Vasilchuk M P and et al. 1997 *Problemy razrabotki metanonosnyh plastov v Kuzneckom ugol'nom bassejne* (Moscow: Akademiya gornyh nauk) p 463
[5] Slastunov S V 1996 *Zablagovremennaja degazacija i dobycha metana iz ugol'nyh mestorozhdenij* (Moscow: MGGU) p 441
[6] Vernigo V M and Kul'chickij V B 2006 *Mining Industry* **4** 6–7
[7] Sikora P, Smyslov D and et al. 2008 Osobennosti zablagovremennoj degazacii ugol'nyh plastov metodom burenija skvazhin s poverhnosti *Gljukauf* **1** 38–45
[8] *Coal mine safety rules* (PB 05-618-03) 2003 (Moscow) 296
[9] *Recommended practice for application of the classification of reserve and possible resources of mineral deposits* 2007 (Moscow: FGU GKZ) 34

IOP Conf. Series: Earth and Environmental Science **24** (2015) 012008 doi:10.1088/1755-1315/24/1/012008

The effect of lightweight agents on the density of cement slurry applied during oil and gas well drilling

A S Bubnov[1], V S Khorev[2] and I A Boyko[3]

[1] Director, Well Engineering Laboratory, Petroleum Learning Center, Institute of Natural Resources, National Research Tomsk Polytechnic University
[2] PhD, Engineer, Laboratory "Drilling Mud and Cement Slurry"
[3] Head of Laboratory "Drilling Mud and Cement Slurry"

E-mail: [1]bubnovas@niitek.ru, [2]vhorev@sibmail.com, [3]Boykoia@niitek.ru

Abstract. The article highlights the principles of cement slurry density changing by adding new lightweight agents (exfoliated vermiculite, aluminosilicate hollow microspheres, tripolite).

1. Introduction

At present, a large number of wells are intensively drilled in hard geological conditions. The regular challenges during well construction process are the cement slurry underlift in a borehole annulus, lost of cement circulation in highly permeable horizons and low cement bond.

All these challenges are somehow connected with inappropriate choice of cement slurry compounds not satisfying geological requirements. For instance, one of the solutions to solve the problem of cement slurry underlift is to use lightweight cement. The most common lightweight agents and their basic characteristics are given below:

Vermiculite – micaceous hydrated silicate mineral with bedded structure; large, golden-yellow or brownish platy minerals. Under high temperature (at heating up to 400-1000°C) vermiculite is split to the worm-like particles, exfoliated and expanded in volume in 15-25 times. Added to cement slurry has a tendency of floating up. At high concentrations loses its spreadability, leads to coagulation and blocks cement squeeze.

Tripolite – soft or low cemented, light, fine-pored opaline sedimentary rock. Colored from white or grayish to dark gray, brownish, red and black. Specific gravity varies from 1,2 to 2,5 g/cm^3. Low water absorbing capacity. Added to cement slurry has a tendency of floating up. Subjected to implosion under high pressure influence.

Aluminosilicate hollow microspheres (AHM) – glassceramic aluminosilicate balls originated at coal combustion. Hollow silicate smooth balls with the diameter varying from 10 to in average 100 micrometers. Particles wall density - 2,5g/cm3. AHM has low density, high mechanical resistance, chemical and heat stability, low thermal conductivity. Added to cement slurry has a tendency of floating up and exfoliating.

2. Methodology

Density of cement slurry samples was measured by picnometer (bottle method). The device is shown in figure 1. The method is approved by GOST (State Standard) 26798.1-96 [2-5,]

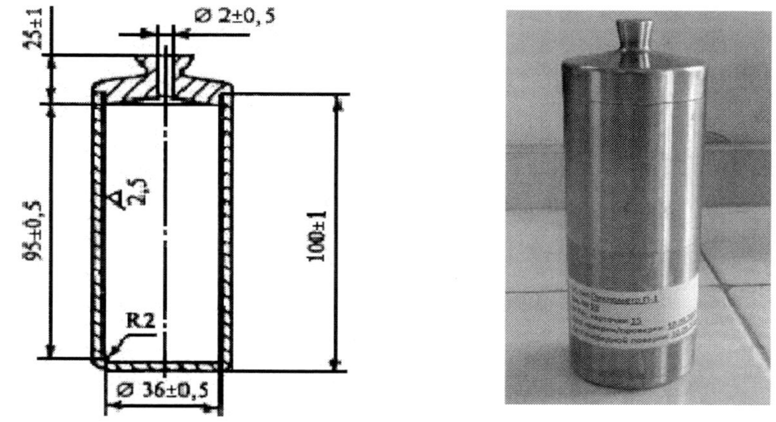

a b

Figure 1. Picnometer
a – Schematic view; b – External view.

Preparation and testing was realized as follows:

- Preparation of experimental cement grout with lightweight agents of definite concentration;
- Picnometer net weight defining;
- Filling in of picnometer with cement slurry. The slurry should fill the channel in picnometer cap;
- Picnometer full weight defining;

Density of cement slurry $\rho_ц$ g/cm^3, was calculated according to formula 1

$$\rho_ц = \frac{m_1 - m_2}{V} \qquad (1)$$

were m_1- picnometer net weight, g;
m_2 - picnometer full weight, g;
V – picnometer volume, cm^3.

The results were rounded to 0,01 g/cm^3. The influence of lightweight agents on the density of cement slurry, at different concentrations, is shown in tables 1-3.

Defining of cement stone stress characteristics was realized in accordance with GOST 26798.1-96. Cement stone bars with different lightweight agents concentration were made for this purpose. Before breaking, the bars were kept in water curing camera during 48 hours at temperature of 75°C. Breaking was realized with the help of special unit "Matest" E160.
Unit external view and the scheme of bar placement are illustrated in figure 2.

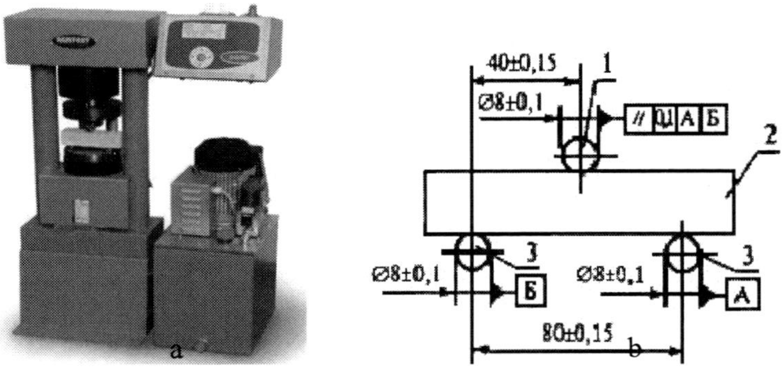

Figure 2. Cement samples stress measuring unit "Matest" E160
a - Stress measuring units "Matest" E160;
b – Scheme of bar placement during breaking
1 – load element; 2 – bar sample; 3 – supporter

3. Results and discussions

Viscosity of slurry was measured by grout flow cone and measuring table. Measuring table – horizontally leveled unit with measuring scale presented in concentric circles having the diameter of min. 70mm and max. 250mm. The volume of slurry was poured into the grout flow cone and raised until the surface of measuring table was covered with slurry. After that, the extreme values of slurry spread boarders were measured. Defining of water loss was realized by pouring the slurry into the measuring cup and keeping it steady for 120 min., than the volume of segregated water was measured. The results of experiments are shown in table 1. [1,3-6]

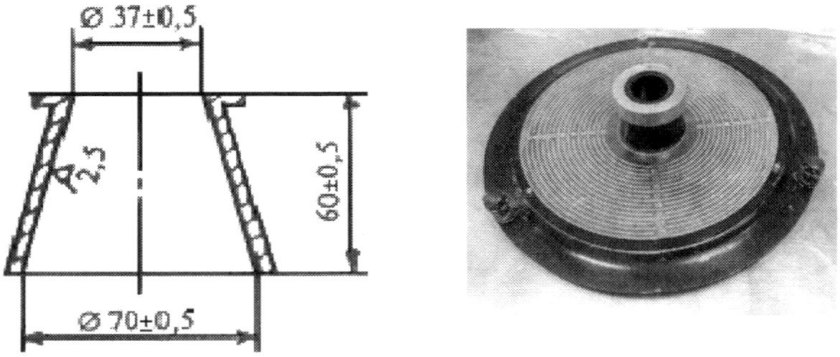

Figure 3. Cone-form

a- Tube filled with cement slurry; b – Grout flow cone external view

Table 1. Lightweight cement density dependence from vermiculite agent.

№	Cement slurry composition	Density g/cm3	Water loss, ml	Viscosity, mm	Cross-bending strength
1	87/13	1,45	0,5	208	2,53
2	89/11	1,49	0,7	205	2,89
3	91/9	1,51	0,7	200	3,07
4	93/7	1,55	0,8	203	3,11
5	95/5	1,50	1	200	3,2
6	85/15	1.40	0,5	180	2,4
7	80/20	1.32	0,5	160	2,0

It is obvious from the table that adding of exfoliated vermiculite positively influences density decrease of prepared cement slurry. Dispersion and bulk density of vermiculite, added during the experiment, also contribute to the change of cement slurry density. Adding of exfoliated mixed-fraction vermiculite results in density decrease, at this, serious density decline is not observed. Sometimes exfoliated coarse-fraction vermiculite particles emerge. This leads to the decline of cement slurry density at the increase of exfoliated vermiculite concentration in the cement slurry.

Table 2. Lightweight cement density dependence from AHM agent.

№	Cement slurry composition	Density g/cm3	Water loss, ml	Viscosity, mm	Cross-bending strength
1	90/10	1,39	0,8	205	2,61
2	92/8	1,42	0,4	207	2,93
3	94/6	1,54	0,5	205	3,19
4	96/4	1,49	0,6	208	3,4
5	98/2	1,52	0,5	205	3,48
6	88/12	1,36	1,4	210	2,2

If compared with exfoliated vermiculite experiments, the application of AHM agent results in density as well as water loss and spreadability values decrease.

Table 3. Lightweight cement density dependence from tripolite agent.

№	Cement slurry composition	Density g/cm3	Water loss, ml	Viscosity, mm	Cross-bending strength
1	86/14	1,43	0,5	215	2,5
2	88/12	1,47	0,5	220	2,61
3	90/10	1,52	0,3	210	2,75
4	92/8	1,52	0	218	3,05
5	94/6	1,53	0	210	3,3
6	88/15	1,40	1	225	2,3
7	97/3	1,56	0	218	3,5

Results of tripolite experiments show good spread ability and water loss characteristics, but stress characteristics are lower than that of vermiculite and AHM.
For clarity, the results of experiments are given in figure 4.

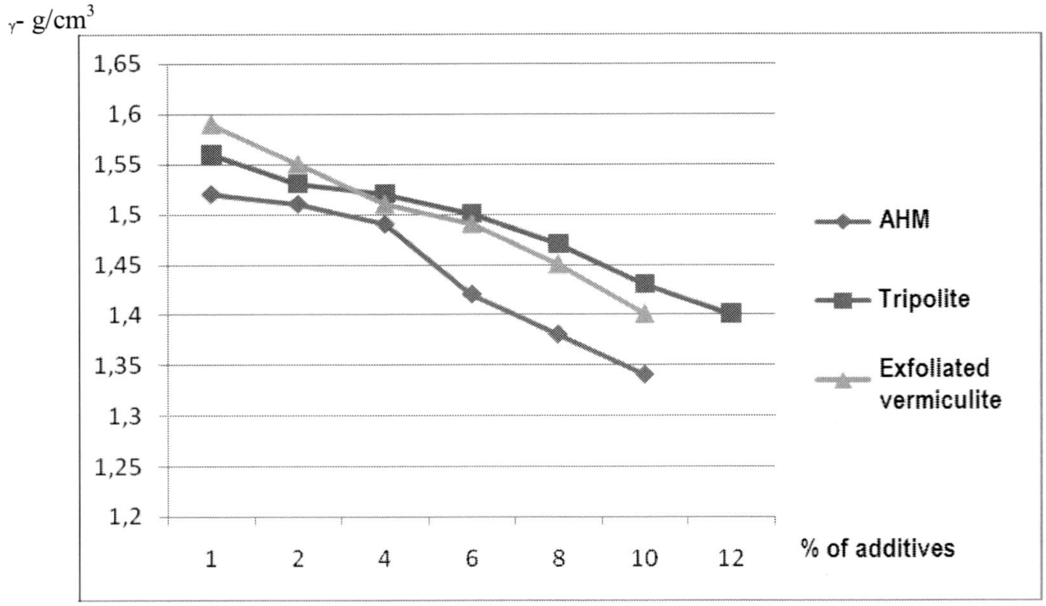

Figure 4. Cement slurry density change from raising of lightweight agents concentration.

4. Conclusion

The results of the experiment allow to make the following conclusions:
- Application of described lightweight agents effectively decreases slurry density from 1.83g/cm^3 (PCT*-I-100 agent free density) to 1.50-1.54 g/cm^3. This allows to mark our cement slurry composition as OTM**-5.
- During application of all the above mentioned lightweight agents it is necessary to use polymer agents to decrease the viscosity as the emerging of interface between cement and lightweight agent is constantly observed.
- For all lightweight agents it is necessary to consider particle-size distribution and bulk density as it greatly influences the density index of cement slurry.
- When concentration of lightweight agents exceeds 10% (for all types mentioned in the article) it leads to the formation of cement grout with low spreadability and stress indexes.

** PCT – oil-well portland cement*
***OTM - Lightweight cement*

References

[1] GOST 12865-67 Exfoliated Vermiculite
[2] GOST 1581-96 Specifications of Oil-well Portland Cement
[3] GOST 26798.1-96 Cement Slurry Testing
[4] Composition and properties of drilling and completion fluids .H.C.H. Darley, George R. Gray
[5] Bubnov A, Boyko A, Horev V and Baryshev I 2014 Research into rheological characteristics and water tapping of lightweight cementing slurry with expanded vermiculite lightweight. *Territory Oil gaz*, **V. 10** p. 18-21
[6] Bubnov A, Boyko I, Horev V and Baryshev I 2014 Research into cement stone strength characteristics and specific weight of vermiculite-based well slurry for oil and gas wells lining. *Territory Oil gaz*, **V. 11** p. 46-49

IOP Conf. Series: Earth and Environmental Science **24** (2015) 012009 doi:10.1088/1755-1315/24/1/012009

Influence of chemical reagents and additives on the rheological properties of lightweight cement slurry with exfoliated vermiculite

A S Bubnov[1], I A Boyko[2] and V S Khorev[3]
[1]Director, Well Engineering Laboratory, Petroleum Learning Center, Institute of Natural Resources, National Research Tomsk Polytechnic University
[2] Head of R&D Laboratory "Drilling Mud and Cement Slurry" Institute of Natural Resources, National Research Tomsk Polytechnic University
[3] PhD, Engineer, Laboratory "Drilling Mud and Cement Slurry"

E-mail: [1]bubnovas@niitek.ru, [2]boykoia@niitek.ru, [3]vhorev@sibmail.com

Abstract. The article is focused on the analysis of such rheological properties and parameters of lightweight cement slurry with exfoliated vermiculite as thickening time, gel strength and water loss.

1. Introduction

The problems caused by inappropriate cement slurry design that does not have required rheological properties occasionally occur during cementing operations. As a rule, inappropriately selected rheological properties of cement slurry result in cement density and strength variation (along hole), poor cement-pipe bond, open and closed pore channels (leading to behind-casing fluid movement), problems associated with cement slurry mixing (high viscosity, cement coagulation) or squeeze cementing (circulation loss, high working pressure, etc.). To avoid all of the above-mentioned challenges, it is required to design cement slurry for desired properties in compliance with nominative documents. Chemical reagents (polymers) or special additives (bentonite, starch, etc.) forming low-viscous structure for cement slurry are basically used to monitor rheological properties. When applying lightweight additives, the influence of polymer additives becomes an important factor which should not be neglected in order to avoid severe consequences.

2. Methodology

The experiments described in this paper are carried out with the application of lightweight additive - exfoliated vermiculite.

Vermiculite is a cleavage-structured mineral of the hydromica group. It is the secondary alteration product (hydrolysis and further weathering) of black phlogopite mica. Vermiculite typically consists of large tabular yellowish-brown crystals. After heat treatment, vermiculite increases 10-15 times in size as in figure 1 [1,3,4].

Figure 1. Appearance of exfoliated vermiculite

When heated, plates transform into worm-like cylinders or golden and/or silver "threads" where extremely fine flakes are observed in the cross-section (exfoliated vermiculite).
Vermiculite parameters are listed in table 1.

Table 1. Vermiculite parameters

Parameters	Vermiculite	Exfoliated vermiculite
Density, g/cm^3	2.4-2.7	0.065-0.130
Vermiculite bulk density after burning, kg/m3	120-200	120-200
Thermal conductivity, W(m,K)	0.05	0.05
Melting point, C	1350	1350
Humidity,%	5	3

Vermiculite is divided into three primary grades based upon grain size:

- Coarse-grained – from 5 to 10 mm;
- Medium-grained – from 0,6 to 5 mm;
- Fine-grained – up to 0,6 mm;

The following vermiculite brands are distinguished:

- Vermiculite 100 – density 100 kg/m3;
- Vermiculite 150 – density 150 kg/m3;
- Vermiculite 200 – density 200 kg/m3;

Only the 0.9 to 0.1 vermiculite fraction was used in the experiments of this study [3].

The spreadability of the cement slurry grout mixed with chemical reagents and additives was measured by means of grout flow cone as in Figure 2 and measuring table. The horizontally leveled measuring table which is equipped with the reference scale basically consists of concentric circles, with minimum diameter being 70 mm and maximum – mot less than 250 mm. After filling the flow cone with grout, the cone was lifted for letting the grout flow out. After allowing the grout to spread over the table surface, the horizontal spread was measured at its widest. Water loss was measured by pouring the grout into the measuring cylinder where it was kept still for 120 minutes. Then, the amount of water accumulated on the grout surface was measured. Experimental results are listed in table 1-4 [1,3].

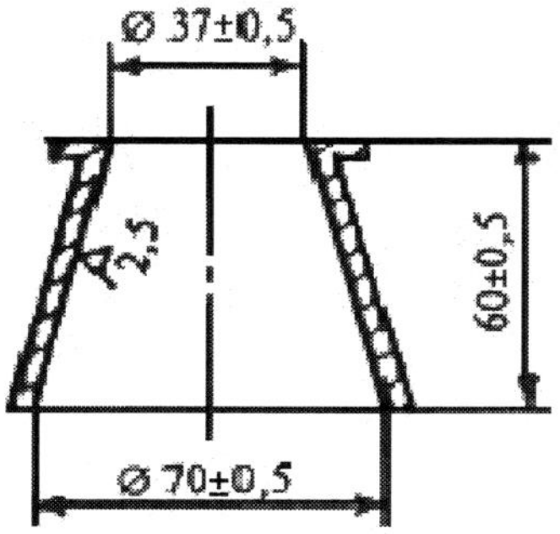

a

b

Figure 2. Grout flow cone
a- grout tube; b- Grout flow cone configuration

Cement slurry thickening time tests were made in atmospheric consistometer Offite 120-80 as given in figure 2. The consistometer stir blade rotated at a speed of 150 r/min ± 5 r/min. Thickening time was evaluated, with temperature ranging from 25°C to 30°C. The test involved mixing the cement slurry and pouring it into the slurry cup up to the inner mark. The test concluded when slurry reached a consistency of 30 Bearden Consistency (BC). The time to reach this consistency value was measured. The temperature which was set in terms of the mean value of moderate well temperatures (75°C) was considered. Test results are listed in table 2-5 [2-6].

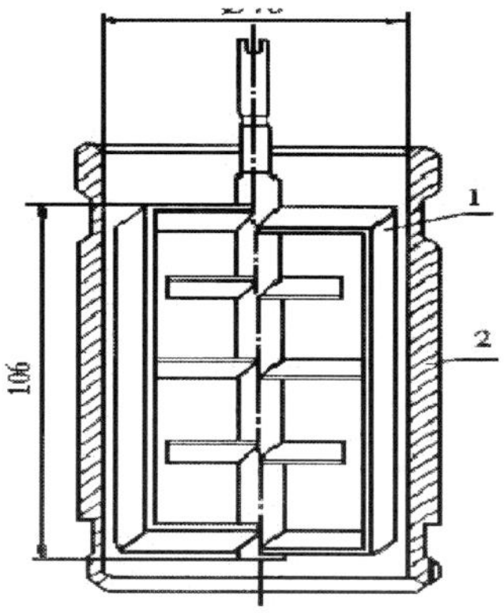

a

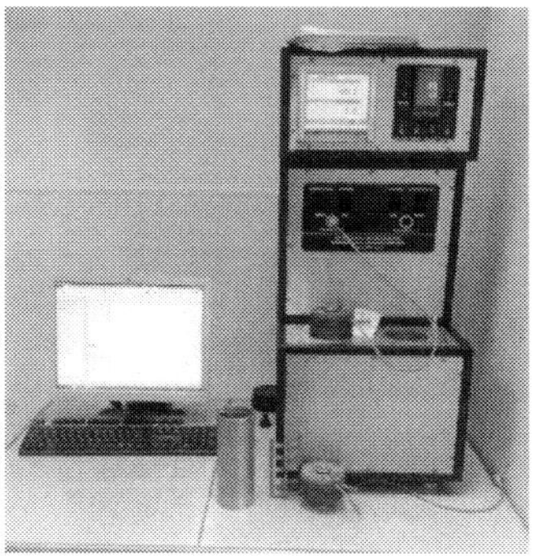

b

Figure 3. Consistometer Offite 120-80 configuration
a- measurement unit 1-blade-type stirrer; 2- cup;
b- configuration.

Gel strength and yield point were measured by viscometer OFFITE 900 given in figure 3. Experimental results are illustrated in table 1-4.

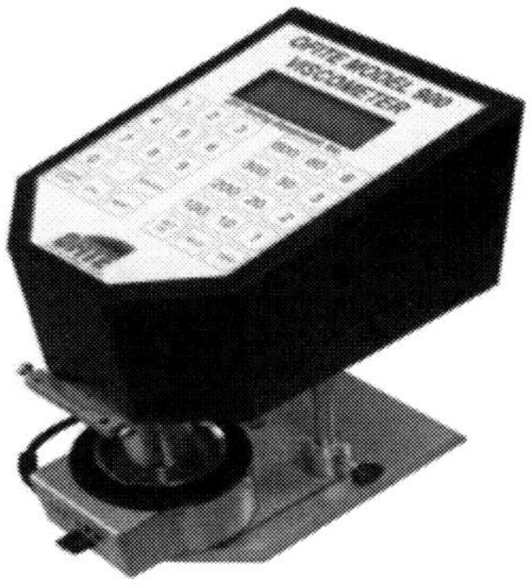

Figure 4. Viscometer OFFITE 900

3. Results and discussions

The concentration of the chemical reagent which was added to the sample was defined before the experiment. GOST (State Standards) and visible binding of vermiculite and cement (i.e. the absence of vermiculite floating effect) were the main criteria to select the concentration values.

Table 2. Influence of chemical reagents on spreadability, mm

Composition	HEC, 0.3%	Bentonite 3%	Servei 1%	Krep, 2%	Starch 0.5%
oil-well portland cement+ vermiculite 95/5	180	245	250	245	more than 250
oil-well portland cement+ vermiculite 92/8	195	250	250	250	more than 250
oil-well portland cement+ vermiculite 93/7	205	more than 250	more than 250	250	more than 250
oil-well portland cement+ vermiculite 95/5	215	more than 250	more than 250	more than 250	more than 250

Table 3. Influence of chemical reagents on water loss, ml

Composition	HEC,0.3%	Bentonite, 3%	Servei, 1%	Krep, 2%	Starch 0.5%
oil-well portland cement+ vermiculite 95/5	0	5	3	4.5	2
oil-well portland cement+ vermiculite 92/8	0	5.5	4	5.5	2.5
oil-well portland	0	6	4	5.0	2.8

cement+ vermiculite 93/7					
oil-well portland cement+ vermiculite 95/5	0	7	5	7	3.5

Table 4. Influence of chemical reagents on gel strength 10sec/10min

Composition	HEC, 0.3%	Bentonite 3%	Servei 1%	Krep, 2%	Starch,
oil-well portland cement+ vermiculite 95/5	17.5/26.5	10/13.9	4.5/328.8	12/15.9	7.0/19.3
oil-well portland cement+ vermiculite 92/8	9/18.5	9.3/12.9	4.8/328.8	11.3/13.5	6.8/16.8
oil-well portland cement+ vermiculite 93/7	8/16.8	9.1/11.7	5.1/328.8	11.1/14.0	6.6/12.2
oil-well portland cement+ vermiculite 95/5	5/10.3	8.5/10.8	5.5/328.85	10.5/13.0	6.3/10.6

Table 5. Influence of chemical reagents on thickening time, hr.

Composition	HEC, 0.3%	Bentonite 3%	Servei 1%	Krep, 2%	Starch 0.5%
oil-well portland cement+ vermiculite 95/5	3:20	2:45	0:08	2:35	0:59
oil-well portland cement+ vermiculite 92/8	3:15	2:40	0:06	2:30	0:45
oil-well portland cement+ vermiculite 93/7	3:15	2:40	0:06	2:25	0:41
oil-well portland cement+ vermiculite 95/5	3:05	2:32	0:05	2:20	0:39

4. Conclusion

Based on the results, the following conclusions can be drawn:

Application of hydroxyethylcellulose (HEC) as a chemical reagent to monitor rheological properties and water loss significantly decreases water loss value, as well as cement slurry spreadability.

Such chemical additives as Servei and Starch considerably decrease the thickening time. This can be explained by the fact that these reagents are considered to be fluid-loss additives which fix free fluid and, by doing so, reduce cement slurry thickening time.

Introducing such chemical reagents as Bentonite and Krep into cement slurry composition is not effective enough to maintain low water loss values.

References

[1] Smolich A and Burlov V 2012 Khimicheskaya stoykost materialov v sredakh neftekhimii i neftepererabotki. Vol **1-2** 390 p.
[2] GOST 12865-67 Exfoliated vermiculite
[3] GOST 1581-96 Specifications of Portland Cements
[4] GOST 26798, 1-96 Experimental Methods for Slurries
[5] Bubnov A, Boyko A, Horev V and Baryshev I 2014 Research into rheological characteristics and water tapping of lightweight cementing slurry with expanded vermiculite lightweight. *Territory Oil gaz* **V 10** p. 18-21
[6] Bubnov A, Boyko I, Horev V and Baryshev I 2014 Research into cement stone strength characteristics and specific weight of vermiculite-based well slurry for oil and gas wells lining. *Territory Oil gaz* **V 11** p. 46-49

Strength properties of cement slurries with lightweights applied in oil and gas wells

A S Bubnov[1], I A Boyko[2] and V S Khorev[3]

[1] Director, Well Engineering Laboratory, Petroleum Learning Center, Institute of Natural Resources, National Research Tomsk Polytechnic University
[2]Head of Laboratory "Drilling Mud and Cement Slurry"
[3] PhD, Engineer, Laboratory "Drilling Mud and Cement Slurry"

E-mail: [1]bubnovas@niitek.ru, [2]Boykoia@niitek.ru, [3]vhorev@sibmail.com,

Abstract. The article is focused on the cement stone strength properties resulted from lightweight cement slurries that meet GOST-1581-96 (state Standards) requirements. Exfoliated vermiculite, hollow aluminosilicate microspheres (HAMs), diatomite and perlite were used as lightweighting additives.

1. Introduction

Nowadays well-drilling is being conducted in complicated geological conditions where conventional cement slurries do not provide required cement-pipe bonding. As a rule, the use of lightweight additives ensures appropriate density of the cement slurry to avoid circulation loss or hydraulic fracture of the formations around the well. However, the cement slurry density decrease results in weakening of cement stone.

2. Methodology

To study the issue a set of experiments was conducted to assess cement stone strength according to GOST 1581-96 (State Standard). Exfoliated vermiculite, hollow aluminosilicate microspheres (HAMS), diatomite and perlite were used as lightweight additives. To study the issue in details the basic parameters of the lightweight additives and their interactions with the cement slurry were initially studied.

Vermiculite is a cleavage-structured mineral of the hydromica group. It forms large tabular golden yellow or fulvous crystals. The mineral expands into worm-like strands and increases 15-25 times in volume when heated (up to 400-1000 ° C). After being added into cement slurry it tends to float up. High vermiculite concentration in slurry leads to coagulation, which makes the slurry grout have poor spreadability.

Diatomite is a weakly-consolidated soft opaline sedimentary rock that can range in color from white and grayish white to dark grey, fulvous, red and black. Its density ranges from 1.2 to 2.5 g/cm^3, porosity value is 50-70%. It has good water absorption. When added into the cement slurry, it mostly floats up. It tends to collapse under high pressure.

HAMS are produced in the process of high temperature coal flame combustion. They have smooth surfaces and diameters ranging from 10 to several hundreds of μm with an average diameter of 100 μm. An average particle density is 2.5 g/cm^3. They are characterized by low density, high mechanical durability, chemical and thermal stability and low thermal conductivity. When added into the cement slurry grout, HAMS tend to float up and generate foamy film on the slurry surface.

Perlite is a glassy igneous rock, the structure of which represents the systems of spheroidal concentric cracks (perlitic structure). It can be of different tones of black, green, reddish fulvous, brown and white colors. Its bulk density varies from 45 to 200 kg/m^3.

Having analyzed the data mentioned above it can be concluded that each lightweight added into slurry grout has particular impact on the strength properties of the cement stone. [1]

A cement compression and flexural machine "Matest", model E160 was used to conduct the experiments. The samples were cement bars with the additives 40×40×100 mm. The bars were kept in water at a temperature of 75°C for 48 hours before being tested. figure 1shows the machine and the lay-out of bars under test.

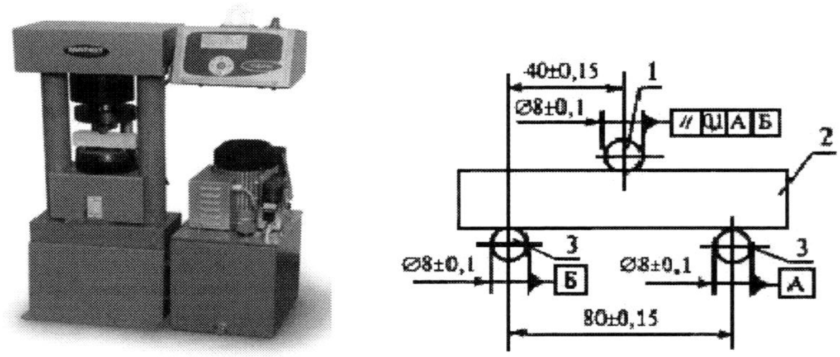

a b

Figure 1. a) Cement compression and flexural machine "Matest", model E160
b) lay-out of cement bars being tested by destruction
1. loading element 2. cement bar sample 3. supports

3. Results and discussions

Tables 1-4 contain the results of strength properties measurement. Minimum 3 samples of each recipe of slurry grout with lightweight additives were tested [2-7].

Table 1. Results of destruction of cement bars with exfoliated vermiculite

№	Composition,%		Density, g/cm^3	Flex. strength mPa
	Oil-well portland cement	vermiculite fraction ≥ 0.6		
1	95	5		2.28
2	95	5	1.36	2.33
3	95	5		2.29
	Oil-well portland cement	vermiculite fraction 0.6-0.3		
1	95	5		2.76
2	95	5	1.44	2.79
3	95	5		2.66
	Oil-well portland	vermiculite		

	cement	fraction 0.3-0.15		
1	95	5		3.47
2	95	5	1.46	3.4
3	95	5		3.23
				max. stress, mPa
	Oil-well portland cement	vermiculite fraction 0.3-0.15	Density, g/cm^3	
	90	10		3.2
1	90	10	1.44	3.13
2	90	10		3.14
3	Oil-well portland cement	vermiculite fraction 0.3≤0.15		
	75	25		2.93
1	75	25	1.49	2.53
2	75	25		2.46
3	85	15		3.91

The results of the cement bar destruction show that the increase in exfoliated vermiculite content decreases strength properties of the samples. It was determined that the samples with finer fraction have more strength properties. Thus, it is reasonable to use appropriate content of exfoliated vermiculite of fine fraction in the slurry grout.

Table 2. Results of destruction of cement bars with diatomite

№	Composition,%		Density, g/cm^3	max. stress, mPa
	Oil-well portland cement	diatomite		
1	92	8		2.08
2	92	8	1.52	2.45
3	92	8		2.23
	Oil-well portland cement	diatomite		
1	90	10		1.98
2	90	10	1.53	2.08
3	90	10		2.23
	Oil-well portland cement	diatomite		
1	87	13		2.46
2	87	13	1.51	2.59
3	87	13		2.5

The experimental data prove that the increase in diatomite content does not significantly influence the slurry grout density. However, it should be taken into consideration that finer fraction of diatomite grains increases bulk density of the lightweight. Thus, the fracture and bulk density inhomogeneity results in slurry density tolerance, which, in its turn, influences cement stone properties.

Table 3. Results of destruction of cement bars with HAMS

№	Composition,%		Density, g/cm^3	Max. Stress, mPa
	Oil-well portland cement	HAMS		
1	97	3		3.41
2	97	3	1.54	3.39
3	97	3		3.31
	Oil-well portland cement	HAMS		
1	96	4		3.31
2	96	4	1.54	3.32
3	96	4		3.29
	Oil-well portland cement	HAMS		
1	94	6		3.21
2	94	6	1.51	3.18
3	94	6		3.18
	Oil-well portland cement	HAMS		
1	92	8		2.16
2	92	8	1.42	2.39
3	92	8		2.3

The data obtained while testing the bars with HAMS show that the most appropriate HAMS content in slurry grout varies from 4 to 6%. With necessary density, the samples cement+HAMS have greater strength properties if compared to the samples cement+vermiculite and cement+diatomite. It should be noted that with 3% of HAMS, the slurry grout is poorly spreadable, which impairs transport property of the grout. More than 8% of the additive content leads to density reduction, which does not meet OB-5 GOST-1581-96 (state Standards) requirements.

Table 4. Results of destruction of cement bars with perlite

№	Composition,%		Density, g/cm^3	Max. Stress, mPa
	Oil-well portland cement	perlite		
1	98	2		3.79
2	98	2	1.53	3.83
3	98	2		3.89
	Oil-well portland cement	perlite		
1	96	4		3.57
2	96	4	1.52	3.46
3	96	4		3.41
	Oil-well portland cement	perlite		
1	94	6		2.98
2	94	6	1.50	3.09
3	94	6		3.10
	Oil-well portland cement	perlite		
1	92	8		2.43
2	92	8	1.49	2.39
3	92	8		2.37

The results obtained by testing cement bar samples with perlite are similar to those of the tests with other additive

4. Conclusion.

It can be concluded that it is reasonable to apply lightweight additives in cement slurries. However, the cost of the additives should also be taken into account since it contributes to the cost of well drilling. Rheological properties are another aspect to be paid attention to. Thus, when HAMS being used as additive, poor slurry spreadability was constantly observed, this impedes proper slurry motion in the annulus. On the contrary, the application of exfoliated vermiculite resulted in sufficient reduction of slurry density with the rheological properties being the same.

References

[1] Smolich A K and Burlov V 2012 *Khimicheskaya stoykost' materialov v sredakh neftekhimii i neftepererabotki* **1-2** p. 390
[2] GOST 1581-96 (State Standard) Specifications of Oil-well Portland Cement
[3] GOST 26798.1-96 Cement Slurry Testing
[4] Bulatov A I and Danyushevskiy V S 1987 *Tamponazhnye materialy Nedra* 280
[5] Ivachev L M 1987 *Promyvochnye zhidkosti i tamponazhnye smesi Nedra* 242
[6] Bubnov A, Boyko A, Horev V and Baryshev I 2014 Research into rheological characteristics and water tapping of lightweight cementing slurry with expanded vermiculite lightweight. *Territory Oil gaz*, **V. 10** p. 18-21
[7] Bubnov A, Boyko I, Horev V and Baryshev I 2014 Research into cement stone strength characteristics and specific weight of vermiculite-based well slurry for oil and gas wells lining. *Territory Oil gaz*, **V. 11** p. 46-49

IOP Conf. Series: Earth and Environmental Science **24** (2015) 012011 doi:10.1088/1755-1315/24/1/012011

Methodological approaches in estimating anomalous geochemical field structure

R. Gavrilov[1] and M. Rudmin[2]
[1 2] Department of Mineral Exploration, Institute of Natural Resources, Tomsk Polytechnic University, Tomsk, Russia

E-mail: [1]gavrilovry@ignd.tpu.ru, [2]rudminma@tpu.ru

Abstract. Mathematical statistic methods were applied to analyze the core samples from vertical expendable wells in Chertovo Koryto gold ore field. The following methods were used to analyse gold in samples: assay tests and atomic absorption method (AAS), while emission spectrum semiquantative method was applied to identify traces. The analysis of geochemical association distribution in one central profile demonstrated that bulk metasomatic aureoles are characteristic of concentric zonal structure. The distribution of geochemical associations is correlated to the hydrothermal stages of mineral formation identified in this deposit. It was proved that the processed geochemical data by factor and cluster analyses provided additional information on the anomalous geochemical field structure in gold- bearing black-shale strata. Such methods are effective tools in interpretating specific features of geochemical field structures in analogous potential ore-bearing areas.

Introduction

Limited reserves of cropping-out deposits significantly impedes the exploration of new ore bodies. Nowadays, the existing processing methods for geochemical information [1-4] for grading geochemical anomalies and interpretation of anomalous geochemical field structure (AGFS) are being updated. These factors enhance for potential undiscovered ore bodies during the geological-exploration stage. It is obvious that an accurate interpretation of anomalous geochemical field structure in early geological survey stages is labor- saving and material resource efficient, and is time--consuming in identifying of ore bodies.

Prognostic prospecting models of mineral deposits plotted in accordance with the analysis and generalization of geological, structural, geochemical, geophysical and other data provide a basis for ore mineralization. Such a profound analysis enables to compare these models to potentially existing ore-bearing areas, ore occurrences and ore zones.

One forecasting element involves the mathematical modeling of ore bodies based on the results of lithogeochemical survey in secondary and primary dispersion aureoles. The method development in identifying the ore mineralization zone erosion and its prospectivity assessment was and is still being promoted (L Ovchinnikov and S Grigoryan [5]). The attempts to evaluate the integral configuration of an ore deposit are not necessarily successful, due to the influence of various natural processes (denudation, weathering, etc.). More critical becomes that obtained information on conventional targets , such as structural accumulations, mineralized zones, stockworks and different morphological bodies, which were investigated within total thickness, along strike and down-dip.

IOP Conf. Series: Earth and Environmental Science **24** (2015) 012011 doi:10.1088/1755-1315/24/1/012011

This research aim is to develop those methodological approaches in interpreting anomalous geochemical field structure of Chertovo Koryto gold -ore deposit that would identify additional exploration criteria for gold-ore targets in black-shale strata. Two principal approaches were applied in the spatial mapping of anomalous geochemical field structure: the first is based on the identification of chemical element associations and their spatial distribution analysis, while the second - the identification of spatial domain according geochemical attributes.

Regional Setting

In previous papers [6] data on the geological position and deposit structure have been presented and described. However, in this present article the author gives only a summary of the target under discussion.

The deposit is situated in north Irkutsk Oblast, in northern Patom highland (Fig. 1), which is a typical example of mesothermal gold mineralization in black-shales.

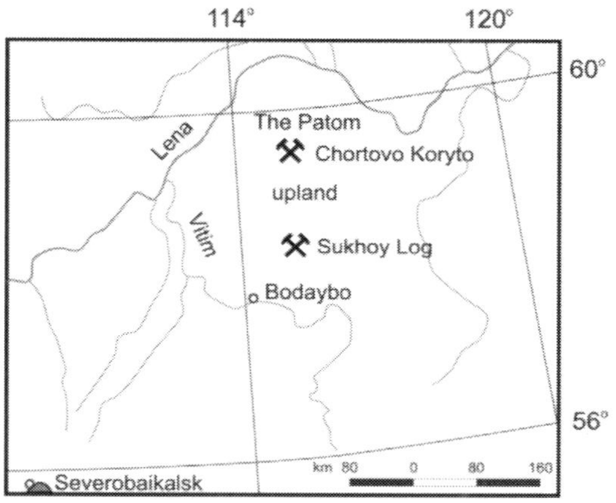

Figure 1. Map of Chertovo Koryto deposit

Ores hosted in Early Proterozoic formation of Mikhailovsk suite carbonaceous terrigenous shales formed a brachysyncline fold with a low angle (from 10 to 20°) of layers, i. e. southward dip to the north changing westward and further to north-NW in the south of the deposit. Brachyfold is dissected by a north -NW-trending tectonic wedge, associated to the Amandraksk deep-seated fault. The suite itself is comprised of metamorphized to epidote-amphibolite facies of sandstones, aleurolites, argillites being subjected to hydrothermal metasomatism effect of propylite-beresite profile. Ore-hosting metasomatites include numerous sulfide-quartz veins and sulfide impregnated pyrite, pyrrhotine, arsenopyrite and other minerals. The ore deposit contour, tracked along the fault up to 1800m., was identified by continuous sampling data. The maximum deposit thickness near the fault is up to 140 meters and gently dipping westward and gradually pinching-out. The contour width of commercial mineralization is about 500 meters.

Research Methods

Exploration drilling involved not only vertical core hole at 50x50m. grid, but also the estimation of C_1+C_2 reserves in Chertovo Koryto deposit. This research is based on the results of drill-hole core sampling: sectional-coring to determine gold content and sectional chip sampling to determine traces of gold mineralization and their behavior within the ore zone. The average length of sectional core samples was one meter, while sectional chip samples vary from 1 to 9 meters, although in average- 3 meters. The first estimation stage of anomalous geochemical field structure (AGFS) included one of the central exploration profiles (BL 15).

Gold samples were analyzed by the assaying method in the laboratories of Lensk Gold Mining Company-subsidiaries: OOO Tonoda and OJSC Pervenets. In this case, the test sensitivity was 0.1 g/t. The gold content was also determined by the atomic absorption method, where the test sensitivity was 0.01 g/t., and this method was used in controlling assaying results. The content of 28 elements was analyzed by the emission spectrum semiquantative analysis in the laboratory of Bodaibin geological survey expedition. The analysis showed results in determining Ag, As, Pb, Co, Cu, Ni and Zn. Other elements were neither detected, nor their content was less than the background concentrations.

The analysis of more than 4000 sectional chip sampling was applied to calculate the statistic factors of gold mineralized traces. The estimation of element background concentrations is conducted by the standard procedure, including the analysis of 170 samples from two drill holes located at the deposit periphery.

To plot the geochemical cross-section zoning statistic calculations of grade correlation, mineralization energy, dispersion and geochemical spectrum variation were applied, as well as cluster and factor analyses.

The differences in the assaying methods for gold and gold traces excluded determining their correlation by the mathematical statistic method. The correlation of the gold content in rocks and statistic index of trace distributions is performed in diagrams.

Results and Discussion

The analysis of AGFS is based on the obtained sampling data of primary litho-geochemical halo dispersion from exploration core wells in one of the central exploration profiles (BL 15). As mentioned above, the difference in analysis methods, chemical element concentration level and assaying methods for gold and gold traces precludes their correlation by the mathematical statistic method. Their correlation could be plotted only by applying content distribution diagrams.

The distribution of ore-genetic elements within the deposit does not correspond with the normal and logarithmically normal laws. The gold content in an ore body varies from milligrams to hundreds of grams per ton.

Depending on the cut-off grade within the ore deposit, the amount of reserves significantly varies. The cut-off grade contouring the ore deposit is 0.5 g/t (Fig. 2). In the central zone there is a maximum ore body thickness with rather enriched remnant ores. At the deposit periphery there are single narrow pinching-out lens of ore with rather low metal content.

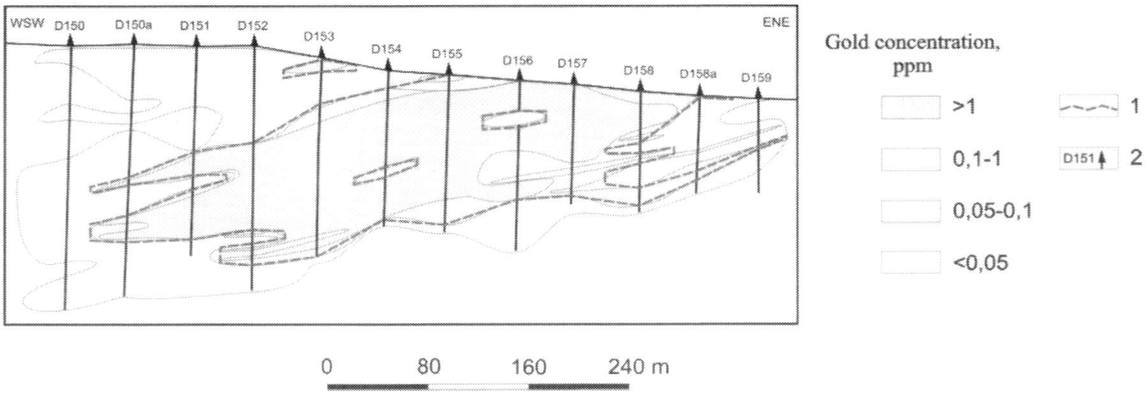

Figure 2. Diagram of gold distribution in the cross-section (BL 15) of the deposit ore-hosting metasomatic aureoles

1) ore deposit contour and lens-satellite; 2) exploration wells and their corresponding numbers

The following gold mineralization traces As, Ag and Pb produce the most contrasting anomalies in bulk metasomatic aureoles. The high-contrasting aureole As is incomparably to that of background values and spatially coincides with the ore deposit contour. Anomalous Ag concentrations partially coincide with the high-contrasting gold aureoles. In this case, contrasting Ag aureoles even exceed the background values hundreds of times and correlate well with remnant ores. The aureole Pb is partially superimposing the ore body contour itself, whereas its maximum value is 300 times higher than the background values.

Investigating the interrelations between ore genetic elements by the grade correlation method, the results indicated that this interrrelation is significantly positive for all the elements. The following interrelation of Ni with Co, Pb, Cu, Zn and Pb-Cu-Zn triads showed the highest values. The contrasting Ag and As anomalies within the deposit have a rather low correlation relationship value (0.21). However, the grade correlation method did not distinctly separate the elements into different groups, which most probably, is connected with the low interrelation threshold value for the enormous processed samples.

To interpret the deposit anomalous geochemical field structure the following indexes characterizing the intensity of hydrothermal-metasomatic mineralization processes were applied: mineralization energy [7], dispersion and variation of the geochemical spectrum [8].

Mineralization energy background values vary from - 1 to 2 units and are rarely detected throughout the ore deposit. The index value reaches 10^4 units and more. So, the mineralization energy depends on the dominant element - As. In the case of excluding this dominant from the calculation equation, the index values incomparably decrease leading to further reduction of the ore-hosting substance volume, which is characterized by the maximum mineralization energy.

The method to estimate the substance differentiation degree according to the geochemical spectrum of one sample was proposed by S Vyborov and I Bystrov. The index calculations of geochemical (GSD) spectrum dispersion and geochemical (GSV) spectrum variations is based on standard dispersion equations and normalized chemical element content variations. Background GSD values do not exceed 0, fraction unit within the surrounding extensive metasomatic aureoles. However, the index values substantially increase reaching the maximum for certain samples up to 10^5 units and even more within the ore zone. The element As significantly influences the anomalous GSD values which is governed by the interrelation in the distribution of their aureoles.

GSV background values at the periphery of extensive metasomatic aureoles do not exceed 40%. These value indexes vary from n to 250% within the ore deposit. In spite of the fact that Despite this variation is the derived function of dispersion, whereas both indexes embrace different genetic information. Anomalous GSV values do not always coincide with the GSD anomalies. Chertovo Koryto deposit is an evident example of this fact. High-contrasting aureole indexes coincide in mineralized zone cross-section, which is obviously conditioned by the sharp difference in element concentrations within the ore body and along its margin. It should be stated that the mineralized deposit zone is characterized by a random content distribution of ore- genetic elements with GSV > 140%. Excluding the GSD and GSV indexes of As from the calculation equation incomparably results in their value decline and significant dimension reduction of high-contrasting aureoles .

Thus, the intensive mineralization indexes are excellent indicators in plotting the volume of metasomatic aureoles within which is located the mineralized zone. However, without additional operations (such as, chemical element breakdown into associations according to a specific principle), these indexes preclude the possible interpretation of the anomalous geochemical field structure of the deposit.

The cluster analysis of mineralized zone dimension identified distinguish 5 sample clusters. There is no distinct interrelation between ore-genetic elements in the first and the fourth clusters. The second cluster unites Cu, Zn, Ag and Pb that characterize a galena-chalcopyrite-sphalerite association. The third sample cluster is characteristic of a considerable Co and Ni content that are involved in the generation of pyrite-pyrrhotine association. In the last (fifth) cluster the dominant is As that is an active element in the arsenopyrite-pyrite-pyrrhotine association. Identified chemical element associations closely match the minerogenesis stage, defined during structural, petro- geochemical investigations in the Chertovo Koryto deposit [6]. Nevertheless, the limited number of chemical elements that are includes in the calculations affected the distribution of clusters within the deposit dimension. A distinct configuration of the zonal geochemical aureole structure was not identified on the basis of the above-mentioned cluster analysis.

The factor analysis was also applied in defining the chemical element associations to interpret the anomalous geochemical field structure. The following three factors incorporate 64% of the total dispersion. The first factor - Pb, Zn and Ag incorporate 30% of total dispersion. The second most significant factor, being 20% of total dispersion, is characterized by Co, Cu and Ni. The third factor includes only one element - As. This proves the fact of the introduction into hydrothermal solutions of elements and their precipitation as native minerals and element- impurities. Identified chemical element associations are also an integral part in the minerogenesis stage [6]. Thus, classifying the close ore-genetic elements into three geochemical associations indicate different conditions and time formation of ore mineralization within large metasomatic aureoles.

Based on processed geochemical data by mathematical statistics methods, a diagram of the deposit geochemical zoning cross-section BL 15 was plotted (Fig. 3).

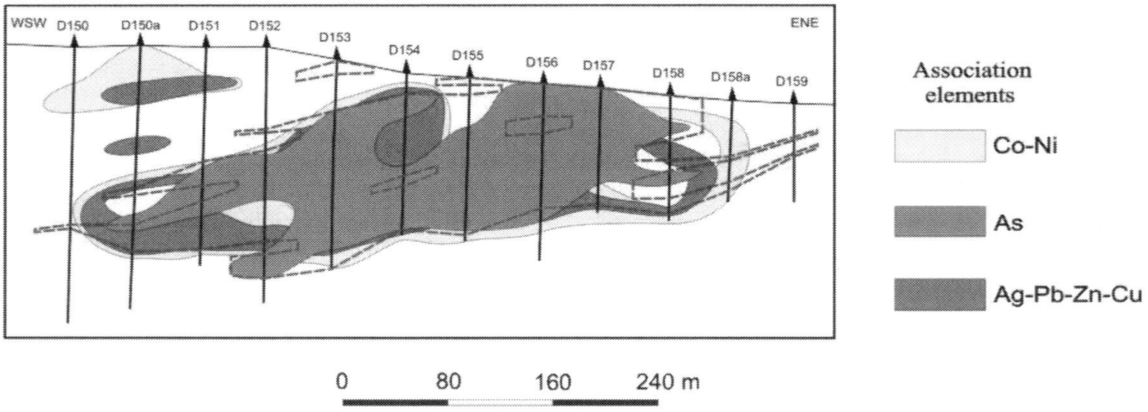

Figure 3. Geochemical zoning of Chertovo Koryto gold deposit in cross- section BL 15 (Ref. to legend in figure 2.)

The high-contrasting Co and Ni aureole correlates to the ore deposit section plane, partially with Ag-Pb-Zn-Cu associations and embraces a rather peripheral location in the mineralized zone. Both these metals have been educed from early hydrothermal fluids accumulated after pyrite-pyrrhotine mineralization. The second group includes only one element - As, found in arsenopyrite , which, in its turn, was a component of late arsenopyrite-pyrite-pyrrhotine association. Due to most extensive spatial superposition of high-contrasting As aureole and mineralized zone, there is limited isolation in the distribution of its aureoles and gold aureoles. The high-contrasting aureoles of the late Ag-Pb-Zn-Cu associations also coincide not only with the ore deposit and its periphery, but also is located more closely to the inner zone, relative to the early Co-Ni geochemical association.

Summary and Conclusions

Methods in interpreting the geochemical field of different targets should be performed by applying all available tools. In this case, the complex analysis of investigated targets are based on those proven methods and approaches that could be used in prospecting-exploration activities. The identification factor for ore bodies is a geochemical field structure. As it has been proven gold anomalies identified in the geochemical studies could not be a precise indicator for ore bodies. An additional criterion could be the zonal structure of the geochemical field itself.

In reality, the application of conventional and non-conventional methods in analyzing the geochemical field can significantly differ depending on the geologic survey stage, goals and objectives of conducted studies. For some cases, it is sufficient to identify the distribution law, background and minimal anomalous content of ore-forming elements and plot their distribution diagrams. However, in most cases to identify ore-genetic element associations and determine combined statistic geochemical zoning indexes, more complex analysis methods and further graphical plotting is a required factor.

Defined geochemical zoning with a rather wide element spectrum of ore associations should be applied for areal lithogeochemical survey in potential deposits and for the data interpretation from mining and single wells.

Grade correlation of chemical element associations is not identified due to the distribution features of ore-genetic elements within the Chertovo Koryto deposit, for example. The intensive mineralization indexes without preliminary identification of element clusters excludes the possibility of obtaining further knowledge on the nature and structure of deposit geochemical field. The geochemical zoning sequence of the deposit metasomatic aureoles was determined by factor and cluster analyses. Three ore-genetic element associations (from the earliest to the latest): Co-Ni, As and Ag-Pb-Zn-Cu were identified. The plotted diagram of geochemical zoning in hydrothermal ore-hosting metasomatic aureoles demonstrates its concentric zoning structure. The diagram shows the correlation of the results obtained in gold fields from other regions (S Grigorov (2007) and V Voroshilov (2009))

Thus, some of the discussed methods in geochemical data processing provide additional information about the anomalous geochemical field structure in gold-bearing black-shale layers. These methods are effective tools in interpreting specific geochemical field structure within ore-bearing areas under similar conditions. The localization of potential areas and the forecasting of ore deposits could be based on the results of areal lithogeochemical mapping and obtained data from mining and single wells. Information on the specific features of the anomalous geochemical field structure and location of ore bodies in such geochemical fields in correlation with other data is significant for accurate gold-ore deposit forecasting.

Acknowledgements
The investigations were supported by National Research Tomsk Polytechnic University, project No. 114-2014.

References

[1] Grigorov S 2007 *J. Home geology* **4** 45.
[2] Voroshilov V 2009 *J. Ore Geology* **51** (1) 3.
[3] Voroshilov V, Savinova O, Ananjev Y and Abramova R 2014 *IOP Conference Series: Earth and Environmental Science* **21** 1.
[4] Timkin T, Lavrov D, Askanakova O and Korotchenko T 2014 *IOP Conference Series: Earth and Environmental Science* **21** 1.
[5] Ovchinnikov L, Grigorjn S and Baranov Je 1973 *J. Izvestiya vuzov* **10** 76.
[6] Kucherenko I, Gavrilov R, Martinenko V and Verkhozin A 2008 *J. Ivestiya* Tomsk Polytechnic University **312** (1) 11.
[7] Safronov N, Mesheryakov S and Ivanov N 1978 *Ore mineralization energy and mineral exploration* (Leningrad: Nedra) p 215.
[8] Viborov S and Bistrov I 1990 *J. Izvestiya vuzov* **4** 102.

Regional stratigraphy and its dependency on tectonic movements (case study: Upper Cretaceous and Paleogene stages in Western Siberia)

V Podobina
Tomsk State University, Tomsk, Russia

E-mail: podobina@ggf.tsu.ru

Abstract. Multiscale and divergent tectonic movements have been identified in Western Siberia of which first - order movements caused transgressions and regressions, as well as the partial formation of sediments. As a result of tectonic movement direction turn, no transgression was observed in the cross-sections of Campanian and Danian central section and in the Priabonian top section. During second-order tectonic movements and undirectional transgression insignificant bed thicknesses and channels were formed. Such movements could have included different tectonic activities within the western and eastern parts of the region limited by the Koltogorsk-Urengoy Rift. Third-order tectonic movement of moderate amplitude promoted either extension, contraction or even depth variations of the marine basin itself.

1. Introduction

The vertical tectonic movements within the West Siberian Plate were both divergent (ascending and descending) and multi-scaled. Wide-range movements resulted in the subsidence or uplift of almost the whole territorial region. These movements (epeirogenic) could be of the first order, i.e. the mostly large-scale and prominent in this region. During the transgression cycles, sedimentary rocks accumulated, which according to the lithological and microfaunal data, differed both vertically and laterally.

2. General characteristics

Ascending tectonic movements generated the uplift of the Western Siberia territory and the destruction of pre-existing out-cropped rocks. Subsequent sediment thicknesses under the influence of further trangression accumulated on the underlying eroded sediments. Moreover, previously accumulated and subsequently destructed stratigraphic units became concealed sequences. Special investigation is vital to determine these free sequences and graded unconformity. These unconformities and concealed stratigraphic units have been observed in the Upper Cretaceous and Paleogene deposits of Western Siberia, which were mainly detected within foraminiferal zones and correlated to adjacent regions. These investigations showed concealed sequences of the Campanian, Danian and Priabonian stages. Accordingly, foraminifer associations were observed in the lowermost or uppermost sequences. In the Late Cretaceous and Paleogene, the amplitude of tectonic movements was gradually increasing as a result of the growing thickness of concealed stratigraphic units. It is well known that Western Siberia became an elevated continent in the Neogene. The accumulation of Neogene continental sediments is mainly observed within the southern latitudinal elongated Omsk depression. Late Cretaceous positive multi-scaled tectonic movements became more apparent in the beginning of the Campanian stage.

Santonian Slavgorodsky suite (lowermost Early Campanian) is widely observed in the marine sediment sequences of Western Siberia [1, 3, 4]. Late Campanian-Maastrichtian (Gankinsk suite) transgression changed its direction – from south to north through the Turgay Straight which, by this time, had widened and significantly deepened. The boreal transgression direction (Santonian – Early Campanian) turned to the south resulting in the outcropping of early Campanian sediments, the bulk of which was destructed by erosion and denudation. This graded stratigraphic unconformity between the Santonian and Maastrichtian sediments,corresponding to the concealed Middle Campanian Stage, can be determined by the correlation of West Siberian foraminiferal zones to those of Central Asia (Kazakhstan) and East-European Platform.

3. Biostratigraphic suite description

The overlying Gankinsk suite (top of Upper Campanian – Maastrichtian) is composed of mainly clays and aleurolites with carbonate impurities, as well as calcareous -secretion foraminifera within three foraminiferal zones: lower *Cibicidoides primus* zone (uppermost Upper Campanian), central *Spiroplectammina variabilis – Gaudryina rugosa spinulosa* zone (lower Maastrichtian) and upper *Spiroplectammina kasanzevi – Bulimina rosenkrantzi* zone (upper Maastrichtian). The analogues of Campanian foraminiferal zones identified in adjacent regions - *Cibicides temirensis, Brotzenella monterelensis,* significant part of *Globorotalites emdiensis* have not been revealed within Western Siberia stratigraphic units.

The transgression direction from south to boreal was again observed in Late Maastrichtian (Gankinsk suite), resulting in almost complete concealed Danian Stage, whereas only the lowermost sequences were preserved from erosion and could be traced in the Omsk depression [2, 4, 10]. Simultaneously, Bakchar iron-ore horizon was forming in the east. The overlying Paleocene Talitsk suite is composed of dark-grey opota clays and includes mainly agglutinated quartz-siliceous foraminifer assemblages with *Ammoscalaria friabilis*.

However, in all marginal shallow-water suite areas calcareous-secretion foraminifera intertwining with agglutinated ones could also be identified [2, 4]. This is also observed in the Selandian Stage (Middle Paleocene, lowermost part of the Talitsk suite). The investigation of the foraminifer assemblages from the top Talitsk suite (Thanetian Stage, Upper Paleocene) demonstrated no uppermost foraminifel zone analogues in this stratigraphic unit. However, the upper planktonic foraminifer zone analogues have been found in the adjacent regions of south Russia: Acarinina acarinata. In Transur Urals the planktonic foraminifer as *Acarinina subsphaerica* with consistent Late Paleocene agglutinated forms have been found in the Talitsk suite (upper horizons) [5]. Therefore, it could be noted that only the upper *Acarinina acarinata* zone (Upper Paleocene) in Western Siberia is concealed. This could be explained by two factors- slight elevation of Western Siberia territory and concealed sequences of upper foraminiferal zone during the Late Paleocene. The Eocene lower subsuite of Lyulinvorsk suite is composed of dark-grey or grey gazes (silicites) and opota clays. The formation of such clays could be associated with the significant hydrological regime changes within the basin itself, which, previously had favourable conditions for both foraminifera and radiolarians in the Early Paleocene. However, deformed sugar-white benthic foraminifera are rare, and this fact makes it impossible to determine their stratigraphic value. The boreal transgression direction turn was not observed here, but stratigraphic unconformity was detected between the Upper Paleocene and Lower Eocene sediments due to the concealed Paleocene upper sequences (planktonic foraminifer zone with *Acarinina acarinata*) (table 1). The following Eocene transgression, partial erosion of previous Thanetian sediments (top Talitsk suite) resulted in graded stratigraphic unconformity [6, 8].

Increasing amplitude of positive tectonic movements has been observed towards the end of Eocene. By this time, Lyulinvorsk suite formation had been completed, which, in its turn, included greenish-grey clay silicite with varying degrees of opoka clays not only in the lower subsuite but also in the middle and upper subsuites. In SW mid-Lyulinvorsk subsuite, Middle Eocene calcareous foraminifera assemblages were detected [4], while in the central and southern areas, Middle Eocene agglutinated quartz-siliceous foraminifera assemblages with *Gaudryinopsis subbotinae* were widely distributed. By the end of Eocene (upper Lyulinvorsk subsuite) the transgression decrease was observed, and the sediments were mainly coarse-grained agglutinated foraminifera. The foraminiferal zone with *Labrospira honesta* was defined. In the uppermost horizons, rare fragments of ferruginized reophacida and haplophragmiida could be found, i.e. layers with *Reophax dentaliniformis – R. subfusiformis* assemblages. The sharp boreal Eocene transgression turn to the southern Tavdinian one (Lower Oligocene) resulted in the uplifting, outcropping and destruction of the Upper Eocene horizon (uppermost layer of upper Lyulinvorsk subsuite). In this case, there is a graded stratigraphic unconformity (between Lyulinvorsk and Tavdin suites), which could be correlated in time to eroded Eocene uppermost foraminiferal zones (most of the Priabonian Stage) [6]. In the overlying Tavdin suite, Lower Oligocene foraminiferal zone with *Cibicidoides pseudoungerianus – Evolutononion decoratum* were defined [2, 4]. Tavdin suite is composed of greenish-grey, sometimes, ferruginated book clays with powered sandy material. This suite encloses benthic calcareous- secretion and planktonic foraminifer assemblages according to which the suite age was determined as Early Oligocene. Tavdin southern transgression comparably to previous boreal Lyulinvorsk one was moderate as it extended practically to the latitudinal Ob river currents. Tavdin suite benthic and planktonic foraminifera are diverse and abundant in the south of the region (near Petukhovo) in comparison with the central and western regions. In the eastern area the Tavdin suite is lithologically altered due to the sandy material increase. There is a lack of foraminifera here. Range of foraminifera in above-mentioned assemblage is similar to Early Oligocene species in West Europe and USA [2]. Integrally, Early Oligocene basin (Tavdin suite) was shallow and moderately lateral within the south-western part of the area. The eastern part of West Siberia, in comparison with the western part, was tectonically more active. Their borderline coincides with the location of the Koltogorsk-Urengoy rift. In the eastern area, the developing 2nd -order tectonic movements are subordinate to 1st -order vertical movements, which, in its turn, are associated with the uplifting and subsidence within the West Siberia territory.

Series	Subsuite	Stages	Zones of planktonic foraminifera		Zones and beds with benthic foraminifera of Western Siberia (Podobina, 2009 [4])		
			Tethyan belt	**Boreal belt**			
			Berggren & Pearson, 2005 [9]	**The Unified Regional Stratigraphic Scheme, 2001 [9]**	**Foraminifer zones**	**Beds with foraminifera**	
PALEOCENE	Upper	Thanetian	M. velascoensis	Acarinina acarinata	unrevealed		
			Ac. soldadoensis – Gl. pseudomenardii				
			Acarinina subsphaerica	Acarinina subsphaerica	Glomospira gordialiformis, Cibicidoides favorabilis	Glomospira gordialiformis, Cyclammina coksuvorovae	
				Igorina djanensis			
	Middle	Selandian	Gl. pseudomenardii – P. variolata	Morozovella conicotruncata	Ammoscalaria friabilis	Cibicidoides proprius	Cyclammina coksuvorovae
			Igorina albeari				
			Igorina pusila	Morozovella			

		i a n		angulata			
	L	D	P. uncinata	Acarinina inconstans	unrevealed		
	l o w	a n i	Gl. compressa – Praemurica inconstans				
	e r	a n	Subbotina triloculinoides	Globoconusa daubjergensis			
			P. pseudobulloides	Euglobigerina taurica	Brotzenella praeacuta	Bathysiphon nodosarieformis, Glomospira charoides	

Table 1. Correlation between Paleocene planktonic and benthic foraminifer zones

From the end of Early Oligocene (Latdorfian Age), Middle Oligocene (Rupelian Age) to Late Oligocene (Chattian Age) a depression zone formed in West Siberia due to 1st -order ascending tectonic movements where continental sediments accumulated (Abrosimovka, Novomikhailovka and Zhuravka suites). At the beginning of Neogene the West-Siberian continent was formed under the influence of tectonic movements, within the territory of which, there was practically no sedimentary environment. Besides high-amplitude 1st -order tectonic movements resulting in the transgression direction turn and the continental uplifting (West Siberia), subordinate 2nd – order tectonic movements existed, which affected both the reduction and expansion of transgression in separate regions of West Siberia. According to lithological evidence these tectonic movements became apparent in the west and east areas and revealed foraminifera in different stratigraphic horizons. From the viewpoint of these factors, the following example should be examined- Ipatovsky horizon. Its western area embraces Sedelnikov suite of compact grey gaze clays (40-60 m in thickness), including single coarse-grained, poorly-preserved agglutinated quartz-siliceous foraminifera (probably, Coniacian Age). Its eastern area embraces similar Ipatovsky suite (Pudino – Napas) with a thickness of 60-270 m. This suite is primarily composed of loose rocks - grey sands and aleurolites. Within the uppermost layer of this suite the Narym iron-ore horizon could be traced, including Late Coniacian assemblage of single brown calcareous-secretion foraminifera with *Dentalina basiplanata – D. tineaformis* [1, 3, 4]. In the west (borehole 23 in Beryozovo) these assemblage layers correlate with sediments containing more abundant species composition of *Dentalina tineaformis – Cibicides sandidgei* assemblages [3].

Rather active 2nd tectonic movements in the eastern area resulted in the significant erosion of previously accumulated rocks. In this case, not only the Danian Stage but also the upper Maastrichtian zone with *Spiroplectammina kasanzevi – Bulimina rosenkrantzi* are concealed. Relatively primitive foraminifera with agglutinated walls, often sugary-white colour, prevailed in the shallowed Early Campanian, Selandian and Thanetian basins of east West Siberia. Within this eastern area Tym iron-ore horizon was detected on the Lyulinvorsk and Tavdin suite boundaries. The rocks of this area include interbedded glauconitic aleurolites and sandstones enclosing fragments of Reophacida and Haplophragmiida.

According to lithological and microfauna characteristics there are a number of examples that confirm the possible 2nd -order tectonic activity within east West Siberia.

4. Conclusion

The channels connecting the West-Siberian basin with adjacent seas opened and disappeared as a result of 2nd -order tectonic activity movements at various times. The most stable and changing was the Late Cretaceous and Paleogene Turgay Channel expanding and deepening during the Turonian, Santonian, Maastrichtian, Eocene (Lyulivorian) and Early Oligocene transgressions. According to foraminifera data, their similarity with adjoining basins, is supposed to have temporarily existed in the

Urals and other West Siberia areas. However, of particular interest is Pre-Yenisey channel formed in the Late Cenomanian when West Siberia was a continent. Thus, Late Cenomanian foraminifera with *Gaudryinopsis nanushukensis elongatus* (*Trochammina wetteri tumida – Verneuilinoides kansasensis* zone) appeared in the south-east of this region (borehole E-150, on the outskirts of Seversk). During Santonian, the Mariinsk channel formed in the south-eastern part through which the characteristic Santonian calcareous- secretion foraminifera migrated from Kazakhstan basin to the West-Siberian basin. These findings in correlation with agglutinated quartz-siliceous foraminifera from central West Siberia showed that the bulk of the Slavgorodsk suite is of the Santonian age . Previously, the Slavgorodsk suite had been assigned to the Campanian Stage excluding such factors as tectonic movements, channel formation, comparison of foraminifer assemblages from different regions, including those foraminifera from northern Kazakhstan and East-European platform [7].

Constant oscillating 3rd -order tectonic movements throughout the Phanerozoic and, in this case, Late Cretaceous and Paleogene resulted in the alternation of relatively deep-water and shallow facies within the sequence, which, in its turn, also influenced the biota composition (foraminifera) in these basins.

References

[1] Podobina V 1989 Foraminiferyi i zonalnaya stratigrafiya verhnego mela Zapadnoy Sibiri Tomsk University Press 232 p
[2] Podobina V 1998 Foraminiferyi i biostratigrafiya paleogena Zapadnoy Sibiri Tomsk "NTL" Press 338 p
[3] Podobina V 2000 Upper Cretaceous Foraminifera and Biostratigraphy of Western Siberia Tomsk (NTL Press) 388 p
[4] Podobina V 2009 Foraminifera, Biostratigraphy of the Upper Cretaceous and Paleogene of Western Siberia Tomsk State University Press 432 p
[5] Podobina V and Amon E 1992 Microfauna and biostratigraphy of Paleogene deposits in the Sarbay section of North-West of Turgay In: *Materials on Paleontology and Stratigraphy of Western Siberia Tomsk* (Tomsk University Press) 88-97
[6] Sokolov B (ed.) 2005 Practical guidebook on microfauna **8** Cenozoic Foraminifera S.-Petersburg (VSEGEI Press) 324 p
[7] Regional stratigraphic schemes of the Mesozoic deposits of the West-Siberian Plain Tyumen 1991
[8] Unified regional stratigraphic scheme of Paleogene and Neogene deposits of the West-Siberian plain 2001 Novosibirsk
[9] Berggren W A and Pearson P N 2005 A revised tropical to subtropical Paleogene planktonic foraminiferal zonation. *Journal of Foraminiferal Reseach* **35** 279-298
[10] Podobina V 2011 The Danian and Selandian calcareous benthic foraminiferal assemblages and biostratigraphy of Western Siberia. *Geologos* **17** 97-110

IOP Conf. Series: Earth and Environmental Science **24** (2015) 012013 doi:10.1088/1755-1315/24/1/012013

Litho-geophysical structure of Paleozoic-Mesozoic contact zones in North-Ostaninsk oil field (Tomsk Oblast)

A Ezhova [1], V Merkulov [2], V Chekanstev [3] and R Abramova [4]

[1] Department of Geology and Mineral Exploration, Institute of Natural Resources, Tomsk Polytechnic University
[2] Department of Geophysics, Institute of Natural Resources, Tomsk Polytechnic University
[3] JSC "Tomskgazprom", Tomsk
[4] Department of Foreign Languages, Institute of Natural Resources, National Research Tomsk Polytechnic University

[1]leav@tpu.ru, [2]merkulovvp@hw.tpu.ru, [3]ChekancevVA@vostokgazprom.ru

Abstract A multidiscipline seismic and gravimagnetic survey, as well as paleomagnetic and lithologic-stratigraphic analysis of pre-Jurassic formations were conducted in North- Ostaninsk oil field within south-east Western Siberian petroleum province. A multidirectional tectonic deformation network merging into the destruction zones of Paleozoic basement was identified. The material composition and age of pre-Jurassic formations within each tectonic block were determined. Fundamentally new geological structure model of the Paleozoic suite was proposed : North-Ostaninsk erosion-tectonic protrusion - a reversed and structurally-complex fracture - deformed tectonic syncline fold where the oil reservoir is confined to the dipping limb. Reversed morphostructures in the erosion-tectonic protrusions could be a prospecting indicator in evaluating the hydrocarbon potential of Paleozoic sediments.

1. Geological setting

North-Ostaninsk oil field is located in the south-east Western Siberian petroleum province and is confined to similar structural high north-west strike. According to seismic survey data (i.e. correlation refraction method- CRM and CDP-method), which has been applied since the 80's of the last century and even nowadays, the structural geometry of the reflection horizon F_2 (Paleozic surface) is characterized by a significant degree of compartmentalization and structural element contrasts [1, 2, 6]. This feature of the basement surface is governed by the extensive development of tectonic deformation systems of two generations (figure 1). The basic one of which is the NW-trending geological strike formed after Hercynian folding, while the latter (Triassic) one – being the NE-trending geological strike. As a result of subduction rocks of different age and composition were isolated on the Jurassic surface.

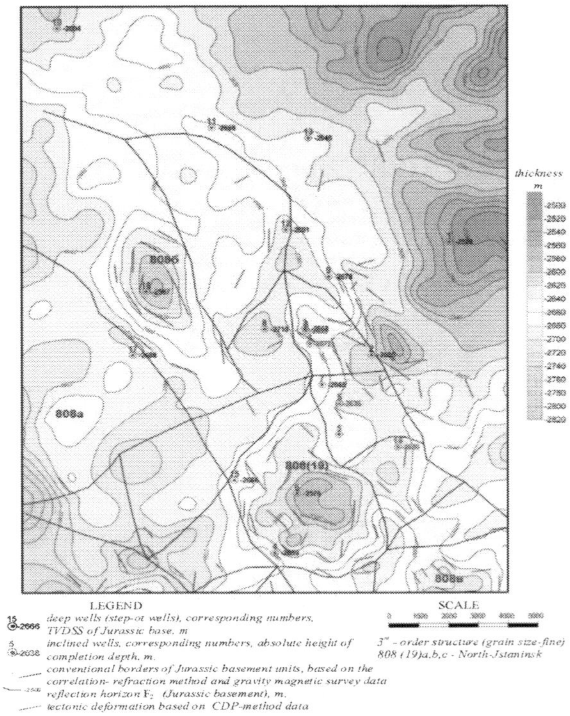

Figure 1. Structural map of the reflection horizon F_2 in North-Ostaninsk oil field based on seismic survey data by CDP-method (according to T.V. Zabug, V.P.Mel'nikov, 2011)

2. Background information

North-Ostaninsk oil fielfd is confined to NW -strike tectonic elevation in the north-eastern Nurolsk sedimentation basin. The structure base includes eroded basement protusion which is laterally restricted and split into separate fault blocks, detected by seismic survey and drilling activities. Heterogeneous Paleozoic formation is composed of complex sequences of limestones, dolomites, magmatic and terrigeneous rocks, embracing depositional processes, whereas, the lithological members could not be traced in correlated well logging.

Research objective- investigation of Paleozoic sediments and rocks in Paleozoic-Mesozoic contact zones within North-Ostaninsk oil field to identify the following factors: void formation, detection of high-productive reservoir zones and bedding orientation by applying the paleomagnetic method for further exploration well distribution pattern.

3. Research data

Geologic-geophysical survey and rock composition study of Paleozoic-Mesozoic contact zones within North-Ostaninsk oil field have been conducted during the 1980's – 1990's [1, 2, 8]. The updated material and data provides practical overviewing of previous survey results in order to define more precisely the tectonic structure of North-Ostaninsk area. Based on core and section description, analytical investigation, age determination and well log interpretation, a lithological characteristic of 19 wells was compiled. According to paleomagnetic investigation results of core samples the bedding elements were determined. Specific features of weathering crust rocks were also studied.

High-precision gravimagnetic survey (scale 1:50 000) was conducted within the North-Ostaninsk area, the data results of which refined such aspects as unit structure, major setting of tectonic elements and material composition of Paleozoic deposits [7].

Correlation of gravimagnetic field anomaly obtained after regional background processing, averaging and subtraction which is depicted in figure 2.

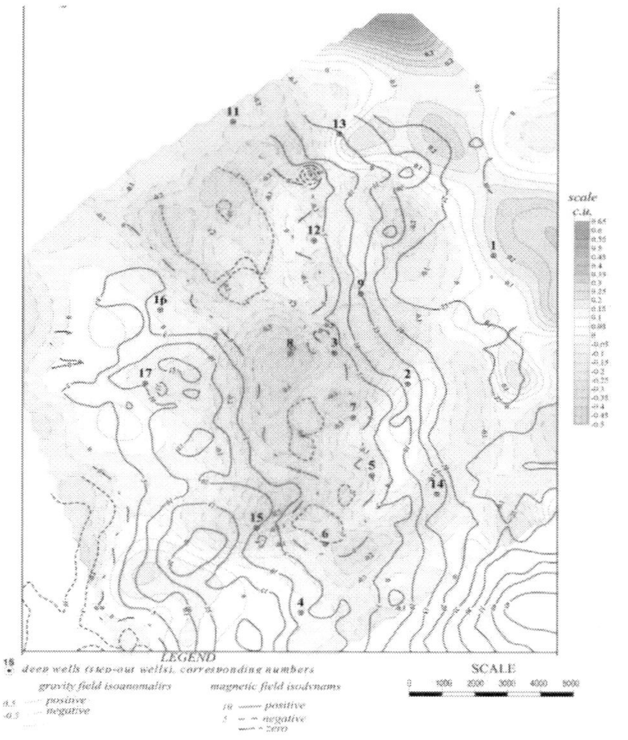

Figure 2. Schema mapping of gravity and magnetic anomalies in North-Ostaninsk area

Regional magnetic field within North-Ostaninsk area is characterized by relatively insignificant field density variation intervals, from + 47 nT to -36nT. The analysis of magnetic field isodynam direction revealed primarily positive and negative NW-trending strike anomalies.

Gravity field anomalies showed a distinct wavy structure, displaying positive and negative anomaly sequences lying from North-east to South-west. Striking isodynams of gravity high and / or low closely matches the behavior of magnetic anomalies.

The analysis of the gravimagnetic field element layout showed that these elements form a characteristic block system with NE and NW-trending strikes, where the predominate one is the NW strike. Such an anomaly localization is associated with the fact that basement tectonics is involved in the formation of the field structure itself as invading fractured zones into the sediment mantle. Most probable existing location of hydrocarbon accumulations is associated with the tectonic structure features and material composition of pre-Jurassic formations. Based on surface magnetic and gravimetric survey data typical micromagnetic anomaly zones and in-situ gravity low develop above these formations within the North-Ostaninsk field, which, in its turn, are considered to be geophysical prospecting indicator.

Application of the paleomagnetic method involving the study of multiple collected samples is based on the rock property which determines individual geomagnetic field direction as a remanence vector [8].

Primary remanent magnetic induction, developing simultaneously with rock formation, has an orientation behavior. Existing magnetic field generates viscous magnetization coinciding with the field direction. Viscous and primary magnetization comprise the vector sum, and, in a first approximation, generates natural remanent magnetic induction, which, in its turn, can be measured in individual collected samples. Paleomagnetic survey results [7, 8] showed that the difference between viscous and primary magnetization is based on their stability level to heating: the more stable the primary magnetization is, the faster the viscous magnetization degrades. The experimental procedure included the following: first, temporary magnetic cleaning was conducted with 2-week duration samples in the "along field" position,and, then within the same time interval-in the "reversed" field to compensate "laboratory" remanent magnetic induction. Further, thermal demagnetization was conducted in V. Aparin thermal oven to compensate the external magnetic field by means of multi-layered permalloy shield, where thermal demagnetization step is 25^0C and final heating temperature - 125^0C. Natural viscous remanent magnetic induction component significantly degrades within this temperature range.

Based on experimental data results vector differences were calculated to determine the viscous magnetization direction, governed by the existing geomagnetic field effect and its corresponding orientation.This data was applied for spatial core orientation. The verification of the obtained directions is the correlation of the dipping viscous magnetization to the existing dipping geomagnetic field in the point of the sampling location, as well as the comparison of the bedding elements to the bedding itself in oriented samples. Orientation angular error of viscous magnetization is $3^0 - 8^0$.

Paleomagnetic survey, involving core samples from wells 3E, 5E, 13R and 15R, was conducted by V. Merkulov in the North-Ostaninsk field. Based on this survey the following bedding elements were revealed: SW layer dip azimuth angle of 60° и 68° (wells 3E and 5E), 34° (well 13R), while in the western area (well 15R)- NE layer dip azimuth angle of 45° (well location ref. to figure1).

Analytical survey was conducted to investigate the material composition of pre-Jurassic formations, involving the following: core and thin section description, chemical and XRD analysis data, determination of reservoir properties and age and log suite interpretation. Based on this data the lithological characteristic of the upper Paleozoic formation was described for exploratory and production wells with defined lithologic member sequences [3, 5].

Due to the lack of core samples, lithological log suite interpretation was conducted, where carbonates showed high-value resistance, gamma-ray intensity (according to neutron logging), low-level radio-activity; silica rocks- low-value resistivity curve, average / low gamma-ray intensity (according to neutron logging); and effusives (volcanic rocks)- induction logging revealing high-value and high-level radio-activity.

According to faunal remain identification, the deposit age in the upper Paleozoic formation was determined: in the east- late Devonian D_3fr (well 16R), Silurian-Lower Devonian S-D_1 (well 2R), Devonian D_{1-2} and D_2gv-D_3 (well 9R); in the south- Lower Carboniferous C_1v (well 6R) and C_1 (well 8R) in the central section.

Permo-Triassic weathering crust rocks underlie Paleozoic- Mesozoic formation contact. Weathering crust products are divided into two large groups: residual and sediment [4]. Residual weathering crust products are structural eluvium, which, in its turn, can be subdivided into three groups depending on the specific weathering crust rock composition: orthoeluvium accumulated as a result of intrusions; para-eluvium accumulated as a result of lithified sedimentary rocks and neo-eluvium accumulated a s a result of the decomposition of losse sediments.

Sediment weathering crust products are those rocks formed as a result of sedimentation of suspension and coarse clastics, as well as transported colliods during the formation of structural eluvium. Both groups of weathering crust products have been observed in the investigated sections. In the elevated sections argillaceous-siliceous sideritized deposits, i.e. para-eluvium accumulated from silica rocks (spongolites), while breccia, as well as shales with enclosed brown iron-ore, siderite and bauxite are confined to declined paleo-relief. Weathering crust rocks revealed high-value electrical conductivity and radio-activity on well loggings, as well as positive anomalies on SP-curves.

4. Results and discussion

To correlate well log in the North-Ostaninsk field, depth formation of Jurassic coal layer was determined. Coal layer U_{10} was considered as a reference horizon to correlate Paleozoic- Mesozoic contact rock zones. Coal layer U_{10} has an unambiguous geophysical characteristic and can be traced throughout the whole territory. To describe Paleozoic- Mesozoic contact zones structural correlations and cross-sections in both sub-meridional (near NS) and sub-lateral (near EW) directions were plotted (figure 3- legend identifying geological sections).

Plotted well lines 11R-12R-3R-7R-5R in the geological cross-section revealed the fact that exposed pre-Jurassic formations in the wells 11R, 3R, 7R and 5R are almost on a level with one another, whereas weathering crust (3.0-4.0 m) embraces siliceous-argillaceous iron-rich rocks, while the upper Paleozoic formation section is formed by dolomites. (figure 4).

At the same time, well 12R cross-section showed quite a different structure. Jurassic sediment thickness from U_{10} top to weathering crust top increases up to 46m. in comparison to 19-23m. in the above-described wells, while weathering crust includes permeable sandy breccia. Palaeozoic formation encloses varied altered effusives (volcanic rocks) with sparse marlstone interlayers.These rocks can be found in the down-dip block. In view of above-mentioned description, both previously mapped fault between wells 3R and 12R and inferred fault between wells 12R and 11R should be considered.

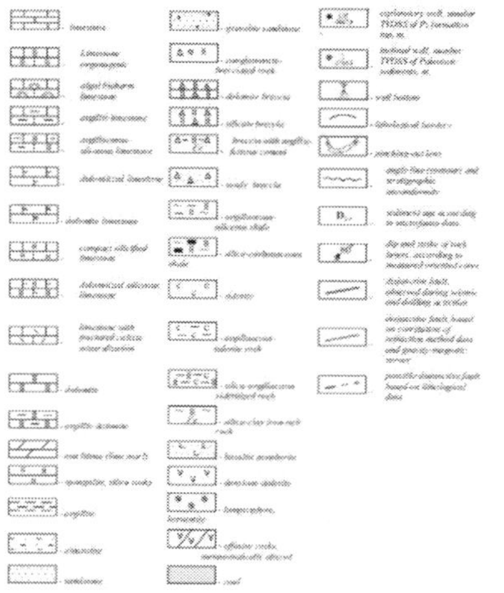

Figure 3. Legend

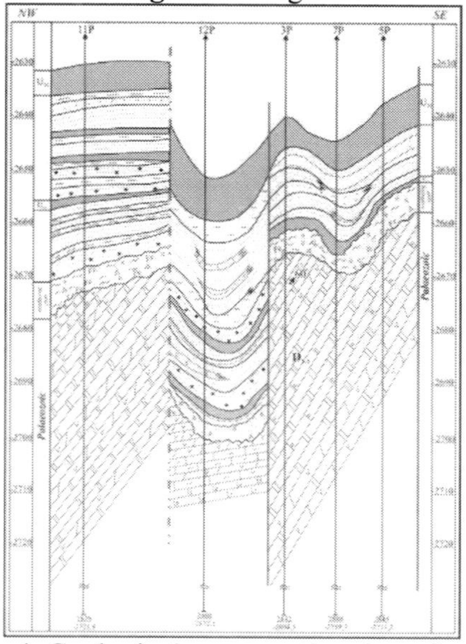

Figure 4. Geological cross-section through plotted well lines 11R-12R-3R-7R-5R depicting Paleozoic- Mesozoic contact zones in North-Ostaninsk field . Horizontal scale 1:75 000; vertical scale 1:500

Plotted well line 15R-6R-5R-14R in the geological cross-section (figure 5) revealed that the Upper Palaeozoic section is comprised of different rock composition: limestone in well 15R, breccia and fractured spongolites in well 6R, replacement dolomite in well 5R and organogenic (algae) limestone in well 14R.

High- hysdometric horizon position (coal layer U_{10}) was observed in the geological cross-section, and, respectively, pre-Jurassic top sediments in well 6R, which was drilled within theblock limited by tectonic faults. In well 6R the rock age was determined as Lower Carboniferous (C_1v), while in 5R - by analogy with well 3R data- Middle-Upper Devonian (D_2gv-D_3).

Palaezoic formation exposed in wells 5R and 14R is different in age and composition: Middle-Upper Devonian replacement dolomites and Sirulian-Lower Devonian organogenic limestone) which, in its turn, confirms the presence of tectonic faults between these wells.

Plotted well line 4R-6R-7R -3R -9R in the geological cross-section (figure 6) revealed that the Upper Paleozoic section in the wells 4R and 6R is composed of silica rocks- spongolites, although in accordance to their thickness (14m. in well 4R and 41m. in 6R) are incomparably. Besides, silica rocks in well 6R are of high porosity and permeability, due to intensive fracturing and leaching which is not observed in well 4R cross-section.

(Legend- Fig. 3)

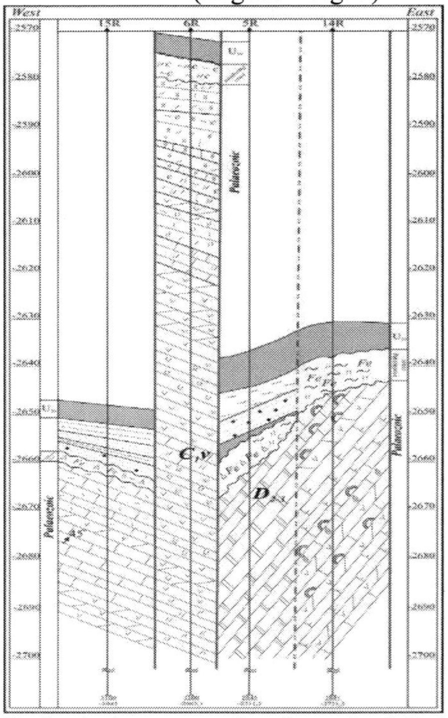

Figure 5. Geological cross-section through plotted well lines 15R-6R-5R-14R depicting Paleozoic- Mesozoic contact zones in North-Ostaninsk field . Horizontal scale 1:75 000; vertical scale 1:500 (Legend- figure 3)

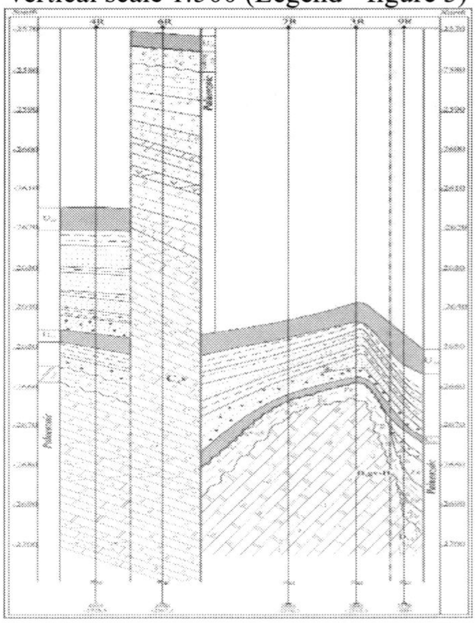

Figure 6. Geological cross-section through plotted well lines 4R-6R-7R-3R-9R depicting Paleozoic- Mesozoic contact zones in North-Ostaninsk field . Horizontal scale 1:75 000; vertical scale 1:500 (Legend- figure 3)

Weathering crust products in both wells are of siliceous- argillaceous sideritized rocks- para-eluvium[4]. High- hysdometric horizon position of pre-Jurassic top sediments in well 6R. Thus, based on the above-described characteristic cross-section feature possible tectonic faulting exists between wells 4R and 6R.

In the block where wells 7R and 3R have been drilled pre-Jurassic sediment sections have identical structure: Middle Devonian replacement dolomites (D_2gv-D_3) and siliceous- argillaceous iron-rich rocks, and in the weathering crust- breccia with agrillo-ferruginous cement.

In the block, NW to the above-described one, in well 9R Lower-Middle Devonian replacement dolomite top is located a bit lower than that in well 3R; while the weathering crust includes red breccias with agrillo-ferruginous cement (16m.). It is assumed that possible tectonic faulting exists between wells 3R and 9R.

A geological model of Paleozoic-Mesozic sediment contact zones was plotted on the basis of an integrated litho-geophysical study of the pre-Jurassic formation. Erosion-tectonic basement protusion underlying the North-Ostaninsk structure is split into separate tectonic fracture blocks. The structural map of plotted reflection horizon F_2 (refer. to figure 1) illustrates not only the block boundaries according to CRM and gravimagnetic data but also numerous random tectonic faults based on seismic survey by CDP-method. Integrated geophysical survey conducted in this studied area provided the contouring of these tectonic faults and their further merging into Paleozoic basement degradation zones (figure 7).

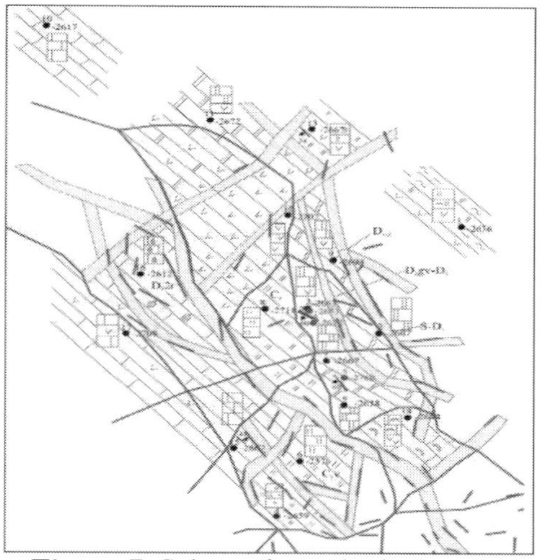

Figure 7. Schematic geological map of upper Paleozoic basement surface in North-Ostaninsk field

To study the outcropping rock cluster distribution on the Palaezoic basement surface the following components were plotted: schematic lithological columns, based on litho-geophysical well logging (upper 30m.); formation age – fauna identification; and strike and dip of rock layers – paleomagnetic investigation of oriented cores.

Summarizing performed litho-geophysical survey the distribution of corresponding rock types within isolated blocks, the boundaries of which correlate with previously identified boundaries has been proposed (figure 7).

According to lithological data, limestones, dolomites, silica rocks and effusives (volcanic rocks) outcrop onto the basement surface.

Limestones found in drilled wells 10R, 16R, 17R, 15R and 14R are in the western, south-western and south-eastern blocks of the North-Ostaninsk structure. They contain fauna remains which are indicators of Late Devonian D_3fr being intensively fractured and brecciated.

Dolomites are widely distributed in the eastern flank, as well as within the well 11R area. They are comprised of limestone interlayers with fauna, being of Silurian- Early Devonian (S-D_1) and Middle-Late Devonian (D_2gv-D_3) age. Dolomites formed as a result of metasomatism during catagenesis under the influence of enriched magnesium water on limestones, i.e. calcite replacement dolomite. The abundance of magnesium ions in the formation waters is the result of the alteration of mafic effusives (volcanic rocks). Generally, dolomites are porous and vesicular. Dolomite replacement is restricted to the oil reservoir of the North-Ostaninsk field.

Silica rocks found in drilled wells 4R, 6R, 8R and 13R are distributed in the south-western, central and north-eastern blocks of the North-Ostaninsk structure. They are basically spongolites, spicula sponges, as well as the remains of crinoids, foraminifera and other fauna which are indicators of Early Carboniferous (C_1v) rocks. These rocks have pores and vugs as a result of leached fauna remains during post-deposition. During well testing of the upper Paleozoic formation section oil bleeding was detected in well 6R.

Effusives (volcanic rocks) include basalt porpyrites, diabase, spilites, i.e. mafic rocks. According to chemical analysis results these post-depositional alterated rocks contain a significant amount of magnesium and ferrum oxides. Effusives can be practically found in all the studied cross-sections as isolated layers or members, but in well 12R these rocks are found in maximum thicknesses. It is assumed that these rocks are distributed throughout the well 12R area and NW to the fault.

5. Conclusion

Thus, in view of lithological data, fauna identification and paleomagnetic survey results, it was defined that the eroded tectonic protusion of the North- Ostaninsk field exhibits distinct inverted morphostructure features, i.e. elevated basement block reflecting structurally complex and fractured tectonic faults within syncline folded block in Paleozoic formations. North- Ostaninsk oil deposit is confined to the steep dipping NE flank of the field itself. Similar correlations as plicative basement structure elements, horizon F_2 erosion surface structures and oil field layout were defined in the following Solonovsk, Kalinov and Severo-Kalino oil fields [8]. It should be noted that above-mentioned oil fields are rather closely-spaced and concentrated in one structural element of the NE Nurolsk depression flange. The evidence of such inverted morphostructures in eroded tectonic protusions could be considered as an exploratory criterion and/or parameter in evaluating potential oil-bearing Paleozoic formations.

References

[1] Belov R, Kondrashov V and Melnikov V 1990 Opyit kompleksnogo seysmofatsialnogo analiza dannyih MOGT i KMPV *Geology oil and gas* **3** 33-36
[2] Belov R 1992 Pereinterpretatsiya seysmorazvedochnyih materialov po verhnepaleozoyskim zalezham UV yugo- vostoka Zapadnoy Sibiri *Geology oil and gas* **4** 5-8
[3] Ezhova A 2007 Genezis pustotnogo prostranstva i emkostno-filtratsionnyie svoystva paleozoyskih kollektorov mestorozhdeniy uglevodorodov Tomskoy oblasti *Geology oil and gas* **3** 53-57
[4] Ezhova A 2007 Genezis pustotnogo prostranstva i emkostno-filtratsionnyie svoystva paleozoyskih kollektorov mestorozhdeniy uglevodorodov Tomskoy oblasti *Geology oil and gas* **3** 53-57
[5] Koveshnikov A 2012 Porodyi-kollektoryi doyurskih karbonatno-kremnisto-glinistyih otlozheniy Zapadno-Sibirskoy geosineklizyi Izvestiya TPU **321** 138-142
[6] Kontorovich V 2002 Tektonika i neftegazonosnost mezozoysko-kaynozoyskih otlozheniy yugo vostochnyih rayonov Zapadnoy Sibiri. *Izd-vo SORAN* 253
[7] Merkulov V 2002 Magnitnyie polya mestorozhdeniy nefti i gaza i vozmozhnosti ih ispolzovaniya pri kartirovanii zalezhey uglevodorodov *Izvestiya Tomsk Polytechnic University* **305** 218-225
[8] Tischenko G, Zyatev G and Merkulov V 1990 Opyit ispolzovaniya elementov zaleganiya porod pri modelirovanii zalezhey nefti i gaza *Geology oil and gas* **10** 15-17

IOP Conf. Series: Earth and Environmental Science **24** (2015) 012014 doi:10.1088/1755-1315/24/1/012014

Stratigraphy and correlation of Lower-Middle Jurassic sediments in SE West-Siberian petroleum-bearing province

A Ezhova[1], A Ostankov[2], E Panova[3] and R Abramova[4]
[1,3]Department of Geology and Mineral Exploration, Institute of Natural Resources, Tomsk Polytechnic University
[2]JSC Tomskgazprom
[4]Department of Foreign Languages, Institute of Natural Resources, Tomsk Polytechnic University
E-mail: [1]eav@tpu.ru

Abstract. Based on the investigation of Lower-Middle Jurassic sediments in SW Siberia, isochron reference clay and coal horizons confined to the roof of chronostratigraphic units were identified; bed series from pre-Jurassic formation bottom to U_{10} coal layer were divided; structural features and distribution of regional cyclites J_{10}-J_{17} in the sequences and throughout the area were determined.

1. Introduction

The studied area is located within the south-east of Tomsk Oblast, whereas, in terms of tectonic relationship - within the SE West-Siberian platform, to the north of Kalgach meso-block, including Boltnoe, Kazan, Ponomarev, Rogalev, West-Somov and Novosomol uplifting (figure 1).

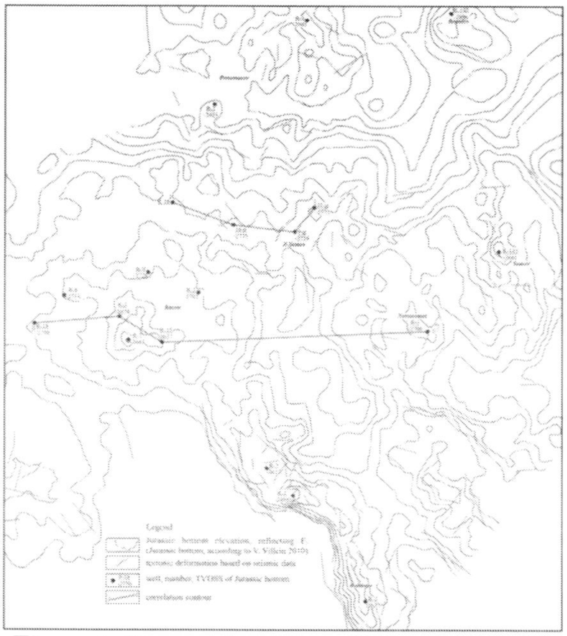

Figure 1. Structural map of reflecting horizon F_2 (Jurassic bottom, according to V Vilkin, 2010)

Hydrocarbon potential of this territory is correlated with Upper Jurassic sediments, where Kazan and Boltnoe oil fields are being developed. In recent years special consideration is being given to Early Middle Jurassic rocks which are potentially hydrocarbon bearing, in view of limited oil flow recovery and flow properties observed throughout thickness sequence.

The major target of the following research includes dismembering and correlation of the Lower - Middle Jurassic sediments in the sequences, as well as profile planning.

2. Initial data

Research data included the following: structural map of reflecting horizon F_2 (Jurassic bottom, according to V Vilkin, 2010); spontaneous potential (SP) diagrams, apparent resistivity, (AR), induction resistivity (IR), natural radioactivity (NR), and neutron loggings and core samples.

Pre-Jurassic formation surface embraces dismembered relief and numerous tectonic deformations (figure 1). Currently, 19 drilled wells have entered pre-Jurassic formation within the studied area. However, these wells are wide-spaced which hampers the possible correlation of geological sections and further investigation of rock distribution in profile plan.

Lower-Middle Jurassic sediments are marked by vertical and horizontal inhomogeneity. Due to the lack of paleontological data, layering is based on the results of geologic-geophysical surveys, spore-pollen analysis, and analogy with contemporaneous rocks containing marine organism remains. The following researchers gave a detailed stratigraphic description of Lower and Middle Jurassic formations in SE West Siberia: F.G. Gurari, V.P. Devjatov, A.M. Kazak, L.V. Smirnov, A.E. Ekhanin, V.I. Moskvin, L.I. Egorov, R.V. Belov, O.S. Chernov, V.A. Kontorovich, et al [1, 2, 3, 4, 6, 8, 9, 10, 11, 12]. Lower Jurassic sediments were tapped in SE West Siberian petroleum province within Ust-Tym, Nurolsk, Barchar and other mega-depressions. However, Middle Jurassic sediments (J_{15} and upwards) have been investigated to a lesser extent.

Dismembering and correlation of Lower-Middle Jurassic sediments (from pre-Jurassic formation bottom to coal layer U_{10}) were based on the principle of sedimentation cycle [7,5], including the following:

1) unified sedimentation sequence, not involving perturbation and concealing of large stratigraphic unit sections, has been observed;
2) horizon markers showing isochrohism features within a relatively limited area and well-defined field-geophysical characteristics have been identified in the sequences.

3. Results and discussion

Horizon markers in studied sequences are defined as coal layers (U_{10}, U_{11}, U_{12}, U_{13}, U_{14}, U_{15}), argillaceous formations (Togur suite and middle subsuite of Urman suite) and Jurassic sediments with crust weathering rocks – Paleozoic sediments contact. Chronostratigraphic units in correlating the sequences are cyclites wherein the top are either underlying coal marker layers of continental sedimentation and / or argillaceous formations formed in marine fresh lake basins.

Basically, regional cyclites are composed of small priority cyclites (zonal and local) which exhibit changing composition orientation from layer to layer and complex texture, i.e. these cyclites are composed of rock layers or beds with gradual and/or sharp boundaries. Accordingly, it should be noted that there is a difference in the two terms: "cyclite" and "layer". The latter is homogeneous three-dimension geological body, bounded above and below by subparallel surfaces. Cyclite – association of at least two layers (beds) recurring in time and space and being the composition reflection of a specific sedimentation cycle. Relating cyclites as chronostratigraphic units of different priorities significantly improves reliable correlation of sedimentary formations, especially for changing lateral continental formations, whereas, reality, sand layers formed at different periods are correlated.

Indexing regional cyclites (as described in previous research [5]) is based on the number of coal layer underlying corresponding formation roof. Genetic relation of sand and coal layers within cyclites can be explained by the facies-phase and facies-cycle sedimentation models: rock unit alteration in sequences, consistent with two phases-tectonic activities in the bottom and tectonic quiescence in the roof. It is well-known that regional continuous coal layers formed during a period of maximum tectonic quiescence, relief leveling and minimum water environment dynamic are relevant to the completion of large sedimentation cycles, which, in its turn, indicates the interstratified rock system of priority regional cyclites.

Integrating all stratigraphic research data of previous years, a conceptually updated stratigraphic chart of Lower-Middle Jurassic sediments (figure 2) was compiled which included the alignment and correlation of the sequences (figures 3 and 4).

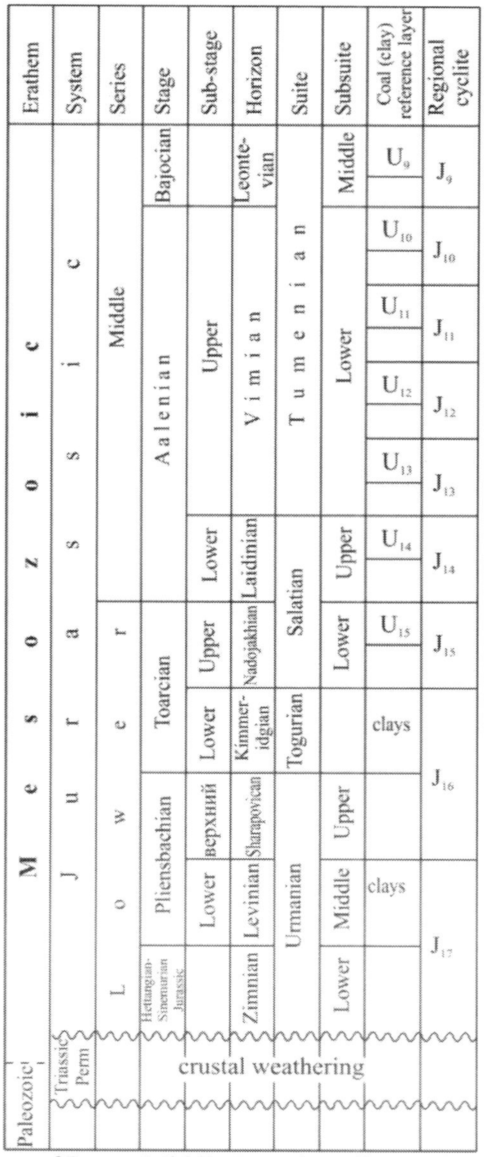

Figure 2. Stratigraphic chart of Lower Middle Jurassic sediments in SE West-Siberian platform (compiled by A Ezhova, 2013)

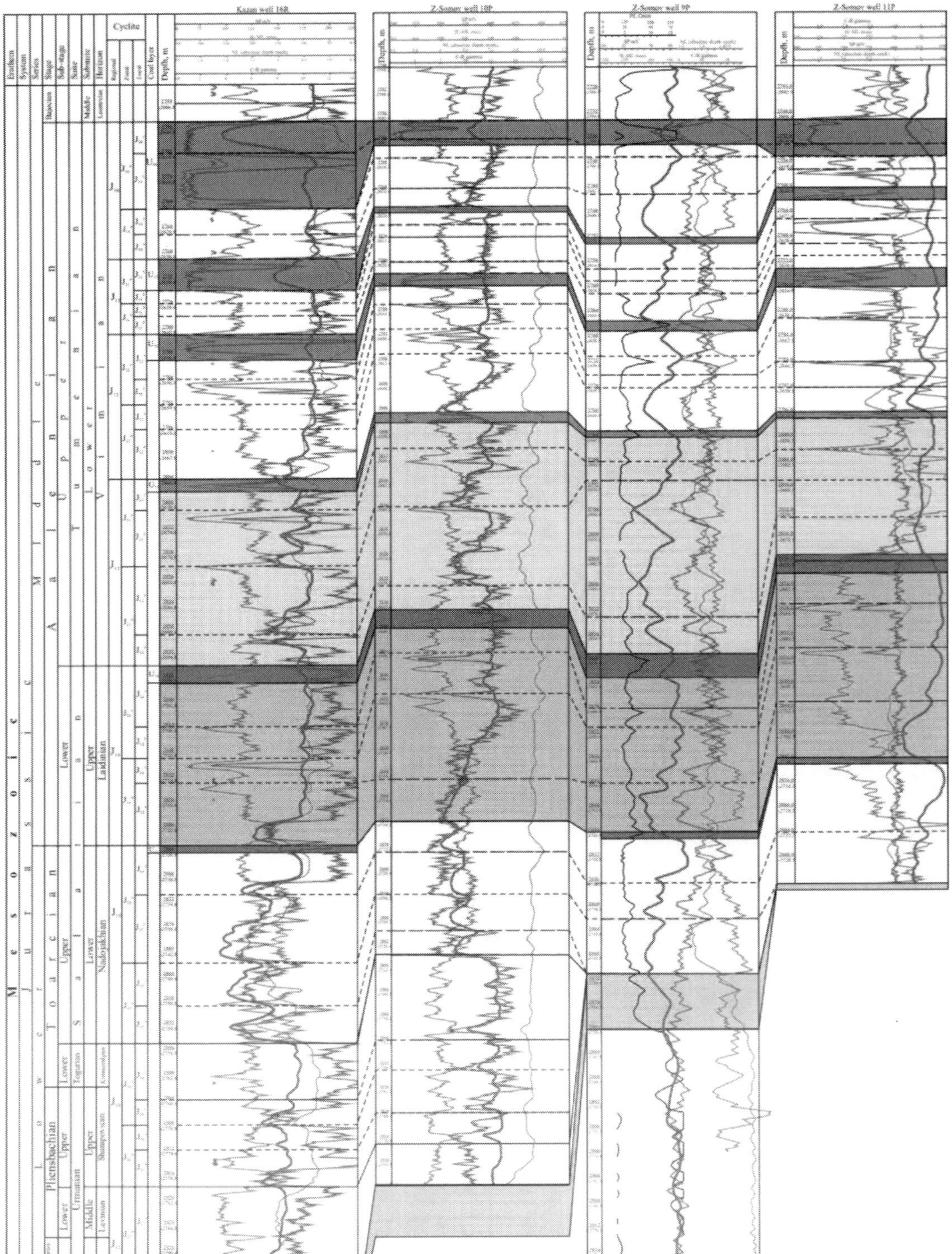

Figure 3. Correlation chart of Lower-Middle Jurassic sediments according to line drilling I-I

IOP Conf. Series: Earth and Environmental Science 24 (2015) 012014 doi:10.1088/1755-1315/24/1/012014

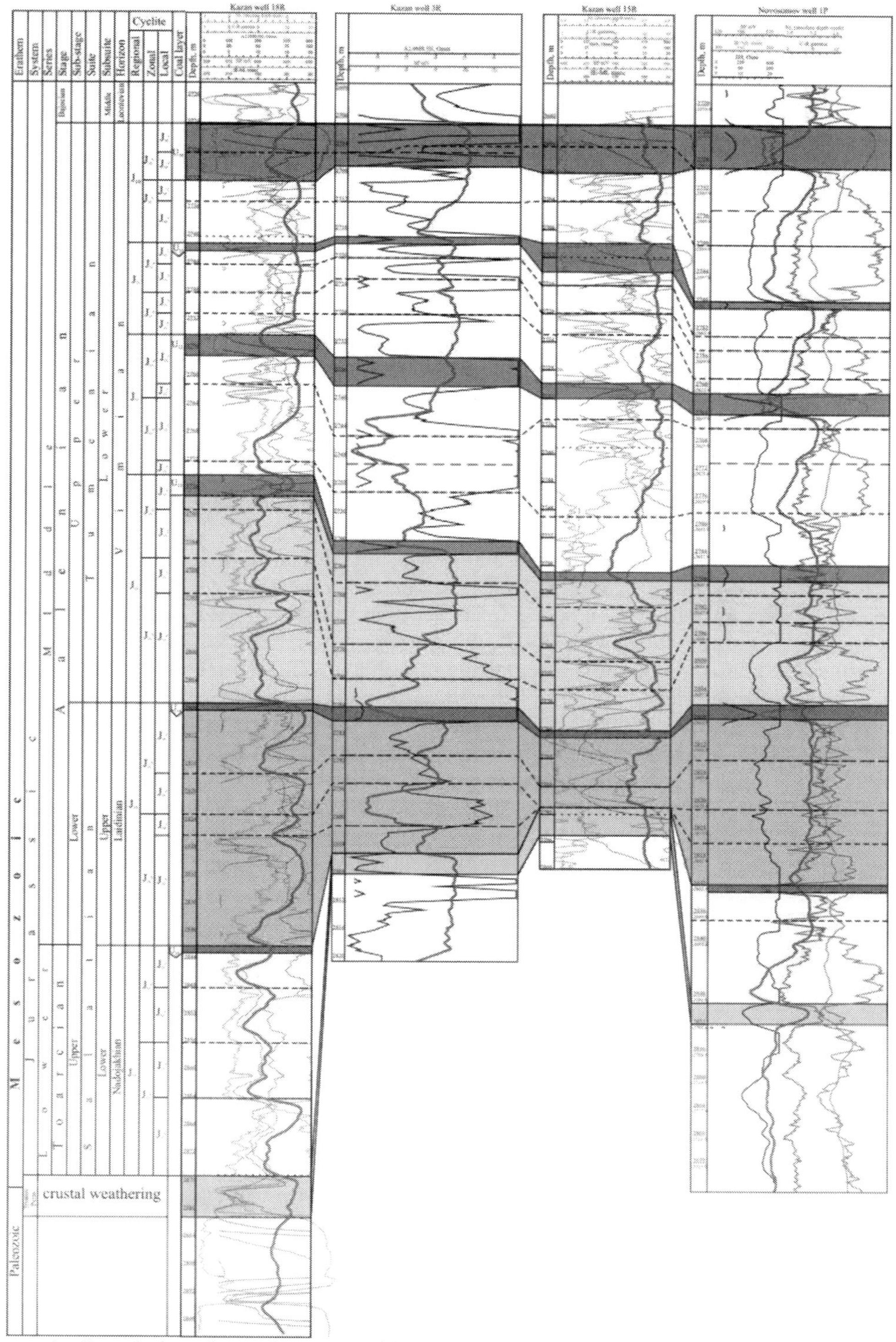

Figure 4. Correlation chart of Lower-Middle Jurassic sediments according to line drilling II-II

The flattening line is the roof of U_{10} coal layer which is considered to be the regional horizon marker in SE West Siberia. According to palynology data U_{10} layer roof is related to Aalenian sediments [3] and is the interface between the lower and middle subsuite of Tumen suite, as well as Vimian stratigraphic horizon roof [6, 2].

Logging response show that U_{10} layer exhibits low values, gamma ray and neutron logging curves, high resistance, low electroconductivity and frequent deep reversed-polarised anomaly. Layer thickness changes from 3 to 14 m, moreover, this layer delaminates and is separated by clay bands.

The regional *cyclite* J_{10} thickness in Kazan uplift embraces 16-22m; in Zapadno-Somov uplift - 12-19m; in NovoSomov and Somov uplifts increases from 25 to 28m; in Ponomarev uplift - sharply increases from 32 to 33m; while in Rogalev and Boltnoe areas decreases down to 19 and 15m., respectively.

The regional cyclite J_{10} is, in some cases, distinctly or conventionally subdivided into zonal $J_{10}^{в}$ (upper) and $J_{10}^{н}$ (lower) cyclites, whereas, further they are divided into local cyclites J_{10}^{1}, J_{10}^{2}, J_{10}^{3} and J_{10}^{4}.

More or less, practically in all sequences of the local cyclite bottoms, especially J_{10}^{4} sandstones, frequently, with carbonate cement, are deposited (pertaining to core sample from Ponomarev well 1P and increased neutron logging values).

Coal layer U_{11} thickness is 1m., increasing up to 5m. in Kazan 16R and Somov 145R wells, while in Ponomarev wells 1P and 2P from 3 to 4m. Regional *cyclite* J_{11}, as described above, is divided into 4 local cyclites, in the bottom of which, sandstones or aleurolites are deposited, while in the top – thin coal interlayers.Cyclite U_{11} thickness ranges from 11 to 20 m., increases up to 22-31 m within Ponomarev and Somov areas, while in Boltnoe well 3-only 6 m.

Coal layer U_{12} with thickness of 2-4 m is rather distinct in the sequences with minimum values of gamma ray and neutron logging, high resistance and low electroconductivity. Regional cyclite J_{12} has inconsistent thickness – from 14 to 33 m and is divided into local cyclites, where sandstones, and even gravelstones are deposited in the bottom (according to core-log data correlation from Zapadno-Somov 9P and Ponomarev 1P wells) and distinct coal interlayers in the top (according to core-log data correlation).

Regional cyclite J_{13}, separated from overlying coal layer U_{13} sediments and having a thickness of 1-3 m embraces a sequence of sand (2-4 m), clay (1-3m) and coal (1 - 1.5-m) interlayers. Due to the existing differential lithological composition, the local cyclites J_{13}^{1}, J_{13}^{2}, J_{13}^{3} и J_{13}^{4} are distinctly identified within the sequences. The deep-seated anomaly is confined to cyclite J_{13}^{4}. The core samples from Zapadno-Somov 11R and Ponomarev 2P wells exhibited coarse-grained sandstones, while the core samples from Kazan 8R well- gravelstones and sandstones with pebbles. Thetre are sandstones, often with carbonate cement in the bottom of overlying local cyclites and clay and coal in the top. The total thickness of regional cyclite J_{13} ranges from 20-37 m., whereas in the cross-sections of Somov 145R and Boltnoe 2P wells, the cyclite J_{13} rocks overlie crust weathering formations having reduced thickness of 15m and 10m., respectively. The bottom of cyclite J_{13} is the lower interface of the upper substage of Aalenian sequence, lower subsuite of Tumen suite and Vimian stratigraphic horizon (figure 2).

The coal layer U_{14} is often splitted into two layers (Kazan 9R, 2R, 16R wells; Zapadno-Somov 10P, 9P and 11P wells; Ponomarev 1P well) having a thickness of 1-4m. Regional cyclite J_{14} is distinctly different from the above-described cyclite J_{13} in its argillaceous composition, and as the overlying cyclites, regional cyclite J_{14} is divided into local ones. However, in this case, sand sediments are mainly confined to the bottom- J_{13}^{4} , whereas, core samples from Zapadno-Somov 9R and Kazan 8R, 9R wells showed the presence of conglomerates, gravelites, sandstone with gravel, while core samples from NovoSomov 1P, Ponomarev 2P and Zapadno-Somov 11P – medium to fine grained sandstones. Fine-grained sandstones, and often aleurolites are deposited in the bottom of local cyclites J_{14}^{3}, J_{14}^{2}, J_{14}^{1}. The roof of these cyclites are apparently evident on interpretated logs and were verified by coal interlayers detected in core samples (Zapadno-Somov 9P, 10P, 11P, Kazan 16R, 18R and Ponomarev 2P wells).

However, the regional cyclite J_{14} has a rather distinct sand sequence as found in some wells, i.e. Kazan wells 18R, 8R, 3R, 9R, 2R and 15R. The layers in these wells have a rather profound SP amplitude and are confined to cyclites J_{14}^1, J_{14}^2 and J_{14}^3 whereas laminating coal interlayers are not well-pronounced. Total thickness of cyclite J_{14} involves 26–38m. , while in Kazan 3R, 8R and 15R wells it decreases to 10–21m., as eroded rocks lie on argillaceous-siliceous crust weathering sediments.

Stratigraphically, regional cyclite J_{14} is correlated to lower substage of Upper subsuite Salatian suite and Ladinian horizon (figure2).

The roof of regional cyclite J_{15} is determined by the coal layer U_{15} defined by core samples from Zapadno-Somov 9P, 11P, Kazan 9R, and NovoSomov 1P wells. According to the Zapadno-Somov 10P well cross-section, this cyclite is poorly-defined and the interface is determined relative to the clay roof. The regional cyclite J_{15} bottom is determined relative to the Togur suite clay roof (Kazan 16R, Zapadno-Somov 10P , Ponomarev 2P and 1P wells) and / or Pre-Jurassic formation roof (Zapadno-Somov 9P, 11P, Kazan 18R, 9R, NovoSomov 1P, Rogalev 145R wells). Erosion contact on the interface of cyclite J_{15} and pre-Jurassic formations (crust weathering) was described: breccia, conglomerate-breccia, gravelites overlie argillaceous-siliceous rocks (NovoSomov 1P, Zapadno-Somov 9P, 11P and Kazan 18R and 8R wells).

Regional cyclite J_{15} thickness is rather inconsistent: from 17 to 40m., while these sediments are absent in Kazan 2R, 3R, 8R and 15R wells. According to production logging data and investigated core samples, well-defined local cyclites J_{15}^1, J_{15}^2, J_{15}^3 and J_{15}^4 ,embracing clastics in the bottom and argillaceous-coal interlayers in the top, are observed in "total" regional cyclite J_{15} sequences (Kazan 16R, 18R, 9R, Ponomarev 2P, 1P wells).

Regional cyclite J_{15} is associated with upper substage of Toarcian stage, lower subsuite of Salatian suite and Nadojakhian horizon, while its roof is along the Lowerr and Middle Jurassic system series boundary (figure 2).

Togur suite (Kimmeridgian horizon) was revealed in Kazan 16R, Zapadno-Somov10P and Ponomarev 2P, 1P wells. Based on core sample description this suite involves dark gray, practically black thin – layered argillites with interlayers and lenses of aleurolites and coal, the age of which was determined as Lower Toarcian by analogy with relevant faunal marine genesis sequences [2,3,4, 6, 8, 10, 11, 12]. The thickness ranges from 7 to 27 m.

Interlaying rocks refer to Urman suite. Upper Urman subsuite has been identified as J_{16} layers [7, 8, 13, 23]. The authors propose to consider this sequence as regional cyclite $Ю_{16}$ being subdivided into local cyclites J_{16}^4, J_{16}^3, J_{16}^2 and J_{16}^1; the roof of the latter cyclite (according to cyclite identification classification [7]) is marine argillite sediments of Togur suite. This structure is well-defined on interpretated loggings and investigated core samples from cross-sections of Kazan 16R, Zapadno-Somov 10P and Ponomarev 2P wells, and probably, Ponomarev 1P well. The sand section of regional cyclite J_{16} (according to the stratigraphic chart) is identified as upper substage of Pleinsbachian stage, upper subsuite of Urman suite and Sharapavian horizon (figure 2). The thickness of regional cyclite J_{16} in the studied territory is 25–46 m.

Underlying sequence is identified as regional cyclite J_{17} based on the description of investigated core samples from Ponomarev 2P and Kazan 16R well cross-sections. According to core sample description these sediments are subdivided into local cyclites, including aleurolites in the bottom and argillites in the top of each cyclite. Upper argillite sequence is identified as lower substage of Pleinsbachian stage and is related to middle subsuite of Urman suite and Levinian horizon. The core sample from Kazan 16R well at a depth of 2930m showed sandstone and along its bottom the lower boundary of local cyclite J_{17}^1 was defined.

Underlying Hettangian-Sinemurian rocks are probably of related to lower subsuite of Urman suite and Zimnian horizon. The total thickness of the regional cyclite J_1 in Ponomarev 2P is 46m and in Kazan 16R- 19m.

Pre-Jurassic formation in the investigated area is described as argillaceous-siliceous shales and probably related to crust weathering rocks. Paleozoic limestones have been identified in the core samples from Kazan 18R well (depth-2882m.), 8R (depth-2855m.) wells, as well as from Kazan 15R (depth – 2791m.) and 3R (depth-2808m.) wells and Zapadno-Somov 9P (depth-2880m.).

4. Conclusion

1. Penetrated thickness in investigated Lower-Middle Jurassic sediment sequence (from pre-Jurassic formation to the top of coal layer U_{10}) changes from 45 to 199m., while the total thickness in Ponomarev 2P well is 276m.

2. Application of sedimentation cyclite principles in studying the Lower-Middle Jurassic formation provided the further identification of chronostratigraphic units and possible observation of their horizontal and vertical sequence development.

3. In terms of hydrocarbon potential the most significant are cyclite J_{13}, J_{14} and J_{15} rocks. The alteration of sandstone reservoir rocks and argillite screen rocks shows oil-saturation characteristic features.

References

[1] Belov R 1998 Zakonomernosti rasprostraneniya zon, blagopriyatnyih dlya akkumulyatsii uglevodorodov v domelovyih otlozheniyah yugo-vostoka Zapadnoy Sibir (PhD paper) Moscow

[2] Geologicheskoe stroenie i neftegazonosnost nizhney – sredney yuryi Zapadno-Sibirskoy provintsii 2005 ed. Gurari F, Devjatov V , Demin V and et al Novosibirsk Nayka p 156.

[3] Gurari F, Ekhanin A and Moskvin V 1988 Markiruyuschie gorizontyi i problemyi korrelyatsii razrezov nizhney chasti chehla tsentra i yuga Zapadno-Sibirskoy plityi *Regional Stratigraphy of Siberian oil-gas bearing provinces* Novosibirsk p 44-53.

[4] Egorova L 1992 Geologiya i kriterii neftegazonosnosti nizhneyurskih otlozheniy yugo-vostoka Zapadno-Sibirskoy plityi Novosibirsk p 25

[5] Ezhova A 2007 Primenenie sistemnogo analiza dlya raschleneniya i korrelyatsii yurskih terrigennyih razrezov na mestorozhdeniyah uglevodorodov Tomskoy oblasti *J. Ivestiya* Tomsk Polytechnic University **311** (1) 59-63

[6] Kazakov A, Devjatov V and Smirnov L 1992 Indeksatsiya plastov gruppyi «Yu» v nizhnesredneyurskih otlozheniyah Zapadnoy Sibiri *J. Aktualnyie problemyi regionalnoy geologii Sibiri* Novosibirsk 64-65.

[7] Karogodin Yu 1990 Vvedenie v neftyanuyu litmologiya *Proceedings of Institute of Geology and Geophysics* Nayka Novosibirsk p 239

[8] Kontorovich V 2002 Tektonika i neftegazonosnost mezozoysko-kaynozoyskih otlozheniy yugo-vostochnyih rayonov Zapadnoy Sibiri *Publishing House SB RAS "GEO"* Novosibirsk p 253

[9] Litostratigrafiya kontinentalnyih otlozheniy nizhney i sredney yuryi yugo-vostoka Zapadno Sibirskoy plityi 1985 / ed. Belozerov B, Brilina N, Danenberg E and Kovalev N *Regionalnaya stratigrafiya neftegazonosnyih provintsiy Sibiri* Novosibirsk p 111-119.

[10] Problemyi stratigrafii nizhney i sredney yuryi Zapadnoy Sibiri 1994/ Devjatov V , Kazakov A, Kasatkina G et al *J. Geology and geophysics* **12** p 3-17.

[11] Stratigrafiya paleogeografiya ranney i sredney yuryi Zapadno-Sibirskoy plityi 1988 / Gurari F, Budnikov I, Devjatov V Ekhanin A, Kazakov A, Moskvin V // *Regionalnaya stratigrafiya neftegazonosnyih provintsiy Sibiri* Novosibirsk p 60-75.

[12] Chernova O 1995 Litologiya i paleogeografiya nizhneyurskih otlozheniy zapadnoy chaste Tomskoy oblasti Novosibirsk p 28

IOP Conf. Series: Earth and Environmental Science **24** (2015) 012015 doi:10.1088/1755-1315/24/1/012015

Pellet impact drilling operational parameters: experimental research

A V Kovalyov[1], S Ya Ryabchikov[2], Ye D Isaev[3], F R Aliev[4], M V Gorbenko[5] and A V Baranova[6]

Institute of Natural Resources, National Research Tomsk Polytechnic University, 30 Lenina Ave., Tomsk, 634050, Russia

[1]arteka011287@mail.ru , [2]kafedrabs@mail.ru , [3]pelletdrilling@gmail.com, [4]faruh_aliev@mail.ru, [5]gmvski@rambler.ru, [6]baranskikh@yandex.ru

Abstract. The article deals with the study of particle-impact drilling that is designed to enhance the rate-of-penetration function in hard and tough drilling environments. It contains the experimental results on relation between drilling parameters and drilling efficiency, the experiments being conducted by means of a specially designed laboratory model. To interpret the results properly a high-speed camera was used to capture the pellet motion. These results can be used to choose optimal parameters, as well as to develop enhanced design of ejector pellet impact drill bits.

1. Introduction

Nowadays the pellet-impact drilling is being actively studied in Drilling Department, Institute of Natural Resources, Tomsk Polytechnic University. This drilling method has a great potential for enhancing the rate of penetration within hard and tough drilling environments and saving time for tripping, which will lead to well cost reduction. Besides, pellet impact drilling can be easily incorporated into the existing mechanical drilling technology without significant reconstruction of a drilling rig.

There are several operational parameters of pellet impact drilling that influence the efficiency of rock crushing, such as: pellet batch weight, diameter and pellet material properties, interval between a drill bit and a hole-bottom, circulation rate, drilling mud type and properties.

The relation between drilling parameters and efficiency of pellet impact drilling was studied and presented in the following works [1, 4-6]. However, the results interpretation seems controversial and unconvincing. It can be explained by the fact that previous testing equipment didn't make it possible to study the pellet dynamics in detail. Thus, the conclusions and recommendations made by the researchers were based on their deduction rather than factual proofs, and may contain inaccurate information.

2. Experiment procedure and statistical data processing

To study the technological processes of pellet impact drilling, a laboratory scale model [3] was designed; figure 1 demonstrates its general layout. A cup (10) is placed on a tray (1) and fixed with a special clinch (3). The cup is used to set a rock sample and observe the process of pellet aspiration and journey. A support (5) is placed on the tray and fixed with a bar (4) in the upper part. To provide the desired distance between the sample and the drill bit and to ensure a smooth descent of the latter into the well model, there is a bit feed mechanism (6) based on friction and rack gear with a flywheel (12). Fluid supply is provided by means of a pumping main (7) and an adapter sub (8) to the drilling string (9). After exerting a drilling action the fluid with cuttings flows from the cup into the tray and then into a drain line (2). A pressure meter (11) is installed above the pellet-impact drill bit for measuring pressure in the pumping main. A cone-shaped arrester (13) ensures forced pellets supply to the ejector pellet impact drill string. Together with a centralizer (14) it lines up the bit in the cup.

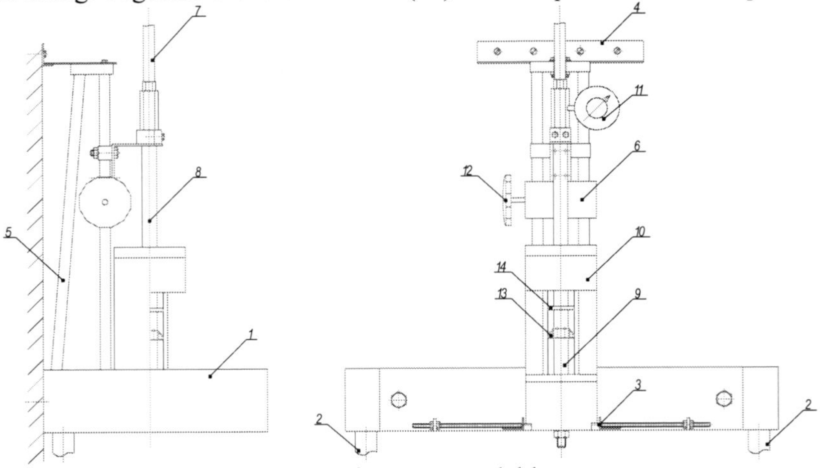

Figure 1. Laboratory model layout:
1- tray; 2- drain line; 3- cup clinch; 4 – bar; 5- support; 6- bit feed mechanism; 7- pumping main; 8- adapter sub; 9 – ejector pallet impact drill string; 10 – cup; 11 – pressure meter; 12 – flywheel for feed process control; 13 - arrester; 14 – centralizer.

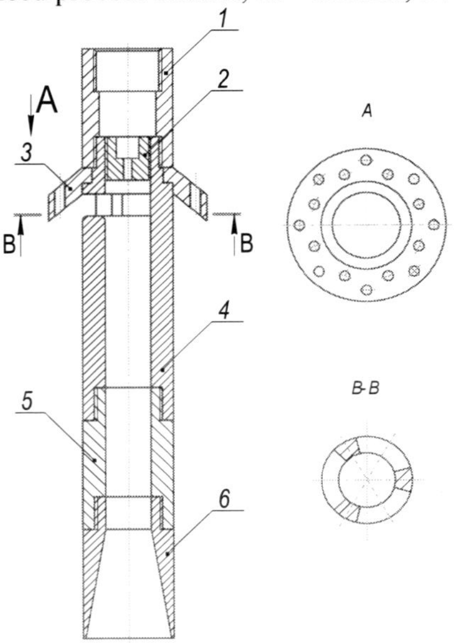

Figure 2. Ejector pellet impact drill bit
1 – discharge sub; 2 – nozzle, 3 – arrester; 4 – inlet chamber;
5 – mixing chamber; 6 – diffuser.

The design of the ejector pellet impact drill bit is shown in figure 2. The drill bit consists of a changeable nozzle (2) and a cone-shaped arrester (3), an inlet chamber (4), a mixing chamber (5) and a diffuser (6). The operation principle of such device is the following: high velocity fluid jet passes through the discharge sub, accelerates in the nozzle and then flows out into the mixing chamber (5). The suction pressure that occurs in the vicinity of the nozzle output ensures aspiration of the fluid with suspended pellets and cuttings from the bore-hole annulus through the holes into the inlet chamber. Then the two-phase mixture passes into the diffuser through the mixing chamber, impinges against the rock and exerts a crushing action. The arrester is used to overcome limitations on circulation rate, as well as to direct the pellets into the inlet chamber perforations.

The experimental research was conducted according to the following methods:

1. Spudding. Drilling was carried out on porcelain tile samples within a set period without lowering the drill bit, well geometrical parameter measurement following this process.
2. Observing. Pellet impact drilling with a transparent bit was filmed with a high-speed camera (3600 frames per second). The video was then analyzed with special computer programs.
3. Reciprocating. Marble samples were drilled. After a definite penetration the drill bit was placed on the hole bottom and then was lifted to keep a set interval between the drill bit and the hole bottom. The reciprocation was carried out at intervals t_{recipr}. The experiment was followed by measuring the parameters: well diameter D_{well}, total well depth h, cylindrical section depth h_{cyl} and bit-hole bottom contact depth h_{bot}.

Minimal number of replications was 3, then arithmetical mean, dispersion, mean square deviation and variation coefficient of the measured parameter were calculated [2]. The variation coefficient did not exceed 10%, which indicates good reproductability.

3. Experimental research of relation between drill bit-hole bottom distance and pellet impact drilling efficiency

The relation between bit-hole bottom distance and pellet-impact drilling efficiency while spudding was studied, other factors being equal. Reference parameters were: d_n=2 мм, d_{mc}=8 мм, r_{nmc}=12 мм, l_{mc}=92 мм, α_{exp}=10°, d^o_{db}=16 мм, D_p=3 мм, M_p=25 г, Q_n=6.7 liter/min, P_n=1.1 MPa, the material under crushing - porcelain tile, the cone-shaped arrester, t=2 min.

It was found out that the increase in distance between the drill bit and the hole bottom results in well diameter and well volume decrease (figure 3). In this case the penetration rate drops.

The well diameter increase can be expressed by the following formula:

, (1)

where R_{well} – well radius, mm; R_{mc} – mixing chmber radius, mm; R_{bb}– distance between the drill bit and the hole bottom, mm; α_{exp} – expansion angle of the diffuser.

As known [4], well volume V_w is calculated according to the formula:

, (2)

where V_p – crater volume resulted from one pellet impact, mm^3; Q_p – pellet flow in the mixing chamber, p/sec; t – drilling time, c; ρ – coefficient that takes into account velocity reduction in real conditions.

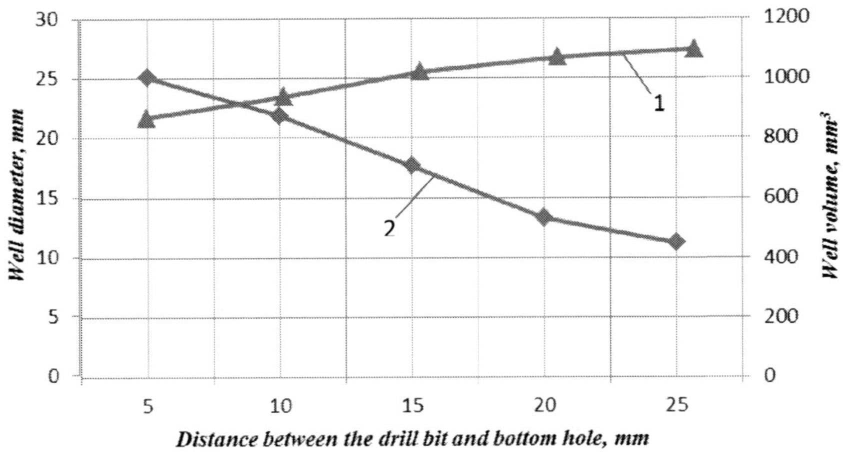

Figure 3. Well diameter (1) and volume (2) dependence on bit –hole bottom distance

The high speed camera made it possible to determine Q_p value with varying the distance between the drill bit and the hole bottom in series 5-10-15 mm (table 1). It is obvious that pellet flow in the mixing chamber decreases on increasing the bit-bottom distance. With the distance increase from 5mm to 10 mm, Q_p value reduces by 10.3% and if the distance increases from 5 to 15 mm, Q_p value becomes less by 29.9%. The well volume should reduce due to pellet flow decrease according to the formula (2), which is proved by the results shown in figure 4.

Table1. Pellet flow in the mixing chamber with bit-bottom distance variations

Distance between the drill bit and the bottom R_{bb}, mm	5	10	15
Pellet flow in the mixing camera Q_p, pellets/sec	583	529	497

While observing the pellet motion we discovered that pellet clouds occurring under the bit cannot leave the area quickly. As the bit-hole bottom distance increases, the number of clouds increases, which results in reduction of pellet flow in the mixing chamber. If the distance is short, the interaction between up-flowing and down-flowing fluids facilitates the pellets to return up the bore-hole annulus after striking against the bottom. If the distance is long, the pellets of the down-flowing fluid bombard the up-flowing pellets, making them accumulate under the bit.

R_{bb}, mm	5	10	15

Figure 4. Interaction between pellet concentration in the bore-hole annulus and bit-hole bottom distance

The research in influence of bit –hole bottom distance on marble drilling efficiency was conducted. Drilling with reciprocation was carried out with the bit-hole bottom distance being 5, 10 and 15 mm.

The well geometrical parameters were measured after the drilling (table 2). The obtained results show that the bit-hole bottom distance increase leads to the increase in well diameter but reduces its depth and volume, thus proving the dependences mentioned above.

Table2. Interaction between well geometrical parameters and bit-bottom distance values

Distance between drill bit and a hole bottom R_{bb}, mm	Well geometrical parameters				
	Well volume V_w, mm^3	Well diameter D_w, mm	Cylindrical section depth h_{cyl}, mm	Drill bit-hole bottom contact depth h_{cont}, mm	Total depth h, mm
5	31800	25.5	52	65.2	69.9
10	29800	26.5	44	57	61
15	27000	27	36	52.1	56.7
Reference data: d_n=2 mm, d_{mc}=8 mm, r_{nmc}=6 mm, l_{mc}=92 mm α_{exp}=10°, d^o_{db}=16 mm, D_p=3 mm, M_p=25 g, Q_n=6.7 liter/min, P_n=1.1 MPa, rock sample – marble, cone-shaped arrester, t_{trip}=4 min, t=40 min.					

4. Experimental study of relation between pellet batch weight and pellet impact drilling efficiency

The first stage of the research was carried out while spudding. The reference parameters were: d_n=2 mm, d_{mc}=8 mm, r_{nmc}=12 mm, l_{mc}=92 mm α_{exp}=10°, d^o_{db}=16 mm, D_p=3 mm, R_{bb}= 10 mm, Q_n=6.7 liter/min, P_n=1.1 MPa, porcelain tile as a sample material under crushing, a cone-shaped arrester, t=2min.

It was found that there is weak dependence between increasing weight of pellet batch and well diameter (figure 5). As it concerns the well volume, it starts to rise with the increasing weight of pellet batch, however, further increase in pellet batch weight results in well volume decrease. It proves the fact that there is an optimal pellet batch weight value for particular drilling environment. At the beginning of the experiment the well volume is less due to smaller number of pellet impacts. Then, the increase in well volume is observed to a definite extend with reaching optimal value of pellet batch weight. Further increase in the value leads to well volume reduction and on reaching the critical pellet batch weight the drilling process ceased due to pellet jamming under the arrester.

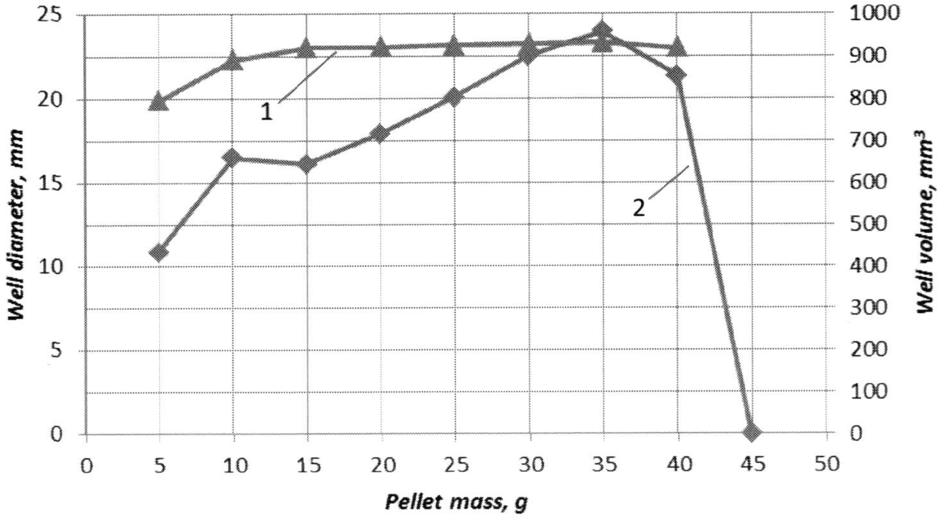

Figure 5. Dependence of well diameter (1) and volume (2) on pellet batch weight

The drilling process with varying pellet batch weight in series 5-15-25 g was recorded on a video. The video data analysis allowed identifying the number of pellets passing the mixing camera per time unit (table 3). According to the data, the pellet flow in the mixing chamber increases with the pellet batch becoming heavier, which results in better drilling efficiency.

Table 3. Interaction between pellet flow in the mixing chamber and pellet batch weight values

Pellet batch (portion) weight M_p, g	5	15	25
Pellet flow Q_p, pellet/sec	176	525	824

5. Experimental study on dependence of pellet diameter on pellet impact drilling efficiency

The research was carried out by drilling marble sample with drill bit reciprocating to discover the relation between pellet diameter and marble crushing efficiency. The results shown in table 4 prove that such dependence does exist for well diameter, depth and volume.

Table 4. Dependence of drilling efficiency on pellet diameter

Pellet diameter D_p, mm	Well volume V_w, mm^3	Well diameter D_w, mm	Well cylindrical section depth h_{cyl}, mm	Drill bit-hole bottom contact depth h_{cont}, mm	Total depth h, mm
2	19000	23	31.2	45	49.9
3	26600	23.9	41.6	55	58.7
3.5	29800	25.8	45.8	58.7	63.3
Reference data: d_n=2 mm, d_{mc}=8 mm, r_{nmc}=6 mm, l_{mc}=92 mm, α_{exp}=10°, d^o_{db}=16 mm, M_p=30 g, R_{bb}=10 mm, Q_n=6.7 liter/min, Pn=1.1 Mpa, rock sample –marble, cone-shaped arrester, t_{recipr}=4 min, t=40 min					

To interpret the obtained data a high speed camera was used. The video data processing allowed identifying pellet flow rate and pellet velocity in the mixing chamber when using pellets of different diameters (table 5).

Table 5. Flow and average velocity of pellets with different diameters in particular sections of the mixing chamber

Pellets diameter D_p, mm	Weight of a pellet m^1_p, g	Pellets flow Qp, pellet/sec	Total number of crushing pellets N_p, pellet	Mixing chamber section	Average pellet velocity V_{av}, m/sec	Average kinetic energy of a pellet E_{kin} on leaving the mixing chamber, mJ
2	0.033	1461	304	initial	7.42	2.682
				final	9.03	
3	0.111	529	90	initial	7.08	7.023
				final	7.95	
3.5	0.175	252	57	initial	6.62	9.816
				final	7.48	

The data analysis showed that the increase in well volume with the increase in pellet diameter occurs due to the fact that a bigger pellet transfers more kinetic energy to the formation (rock sample) causing more destruction.

6. Conclusions

The experimental research discovered the influence of such factors as bit-bottom distance, pellet batch (portion) weight and pellet diameter on the efficiency of pellet impact drilling. These relations can be used for designing new sophisticated drill strings, and calculating optimal operational parameters. The obtained relations are well grounded by using high speed video.

The following challenges can boost further investigation:

- to develop optimal pellet batch weight calculation methods;
- to find the reasons and solve the problem of pellet jamming in bore hole annulus under the arrester;
- to study the rate of impact of different parameters on the well diameter;
- to conduct theoretical and experimental research on relations between fluid properties, flow rate, as well as pellet material and pellet impact drilling efficiency.

- **Abbreviations**

d_n	Nozzle diameter [mm]	T_{recipr}	Reciprocation frequency [min]
d_{mc}	Mixing chamber diameter [mm]	D_w	Well diameter [mm]
r_{nmc}	Distance between the nozzle outlet and the mixing chamber [mm]	h	Total depth of the well [mm]
l_{mc}	Mixing chamber length [mm]	h_{cyl}	Cylindrical section depth [mm]
A_{exp}	Expansion angle of the diffuser [°]	H_{cont}	Bit-hole bottom contact depth [mm]
d^o_{db}	Drill bit outer diameter [mm]	V_w	Well volume [mm^3]
D_p	Pellet diameter [mm]	Q_p	Pellet flow in the mixing chamber [p/m]
M_p	Pellet batch weight [g]	N_p	Total number of crushing [p]
R_{bb}	Distance between the drill bit and the hole bottom [mm]	m^1_b	Pellet weight [g]
Q_n	Fluid flow from the nozzle [l/min]	V_{av}	Average pellets velocity in the mixing chamber [m/c]
P_n	Upstream nozzle pressure [MPa]	E_{kin}	Average kinetic energy of a pellet on leaving the mixing chamber [mJ]

t Drilling time [min]

References

[1] Zaurbekov S A 1995 Povyshenie effektivnosti prizaboynykh gidrodinamicheskikh protsessov pri sharostruynom burenii skvazhin *Author's abstract* p.18
[2] Kvesko N G and Chubik P S 2010 *Metody i sredstva issledovaniy* (Tomsk: TPU) p. 112
[3] Kovalyov A V, Ryabchikov S Ya, et al 2014 Patent RU 143090 U1
[4] Uvakov A B 1969 *Sharosruynoyue bureniye* (Moscow: Nedra) p. 207
[5] Shtrasser V V Issledovanie protsessov razrusheniya gornykh porod udarami sharov (k teorii sharostruynogo bureniya) *Author's abstract* p. 217
[6] Eckel I E, Deily F H and Ledgerwood L W 1956 Development and testing of jet pump pellet impact drill bits *Transaction AIME* **207** 15.

Designing the ejector pellet impact drill bit for hard and tough rock drilling

A V Kovalyov[1], S Ya Ryabchikov[2], Ye D Isaev[3], F R Aliev[4], M V Gorbenko[5] and A B Strelnikova[6]

[1]Teaching assistant, Institute of Natural Resources, National Research Tomsk Polytechnic University, 30 Lenina Ave., Tomsk, 634050, Russia
E-mail: arteka011287@mail.ru
[2]PhD, Professor, Institute of Natural Resources, National Research Tomsk Polytechnic University, 30 Lenina Ave., Tomsk, 634050, Russia
E-mail: kafedrabs@mail.ru
[3]Engineer, Institute of Natural Resources, National Research Tomsk Polytechnic University, 30 Lenina Ave., Tomsk, 634050, Russia
E-mail: pelletdrilling@gmail.com
[4]Student, Institute of Natural Resources, National Research Tomsk Polytechnic University, 30 Lenina Ave., Tomsk, 634050, Russia
E-mail: faruh_aliev@mail.ru
[5]PhD, Associate Professor, Institute of High Technology Physics, , National Research Tomsk Polytechnic University, 30 Lenina Ave., Tomsk, 634050, Russia
E-mail: gmvski@rambler.ru
[6]PhD, Associate Professor, Institute of Natural Resources, National Research Tomsk Polytechnic University, 30 Lenina Ave., Tomsk, 634050, Russia
E-mail: annas24@yandex.ru

Abstract. There are many types of ejector pellet impact drill bit providing impact rock drilling. Basic types of drill strings have been regarded, the essential requirements for the most efficient facilities to drill hard and tough rocks are formulated. With regard to these requirements, the ejector pellet impact drill bit design appropriate for operating under given conditions has been proposed

1. Introduction

Pellet impact drilling is provided by pellet drill bits, which produce the continuous circulation of pellets in the bottom-hole zone. This method of drilling was proposed in 1955 by Carter Oil Company's scientists (USA). Initially, the method had inherent advantages over rotary drilling [1]:

- the length of the round trip increases as soon as it is limited by wear-resistance of pellet drill bits only, which is caused by the ability to replenish the pellets without pulling the drill string from the hole;
- the construction of the drill bit is simple;
- bit weight is not required;
- rotation is not required, which reduces drill pipe failures and at the same time allows the use of lighter drill pipe.

In the description of field work results, the term of "gravity-aspirator" (figure 1) was used to name the drill bit. In this construction, the drill string (1) with the primary nozzle (2) at the end is connected with the secondary nozzle| (3) by bars (4). The construction is also supplied with bit feet (7), which contact with the hole bottom to space the drill bit off the hole bottom and to prevent it from destroying. When the pellets are at the hole bottom, the jet pump is started. The fluid accelerates in the primary nozzle (2) and runs with high velocity into the secondary nozzle (3). This results in discharge zone between the primary nozzle (2) and the secondary nozzle (3), which aspirates annulus fluid with suspended pellets (5) and cuttings (6). Then the two-phase mixture goes through the secondary nozzle and breaks the rock.

After that, the fluid draws pellets (5) to the annular space between the drill bit and the wellbore, which rise, partially recirculating, above the primary nozzle (2) and then slow down because of annular space extension and decrease in the amount of fluid in the result of injection to form the pellet cloud (8). Being subjected to gravitation, pellets then fall to be aspirated. The process repeats many times.

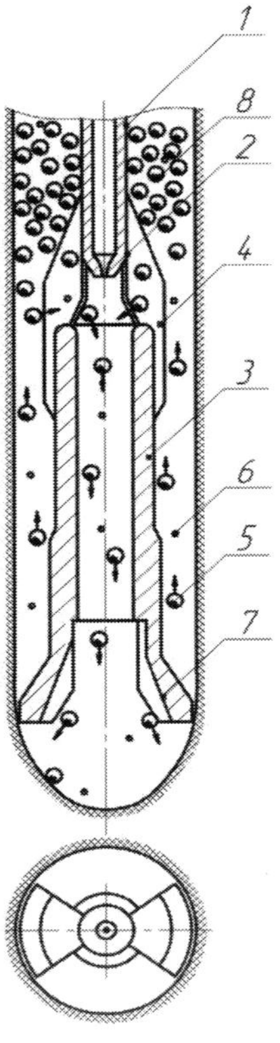

Figure 1. Gravity-aspirator drill bit
1 – drill string; 2 – primary nozzle;
3 – secondary nozzle; 4 – bars;
5 – rock-breaking pellets; 6 – cuttings;
7 – feet; 8 – pellet cloud.

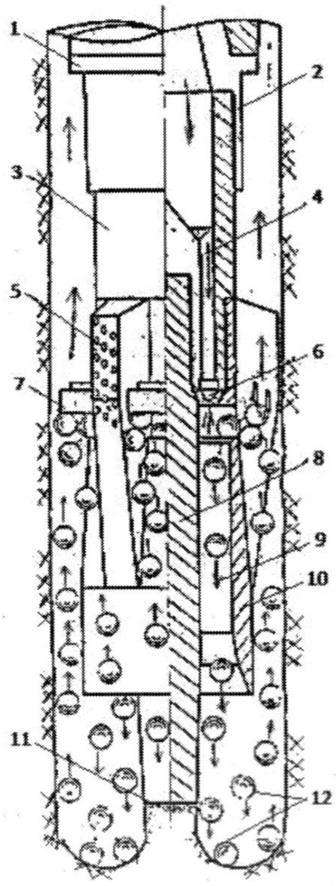

Figure 2. Ejector pellet impact drill bit with the nozzle and circular mixing chamber
1 – calibrating device; 2 – sub;
3 – connector end;
4 – fluid delivery channel;
5 – calibrating and centralizing bars;
6 – circular nozzle; 7 – arrestor;
8 – drill bit holder;
9 – circular mixing chamber;
10 – drill bit body; 11 – hard alloy teeth;
12 – pellets.

IOP Conf. Series: Earth and Environmental Science **24** (2015) 012016 doi:10.1088/1755-1315/24/1/012016

The disadvantages of the construction are the following:

- high rate of feet wear because of their contact with rocks and circulating pellets impact;
- drill bit rotation required to break the hole bottom zone overlapped by the feet;
- large zone of the hole bottom overlapped by the feet;
- necessity to maintain the set flow rate: otherwise, if the flow rate decreases, the pellet cloud is located below the nozzle, and if the flow rate increases, the pellet cloud is above the nozzle, which results in fall of the amount of pellets in the secondary nozzle and drop of drilling rate.

These disadvantages of the ejector pellet impact drill bit and departure from the initial concept of pellet drill bit resulted in further research carried out by both scientists and production workers. Today, there are many different types of the ejector pellet impact drill bits [2] that provide:

- required well diameter throughout drilling process;
- necessary spacing between the drill bit and the hole bottom;
- optimal track of pellets race within the hole bottom zone.

To prove the choice of the ejector pellet impact drill bit to drill hard and tough rocks is an important objective for a scientist.

2. Analysis of ejector pellet impact drill bit constructions

S. Zaurbekov in his work [3] pointed at the small number of pellets in the mixing chamber (called “secondary nozzle” in [1]) of the impact drill bit with the nozzle and the tubular mixing chamber arranged consequently in line. Pellet loss is no more than 140 pellets per second. As a result, the drill bit depicted in figure 2 was proposed. The construction is supplied with the nozzle (6) and the circular mixing chamber (9) and breaks the rock in the central part of the hole bottom with the drill bit holder (8) having hard alloy teeth (11) (figure 2). This construction provides the necessary spacing between the drill bit and the hole bottom. In addition, the arrestor (7) allows reaching the maximum flow rate and drawing the pellets into the inlet ports.

However, the construction is characterized by the following disadvantages:

- the slot nozzle makes the process of drill bit constructing more complicated;
- fast drill bit holder wear, especially when drilling hard and tough rocks;
- rotation and bit weight are required, which is in contradiction to the initial concept of pellet drill bit and causes challenges to drilling process;
- the combined method of breaking the hole bottom rock being applied (pellet impact drilling in border zones and rotary drilling in the central part), different time periods are required for rocks failure, which, according to the works [1, 4, 5], gives less drilling rate;
- the maximum diameter of pellets is less than that of pellets used in the impact drill bit with the nozzle and the tubular mixing chamber, which, according to the works [1, 4, 5], gives less drilling rate.

Some scientists [6, 7] consider the ejector pellet impact drill bit the most efficient when there are the nozzle and the tubular mixing chamber arranged consequently in line while the border zones are broken with blade (figure 3) and shot mechanical devices (figure 4). However, there are limitations connected with combined rock failure in the hole bottom zone: mechanical devices wear, need in rotation and bit weight, challenges of drilling control.

3. Requirements for drill bits appropriate for hard and tough rocks drilling

To sum up, ejector pellet impact drill bit constructions described above are efficient to drill soft and medium-hard rocks. To construct the ejector pellet impact drill bit for drilling hard and tough rocks, the following requirements are essential:

- there should be no constant contact of the drill bit with the hole bottom, which means that drill bits with feet, holders and shoes are unacceptable;
- rotation and bit weight are not acceptable as soon as they make the drilling process more complicated;

- the construction should be designed on the basis of the drill bit with the nozzle and the tubular mixing chamber arranged consequently in line;
- there should be a diffuser at the edge of the mixing chamber which increases the efficiency of pellet impact in the border zones of the hole bottom; it also helps to provide the minimal spacing between the drill bit and the hole bottom, which gives higher impact drilling rates;
- the arrestor is necessary to provide high and sustainable flow rate essential to break hard and tough rocks;
- spacing between the drill bit and the hole bottom and the well diameter should be under control throughout the drilling process;
- pellets should be replenished during the round trip without pulling the drill string from the hole;
- the facilities should be simple to construct and operate.

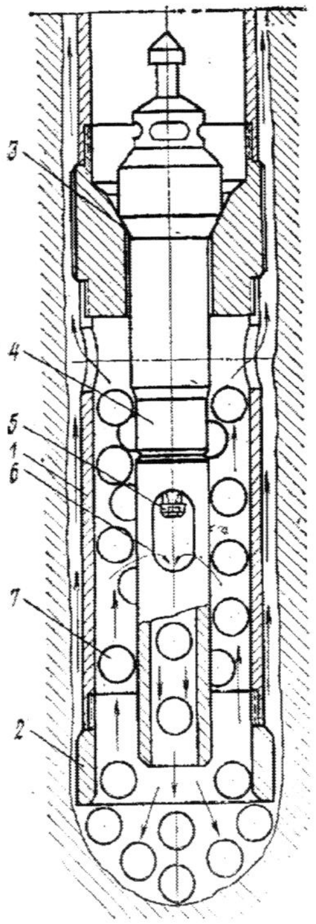

Figure 3. Blade ejector pellet impact drill bit
1 – body; 2 – rock-breaking piston shoe;
3 – seat; 4 – jet apparatus; 5 – nozzle;
6 – mixing chamber with inlet ports;
7 – pellets.

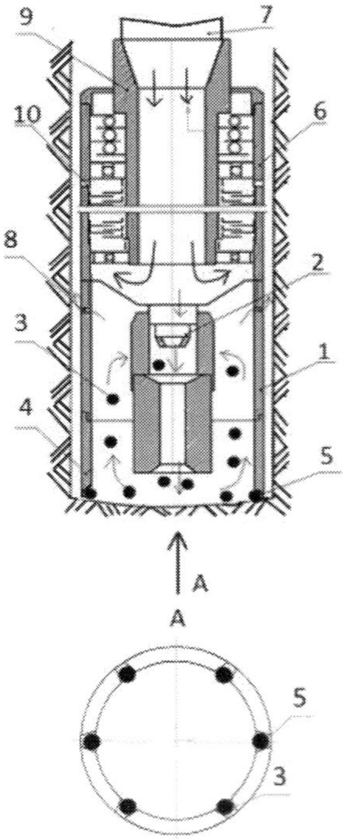

Figure 4. Shot ejector pellet impact drill bit
1 – body; 2 – jet apparatus;
3 – rock-breaking pellets;
4 - rock-breaking circle; 5 – seats;
6 – turbine driver; 7 – drill string;
8, 10 – holes; 9 – axis.

4. Ejector pellet impact drill bit for hard and tough rocks drilling

The authors of this work propose the construction of the ejector pellet impact drill bit (figure 5) deprived of constant contact with the hole bottom. The construction includes the cone-shaped arrester

(3), which provides the necessary flow rate and directs the pellets to the inlet chamber (4). There is a diffuser at the end of the drill bit to provide the required well diameter throughout drilling process in case of less space between the drill bit and the hole bottom.

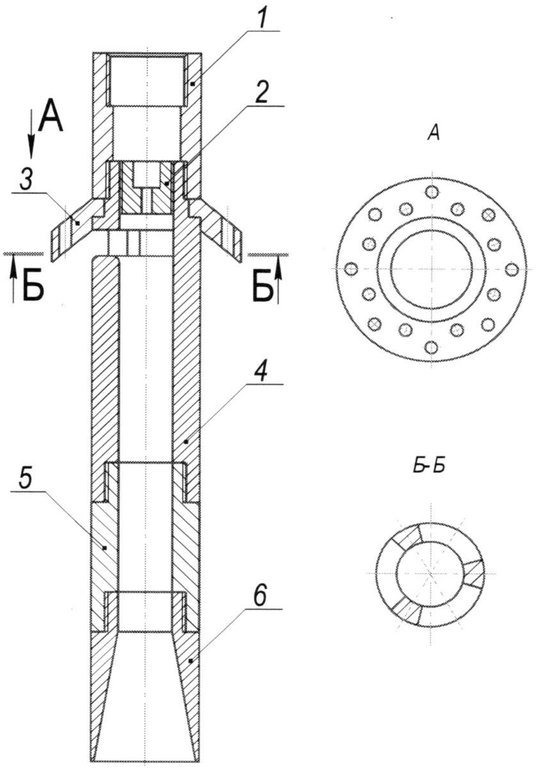

Figure 5. Ejector pellet impact drill bit
1 – discharge sub; 2 – nozzle, 3 – arrester; 4 – inlet chamber;
5 – mixing chamber; 6 – diffuser.

Today, the authors are investigating the problem of control methods to provide optimal spacing between the drill bit and the hole bottom and to maintain the set well diameter while drilling with these facilities. The pellet supplier is also being designed for the pellets to be replenished within the round trip.

References:

[1] Eckel I E and Deily F H 1956 Development and testing of jet pump pellet impact drill bits Transaction AIME **207** 15

[2] Kovalyov A V and Yakushev D A 2014 Problems of Geology and Subsurface Development: The materials of XVIII International Scientific Symposium in honor of Academician M.A. Usov V **2** 412-14

[3] Zaurbekov S A 1995 *Povysheniye effectivnosti prizaboynykh gidrodinamicheskikh protsessov pri sharostruynom burenii skvazhin* 18GOST 26798.1-96 Cement Slurry Testing

[4] Uvakov A B 1969 *Sharosruynoyue bureniye* Nedra 207

[5] Shtrasser V V 1966 *Issledovaniye protsessov razrusheniya gornykh porod udarami sharov (k*

teorii sharostruynogo bureniya) 217
[6] Davidenko A N and Ignatov A A 2013 *Abrazivno-mekhanicheskoye udarnoye bureniye skvazhin* 110
[7] Uvakov A B and Shtrasser V V *Sharosruyniy snaryad dlya bureniya skvazhin* Patent № 417599, E 21 B 7/18. № 1451266 Filed: 15.06.1970 Published: 28.02.1974

Production of pure raw quartz by enriching quartzites

M Korovkin [1], L Ananyeva[1], V Ginsar [2] and O Razva[1]

[1]Institute of Natural Resources, Research Tomsk Polytechnic University
[2]Federal State Budgetary Institution of Science Institute of Monitoring of Climatic and Ecological Systems of Siberian Branch of the Russian Academy of Sciences

E-mail: mvk@tpu.ru,

Abstract. The sedimentary metasomatic quartzites of Antonovsky deposits (Western Siberia) almost completely consist of high-density α-quartz micrograins. The grain purity is conditioned by self-purification of quartz microgranules in the process of lithification. The work proposes an effective aeromechanical technology for deep purification of quartz microgranules by applying high-temperature chemical exposure to produce commercial raw quartz.

1. Introduction

Quartz-containing rocks (quartzites, quartz sandstone, quartz sands and sand-gravels) are traditionally used as fluxing agents and raw materials for producing ferroalloys and refractory materials in chemical, glassware, building and other industries. The global production and consumption market of raw quartz is currently characterized by a consumer boom in chemically pure quartz varieties. Such an interest is associated with high technology application in producing items for semiconductor microelectronics, special materials for optical and acoustic electronics, optical fiber systems, nanomaterials, unique ceramic materials and silicon carbide (SiC) and silicon nitride (Si_3N_4)-made products for ferrous and nonferrous metallurgy, chemical industry. However, of great commercial importance is the increasing demand for crystalline silicon conditioned due to the development of helioenergetics and establishment of photoelectric plants with implemented silicon-based elements that transform solar radiation into electric energy. The principal limitation of the wide-scale application of ground solar energy systems is the high price of "solar-quality" silicon. Therefore, the current targets are both to develop new production technologies for "solar-quality" silicon, which, in its turn, would drastically reduce its cost, and to select such refined industrial silicon that would be a carbothermal renewable natural quartz, i.e. rock crystal. With due consideration Due to the limited distribution and depletion of conventional pure raw quartz (rock crystal) deposits, there is no doubt in the relevance of the problem in estimating the quality and industrial application potential of abundant quartz rocks. As quartzites have similar properties, they could be a promising source of highly-pure raw quartz materials.

In this case, quartzites of Antonovsk deposits (Kemerovo Oblast, Russia) with estimated reserves of 660 mln tons and currently used as raw source for ferrous metallurgy is of great importance. [1-3] Due to specific formation conditions, Antonovsk quartzite deposits (Kemerovo Oblast) could be a potential high-quality raw quartz source [1, 2].

2. Research Methods and Results

The results of mineralogical and geochemical studies showed that these quartzites are monomineralic rocks with rather high quartz content. Quartzites of sedimentary metasomatic origin almost completely consist of high-density α-quartz micrograins varying from 0.01 to 50 microns, the purity of which is conditioned by quartz microgranule self-purification during lithification in the early metagenesis of quartz-hydromica-sericite series [4]. Within local areas, particularly in significant crushing zones, the hypergene processes destroy the quality characteristics of chemically pure quartzites which determine the primary industrial grades of quartzites. However such rocks do not contain poisonous or toxic elements (table 1).

Grade / element content [10^{-4} %]	U	Th	Yb	Sm	La	Hf	Au	Ag	Tb	Sc
KR-2	0.2	-	-	0.5	0.3	1.3	0.005	-	-	0.8
KF	0.4	0.6	0.2	0.7	0.6	1.1	-	-	-	1.0
KSh	0.4	0.6	0.2	0.8	0.8	1.2	-	-	-	2.4

Table 1. Average content of rare, radioactive and rare-earth elements in quartzites of different industrial grades

The most pure quartzites exhibit a low quartz microgranule crystallinity degree [5, 6], lack of isomorphic replacements and impurity ions (Al, Fe, Ti, Na and other) in the α-quartz crystal phase.

The estimation of boron and phosphorous concentrations of more than 1 ppm is of particular importance and, which, in its turn, restricts the production application of raw quartz for "solar-quality" silicon, where they are doping additives changing the conductivity type of semiconducting silicon, and quartz and special borosilicate and borophosphate glass products, whereas such elements are modifying cations.

Secondary ion mass spectrometry (SIMS) with sensitivity to practically all isotopes of ~ 0.1 ppm detects boron and phosphorous in quartz grains. The isotope spectra were measured by a focused ion beam from the surface of specially prepared quartzite samples on molybdenum support, and from a quartz grain. Preliminary sample surface argon etching was applied to remove surface defects.

The panoramic SIMS spectra registered the isotopes of those chemical elements that comprise the molybdenum support (Mo^{90}...Mo^{100}, Fe^{56}, Sc^{45}, C^{12}, N^{14} and others) and in the studied quartzites (Si^{28}, Al^{27}, K^{39}, O^{16}, Na^{23} and others). The content of such elements in initial raw quartzites is 10 times higher than in granulated quartz (Malo-Kutulakskoe deposit, Yakutia). However, after crushing and purification by proposed aeromechanical separation method, the mpurity element content in quartzites decreases, while in granulated quartz samples it remains unchanged. It could be assumed that the chemical impurity elements in quartzites of accessory minerals and authigenic films within intergranular space can be removed by such methods as mechanical separation and chemical etching, whereas their extraction excludes special purification methods in alloy.

Special experimental SIMS methods in relevant recording range (figures 1 and 2) were employed to detect boron and phosphorous in quartzite grains from Antonovsk. Figure 1 depicts a detailed spectrum of detected ions within the range of 9 .... 17 AMU (atomic mass unit). The spectrum clearly showed the isotopes of carbon C^{12} and oxygen O^{16}. The attempts to detect the presence of ions of boron B^{11} by mathematical modeling failed

IOP Conf. Series: Earth and Environmental Science **24** (2015) 012017 doi:10.1088/1755-1315/24/1/012017

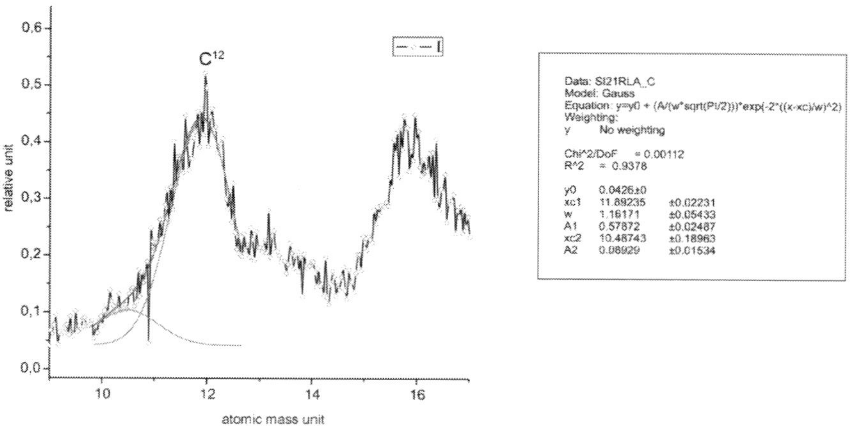

Figure 1. Fragment of isotope spectrum within the range of 9 ... 17 AMU (atomic mass unit)

The presence of phosphorous P^{31} ions (figure 2) in enriched quartzites was not detected, in spite of the fact that certain quartzites were identified in phosphorous pentoxide by chemical analysis. Atomic emission analysis and Rutherford back-scattering method also indicated the absence of phosphorous ions in enriched quartzites.

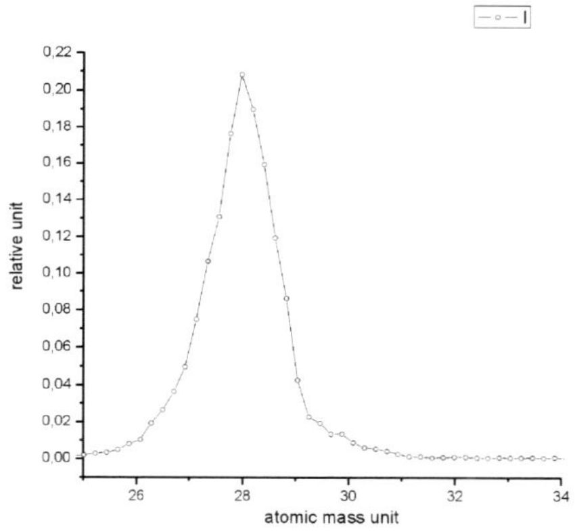

Figure 2. Fragment of isotope spectrum within the range of 25 .... 34 AMU (atomic mass unit)

The obtained results showed that under such quartzite formation conditions the self-purification of quartzites (autolysis-accompanied crystallization) and the reverse process - infiltration of other chemical element ions into the quartzite crystal lattice did not occur.

Veined and granulated quartz due to close ion radii demonstrate the isomorphic silicon ion replacement effect by aluminum and phosphorous ions, while isomorphism was not observed in studied quartz grains in sedimentary metasomatic quartzites.

It should be noted that neutron activation analysis, spectral analysis and ICPMS method indicated different chemical element impurity content in the initial quartzite samples which is typical for accessory minerals. Ferrum oxides, feldspar, hydromica, illite and chlorite were identified along quartz recrystallizated cracks, boundary fractures and in intergranular space.

All these impurities (besides ferrum oxides) are easily removed by vibrational screening and further flushing. Ferrum oxides as authigenic films developing on the surface of quartz grains are difficult to process. In some cases, in order to satisfy the high requirements for raw quality, deep enrichment of quartzites by a complex of mechanical, chemical, thermal and other methods should be performed. Highly pure fine-grained Antonovsk quartzites, slightly containing harmful structural impurities, could be easily enriched in contrast to monocrystalline raw materials which that require special purification in alloy.

Considering the uniqueness of the Antonovsk microquartzites formation conditions an effective aeromechanical technology was developed for profound purification under high-temperature chemical exposure. This technology has embraced the best features of existing technologies [7]. Fine raw material purification is performed for 5 …... 200 mcm grains in an aeromechanical reactor, where the material is further milled simultaneous deslurring and thermochemical treatment, which, in its turn, produces high quality quartzites with SiO_2 content - 99.96% and more (table 2).

Sample name	Content [wt%]				
	SiO_2	**Fe_2O_3**	**Al_2O_3**	**TiO_2**	**ignition loss**
Initial quartzite	98.89	0.45	0.3	0.01	0.35
Fraction of 1 mm (enriched)	99.58	0.1	0.3	0.025	-
Fraction of 0.4 mm (enriched)	99.79	0.008	0.3	0.022	0.04

Table 2. Content alteration of certain oxides in quartzites as a result of aeromechanical separation

3. Conclusion

The conventional methods for enriching raw quartz decrease several-fold the total mass fraction of impurity element content, for example, electromagnetic separation – 4 -fold, electrostatic separation - 10-fold, chemical treatment-25-fold, and foam flotation - almost 100-fold. However, preliminary enrichment of rough material during extraction regional enterprises is recommended, while fine enrichment should involve the following high-tech operations: electric-impulse destruction of raw materials, high-intensity magnetic separation, ultrasonic purification of disintegrated particle surface, with further express-control of product quality and raw material loss during each technological stage.

Proposed advanced technology including aeromechanical separation under high-temperature and chemical exposure was developed for fine purification of unique Antonovsk quartzites, producing commercial-grade raw quartz after the first enrichment cycle (table 3).

Purpose	Grade	Component content [%]	Coarseness [mm]		
Production		SiO_2 not less than	Fe_2O_3 not more than	Al_2O_3+CaO not more than	
green SiC	top	99.5	0.1	0.3	5.0 – 0.5
black SiC	first	99.0	0.2	0.5	5.0 – 0.5
Antonovsk quartzite (after the first enrichment cycle)		99.58	0.100	0.30	1.0

Table 3. Technical conditions for raw quartz for the production of silicon carbide

References

[1] Ananyeva L, Ananyev Yu, Dolgov I, Korobeynikov A and Korovkin M 2001 *J. Izvestaya* Tomsk Polytechnic University **304** (1) 123
[2] Ananyeva L and Korovkin M *J. Izvestaya* Tomsk Polytechnic University **306** (3) 50
[3] Korovkin M, Ananyeva L and Antsiferova A 2012 *J. Izvestaya* Tomsk Polytechnic University **320** (1) 16
[4] Korovkin M and Ananyeva L 2012 Mineralogy, geochemistry and mineral resources of Asia **B 2** 139
[5] Korovkin M, Ananieva. L and Antsiferova A 2011 Proceedings of the 10th International Congress for Applied Mineralogy (ICAM), Trondheim, Norway 403
[6] Razva O, Anufrienkova A and Korovkin M 2014 *J. Modern knowledge-intensive technologies* **7-2** 27
[7] Korovkin M, Ananyeva L, Antsiferova A and Ginsar V 2010 Proceedings of the 10th International Research and Practice Conference for Study, Development and Application of High Technologies in Industry, Saint-Petersburg, Russia 245

IOP Conf. Series: Earth and Environmental Science **24** (2015) 012018 doi:10.1088/1755-1315/24/1/012018

Litho-mineralogical characteristics and facies formation conditions of oil-reservoir rocks J_1^1 and J_1^2 in Kazan oilfield (Tomsk Oblast)

L Krasnoshchekova [1], D Cherdansteva[2] and R Abramova[3]

[1] Department of Geology and Mineral Exploration, Institute of Natural Resources, Tomsk Polytechnic University
[2] Department of Oil and Gas Industrial Facilities Design Construct, Institute of Natural Resources, Tomsk Polytechnic University
[3] Department of Foreign Languages, Institute of Natural Resources, Tomsk Polytechnic University

E-mail: [1]krasnl@tpu.ru, [2]dannia@tpu.ru

Abstract The detailed investigation results of textural rock characteristics indicated the predominate lithological facies in formation layers J_1^1 and J_1^2 (Kazan oil/gas-condensate field), their sedimentation sequence and further alteration of environmental formation conditions. The stage processes and secondary alterations in rocks significantly influencing the reservoir properties were determined.

1. Introduction

Kazan oil/gas-condensate field is located in the south of Tomsk Oblast. Structure-tectonically this field is confined to the Kazan dome-shaped uplifted Nurolsk depression. The geological profile embraces metamorphosed rocks of Paleozoic folded basement and argillo-arenaceous sediments of Mesozoic-Cenozoic sedimentary mantle. In terms of field development, the Upper Jurassic layers -J_1^1 and J_1^2 (Vasugan suite) extending horizontally and vertically and tapped in all drilled wells are of principal interest.

2. Research methods

The mineralogical and structure-texture description of core samples and thin sections (core sampling depth 2481,3-2511,2m.) was based on the investigated strata data from one Kazan field appraisal well. The lithological rock classification involved sample splitting according to well logging.

Determining structure types and lamination of cored layers was performed with respect to the suggested Botvinkina (1962; 1965) and Alekseev (2007) methods, whereas mineralogic-petrographic analysis techniques are described in the works of Logvinenko and Sergeeva (1986).

3. Results and discussion

Net pay formation is composed of medium/fine-grained light-grey sandstones interlayered by silty-clay material. According to a detailed lithological core description the constant texture diversity of layer rocks was identified (table 1).

Type	Description
Non-stratified (a)	Homogeneous (massive)
Indistinctly laminated (b, c)	Graded
Disrupted (d) (burrows)	Partially disturbed
	Completely disturbed
Bedding (bedded texture)	
Cross-bedded (e)	Slightly sheared
Lenticular (swaley bedding) (f)	Slightly / severely disrupted
Wavy cross-bedding (g,h)	Parallel
	Non-parallel slightly / severely disturbed
Horizontal bedding (i)	Non-uniform (heterogeneous)

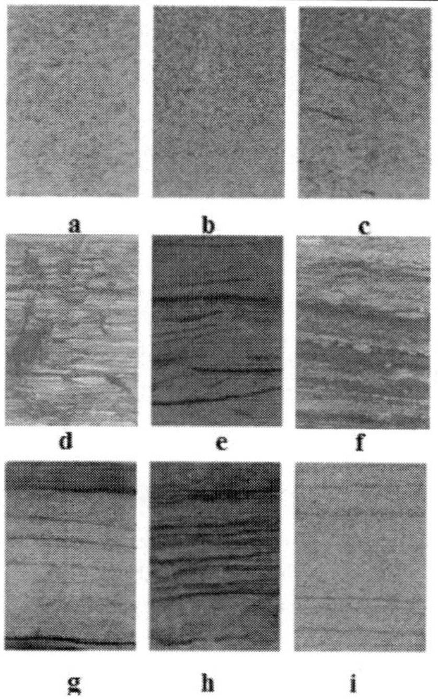

Table 1. Classification of J_1^1 and J_1^2 layer texture (Kazan oil field)

The rock structure-texture analysis results of lengthwise coring from the appraisal well indicated the rock formation conditions within the sedimentary sequence of studied layer (table 2).

Facies	Sub-facies	Macrofacies
Continental	Alluvial	Floodplain deposits in river valleys
	Lacustrine	Deposits in stagnant and marshy lakes
		Deposits in open lacustrine waters
Transition (Coastal-continental)		Deposits in bay-lagoon coasts
Basin	Shallow basin	Deposits in open moving basin shoals

Table 2. Facies dismembering pattern for J_1^1 and J_1^2 layers (Kazan oil field)

Besides the above-mentioned textural core sample description of rock sections, the following characteristic features were investigated: lithological varieties and further identification of the mineral - petrographic composition of sandstone and siltstone layers, cementing material composition and content, grain-cement relationship and secondary alteration of investigated samples.

Terrigenous constituent of rock included quartz grains (60-75%) and feldspar (including potassium feldspar and plagioclase – up to 15-35%), rarely, fragments of quarzites, micro-quarzites, meta-andesites, meta-basalts, meta-granites and pegmatites, argillic/ argillic-hydromica sericite and sericite shales (up to 10-15%). According to Shvanov classification (1987), rock alteration is classified as mesomictic, to a lesser extent, oligomictic (figure1).

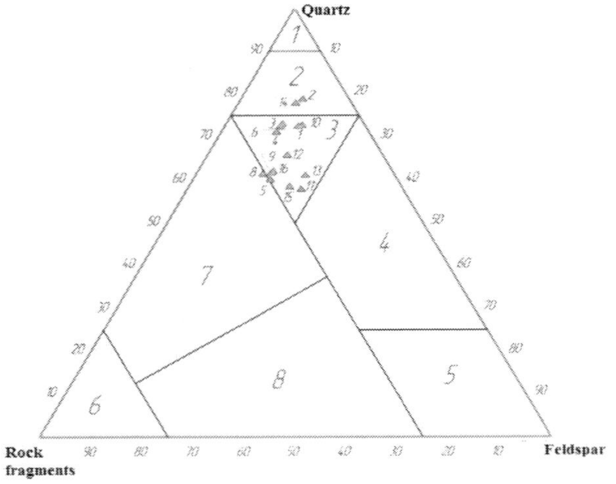

1- quartz
2- oligomictic
3- mesomictic
4-arkose
5- feldspar- arkose sandstones
6- proper greywacke
7- quartz greywacke
8- feldspar greywacke

Figure 1 Classification diagram of sandy rocks (Shvanov, 1987)

The bulk of the fragment material included fine-grained sand fraction (0.1-0.25mm). Cement content in sandstones ranges from 3-5% to 20%, while in the layer roof – up to 35%. Cementation type- filmy, porous and sometimes basal.

The thin sections exposed angular, subangular, as well as subrounded and rounded fragments. Degree of grain roundness ranges from 0.2-0.5. The sandstones are well-sorted.

Competent fragments are mica - muscovite dominated (up to 3-5%), often alternating to hydromuscovite with decreased interference color and cracking on grain tips (figure 2). Chlorite replacement is found in separate muscovite horse.

The secondary rock alterations showed mechanical and plastic deformation characteristics of clastic grains as cracking and formation of random-oriented cracks or as formation of composition plane in the folds of feldspars. The cracks are localized in the central grain sections or peripherally in grains, as well as persistent throughout the fragments, partially forming hydromica (figure 3).

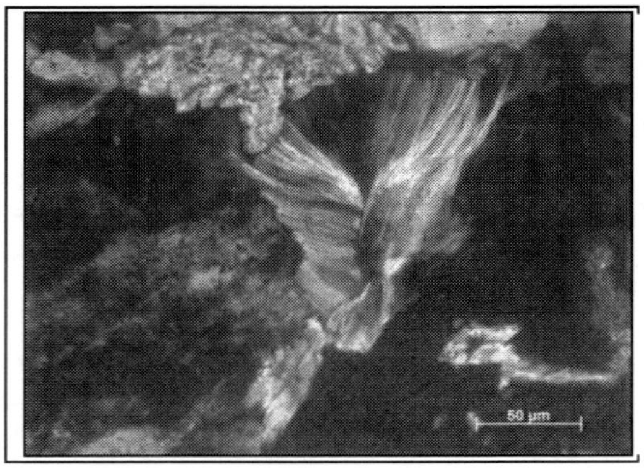

Figure 2. Formation of hydromica from muscovite (detected by analyzer)

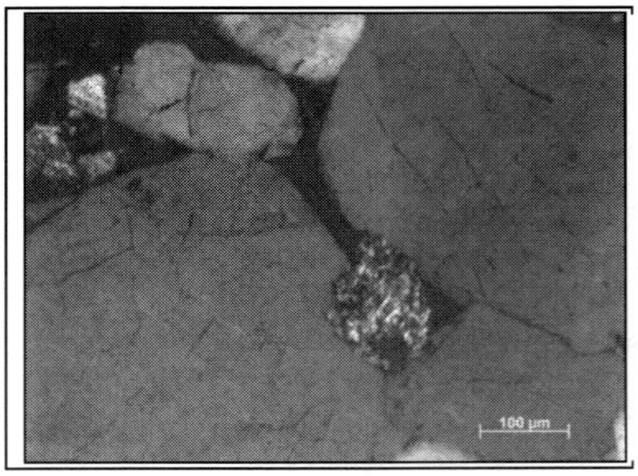

Figure 3. Cataclasis cracks in terrigenous sandstone material (detected by analyzer)

Coupled with pressure solution and generation of orthomorphic and incorporated grain contact textures, the regeneration of grains by quartz cement was observed (figure 4). The grain fringe shape is governed by the volume of free pore space and its dimension varies from a thousandth mm to 0.1mm. Inclusions of numerous dispersed clay flakes are predominately observed in the central section of the fringes, embracing the possible contours of earlier regenerated growths.

Leaching zones with further formation of intragranular porosity were partially revealed in feldspars where sericite, calcite and/or saussurite aggregates (in some cases) also formed parallel (figure 5).

Figure 4. Regenerated and orthomorphic - incorporated texture in quartz (detected by analyzer)

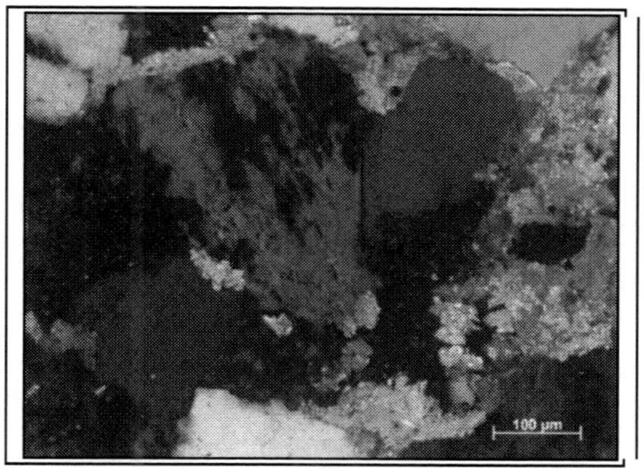

Figure 5. Formation of leached pores in clastic material- feldspar (detected by analyzer)

Well-decrystallized kaolinite as vermicular growths (figure 6) was identified in closed intergranular pore spaces. Segregation of reddish-brown and/or brownish-black paleo-organic material is frequently restricted to kaolinite pores (figure 7).

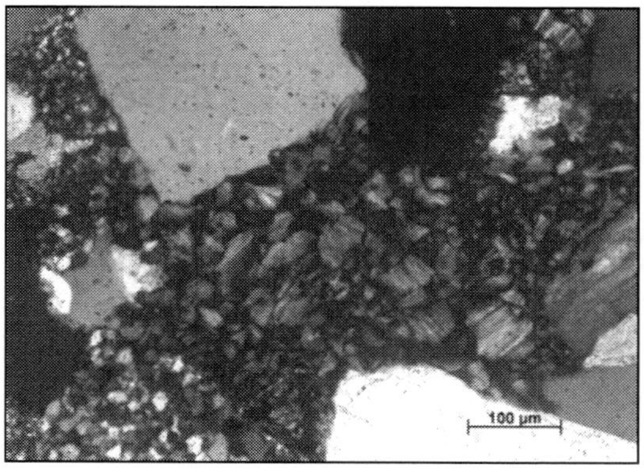

Figure 6. Decrystallization of kaolinite aggregates in pore spaces (detected by analyzer)

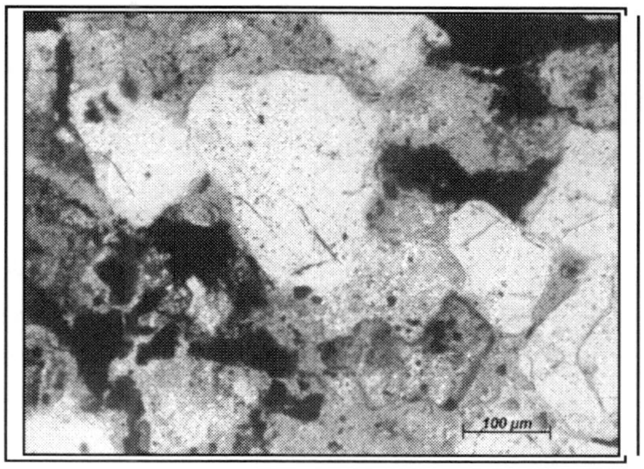

Figure 7. Segregation of bitumen in sandstone pore spaces (detected without analyzer)

Carbonate cement corrodes terrigenous material and expands into grains as spots and/or into the pore space, forming pore-filling, porous-basalt or basalt cement (figure 8).

Carbonaceous, argillous and sideritic lamina were observed in the rocks. Paleo-organic growths of red-brown to black, sometimes as detrital inclusions, were commonly found in the thin sections, indicating the micro-lamination of rocks (figure 9). Paleo-organic material occurs in different alteration stages – from gelification to fusainisation.

IOP Conf. Series: Earth and Environmental Science **24** (2015) 012018 doi:10.1088/1755-1315/24/1/012018

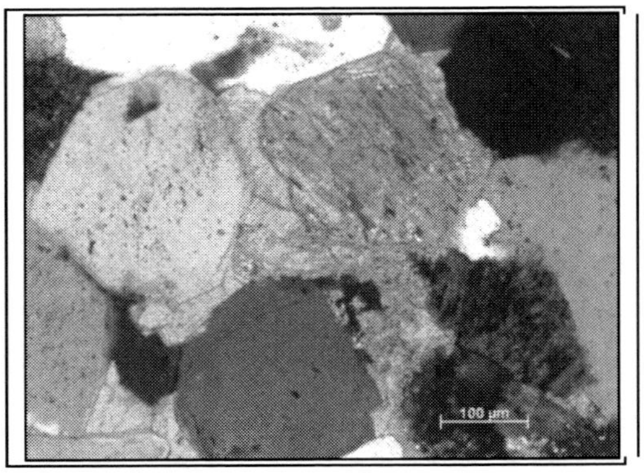

Figure 8. Carbonate replacement in terrigenous material (detected by analyzer)

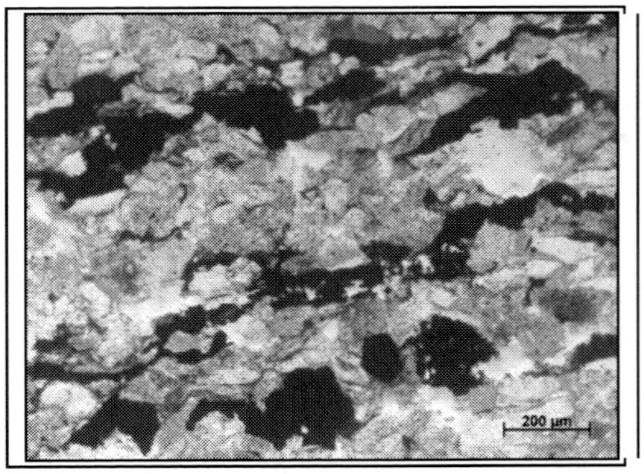

Figure 9. Sandstone with vegetative detritus remains (detected without analyzer)

Process development of recessive and superimposed lithogenesis in rock-samples from the appraisal well is depicted in table 3.

Sample №	depth,m.	Stage processes								Superimposed (epigenetic) processes				Bitumen formation	Pyritization Oxidizing petroliferous substance
		Compaction		Deformation		Paleo-organic material	Pyritization	Hydromication	Regeneration	Carboni-zation		Pore leaching	Kaolinitization		
		Orthomor-phism	Incorpo-ration	Plastic deformation	Shattered cracks					Calcite	Siderite				
1	2482,2														
2	2483,0														
3	2483,3														
4	2484,18														
5	2487,26														
6	2488,19														
7	2492,8														
8	2494,06														
9	2494,52														
10	2495,99														
11	2496,88														
12	2476,28														
13	2498,16														
14	2500,3														
15	2504,7														
16	2508,14														

Table 3 Characteristic features of lithogenetic processes in sandstones of J_1^1 and J_1^2 layers (Kazan field)

Note: Thick vertical lines reflect the qualitative occurrence frequency of mineral and texture regeneration

4. Conclusion

Textural feature rock classification of sandstone reservoir layers promoted the identification of corresponding reservoir facies. The transition sedimentation environment characterizing the shift of continental facies to basin facies was identified in the cross-sections. Thus, in the studied well the continental sediments included lacustrine aleurolites and fine-grained sandstones of surface lacustrine

IOP Conf. Series: Earth and Environmental Science **24** (2015) 012018 doi:10.1088/1755-1315/24/1/012018

macrofacies sediments. Medium-fine-grained sandstones, with well-sorted material and high carbonate content accumulated in the basins (i.e. extensive fresh-water intracontinental basin with levelled floor and depth up to the first ten meters). This subgroup includes macrofacies sediments outermost from basin coastline.

The predominate stage processes – compaction (orthomorphism, incorporation) and deformation (cataclasis) were determined. It should be noted that mechanical rock deformation expands more potential hydrocarbon migration paths. Superimposed (epigenetic) alterations involve those processes that both positively and negatively influence the reservoir properties. Both pore leaching and kaolinitization improve the reservoir properties of sandstones, consequently, carbonization, cementing rock pore space and "healing" cracks decrease the reservoir pore space and permeability.

References

[1] Alekseev V 2007 *Atlas fatsiy yurskih terrigennyih otlozheniy (uglenosnyie tolschi Severnoy Evrazii)* (Ekaterinburg Izd-vo : UGGU) pp 209
[2] Botvinkina L 1965 *Metodicheskoe rukovodstvo po izucheniyu sloistosti* (Moscow :Nedra) pp 259
[3] Botvinkina L 1962 *Sloistost osadochnyih porod* (Moscow: Izd-vo AS USSR) pp 542
[4] Logvinenko N and Sergeeva E 1986 *Metodyi opredeleniya osadochnyih porod Guidebook for universities* (Leningrad: Nedra) pp 240
[5] Shvanov V 1987 *Petrography of sandstones (Composition, classification, description of mineral types)* (Leningrad: Nedra) pp 269

IOP Conf. Series: Earth and Environmental Science **24** (2015) 012019 doi:10.1088/1755-1315/24/1/012019

Reliability assessment of tunneling flow charts

N A Antropova[1], V G Krets[2], V G Luk'yanov[3] and A V Baranova[4]
Tomsk Polytechnic University, Tomsk, Russia

[4]baranskikh@yandex.ru

Abstract. When driving horizontal workings of different kinds it is typical to reserve excessive redundant equipment. However, the lack of resources (facilities, people) and poor management don't ensure goal achievement [3]. The paper presents reliability assessment method of tunneling with the help of mathematical modeling. The mathematical model is developed on the basis of factual data of tunneling conditions and no-failure operation probabilities of the flow chart elements are estimated. The method of reliability assessment allows identifying organizational, technological and technical elements of the reserves to increases drive efficiency and improve feasibility characteristics.

1. Introduction

Tunneling is used while laying pipes and communication lines, prospecting works and etc. The reliability of tunneling flow chart is ensured by specification documents, experience of advanced crews and enterprise's traditions. Thereby, the use of excessive redundant equipment and high amount of general labor often occurs. If there are insufficient resources (facilities, staff, number of faces and etc.) and poor management, the aim can hardly be reached.

The increase in exploration efficiency being one of the main challenges nowadays, tunneling flow chart reliability assessment becomes a relevant issue. It is also necessary to evaluate to what extend the flow chart reliability impacts the final results of tunneling.

Factual statistic data of tunneling show that there is great loss of working hours caused by several factors: shortage of equipment in productive workings (such as trolleys, perforators and electric locomotives), non-productive works to be fulfilled (equipment repair, derailed trolleys and electric locomotives replacement) and technical-organizational reasons (the absence of electric power or/and compressed air, high gas content). Thus, redundant equipment can't guarantee continuous production process due to high breakdown rate of the machines. Therefore the method of reservation which is widely applied to increase reliability can't be regarded as crucial.

2. Methodology

Mathematical modeling is one of the methods to represent reliability of tunneling flowcharts, estimate and improve failure-free operation. It is acceptable to describe reliability of any type of mine workings.

Let's represent tunneling technology as a complex system of j-elements (j= 1, 2, k,..., n). The elements can be substituted by the machines to drill blast-holes, underground loaders, electric locomotives, fans, trolleys exchange schemes, number of shifts a day, salary system, mining and geological conditions and etc. The choice of elements is in direct relation to tunneling conditions and aims of exploration so that the results to be obtained will be easy to use.

The elements are united to perform a net-graph G (Ej), where Ej is a set of graph arcs [1]. According to the tunneling flowchart all the arcs are connected in such a way that parallel arcs represent independent network operation while successive arcs represent dependent one (figure 1). The most likely system elements are:

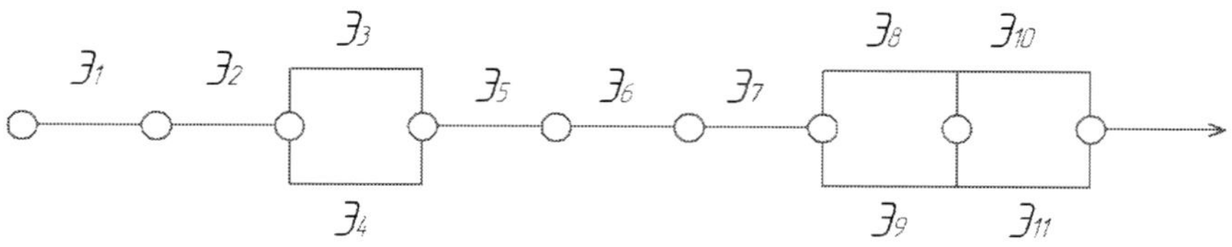

Figure1. Reliability graph G(Ej),where E_1 is electric power; E_2 – compressed air; E_3 – drilling process; E_4 - fixing; E_5 – charging and blasting E_6 - ventilation; E_7 – electric locomotive operations; E_8 – loading and exchange; E_9 - empties; E_{10} – auxiliary operations; E_{11} – temporary track laying.

The graph shows 11 arcs representing interconnected elements and technological processes. Each arc is estimated by failure distribution function $P_j(t)$, j= 1, 2, k,…,n).

The research [2] proved that the function $P_j(t)$ may be expressed in terms of the exponential distribution of probability density function:

$$P_j(t) = 1 - \alpha_j e^{\beta_i t} . (1)$$

Function (1) is deduced by statistical manipulation of elements E_j failure frequency during the observed period of time [0,t].

By the exhaustion of arcs from E_I to E_{II} along path η_j and basing on the hypothesis about independent random variables, we can express the probability of failure-free flowchart operation as all arcs paths product: η_j:

$$P_{\eta_j} = P_1 \cdot P_2,..., P_k = \prod_{j=1}^{k} P_j . (2)$$

For technological system represented in Figure 1 these paths are the following:

$\eta_1 = (Э_1 Э_2 Э_3 Э_5 Э_6 Э_7 Э_8 Э_{10})$,

…………………………………

$\eta_8 = (Э_1 Э_2 Э_4 Э_5 Э_6 Э_7 Э_9 Э_{11})$.

As a result, we have 8 values of the reliability assessment of tunneling flowchart:

$P_{\eta_1} = 0{,}535, P_{\eta_2} = 0{,}528, P_{\eta_3} = 0{,}518, P_{\eta_4} = 0{,}512, P_{\eta_5} = 0{,}51, P_{\eta_6} = 0{,}54, P_{\eta_7} = 0{,}521, P_{\eta_8} = 0{,}529$..

Let's take $\eta_{j\min} = \min(P_i,..., P_k)$ as a minimum reliability evaluation among all the paths of tunneling, while path $\eta_{j\max} = \max(P_i,..., P_k)$ provides possible reserve for technological system reliability increase by value $\Delta p = P_{\eta_j} \max - P_{\eta_j} \min$ as long as it is impossible to increase the reliability along path $\eta_{j\max}$.

Thus having identified elements Ej with low reliability of failure-free operation (table 1) we take measures to increase their reliability. New evaluation is $P'_{\eta_j} > P_{\eta_j}$ and this increase can reach the level of $P_{\eta_j} \max$ which is to become new reliability assessment of the path.

3. Results and discussions

Statistic data over the period of 7.6 months were collected to test the reliability model. Within this period there were 4788 man-hours at Magianskaya prospecting expedition. The time spent for different operations of technological cycle is the following: rock loading (E_8) – 1180 h.; rock haulage (E_9) -702 h.; track laying by auxiliary workers (E_{11}) – 540 h.; blast-hole drilling (E_3) – 1252 h.; charging and blasting (E_5) – 270 h.; ventilation (E_6) – 250 h.; auxiliary operations (E_{10}) – 324 h.

The downtime at Magianskaya prospecting expedition makes about 22% or 1053 h. due to the following reasons: absence of compressed air or electric power – 311 h.; lack of materials – 105 h.; equipment failure – 242 h.; absence of transport – 134 h., bit and crown bit breakage – 37 h., derailed trolleys readjustment – 56 h.; other reasons – 169 h.

The reliability values of the system elements are given in table 1.

Table 1.

E_j	1	2	3	4	5	6	7	8	9	10	11
P_j	0.94	0.94	0.99	0.98	0.85	0.92	0.95	0.97	0.98	0.85	0.84

4. Conclusion

Analyzing the table we come to the conclusion that in most cases the downtime is caused by auxiliary operations, trolleys loading and exchange, equipment failure and idle time in working shifts which is caused by technical-organizational reasons. Reliability of these elements is low but there is potential to increase it. Thus the reliability of the system within element E_9 (auxiliary operations) can be significantly improved if a shaft crew is exempted from nonproductive work (charging up the locomotive battery, rails and pipes delivery, perforators repair). The reliability of element E_5 can be improved as well by using enhanced constructions of tracks in the face zone and improved schemes of trolleys exchange (loading elevators, cherry pickers). It is also possible to increase reliability within element E_6 without any significant costs by supplying the face with certain service parts and improving maintenance service. Effective measures in this case are to streamline supply, stock the service parts which are often broken down, improve production standards and ensure continuous power and compressed air supply.

Calculation results demonstrate that it is possible to increase working efficiency by 29% at Magianskaya prospecting expedition only by managing logistic factors on the scientific basis without any additional investments. Thereby, the method of flowcharts reliability assessment let us identify element by element organizational, technological and technical reserves to increase tunneling efficiency and to influence engineering-and-economical performance in the whole.

References

[1] Venttsel' E S 1988 *Issledovanie Operatsiy: Zadachi,Pprintsipy,Mmetodologiya* (Moscow: Nauka (Science)) p 208
[2] Gmurman V E 2003 *Teoriya Veroyatnostey i Matematicheskaya Statistika* (Moscow: Vysshaya shkola (Higher school) p 479
[3] Luk'yanov V G and Krets V G 2014 *Gornye Mashiny i Provedenie Gorno-razvedochnykh Vyrabotok* (Tomsk: TPU) 342
[4] Roginskiy V M 1975 Nadezhnost' tekhnologicheskikh sistem i rezervirovanie oborudovaniya na podzemnykh gorno-razvedochnykh rabotakh *Razvedka i Okhrana Nedr* **5** 34
[5] Burkov P V, Chernyavsky D Yu, Burkova S P and Konan A 2014 Simulation of pipeline in the area of underwater crossing *Proc. IOP Conf.: Earth and Environmental Science* [Electronic resources] (Bristol: IOP Publishing Ltd.) **21(1)** 1-5

IOP Conf. Series: Earth and Environmental Science **24** (2015) 012020 doi:10.1088/1755-1315/24/1/012020

Functional-analytical capabilities of GIS technology in the study of water use risks

O G Nevidimova[1,a], E P Yankovich[2,b] and K S Yankovich[2]

[1]Institute of Monitoring of Climatic and Ecological Systems SB RAS, Tomsk, Russia
[2]National Research Tomsk Polytechnic University, Tomsk, Russia

E-mails: [a]olga-nevidimova@mail.ru, [b]yankovich@tpu.ru

Abstract. Regional security aspects of economic activities are of great importance for legal regulation in environmental management. This has become a critical issue due to climate change, especially in regions where severe climate conditions have a great impact on almost all types of natural resource uses. A detailed analysis of climate and hydrological situation in Tomsk Oblast considering water use risks was carried out. Based on developed author's techniques an informational and analytical database was created using ArcGIS software platform, which combines statistical (quantitative) and spatial characteristics of natural hazards and socio-economic factors. This system was employed to perform areal zoning according to the degree of water use risks involved.

Introduction.

The problem of studies of water use risks is closely connected with the problem of comprehensive knowledge about the processes which occur in natural and socio-economic environments. There is a need for precise and geo-correlated information about the ecological situation of geosystems located in a particular region. Therefore, in order to provide an integrated assessment of water use conditions, which, to a great extent, are the result of intensified human activities within the floodplains of river channels, we should incorporate results of separate studies considering either climatic, hydrological or socio-environmental conditions in the area [1].

Water use is a key factor which determines the society functioning and development of its economic potential. Tomsk Oblast possesses significant surface water resources which by far exceed the needs of industry, agriculture, housing and communal services. Surface water resources in Tomsk Oblast are concentrated in 131 000 water bodies, including 8 100 rivers which are 95 thousand km long in total. River network density ranges from 0.39 to 0.29 km/km^2 [2]. Tomsk Oblast is characterized by extensive methods of water use and there are local areas of increased intensity of water use due to the peculiarities of their economic structure.

Water resources in Tomsk Oblast are used for oil and gas production, manufacturing, agriculture activities, water supply of cities and towns, timber floating, fisheries. In a number of areas within the region, rivers are practically the only means of transportation. Population of the region is concentrated within the valleys of the rivers Ob, Tom, Vasyugan, Chulym and Ket [2].

However, climate conditions of Tomsk Oblast do not only shape and regulate the river system dynamics, but also determine the direction, intensity and risks in the use of its water bodies – rivers. The commercial use of water is made under the influence of multiple adverse climatic factors: low winter temperatures, considerable weather variability, sharp fluctuations of daily and annual temperatures, unfavourable hydrological situation in the area, strong and persistent winds, and large swampy areas. Based on many environmental indicators, the region can be referred to as an area extremely uncomfortable for living. A high number of natural hazards is observed here (from 180 to 240 cases of natural risks per year), increasing by almost 6% annually [3,4]. When making analysis and forecasts of these hazards, there is a need for precise and geo-correlated information about events and processes, based on which a detailed assessment of the area can be made. This task can be successfully accomplished using modern GIS (Geographic Information System) technologies. Development of GIS technologies demonstrates benefits of the integrated use of varied data to receive new information about geographic features. Systematic and comprehensive approach is the most important characteristic of GIS [5]. These capabilities make GIS technology the most effective method for spatial-temporal assessment of existing climatic hazards for the purpose of water resource management [6].

Materials and methods

The data from the most representative meteorological and gauging stations over the last 40 years were used as GIS modeling base for spatial distribution of climatic and hydrological conditions and water use risks in Tomsk Oblast. ArcGIS software was used as a tool for processing, analyzing and reporting data [7]. The advantage of this GIS-based software package is that it can create an extensive information base, combining digital, cartographic characteristics of natural hazards and their dynamic state. Basic functions of ArcGIS software within the framework of the created geographic information system contribute to comprehensive analysis of natural hazards by adding statistical information from standalone separate tables to the tables of spatial object classes within the system. This, in turn, provides an opportunity to combine diverse parameters and produce multi-component maps for individual tasks of environmental management, thus visualizing the received results.

In the present study, development of a spatial dynamic model of hazardous natural processes based on GIS technology has involved several stages. The first stage was the formation of a geodatabase, which consists of standalone tables, vector and raster data, geographic feature behavior rules. Spatial data were divided into two blocks representing feature datasets. One feature dataset includes a topographic base. Another dataset contains thematic data, including information about location of weather and gauging stations presented as point features.

To obtain correct results, all spatial data were arranged into a single coordinate system with regard to projections, distortions and generalization rate.

Standalone tables were formed on the basis of statistical data comprising climatic and hydrological parameters. From the entire set of natural factors, only those were selected that could be hazardous to water use processes.

Data on natural risks which were included into the standalone tables were divided into 3 groups: climatic, hydrological and socio-economic.

When calculating the integrated climatic hazard index, $NC = \frac{\sum_{i=1}^{7} n_i \times 100}{7}$ seven climatic factors were considered among which there are those producing the most negative impact on the natural resource management activities in the studied area:

1. number of windy days (wind speed of 15 m/s or more);
2. snowmelt runoff /melt water formation (average daily intensity of snowmelt in mm);
3. temperature fluctuations (maximum temperature difference at the soil surface in April - a period of extreme instability of temperature);
4. number of days with excess rainfall (per year);
5. exposure to storms (maximum number of storms per year);
6. risk of drought (maximum number of days per year with humidity below 30%);

7. low winter temperatures (maximum number of days per year with the temperature of -35°C and below).

Hydrological stress index was calculated as $NG = \frac{\sum_{i=1}^{5} k_i}{5}$, where k_i – a coefficient corresponding to the value of five major hydrological factors characteristic of Tomsk Oblast:

1. low water level (number of days per year with a low water level);
2. ice phenomena (complex index considering ice-drift and ice-cover duration and early icing;
3. high water level (number of days per year with a maximum level of high water);
4. difference between low and high water levels;
5. intensity of ice phenomena and meltwater accumulation;
6. sediment runoff rates.

Basic information included socio-economic indicators characterizing the condition of water resources management and possible damages from the natural hazards. Social and environmental stress index NR within a separate river channel borderlines was calculated by the following formula $NR = \frac{\sum_{i=1}^{3} g_i}{3}$, where g_1 – population density score (within riverbed area); g_2 – bridge crossing frequency score; g_3 – score indicating the number of environmentally hazardous oil-gas pipeline and river channel intersections.

The quantitative values listed above were used as input data and were correlated to the location of the selected weather stations, gauging posts and water bodies, with the help of Join – a specialized tool ArcGIS. Each indicator was included into the database with reference to observation year. The database structure and functions provide the possibility to add and modify information.

The next stage involved the spatial analysis of data, including areal zoning, creating a series of thematic maps and visualization. To perform regional variability assessment of separate hazard parameter, its values were interpolated using an additional ArcGIS Spatial Analyst Module.

Results and discussion

Based on the conducted analysis, the areal zoning has been performed according to the degree of water use risk. In addition, the areas with the greatest frequency of dangerous climatic phenomena as well as areas with different variability degrees of meteo-climatic parameters were plotted in a series of thematic maps.

1.1. Major trends of climate change in Tomsk Oblast

According to the received estimates, over the past 10 years the temperature has been rising at an average rate of 0.34°C per year in the entire Tomsk Oblast area. For the vast plain areas, actually obtained values of average temperature increase ranged from 0.68 to 0.92°C over the studied period. Most intense warming occurred in the northern and south-eastern parts of Tomsk Oblast, the least intensive warming – in the western parts. Of particular interest is submeridianally oriented area, located along the main waterway of the oblast – the Ob River and its major tributaries – rivers Tom, Parabel, Vasyugan and Tym. This zone is characterized by relatively high values of the average temperature increase (up to 0.85°C) (figure 1).

The rainfall variability analysis has shown that the increase in precipitation is the main trend throughout the region. Moreover, the growth rates of average annual rainfall exceed by 2.5 times the rainfall reduction rates, although both values are small and do not exceed 2.8 mm in the absolute value over the study period (figure 2).

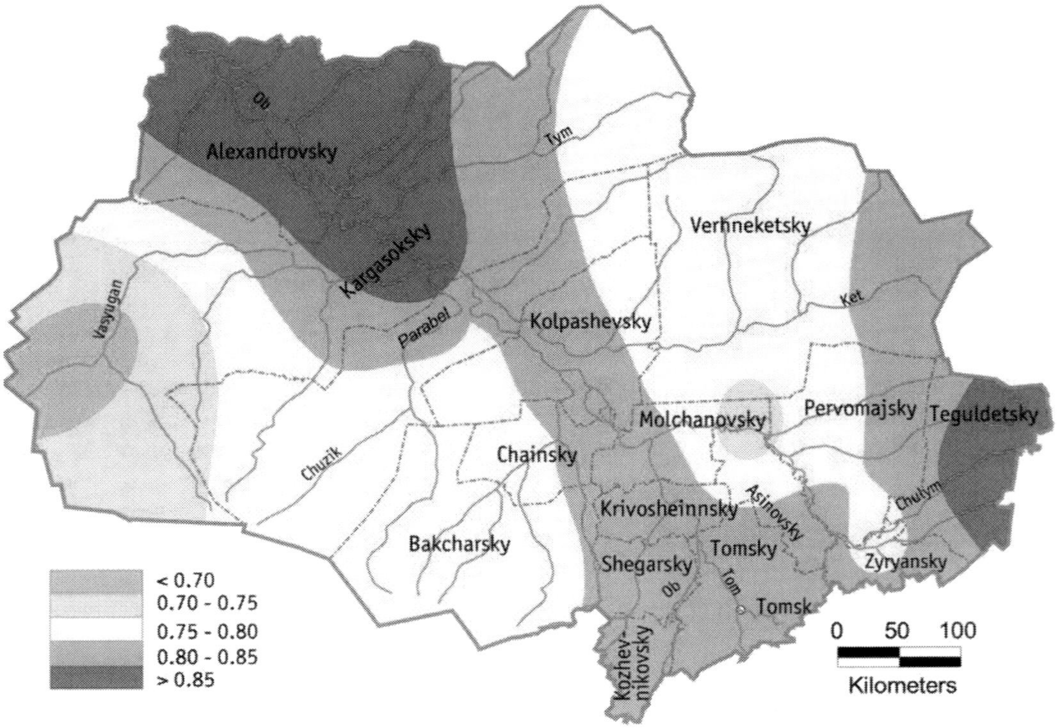

Figure 1. The average annual temperature changes in the years 1976-2006.

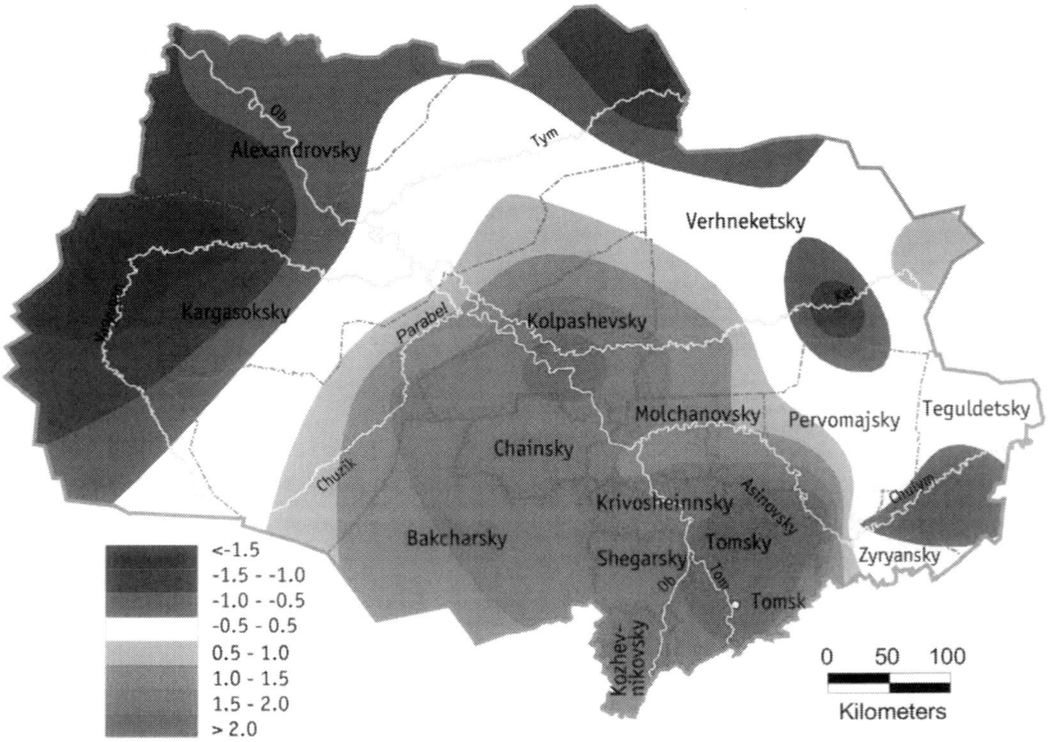

Figure 2. The average annual precipitation changes in the years 1976-2006.

The precipitation growth trend is observed in the central and southern areas. Furthermore, it should be noted that most of these areas are subjected to increasingly strong local effects of dangerous exogenous processes including coastal and gully erosion as well as landslides.

Another important result of the rainfall variability analysis is the distinguished reduction of rainfall in the western and northern parts, as well as significant reduction in precipitation levels in some eastern parts of Tomsk Oblast. With the increase of average annual temperatures, especially during the coldest part of winter, reduced rainfall can cause degradation of relict permafrost and contribute to thermokarst phenomena.

These findings were used to perform the spatial assessment of the studied climatic factor risks. The wind regime in Tomsk Oblast is characterized by significant spatial homogeneity of strong gusty winds (figure 3).

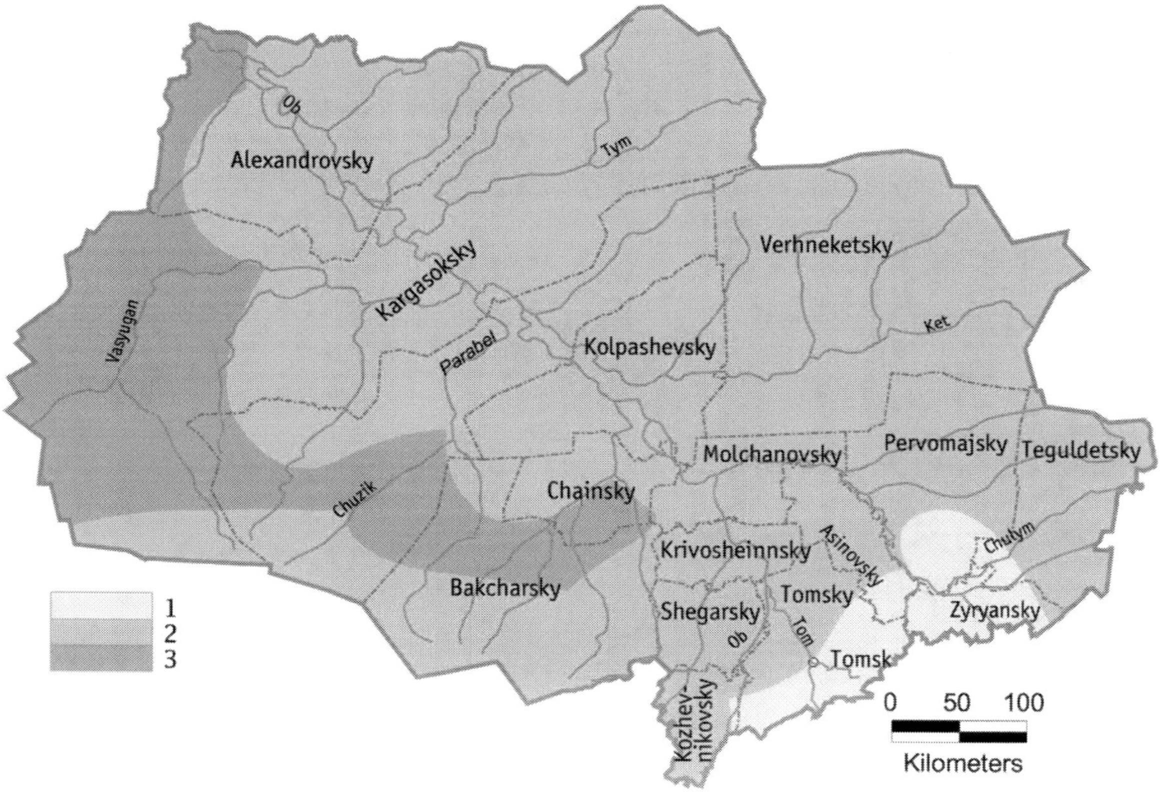

Figure 3. Evaluation of wind regime risk level: 1 - weak; 2-average; 3-high.

Since threshold wind velocity resulting in soil erosion ranges from 5 to 12 m/s in Tomsk Oblast, the increased wind erosion hazard rate is characteristic of almost entire Tomsk Oblast area.

1.2. Hydrologic vulnerability of Tomsk Oblast

These trends are changing the heat and moisture ratio, which determines the dynamics of climatic processes, including the mode of drainage system operation in the studied area [8]. Based on the calculations of hydrological stress, the territory was ranked by the degree of hydrological hazards. Thus, it was found out that the major risk factors for the Ob River are the intensity of snowmelt water accumulation (in southern areas) and sediment runoff rate fluctuations. The growth of these factors leads to increased soil erosion and riverbed deformation.

IOP Conf. Series: Earth and Environmental Science **24** (2015) 012020 doi:10.1088/1755-1315/24/1/012020

1.3. Water use risk assessment

Analysis of climatic and hydrological characteristics of the area provides assessment of water use risks, occurring as a result of human activities within the river floodplain complexes. Based on comprehensive analysis of climate, hydrological and socio-economic stress parameters, the integrated water use risk assessment was conducted. Water use risk calculations carried out within Tomsk Oblast introduce a wide assessment range (figure 4). The map shows the water stressed areas ranked according to risk level: high and heightened risk water stress zones constitute over 66.5% of total area, medium risk zones account for 9,89 %; low and reduced risk water stress zones comprise 23,61 % of Tomsk oblast area.

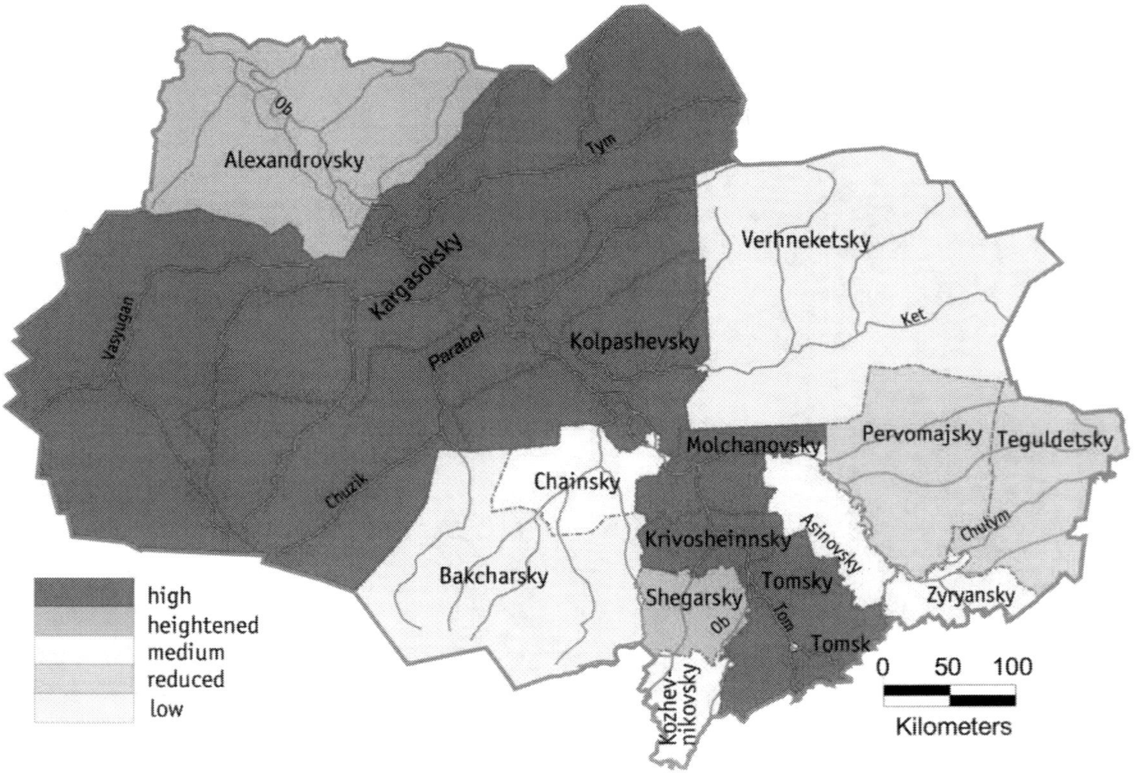

Figure 4. Water use risks in Tomsk oblast

Thus, considering high risk area prevalence, the water use hazards can be referred to as mass phenomena in Tomsk Oblast. The areas adjacent to the Ob River are of particular concern. Despite the obvious appeal of these areas, economic development of the region results in risk increase during high water periods. Population pressure in the south of the region (Tomsky, Krivosheinnsky, Molchanovsky raions), oil and gas pipeline transportation development in oil-bearing parts of the region (Alexandrovsky, Kargasoksky, Parabelsky raions), enhanced by high hydrological stress and active river channel formation processes account for high levels of water use hazards in these areas. In this respect, the eastern parts of the oblast (Verhneketsky, Pervomajsky, Teguldetsky raions), where social infrastructure and pipelines transportation are least developed, are less stressed. Relatively low values of climatic, hydrological, socio-environmental stress indexes in Bakcharsky Raion determine relatively low natural hazard potential of the area.

Conclusion

The conducted studies have revealed a set of features which characterize natural and anthropogenic interconnections and meteorological conditions in the studied area. The collected factual material was used as the basis for the analysis of climatic, hydrological and socio-economic impacts on the spatial distribution and degree of water use risks. Interactions of climatic and socio-economic processes define both the nature and the amount of water use risks. Territorial differentiation of these risks is determined by hydrological and socio-economic characteristics, rather than by meteorological and climatic conditions of the area. The degree of risk is proportional to the territory inhabitance level and the level of its economic development. The proposed approach to water use risk assessment can be used to perform the areal zoning according to different climatic, hydrological parameters and social factors, based on local conditions.

A large number of location-based and temporally distributed data on the climatic, hydrological and socio-economic conditions of water use determines the need for their integration into a single data storage and processing system, which can serve as an analytical tool for making decisions. Due to GIS capabilities, which allow accumulating a wealth of information, it is possible to conduct rapid, real-time analysis and visualize distribution of dangerous natural processes and water resources management risks.

References

[1] Hall JW, Watts G, Keil M, de Vial L, Street R, Conlan K, O'Connell PE, Beven KJ and Kilsby CG 2012 Towards Risk-Based Water Resources Planning in England and Wales Under a Changing Climate *Water and Environment Journal* **26** **118** 118–129

[2] Rakovskaya EM and Davydova MI 2003 *Physical Geography of Russia*: *Part 2: Asian region, the Caucasus and the Urals* (Russia)

[3] Nevidimova OG and Yankovich EP 2009 Nature-climatic Hazard Analysis by GIS-Technology *Mining informational and analitical Bulletin (Scientific and technical journal)* **17** 305-309

[4] Volkova ES, Nevidimova OG and Melnik MA 2011 Comprehensive risk analysis in the area of nature management in Tomsk Oblast *Geography and natural resources*. **2** 39-46

[5] DeMers MN 2000 *Fundamentals of Geographic Information Systems*. (New York (NY): John Wiley & Sons)

[6] Nagarajan M and Basil G 2014 Remote Sensing- and GIS-based Runoff Modeling with the Effect of Land-use Changes *Natural Hazards* **73** **2023** 2023-2039

[7] Harder Ch, Ormsby T and Balstrom Th 2013 *Understanding GIS: An ArcGIS Project Workbook* (California, USA)

[8] Nevidimova O 2011 *Hydropower Resources and Hydrological Hazards in Western Siberia* In: 11-th International Multidisciplinary Scientific GeoConference SGEM 2011 (Bulgaria Albena) Vol.II pp. 867-872.

IOP Conf. Series: Earth and Environmental Science **24** (2015) 012021 doi:10.1088/1755-1315/24/1/012021

Textural and structural features, composition and formation conditions of arenaceous rocks in PK_1 horizon, Pokursk suite in south-eastern Pur-Tazovsk area (Yamalo-Nenets Autonomous Territory)

N Nedolivko[1], T Perevertailo[2], A Barkalova[3] and T Dolgaya[4]
[1, 2, 3]Department of Geology and Mineral Exploration, Institute of Natural Resources, National Research Tomsk Polytechnic University Tomsk, Russia
[4]Department of Foreign Languages, Institute of Natural Resources, National Research Tomsk Polytechnic University Tomsk, Russia

Email: [1]nedolivkonm@yandex.ru, [2]ptg@tpu.ru, [3]b.alexandra1992@yandex.ru

Abstract. Terrigenous deposits of the productive PK_1 horizon in Pokursk suite of Pur-Tazovsk area (Yamalo-Nenets Autonomous Territory) were studied to specify the structure and identify the formation features. Complex horizon structure of sequential silt, mudrock interbedding and alternation has been identified. Littoral-marine type of sedimentation (occurring during the total marine transgression increase) by the help of genetic rock features identified during the core sampling and by the granulometry and X-ray-phase results analysis has been determined.

Introduction

Identification of arenaceous reservoir rock genesis and mode of occurrence is one of the major Petroleum Geology targets, particularly when the non-structural traps should be explored. The target of this research is to specify productive PK_1 (Upper Cretaceous, Cenomanian) horizon structure of Pokursk suite in south-eastern Pur-Tazovsk area (Yamalo-Nenets Autonomous Territory) and identify formation features on the basis of rock genetic features determined by the core microscopic analysis, granulometry and X-ray-phase analysis. Relevance of the research is explained by the Cenomanian complex structure characterized by the polyfacial sedimentation and the absence of data about its formation. Thus, the Cenomanian deposits have been characterized as continental ones in the performance reports on some oilfields. In fact, the performance reports and published data consider the genesis of the Cenomanian deposits of Western Siberia to be related to transition zones of sedimentation (deltaic facies) [1, 4, 5] or desalinated sea basins [11, 7, 8]. According to our research data, PK_1 horizon deposits of Pur-Tazovsk area were formed along the coast of the shallow sea basin against the sea floor down warping and marine transgression increase.

Structure, lithology characteristic features, genetic features and formation conditions of PK_1 horizon by the microscopy data

Studied PK_1 horizon deposits encompassing gas deposit, occur at the depth intervals of 1307.5–1353.0 m in the upper layers of the productive Cenomanian formation, overlapped by the Upper Cretaceous marine clayey rocks of Kuznetsov suite (Turonian Stage). According to the lithology characteristic features and formation conditions (profile description and genetic features identification were made by

[3, 6], they are divided into five arenaceous interlayers, indexed as PK_1^5–PK_1^1 from the bottom to the top and separated by thin clayey and sometimes clayey-coaly interlayers (figure 1).

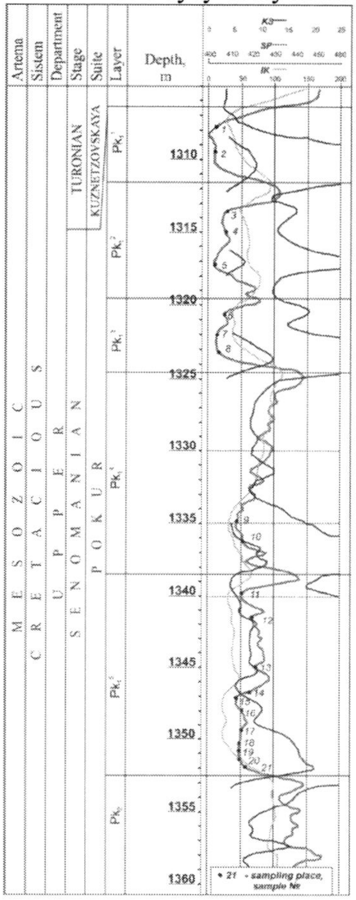

Figure 1. Dissection chart, PK_1 horizon geophysical characteristics, sample numbers and sampling areas.

PK_1^5 lower layer (1353.0–1339.0 m) is discordant to the washout and lies on the lumpy mudrock pierced by plant roots. It is characterized mostly by fine- and medium-grained sandstones with horizontal, oblique-low-angle and diverse-directional cross bedding oriented angularly 30–35° to the core axis (figure 2, A). The bedding is specified by the inwash of clayey material and plant detritus on the bedding planes. Rare clay bands with benthonic detritovore ichnofossils (*Chondrites*) as the holes and short channels occur. Intraclast layers (clasts of water-worn and re-deposited sedimentary rocks) of argillaceous siderite and clayey composition are identified. The crowned layers of the deposit are of clayey, dark grey, unclear-wavy-stratified rocks comprising small impurities of plant detritus and pyrite concretions.

PK_1^4 layer situated above (1339.0–1325.3 m) overlies PK_1^5, which can be observed in the water-worn bedding, uneven contact and rounded intraclasts of the argillaceous siderite and clayey composition. The layer is characterized by fine-grained aleurite sandstones with carbonaceous plant detritus inwash, that shows the cross (the angle of gradient is 35–40° the core axis) unidirectional and diverse-directional lamination, which are deformed by turbidity, washout and channels of small benthonic animals (for example, *Chondrites*) (figure 2, B). Upwards in the profile, the sandstone layers are replaced by mudrock gradually transferring into coal.

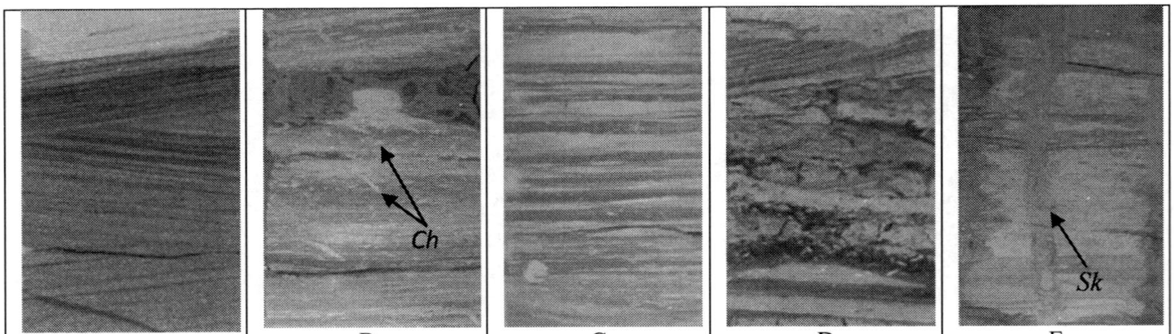

A	B	C	D	E
Diverse-directional cross bedding, PK_1^5, 1348.1 m	Traces of benthonic organisms type *Chondrites (Ch)*, PK_1^4, 1331.6 m	Wavy bedding, traces of detritivore type *Chondrites*, PK_1^3, 1320.3 m	Wavy cross diverse-directional bedding, bioturbation, PK_1^2, 1317.5 m	Sandstone with bioturbation types – *Chondrites and Skolithos (Sk)*, PK_1^1, 1312.0 m

Figure 2. PK_1^5–PK_1^1 layer rock features.

PK_1^3 layer (1325.3–1320.0 m) underlies unevenly coal seam It is characterized by the fine-grained aleurite sandstones, poorly cemented by clayey cement with primarily wave-like (cross-wave, undulating, wave, lenticular-wave), rarely, horizontal lamination. The bedding is fine and thin, formed as the result of carbonaceous plant detritus inwash and clayey material. At the top of the bed, the lamination is deformed by the root systems. The crowned layers of the profile are of thin lenticular wave-like alternation of light-grey aleurites and grey clayey rocks where benthonic detritovore ichnofossils (*Chondrites*) (figure 2, C) are found.

PK_1^2 layer (1320.0–1312.5 m) has unstable lithological composition. Continuous alteration of fine-grained light-grey aleurite sandstones (from light-grey coarse-grained arenaceous to grey fine-grained and clayey) and dark grey clayey rocks with the changing through the profile: from cross wave (unidirectional and diverse-directional) (figure 2, D) to wavy, lenticular wave and rarely horizontal. Layers often have multiple washouts, turbidity and intensive bioturbation traces (*Chondrites*). Bioturbation traces and root systems remains are often found in the rocks.

PK_1^1 layer (1312.5–1307.5 m) occurs on the underlying deposits with sharp contact. It is characterized by the light-grey aleurites and sandstones, arenaceous aleurites with wave-like bands of clayey material, benthonic "sand eaters" (*Skolithos*) and detritovore ichnofossils (*Chondrites*) (figure 2, E). Upwards in the profile, the layers are replaced by clayey with unclear bedding rocks characterized by turbidity traces of precipitation, holes and channels of small benthonic organisms (*Chondrites*).

The lower part of the profile (layers PK_1^5 and PK_1^6) are composed of more coarse-grained rocks with cross, diverse-directional and wavy lamination, coal bands, root remains and benthonic ichnofossils was formed under conditions of alternating wave motion of water with relatively high dynamics; the upper part (layers PK_1^3, PK_1^2 and PK_1^1) with coarse-grained arenaceous aleurite rocks, mainly wavy lamination and intensive bioturbation was formed under conditions of lower dynamics and alternating wave motion of water. Features of marine (wavy lamination and multidirectional cross-bedding characterize wave character of the aquatic sedimentary environment, ichnofossils type *Skolithos* and *Chondrites* are formed in sublitoral [2]) and continental origin (coal bands, root remains) show the alteration of marine\continental conditions and coastal-marine regime of sedimentation. In general, the sediment accumulation in PK_1 horizon occursin the setting of the sea floor downwarping of the sedimentation basin and is related to Cenomanian-Turonian marine transgression.

Deposits genesis identification for horizon PK_1 according to analytical results

Marine origin of sediments is indicated by the cementing mixed-layer minerals like hydromica-montmorillonite, montmorillonite, chlorite (according to the X-ray-phase analysis) (figure 3). The components content increases through the profile from bottom to top, but kaolinite content substantially decreases in the same direction.

Granulometric spectrum of arenaceous aleurite rocks is dominated by arenaceous and aleurite fractions; pelitic fractions the content of which is not higher than 10 % (figure 4), are dramatically subjected to good deposit elutriation. The rocks are poorly cemented by the clayey cement; these can be easily deformed and, as a result, provides informative granulometric characteristics.

According to the granulometric analysis data (table 1) and calculations made by Falk's equation [3], granulometric spectra of rocks have the following characteristics: positive values of asymmetry – prevalence of coarse-grained fractions and high dynamic activity of the sedimentation environment; positive values of excess reflect the sediment resorting stability [9, 10]; sorting coefficient increase (S_o) through the profile from the bottom to the top – sediment resorting improvement from the medium – to well-sorted grains.

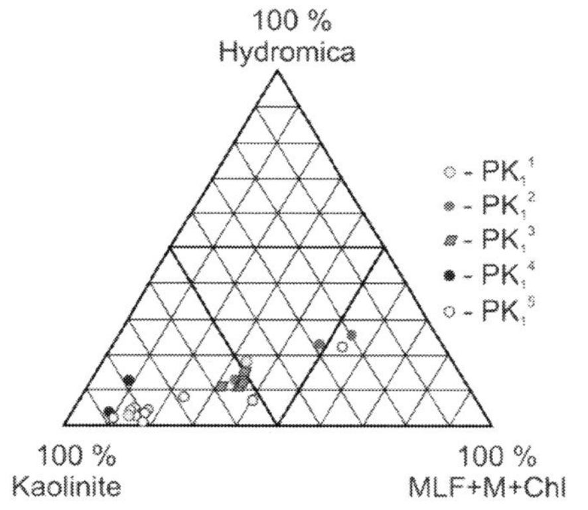

Figure 3. Triangle diagram of clayey cement content in arenaceous and aleurite horizon PK_1 rocks.

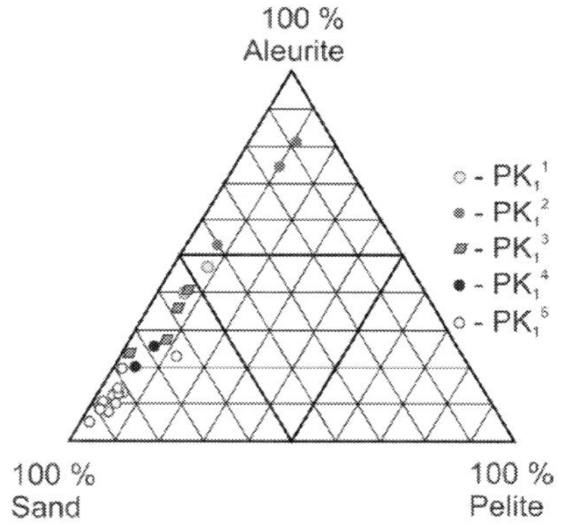

Figure 4. Figurative points position for granulometric composition of arenaceous and aleurite horizon PK_1 rocks.

Notes: MLF – mixed layered minerals type of hydromica-montmorillonite, M – montmorillonite, Chl – chlorite.

Table 1. Granulometric parameters of arenaceous rocks in horizon PK_1

Layer	Average content (%) and fraction size (mm)					Granulometric coefficients			
	1–0.5	0.5–0.2	0.2–0.1	0.1–0.01	<0.01	Md, mm	Sorting (S_o)	Asymmetry (from-to)	Excess (from-to)
PK_1^1	0,08	2.28	47.71	43.83	6.10	0.08–0.09	0.39–0.48	0.22–0.34	1.07–1.15
PK_1^2	0.05	2.45	19.43	69.04	9.03	0.05–0.08	0.36-0.43	0–0.08	0.7–1.04
PK_1^3	0.04	6.69	55.63	31.50	6.15	0.09–0.12	0.40–0.53	0.19–0.37	0.89–1.6
PK_1^4	0.08	5.06	67.19	22.11	5.55	0.10–0.11	0.49–0.53	0.06–0.32	1.14–1.61
PK_1^5	0.17	40.14	42.74	12.14	4.81	0.12–0.21	0.48–0.77	0–0.68	0.74–2.62

Imaging points of granulometric spectra of arenaceous and aleurite rocks in the range of VIII and VII fields (figure 5) on Rozhkov dynamogenetic diagram (asymmetry – excess) is relevant to the shallow water wave production, powerful wave run-up and breaking wave of the coastal facies in great open water areas (layers PK_1^5, PK_1^4 and PK_1^3) and shallow water wave processes, neutral coastal zone of the coastal marine facies (layers PK_1^3, PK_1^2, PK_1^1). Sandstones were formed in stable high dynamic environment where active reworking of the material dominated over its supply [9, 10]. Formation conditions together with the core data taken into consideration may be related to the shore bars, as the sand material accumulated, they emerged and were covered by vegetation.

Coastal layers PK_1^5 and PK_1^4 have been identified in the sediment formation and coastal marine formation conditions (layers PK_1^3, PK_1^2 and PK_1^1).

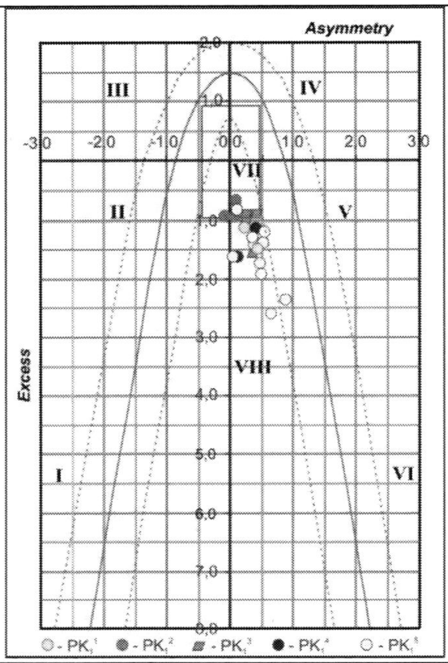

I – stagnant sedimentation conditions at the water areas bottom of various depths. Marine facies. II – bottom currents or turbidity flows. Marine facies. Hydromechanical or physical destruction of igneous rocks, marine rock erosion. Continental facies of the source areas, crust of weathering. III – weak, mostly river currents. Continental fluvial facies. IV – strong river or alongshore currents. Continental fluvial or coastal marine facies. V – shallow water wave production, strong alongshore currents. Wave run-up. Coastal marine facies, continental beach microfacies of big plain rivers. VI – shallow water wave production, strong wave-up – upper part of the area, marine sand beach eolation – lower part of the area (coastal dune microfacies). In general coastal facies of the water areas. VII – eolian rewarking of the river sediments – upper half of the rectangle. Continental desert facies (continental dunes). Low right quarter of the rectangle – shallow water wave processes, neutral coastal zone. Coastal marine facies. VIII – shallow water wave production, powerful wave run-up and breaking wave. Coastal facies of great open water areas.

Figure 5. Rozhkov dynamogenetic diagram (asymmetry – excess) [9].

Conclusions

Core and granulometric analysis data have identified the complex structure of the productive layer PK_1, 5 reservoirs are separated by coaly-clayey (in the lower part) and clayey bands have been determined.

Lithological composition has been described, genetic features have been identified and coastal (layers PK_1^5 and PK_1^4) and coastal marine (layers PK_1^3, PK_1^2 and PK_1^1) conditions for deposits formation have been determined.

References

[1] Aleksandrova G, Kosmynin V, Postnikov A 2010 Stratigraphy and sedimentary environments for Cretaceous deposits in the southern part of the Var'egansk megabar (Western Siberia) *Stratigraphy and Geological Correlation* **18**, 4, 65–91

[2] Frey R W, Seilacher A 1980 Uniformity in marine invertebrate ichnology *Lethaia* 13, 183–207

[3] Grossgeim V, Beskrovnaya O, and Gerashenko I 1984 *Methods of paleogeographic reconstruction (during oil\gas exploration)* (Leningrad: "Nedra"), 22–24

[4] Kosmynin V 2012 *Analysis of lithological heterogeneity of oil and gas deposits of Pokur suite in southern Var'egansk megabar* (Russian Edition LAP LAMBERT Academic Publishing), 264

[5] Lebedeva N, Agalakov S, Beizel A 2004 Palynostratigraphy and structure of the Upper Cretaceous section based on well 113 of the Yuzhno-Russkaya area (Pur-Tazovsk area, Western Siberia) *Geology and Geophysics* **45**, *6–7: News of Paleontology and Stratigraphy*, 191–207

[6] Nedolivko N 2008 *Core research for oil/gas wells. Training workshop for the students of Petroleum Geology as a speciality* (Tomsk: TPU), 158

[7] Podobina V 2000. *Upper Cretaceous foraminiferas and biostratigraphy of Western Siberia* (Tomsk. NTL Press), 388

[8] Podobina V 2012 New data on Upper Cenomanian foraminiferas and biostratigraphy of West Siberia northern district. *Bulletin of Tomsk state University* 361, 182–187

[9] Rozhkov G 1978 Geologic interpretation of granulometric parameters according to fractional sieve analysis *Granulometric analysis in geology* (M. "Nedra") 5–25

[10] Rozhkov G 1986 Fragmentary material differentiation and granulometric diagram $\alpha - \tau$ according to indirect grains count *Mechanical differentiation of the solid substance on the land and offshore* 97–117

[11] Zakharov V, Beizel A, Pokhialainen V 1989 Discovery of the marine Cenomanian in Siberia north *Geology and Geophysic,* **6**, 10–13

Forest fire situation analysis over forest reserve land in Tomsk petroleum province

O A Pasko[1] and A V Baranova [2]
[1 2]Natural Research Tomsk Polytechnic University, Tomsk, Russia.

[1]oap@sibmail.com, [2]baranskikh@yandex.ru

Abstract. The paper delivers the analysis of space-time characteristics of forest fire ignition and spread in the North of Tomsk oblast, i.e. petroleum production area (Kargasok, Parabel and Teguldet districts). It also presents long-term and seasonal forest fire behavior including fire ignition and spread frequency (annual and seasonal), the fire season duration and their zonality. The main driving factors of forest fire ignition both human and natural ones are revealed.

1. Introduction

Experts estimate the total hydrocarbon resources of Tomsk petroleum province at 2.95 billions of tons, which makes the province one of the most prospective areas of Russian fuel and energy complex. The production and infrastructure development including transport accessibility contributes to the increasing number of forest fires on the one hand, but facilitates their discovery and suppression [1]. The fire destructs productive woodlands that could have been reclaimed and is of serious hazard to the oil and gas complex. The degree of fire danger is determined by oil and gas flammability. Among technogenic factors are oil spills, atmospheric gas pollution, failed electrical equipment [2], left wood, assarted trees and wastes [3]. Another aggravating factor is the fact that it is pine forests that are typical for the northern area of Tomsk oblast. These forests have the most fire risks potential (the geographic range of birch forests is limited up to the 57th degree of northern latitude).

The number of fires in European and North American countries has been reduced lately. It can be attributed to satellite monitoring, automatic remote sensing data processing and effective use of firefighting resources [4]. On the contrary, the number of fires in the European part of Russia has increased many-fold due to forest management reorganization, the abolition of federal fire management body and forest service staff reduction [5].

The aim of the work is to study space-time characteristics of forest fire ignition and spread in the North of Tomsk oblast, i.e. petroleum production area (Kargasok, Parabel and Teguldet regions).

The research object is the pattern of forest fire ignition and its spread over the forest reserve lands.

To achieve the aim the following tasks are to be solved:

1. to study fire behavior in the North of Tomsk oblast including fire ignition frequency (annual and seasonal), fire season duration and the zonality;
2. to forecast the rate of fire danger and compare it with factual data;
3. to distinguish natural and human factors facilitating fire ignition and spread.

2. Research object, tools and methods

The research object is forest reserve lands of Tomsk oblast that are exposed to forest fires. It is the area where the main oil and gas field facilities of the region are situated. Long-term data of Tomsk oblast forestry department were analyzed (1991-2013). To predict the fire danger in 2013 Nesterov's method was applied with the obtained data being compared with factual ones according to the following formula:

$$G = \sum_{1}^{n} (t-\omega)\, t,$$

where G is burning index of ground cover, °C; n – the number of dry days (days with precipitation less than 2.5 mm are considered to be dry); t – atmospheric temperature for 12 hours, °C; ω – dew point,°C [6].

Statistical data processing was carried out by means of software STATISTICA.

3. Results and discussions

The research results prove that the number of forest fires varies by location and from year to year. Changes of forest fire danger in Tomsk oblast have a cyclical pattern (figure 1). The annual average number of forest fires over the period from1990 to 2013 is 276. The fire frequencies are: about 6 years - for maxima (more than 400 fires per year) and about 4 years for minima (less than 200 fires). The intermediate fire activity peaks make a 4 to 6-year cycle. The first cycle stage, which is characterized by the increasing number of fires, lasts 1-2 years. The next is a peak followed by the second cycle stage (the decrease of fires to the minimum) which continues 1-3 years. Sharp contrasts of the graph curve and no stabilization period are particular features of the graph.

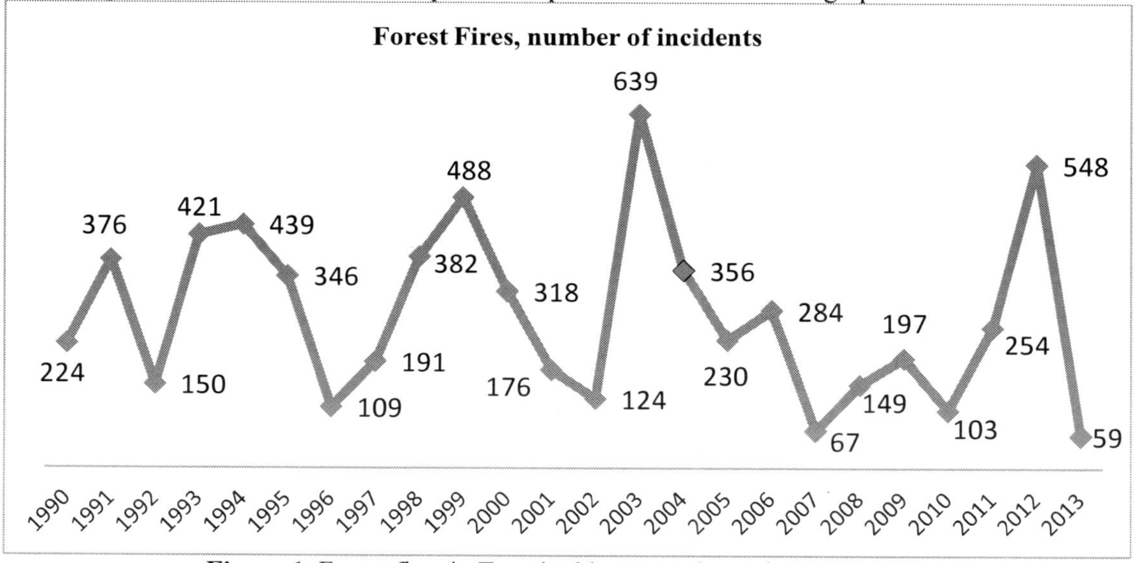

Figure 1. Forest fires in Tomsk oblast over the period 1990-2013

In general, there are three peaks of seasonal fire activity in Tomsk oblast: spring, summer, with stable fires, and autumn. However, Aleksandrovskiy, Kargasokskiy and Parabelskiy districts are characterized only by stable summer fires, mostly in June (figure 2). The areas of these districts refer to the second and third danger classes of the fire danger rating, which means there is high probability of surface fire during a whole fire season. The fire frequency is quite low but the fire areas are significant, which is attributed to poor area development and difficulties in fire spot detection. The factual fire season lasts no more than 87 days.

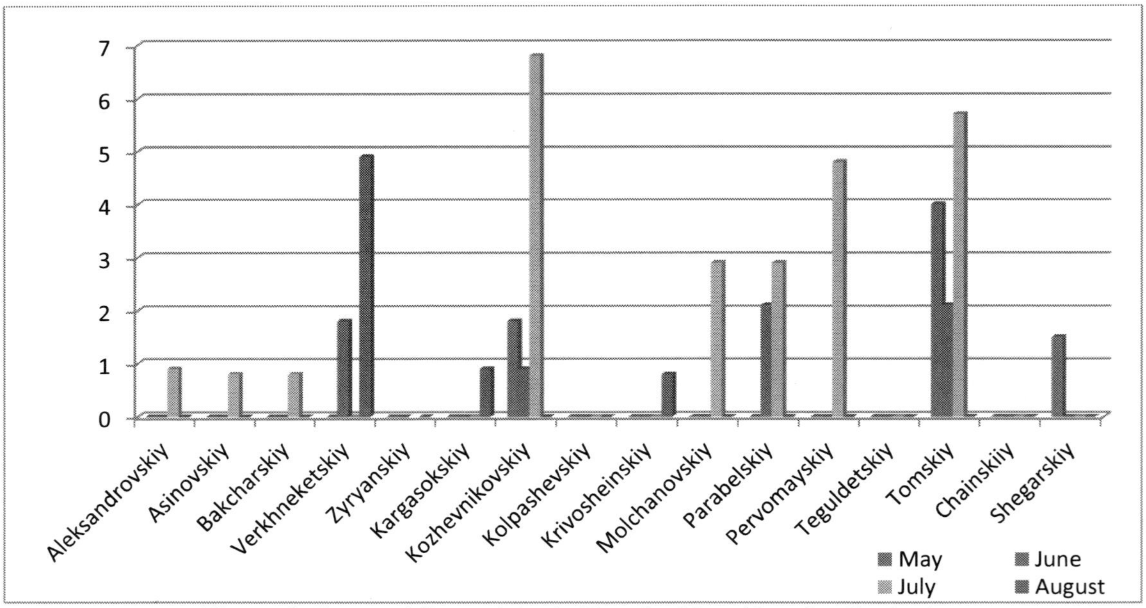

Figure 2. Seasonal fire activity in Tomsk oblast (data for districts)

Nesterov's method was used to predict fire situation in 2013. The factual and predicted data show high degree of repeatability (figure 3). In northern part of Tomsk oblast the fire season of 2013 started on 10 May. 59 fires were registered over the fire season. The fire area covered 2,211 ha including 2,164.3 ha of forest area (figure 4). The figures show a decrease of 548 fires in comparison with the same period a year earlier as well as a 295,586-ha reduction of fire area.

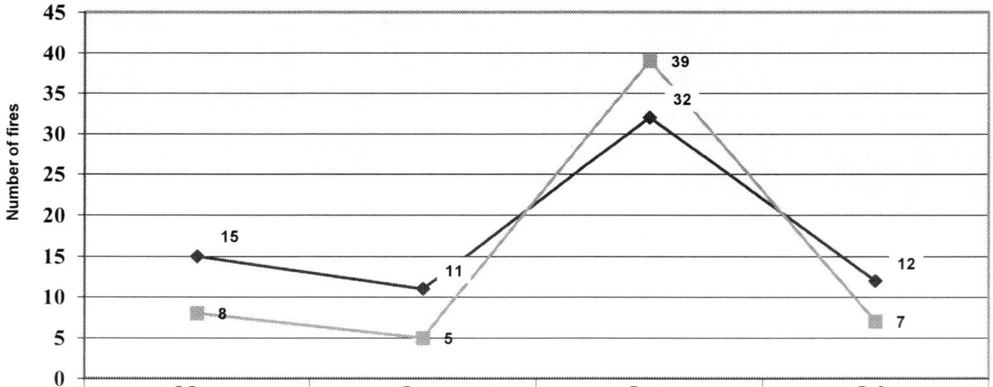

	May	June	June	July
Prediction	15	11	32	12
Fact	8	5	39	7

Figure 3. Accuracy of fire danger forecast in 2013 year in Tomsk oblast

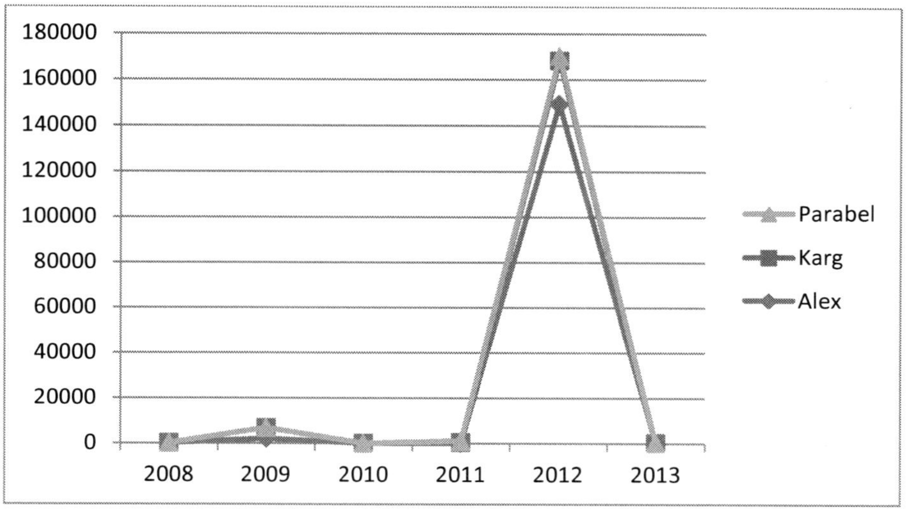

Figure 4. Changes in forest fire in Parabelskiy, Kargasokskiy and Aleksandrovskiy districts of Tomsk oblast from 2008 to 2013 years

The variables of Nesterov's formula reflect natural factors of fire ignition and spread, which are typical of the North of Tomsk oblast. High moisture of forest fuel prevents the forest from burning most part of the fire season. However, if a long dry period occurs, moss -ground cover and debris layer become dry, which makes them the basic source of fire danger. As a result, 98.5% of all the wildfires in Tomsk oblast is surface fires, 1.4% is crown fires and 0.1% is ground fires. Another factor associated with fire ignition is coniferous (pine) forests that dominate in the region. Up to 70% of all the wildfires occur in grass pine forests. Other factors of seasonal fire activity peaks are dry periods at the end of spring or beginning of summer, hot temperatures in summer, low humidity, thunderbolts and winds bringing dry and warm air from Northern Asia, Mongolia and Eastern Siberia. Such climatic pattern facilitates all the forest areas having high level of fire readiness by the middle of summer.

The intervals between fires and their frequency determine the dynamics and state of forest cover and vice versa. Thus, their study should be the foundation for the fire management. Every forest cover has its fire pattern (type, intensity, size, frequency interval) and dynamics of after-fire forest reproduction. It is obvious, that fire modeling allows revealing the fire patterns that are typical for a particular area and forecasting development trends of forest ecosystems.

Human factors should be also regarded as the reason of forest fires in the North of Tomsk oblast. These are the following factors: in-situ burning of oil spills, exhaust of waste gas, falls of rocket stages, activities of exploration companies, local people's activities and power line breakdown. It is not possible to classify these factors nowadays, as there are no factual data on this point.

The forest fires can threaten settlements, impede economical activities and destroy vast areas of wood resources, which lead to economic losses. On the other hand, wildland fires are a natural part of the forest ecosystem and important valuable tool of its self regulation, sustainable development and excessive fuel disposal. Fire danger forecasting, GIS and video monitoring of fire situation and its development, new technologies used for firefighting as well as land use works (control lines, lanes and water sources) allow reducing the number of fires and fire risks in Tomsk petroleum province. It should be noted that the stable decrease in fire incidents in Tomsk oblast is not registered. Thus, the Russian scientists' idea of the Forest Service reorganization having negative impact has not been proved yet. New fire detecting and fighting technologies, which are effectively used in European and North American countries, have not improved the fire situation in Tomsk oblast either.

4. Conclusion

1. Zone fire characteristics in the petroleum production areas of Tomsk oblast are conditioned by high natural fire danger of pine forests and climatic peculiarities of the region.
2. The seasonal and annual changes in number of fires in Tomsk oblast are of cyclical character.
3. Nesterov's method of fire danger forecasting shows high repeatability with the development of real situation.

References

[1] Pasko O 2013 The forest resource of Tomsk Oblast and its use *Applied and Fundamental Studies* 102-110
[2] Fujioka F 2009 Fire danger and fire behavior modeling system in Austrelia Europe and North America *Developments in Enviromental Science* **8** 471-497
[3] Bytnerowicz A, Arbaugh M, Riebau A, Andersen and C Elsevier 2009 Wildland fires and air pollution 643
[4] Bleken E and Mysterud I 2007 Forest fire and environmental management *A technical report on forest fire as an ecological factor* University of Oslo 266
[5] Khodakov V and Zharikova M 2011 Lesnye pozhary i metody ikh issledovaniya *Kherson Grin' DS* 470
[6] Nikishchenko N 2011 *Kompleksnaya otsenka prirodnykh i antropogennykh faktorov vozniknoveniya pozharov na zemlyakh lesnogo fonda Voronezhskoy oblasti* Author's abstract 22
[7] Kudryavtsev M, Lukin V, Malinetskii G, Mitin N, Naumenko S, Podlazov A, Rumyantsev A, and Toropygina S 2008 Forest conflagration extinguishing problems on Russian Federation territory *Preprint, Inst. Appl. Math., the Russian Academy of Science* 40

Vasyugan horizon structure features within junction zone of Ust-Tym depression and Parabel megaswell (Tomsk Oblast)

T Perevertailo[1], N Nedolivko[2], and T Dolgaya[3]
[1,2] Department of Geology and Mineral Exploration, Institute of Natural Resources, National Research Tomsk Polytechnic University Tomsk, Russia
[3]Department of Foreign Languages, Institute of Natural Resources, National Research Tomsk Polytechnic University Tomsk, Russia

E-mail: [1]ptg@tpu.ru, [2]nedolivkonm@yandex.ru

Abstract. The research area includes Upper Jurassic deposits of Vasyugan horizon. To identify the territorial relationship between Vasyugan and Naunak suites, rugged topography and well profile figurecorrelation have been made; detailed core lithology analysis (Gurarinskaya, Sobolinaya and Yasnaya areas) has been carried out, conclusions on paleoenvironment have been made. It has been identified that sedimentation occurred in the range of the studied territory where continental facies played a more significant role than it was for Vasyugan suite deposits. Thus, the conjunction zone of Ust-Tym depression and Parabel megaswell are spatially the transition zone of lithologo-facial Vasyugan suite replacement into Naunak one.

1. Introduction

Oil-bearing capacity of Upper Jurassic deposits in the West Siberia territory has been proved by the range of oil field discoveries. A number of contemporary researchers state that “the Mesozoic stratigraphy in the subsurface of the West Siberian Basin contains prolific hydrocarbon accumulations, and thus the depositional environments of marine and marginal marine Jurassic and Cretaceous age sediments are well-established” [1]. Besides, paleogeographic reconstruction of the West Siberian basin during the Jurassic has been proposed by A.E. Kontorovich [2]. However, despite the long history of research and useful know-how material, there are some problems left to be solved. The regional stratigraphic scheme of the Upper Jurassic south- east West Siberia territory shows that Vasyugan horizon includes Vasyugan and Naunak suites. But, so far, the correlation of these territories has been obscure.

The detailed lithologo - stratigraphic separation of Vasyugan suite profile in Tomsk Oblast was made by a number of researchers [3, 4], and, Vasyugan suite, according to widely used concepts is subdivided into 2 subsuites: the lower- mostly clayey with the sandstone strata J_1^5 and J_1^6 locally distributed, and the upper - characterized by sandstone, siltstone, clay and coal interlayers, comprising J_1^4, J_1^3, J_1^2, J_1^1 strata. Upper Vasyugan subsuite has undercoal (regressive), intercoal (continental) and overcoal (transgressive) layers. In the south-east of the territory the intercoal deposits merge with Naunak suite deposits. Naunak suite development area borders on Ust-Tym depression and eastern territories [5].

2. Methodology

To identify the structure features of Vasyugan suit transition into Naunak suite, a detailed lithological study of the well profiles was made. The wells were drilled in the junction zone of Ust-Tym depression and Parabel megaswell (Double bench – Yasnoye oil field) and in Ust-Tym depression southern part (Sobolin swell – Sobolin and Gurarinsk oil fields) (figure 1). Well profiles separation and correlation were made with the allowance for system analysis of rock-strata associations [6], on the basis of which the studied deposits of the stratigraphic interval are incorporated into regional cyclite J_1 (figure 2), where local cyclites are identified (from J_1^5 to J_1^1).

IOP Conf. Series: Earth and Environmental Science **24** (2015) 012023 doi:10.1088/1755-1315/24/1/012023

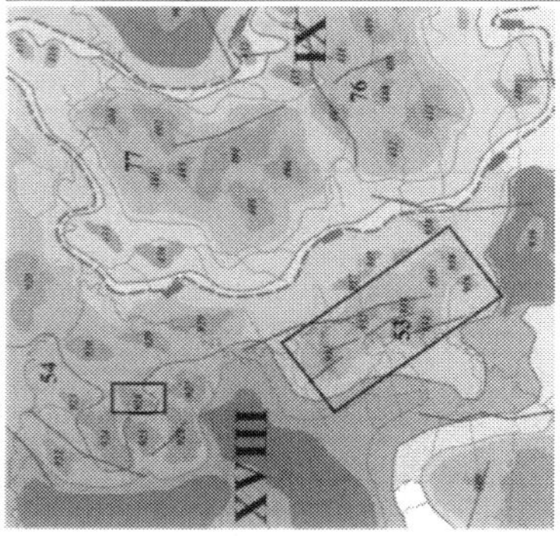

Legend:
Structures I order
IX Parabel megaswell
XVIII Ust-Tym depression
Structures III order:
53 Sobolin swell
54 Double bench

Figure 1. Jurassic structural stage of the sedimentary cover for Tomsk Oblast western territories. Extract from Tectonic map. (Editor: A Kontorovich, 2002)

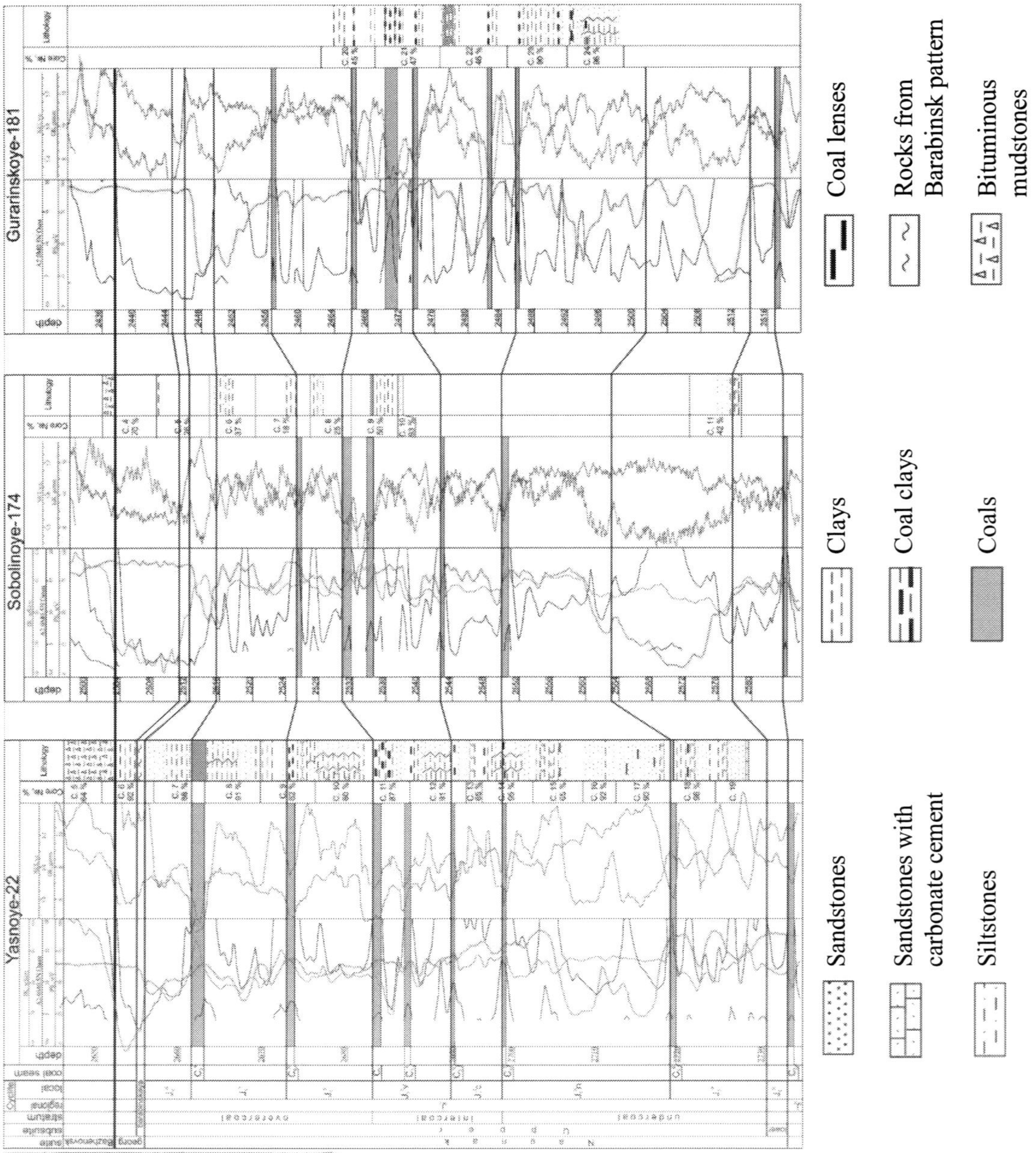

Figure 2. Correlation scheme for Yasnoye, Sobolin, Gurarinsk oil fields Upper Jurassic deposits

3. Results and discussion

On the stratigraphic schemes of Yasnoye, Sobolin, Gurarinsk oil fields the Upper Jurassic profile is characterized by deposits of Naunak, Georgiyev and Bazhenov suites. *Cyclite* J_1^5 in the lower part of the studied profile (according to the core analysis) is characterized by grey and brown-grey homogeneous with thin horizontal lamination sections siderized by clay with benthonic ichnofossils (*Palaeophycus),* which characterize a low dynamic marine environment. Clayey bottom deposits in Sobolin and Gurarinsk areas become arenaceous and form J_1^5 stratum characterized by light-grey fine-grained aleurite sandstones comprised of carbonized and siderized macro-organic inclusions, channels and holes of benthic animals. Sandstone thickness increase is marked sideways from the local uplifts arches to their limbs.

Cyclite J_1^4 deposits (undercoal thickness) mostly cover the complex of arenaceous silt rocks interbedded by clays (Gurarinsk, Sobolin areas) or by stratum C_1^g (Yasnoye area) characterized by coal or coal clays. The cyclite thickness ranges from 8m to14 m.

In Yasnoye area the cyclite in the lower part has a heterogeneous structure and is characterized by continuous and frequent coal seams, siltstone and fine-grained sandstones interbedding the formation of fine, diversidirectional cross and low-angle lamination, sometimes the laminae cutoff (figure 3 A) and specified by environment hydrodynamic activity. The rocks are intensively subjected to bioturbation and contain various benthonic ichnofossils (*Chondrites, Scolithos, Teichichnus*) (figure 3 B, C) that attest to good aeration and water area heating.

The upper cyclite part constitutes clayey rocks with interbeds of clay-siderite, clay-coal and coal composition, fine horizontal and wavy stratified structure. The rocks contain plant detritus of various preservation stages, fragments of siderized wood, root systems remains, pyrite. Flood-coast -lagoon environment to the final cyclite formation was likely to present in the studied area.

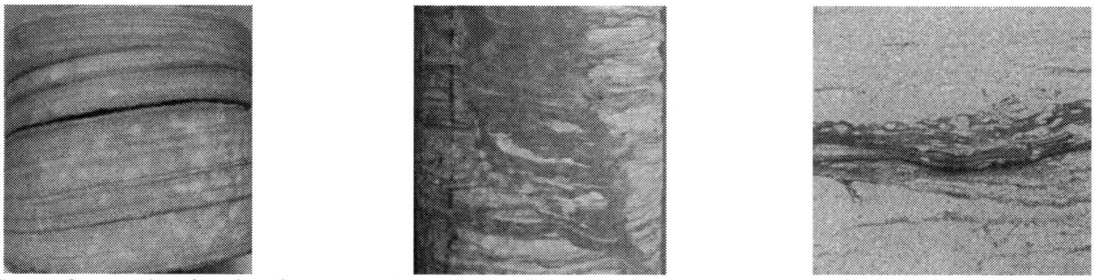

A) Sets of cross lamination in sandstone (well № 20) B) Ichnofossils Teichichnus (well № 20) C) Ichnofossils Chondrites (well № 22)

Figure 3. Cyclite J_1^4 rock features (Yasnoye area)

Cyclite J_1^{3n} with the thickness of 12–20 m is overlaid at the top by the coal seam C_1^{c}. Well profiles in the studied areas are different in structure. Yasnoye area profiles contain light-grey sandstones, from fine-grained to coarse-grained. Frequent continous interbeds with various granulometric composition and graded bedding occur. Rocks have fine low-angle, unidirectional and diversidirectional lamination (figure 4 A) deformed by wash-out and bioturbation, comprising intraformational pebbles and coal inclusions. Sandstones are overlaid by homogeneous horizontal, wave lense -like, stratified clays, partly siderized ,while on the top - by coal clays.

Thus, sediment accumulation in high dynamic environment was preserved during the whole cyclite formation process and is related to coastal bars and barrier islands exposed to the surface and periodically subjected to wash-out.

Gurarinsk area profile (well № 181) is characterized by the frequent interbedding of various lithological composition rocks: sandstones, siltstones, clayey-aleurite and coal clays. Bioturbation structures occur everywhere (figure 4 B), siderite concretions, fragments of wood, root remains (figure4,C) coal interlayers.

The described features relate to the marginal lagoon sections where sea water continuously penetrated and introduced sand and aleurite material. In the relatively calm periods, fine-grained sediments accumulated. Presence of coal interbeds and lenses in cyclite J_1^{3n} profile shows the sea basin shallowing, plant-growing and swamping of the area.

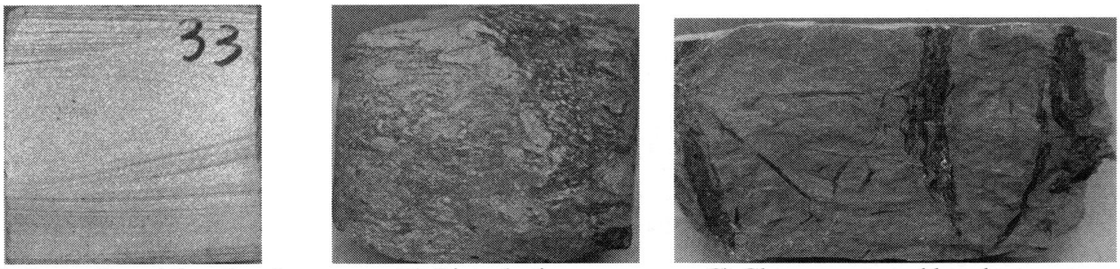

A) Cross diversidirectional lamination (well Yasnaya 22) B) Bioturbation (well Gurarinsk 181) C) Clays penetrated by plant roots (well Gurarinsk 181)

Figure 4. Cyclite J_1^{3n} rock features

In general, cyclites J_1^4 and J_1^{3n} have a regressive structure and relate to undercoal thickness.

Cyclite J_1^3c *and* J_1^3v formation (intercoal thickness) relate mostly to the continental regime of sedimentation.

According to the identified genetic features, it may be assumed that during the process of J_1^3c deposition the fine-grained sediments of marine origin accumulated at the foot, while in the continental -at the top. The process occurs at the place of bar sandstones distribution (Yasnaya area). Sand material deposition spatially occurs in the areas where under water shoals and lagoons predominated (Gurarinsk area). Rocks are characterized by increased siderization, clays have root system remains. Black coal clays, grey siltstones and black coals interbedding occur in the upper part of the cyclite.

J_1^3v sediments accumulation took place during the continental regime within the lacustrine alluvial plain. The coal seam C_1^a is determined in J_1^3v composition. The continental sedimentation regime was concluded by the coal seam C_1 formation, characterized ,according to the core analysis, by coals and black coal clays with rare thin aleurite bands. The cyclite thickness is 8-20 m.

Based on core samples from well № 20 (Yasnaya area) channel shoal facies were identified. Sandstones filling the channel refer to fine-grained varieties with intraformational clay pebble inclusions. The bedding is rare, capillary, cross and cross wave-like, formed by siderite, clayey material plant detritus inwash. Upwards in the profile, the channel deposits are overlaid by back marsh sediments and characterized by siltstone and clayey interbedding with carbonized root remains, concretions and pyrite psuedomorphs along plant remains, siderite concretions.

Deposits uncovered by wells № 22 (Yasnaya area), № 181 (Gurarinsk area) and № 184 (Sobolin area) were formed in temporarily living flood plain environment conditions. These are characterized by repeated interbeddings of sandstones, siltstones, complex clayey aleurite rock composition, clayey and coal rocks, coal. Rocks are irregularly siderized, they have clay detritus, remains of surface and root plant fragments (figure 5).

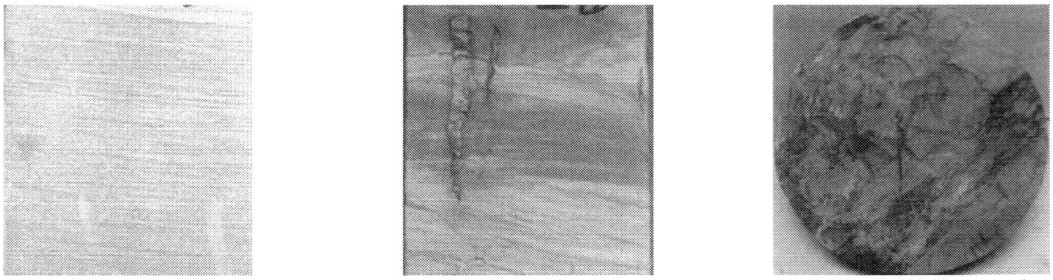

A) Thin stream lamination of flood sands (well Yasnaya № 22) B) Root remains (well Yasnaya № 22) C) Surface plant remains (well Gurarinsk № 181P)

Figure 5. Facies of temporarily living flood plain areas (cyclite J_1^{3v})

Well- isolated cyclites J_1^2 J_1^1, J_1^0 are marked out in the upper J_1 horizon (overcoal thickness). Alteration and sequence of lithological rock variety stratification illustrate the transgressive-regressive strata character. Coal bands characterize the continuous sea water-level decline and continental sedimentation conditions restoration in the studied area.

Cyclite J_1^2 of the studied area is characterized by the clayey-aleurite strata and separated from the overlying deposits by the coal shed indexed $C_0^{6.}$ The thickness of the deposits is not more than 16m.

Various multiple ichnofossils,that are found in the cores during drilling, provide basic evidence of the fact that at the beginning of the deposit formation the whole territory was covered by sea. Besides, regular continuous aleurite and clayey material interbedding, wave-like and cross wave-like alternation with thin low-angle and horizontal lamination show regular variability of aqueous environment dynamics.

Follow-up sea regression resulted in the territory proximity to the shore line, bog and vegetation formation that became rock features of continental origin. The upper part of the profile studied by coring includes clayey rocks comprised of plentiful carboniferous plant detritus, large well- preserved plant fragments, root system remains, black coal clays and coals interbeds.

Sediment accumulation of the *cyclite* J_1^1 was made during transgressive-regressive stage. The thickness of these deposits varies from 4 no 11 m.

Well profiles in Gurarinsk area show distinct regressive sedimentation nature with fragment size increase from bottom to top. Ichnofossils *(Chondrites и Terebellina),* structural and textural rock features, carbonate sediment formation indicate that the process occurred in desalted marine water area in the places close to the shore line and relatively calm hydrodynamics.

In Yasnaya area wells, the sandstones occupy the midsection of the profile, having cross and cross wave-like lamination and are fine-grained, irregularly calcitized and siderized. Upwards the profile, they are changed by lagoon sediments characterized by siltstone and clayey material interbedding with the pyrite and siderite concretions abundant, worm channels, remains of surface plant fragments and roots. At the top, the deposits are overlaid by the coal seam $C_0^{a.}$

Overlaid deposits are of the cyclite J_1^0 , sedimentation of which occurred in conditions of low shallow water mobility. The process resulted in the formation of irregular arenaceous aleurite clayey rocks of deformed bioturbidity (figure 6, A). In general, the cyclite thickness varies from 3 to 7 m.

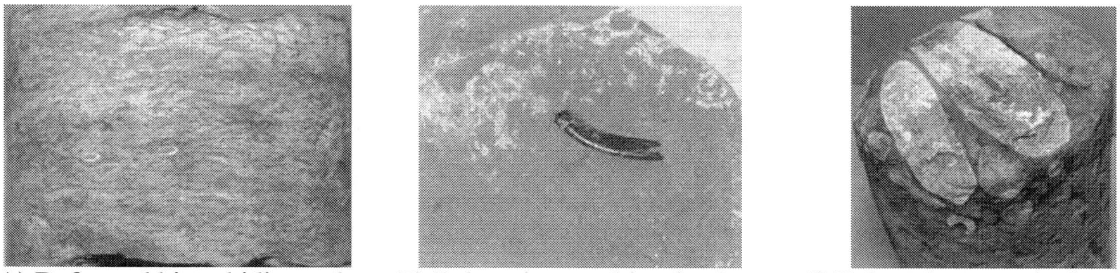

A) Deformed bioturbidity rock, Terebellina ichnofpssils (well № Yasnaya 20) B) Belemnite onychites in clays (well Gurarinsk № 182) C) Belemnite rostra in the rocks of Barabinskaya pattern (well Gurarinsk № 182)

Figure 6. Features of cyclite J_1^0 rocks and Georgiyevk deposition series

On the log Georgiyevk deposition series are positively marked out according to(showing) specific resistance decrease and natural radioactivity increase. These are characterized by dark-grey clays with *ro*stra remains and belemnite onychites (figure7, B). Barabinsk pattern rocks are marked out at the base of the series. These occur washed out on the underlying rocks of horizon J_1 and are quite different in appearance and substance composition. These are heterogeneous non-sorted aleurite-clayey rocks with green-bluish tinge, breccia-like with phosphate inclusions, great amount of pyrite and glauconite, numerous belemnite rostra characterized by calcite (figure 7, C).

The total deposition thickness of Georgiyevk series within Yasnaya area is almost 3 m, increasing in the southern part of Ust-Tym depression to 12 m.

4. Conclusion.

1. Upper Jurassic deposits widely spread in the profiles of Gurarinsk, Sobolin and Yasnaya fields represent spatially the transition zone of lithologo-facial Vasyugan suite replacement into Naunak one.

2. The lower part of the studied profile (cyclite J_1^5) according to the lithological characteristics and formation conditions are related to Low Vasyugan subsuite.

3. Cyclites J_1^4 and J_1^3n formed during sea regression under conditions of progressive transition process from coastal marine environment to flood plain coast environment are related to the deposits of the Upper Vasyugan subsuite undercoal stratum.

4. Cyclite J_1^3c and J_1^3v sediments accumulating in continental conditions are also related to temporary analogue – the intercoal strata of the Upper Vasyugan subsuite, but are notable for its increased thickness and coal – bearing capacity.

5. The most significant differences are in the overcoal stratum structure where, according to the core analysis, marine and continental features of sedimentation combine; coal seams C_0^a and $C_0^{б}$ are marked out in the profile. Cyclites J_1^2 J_1^1 and J_1^0 formation regime are of transgressive-regressive (pulsating) character and continuously changed from marine to continental.

6. In general, the studied profile is comparable to Vasyugan suite profile, but it is different due to the thickness increase, higher content of coal material including either the overcoal stratum. This shows the more significant role of continental deposits.

7. The fact that the territory is related to the transition zone is also due to Georgiyev and Bazhenov suites deposits that overlie the overcoal stratum. These are not Maryanov suite deposits that are thought to be in the profiles of Naunak suites.

References

[1] Le Heron D, Buslov M, Davies C, Richards K and Safonova I 2008 Evolution of Mesozoic fluvial systems along the SE flank of the West Siberian Basin. *Sedimentary Geology*. **1-2 Vol. 208** 45-60
[2] Kontorovich A, Kontorovich V, Ryzhkova S and et al. 2013 Jurassic paleogeography of the West Siberian sedimentary basin. *Russian Geology and Geophysics*. **8 Vol. 54** 747-77
[3] Danenberg E, Belozerov V and Brylina N 2006 *Geological structure and oil-bearing capacity of Upper Jurassic-Cretaceous deposits in the West Siberian south-east (Tomsk Oblast)* p 291
[4] Ezhova A and Nedolivko N 2003 Stratigraphy and deposits correlation of Mid - Upper Jurassic western part of Nurolskaya depression // Mesozoic stratigraphy problems for West Siberian Plate (Interdepartment materials on Mesozoic Stratigraphy Conference of West Siberian Plate): Collected scientific articles/ edited by Gurary F.G., Mogucheva N.K. Novosibirsk. Research Institute of Geology and Geophysics (SNIIGGIMS) 107-117
[5] Shurygin B, Nikitenko B and Devyatov V et al. 2000 *Stratigraphy of oil/gas fields in Siberia. Jurassic Period* (Novosibirsk. RAS "GEO") p 480
[6] Karogodin Y and Gaideburova E 1989 *Layered associations research for oil and gas deposits (Field geophysics data complex)* (Novosibirsk. "Nauka" Siberian department) p 108

IOP Conf. Series: Earth and Environmental Science **24** (2015) 012024 doi:10.1088/1755-1315/24/1/012024

Metallogenic problems of hydrothermal gold deposit formation: facts and arguments

I V Kucherenko[1] and Zhang Yuxuan[2]
[1,2] Department of Geology and Mineral Exploration, Institute of Natural Resources Tomsk Polytechnic Unversity

E-mail: [1]kucherenko.o@sibmail.com

Abstract The research problem of substituting existing hypotheses by a well-defined re modeling of the geological processes, which, in its turn, initiate and govern the formation of hydrothermal gold-ore deposits, is becoming more and more relevant. The four existing hypotheses - granitogene, basaltogene (magmagene), metamorphogene, and polygene have been considered and discussed. The interpretation of different hypotheses and analysis of substantiating facts showed the inconsistency of above-mentioned hypothesis-approaches. Up-dated petrological, petrochemical, geochemical data proved the following: generation of metal-bearing fluids in sources of moderate alkali-basalt melts during the late basalt formation stages show recycling Late Riphean to Late Paleozoic convergent antidromic granite diorite dolerite magmatic complexes.

1. Introduction

The attained advanced investigation of ore formation conditions within hydrothermal gold-ore deposits could be associated with such an unsolved problem as metallogenic mineralization dependence to global geological processes.

It should be mentioned that such questions as the sources of ore-gold and foreign- metal impurities, the generation conditions of metal-bearing fluids and the ore path to future deposits could be the evidence base in plotting a geologic-genetic model of fluid-ore-magmatic gold-producing systems, as well as being an effective propspecting criterion in evaluating those commerical gold-ore deposits buried at an accessible depth while excluding eroded deposits.

Granitogene, basaltogene (magmagene), metamorphogene, polygene hypotheses of hydrothermal gold-ore deposit formation were proposed at the turn of 21st century. Probable metal and fluid sources could be granite and/or basalt magma reservoirs (according to magmagene hypothesis); rocks in dynamo-zonal metamorphism areas of hydration and dehydration (according to metamorphogene hypothesis); and/or as numerous sources having been subjected to regional metamorphism , such as sedimentary rocks and silicate melts (according to polygene hypothesis). The diversity of existing hypotheses, being relevant to both gigantic and unique deposits, excludes such a fact as the convergent ore-formation itself, while the existence of such hypotheses only attests either the lack of empirical data or unreliable data.

2. Analysis of hypothesis evidence base

The disadvantages of previous research methods focused on the ore-forming potential in dynamo-zonal metamorphism have been discussed [1]. The decennial integrated analytical data showed a rather apparent tendency, i.e. fully-developed "chaos" of the hypothesis evidence base itself, resulting in this diversity. This "chaos" could be described as different definitions by different authors concerning the pre-gold ore grade (from the first mg/t to multi gr/t) in one and the same ore-hosting black-shale formation, in ultrametamorphic Archean basement complexes and in rather young regional dome structures. Whether sharp positive geochemical anomalies in near deposit wallrock zones formed in metamorphic and/or metamorphosed rocks could be either the cause (assumption) or result of ore formation is even today an open question.

This "chaos" could be explained by the fact that both research approaches and methods were and are irrelative to the stated investigation objectives and tasks. According to V.I. Vernadsky and A.E. Fersman, the aim of these investigations was to show the geological history of chemical elements on the planet and estimate the concentration or dispersion of metals during every formation stage on the basis of quantitative process, as well as to evaluate the subsequent alteration of sedimentary rocks subjected by metamorphic processes and associating mineralization of hydrothermal metasomatic processes. In most cases, host rock metasomatic alterations accompanied by beresite and propylite formation metasomatites are classified as metamorphic processes of hydration, while hydrothermalites are related to biotite-chloritic, chlorite-sericitic, albite-sericitic and other "subfacieses" of greenschist facies [2, 3]. A matter of endless arguments still remains the fact how to determine the total phased gold content.

It is obvious that the out-dated research paradigm should be revoluntionized, i.e from a primitive understanding of geochemistry to a more defined conception of its objectives and tasks. The proposed geological history reconstruction of petrogenic and oregenetic elements (gold and its geochemical accessory minerals – silver and mercury) in the Early and Late Proterozoic black shale formations and in ultrametamorphic complexes within wallrock areas of the following deposits: Kuznetsky Ala Tau (Berikul), North Transbaikal (Irokindinski, Kedrovski, Karalonski), Lensky district (Sukholozhsk, Verninsk), Patom plateau (Chortovo Koryto) indicated that host rock alteration is pertained to zonal metasomatic beresite areolas and propylite metasomatite formations, while the bulk pre-ore metal content shows the syn-ore origin of their anomalies. The obtained data showed the composition-genetic mineralization homogeneity developed in Late Riphean to Late Paleozoic black shales, ultrametamorphic complexes and other crystalline rocks [4].

The evidence for the granitogene hypothesis is based only on the data of single-version interpretation [5–6]. Bulk magma pockets including granite melts and inherited plutons, as well as granitoid massifs generate sufficient metal-bearing fluids for future commercial gold-ore deposit development.This could be a reason-based choice in considering the possible formation of comparable and/or significant volumes of basalt melts and metal-bearing fluids in the mantle. Water content in granite melts embracing aqueous metasomatic minerals (biotite, amphibole, and other minerals) in mantle xenoliths combines with the lack of water in mantle magmas. The spatial-temporal proximity of deposits to granitoid massifs, mature focused dome-shaped blocks and associations with early pre-ore dikes of acid rocks could be indicators of either genetic or paragenetic interrelationships. Granitoid plutons, blocks, siliceous dyke belts and deposits in the crust are controlled only by deep faults, which are channels for injected early hyperthermal fluids governing crust bedding palingenesis and forming large granitoid-bodies and late metal-bearing fluids, which, in its turn, generate deposits. Therefore, the identified near-pluton zoning of gold-ore mineralization is comparable to near-fault zoning, indirectly associated with plutonic intrusions. The concept that ore-forming fluids are silicate brines (magma) is inconsistent with the existing fact of relatively low fluid concentrations (up to 15…20 % wt of NaCl) within mineral ore vacuoles. Silica debris during host rock metasomatism is up to 50 % wt [7] which indicates the sterility of inflowing metal-bearing fluids in respect to SiO_2. However, the feldspar content in gold ore-bearing quartz veins was probably not the result of granite-inherited veins by generation of metal-bearing fluids in granite melts, but could have been the determined balance of petrogenic components in wallrock metasomic columns by silica, aluminum silicate and sodium diffusion from porous wallrock solutions in the interflowing potassium-calcium fractured solutions during metasomatism and mineralization. In the 1970s the agrument matching of commercial gold-ore deposits to granitoids,which are either metal

IOP Conf. Series: Earth and Environmental Science **24** (2015) 012024 doi:10.1088/1755-1315/24/1/012024

enriched or metal depleted, introduced the concept of substituting the granitoid metallogenic trend for the potential granitoid mineralization [8].

This substitution furthered the study of applying granitoid minerals in the thermadynamic and physico-chemical regimes of active granite melts and the distribution and evolution of crystallized solid gold content in melts,which show a significant affinity of sulphur, chloride and other elements. In this case, another important factor is to estimate the gold-bearing melt potential, to evaluate the conditions and metal-bearing fluid generation in above-mentioned melts based on petrochemical parameters, as well as other parameters [9–12]. The obtained results are compared with the studied massif granitoids (with or without areal-temporal gold ore associations) in hydrothermal gold deposits and classified as unique, large, average or small. However, there are some emerging fact-based conditions that exclude the possible relevant formulated conclusions concerning the above-mentioned. These are the following: (1) high analyzed data uncertainty that granitoids are the bulk source of residual melts within which metal-bearing fluids could have really generated; (2) no criteria to effectively evaluate that amount of metal-bearing fluids which would be sufficient to produce commercial deposits including differentiated deposits with respect to contained metal mass. In this case, the genetic relation of hydrothermal gold deposits to granites still remains unproved.

Substantiation of magmegene (basaltogene) hypothesis is based on the factual data describing the isotopic content of sulphur in sulphides and carbon in carbonates responsible for meteorite probable cause blast (primeval mantle), associations of gold ores and platinium. There are known factual evidences such as isotopic constitution of sulfides and carbon of carbonate ores associated with meteoric standard (primitive mantle), metal platinum group associations in gold ores, presupposed mantle element impurities - mercury in gold and nickel wallrock metasomatite in chlorites, spatial-temporal association of gold ores with mantle plumes, small-sized basalt bodies– dykes, sills.

3. Results and discussion

Up-dated information was correlated from eighteen deposits of Yenisei, Kuznetsko-Ala Tau, East-Sayan, Northern Transbaikal (Muisk), Lenski gold ore areas within south mountain-folding of Siberian craton margin, including the above- mentioned ones. According to the basaltogene hypothesis, these facts show the principally relative frequency and sequence of events in Late Riphean, Early-, Middle- and Late Paleozoic mineralization, westward-eastward younging deposits. Plutons of early palingenic granitoids or mature ultrametamorphic dome structures accompanied by acid rock dikes (aplites, pegmatites, granite-porphyry, felsitic microgranite-porphyry and etc.), beds and dykes of granite diorites (microdiorites, diorite-porphyry), associations of late pre-ore dykes, inter-ore dykes, post-ore dykes of medium-alkali dolerites (figures 1 and 2) were sequentially formed under geodynamical collision conditions in active continental margins or intracontinental rifting in all the areas during tens of millions of years (according to radiological methods). Specified magmatism derivate is controlled by deep faults and combined into convergent antidromic magmatic complexes, the latter of which recurring in space and time indicates their mineragenic type.

Intensifying anisotrophy of mechanical granitoid properties in pre-ore dolerite dykes is often accompanied by quartz veins, as well as early accumulated ore-mineral complexes. Dykes in vein exocontacts embrace beresite growths alternating to porpylites on the periphery. In areas of weakly alternated massive dykes the mineral-chemical composition of these dykes is comparable to similar normative rocks. As this could be found in numerous deposits, confined to granitoid massif complexes, it indicated that residual granite melts did not exist before the injection of early basalt melts; otherwise, if the melts mixed then "mottled" hybrid dykes could have formed and which were undetected in the investigated deposits. It is obvious that in this case non-existinggranite melts could not have been the source of metal-bearing fluids, the early metal-bearing fluid injections after intrusion of early basalt melts into granitoids.

Inter-ore dykes of dolerites cross-cut not only early ore-mineral complexes with decompressed quartz vacuoles in dyke exocontacts, but also veins formed in late ore-mineral complexes with beresite growths in their casings. Black, greenish-black massive, usually, fine/medium -grained,

IOP Conf. Series: Earth and Environmental Science **24** (2015) 012024 doi:10.1088/1755-1315/24/1/012024

being confined to welded host rock exocontacts, these dykes are converted into metasomatities underlying the host rocks, to a certain extent, either hydrothermally altered or unaffected in their exocontacts. In rare “logan rocks”, weakly altered thick dykes, their structure and mineral-chemical composition are related to moderate alkali dolerite, labradorite and aurite with olivine, grothite, titanite impurities (table 1, sample 2).

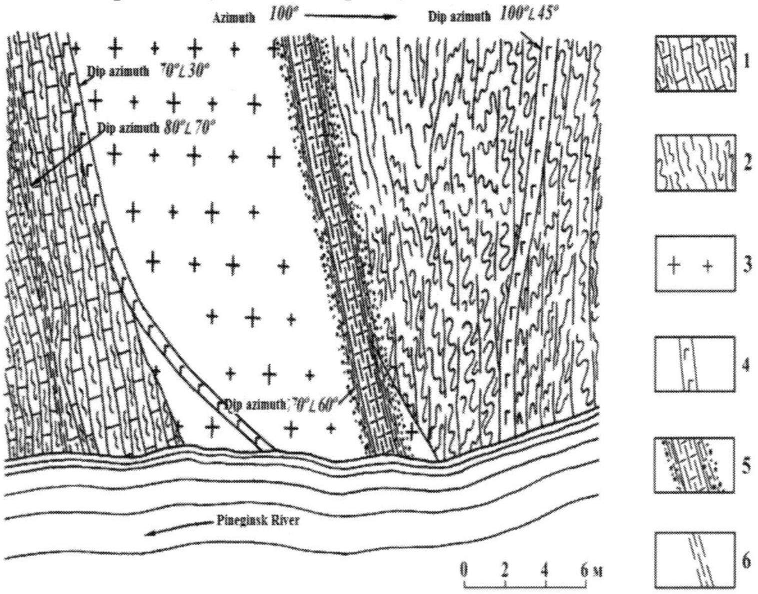

Figure 1. Kedrov deposit: (1) structure-temporal relations of sequentially formed gneiss limestones, (2) carboniferous shales, (3) mature dome structure,diorite-porphyry dykes, (4) dyke of medium-alkali dolerite, (5, 6) sulphide-quartz gold vein with beresite growths and bedded rocks.
Beresite growths, indicating their pre-ore age, are visible on river beds in the vein exocontacts within dolerite dykes

Based on the X-ray spectrography method, the following metasomatic minerals of ranging %wt were identified not only in one and the same dykes, but also in different ones: reddish-brown, pinky, earthy-green biotite (up to 60 %wt), ferro-magnesian hornblende (up to 20%wt), actinolite, tremolite, talc, serpentine, zoisite, epidote, magnesian iron chlorite, calcite, dolomite, ankerite, albite, sericite, quartz, apatite, rutile, leucoxene, celestine, barytocelestite, magnetite, pyrite. Two generations of biotite were identified: early generated biotite replaced by chlorites and other metasomes, and late generated biotite forming elongated flakes without replacement features and framing former augite crystals, which, in its turn, have been completely converted into fine- grained chlorite, sericite, carbonate aggregates and other minerals.

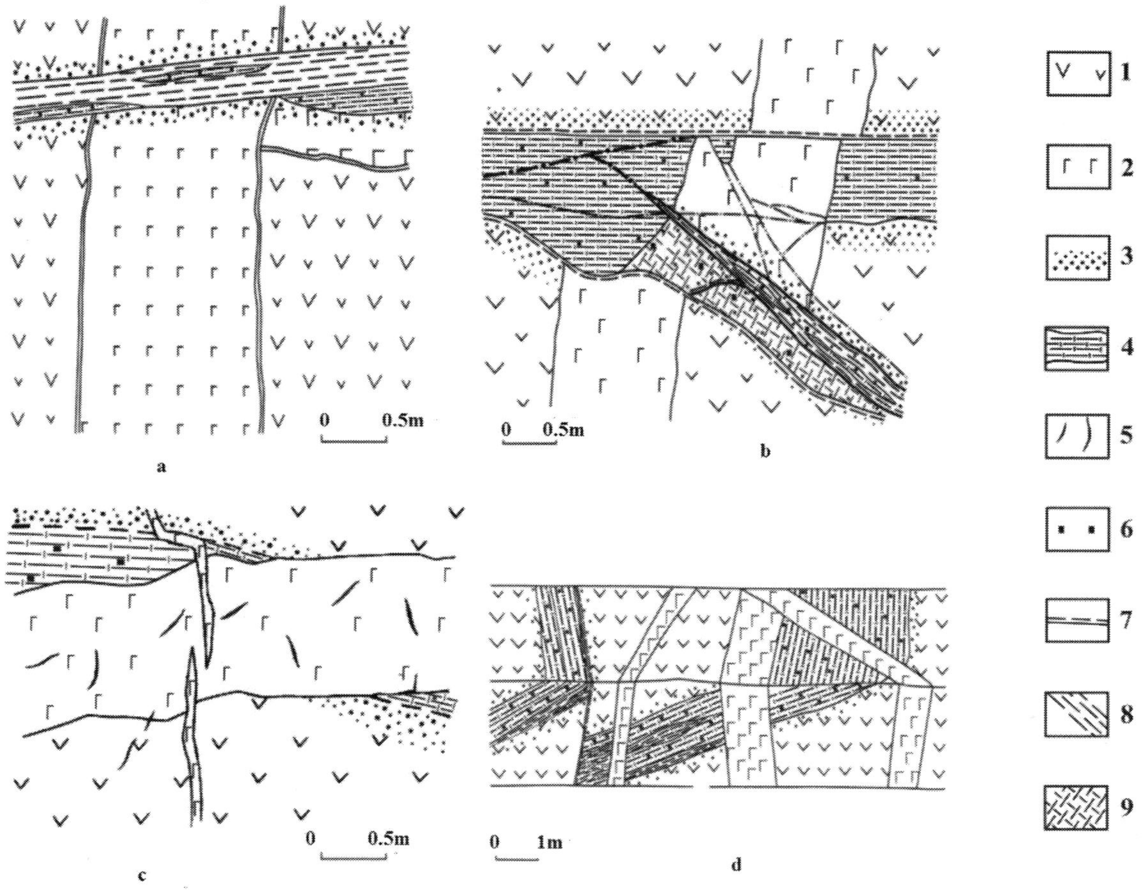

Figure 2. Berikulsk deposit. Structure-temporal relations of gold ores with pre-ore (two generations, *a*) inter-ore (*b,c*) post-ore (two generations *d*) dykes of moderate alkali dolerites.
(1) Berikulsk suite blanket basalt porphyries ($€_1$), (2) dykes of moderate alkali dolerites, (3) beresites, (4) quartz gold veins, veinlets, lenses, (5) carbon-quartz veinlets, (6) pyrite, (7) tectonic wedges, (8, 9) schistosity and rock breaking zones

Based on balance calculations, dolerite chemical compositions during internal dyke metasomatism were identified – silica, sodium output and potassium, carbon dioxide input, sometimes partial replacement of sulphur, phemophile element associations (P, Ti, Mg, and not always Mn, Fe, Ca) (table 1) and gold (up to 3 g/t).

Stated data provides accurate reconstruction of geological processes during the late basaltoid stage of magmatic complex formation.

Metasomatic origin of biotite is proved by its weight (tens %wt) which is incomparable to the late-magmatic biotite wt in moderate alkali dolerite (up to 5 %wt). Hornblende crystals in comparison to late-magmatic hornblende are found in apodolerite metasomatites in the areas of rather intensive metasomatic alteration of dolerites.

Numbers of samples	Oxide content in % wt (first line). Input-output value of element atoms in standard geometric volume 10000 Å^3 in % to number of their atoms in standard geometric volume of initial dolerite (second line)															
	SiO_2 Si	Al_2O_3 Al	K_2O K	Na_2O Na	CaO Ca	MgO Mg	FeO Fe^{2+}	Fe_2O_3 Fe^{3+}	TiO_2 Ti	P_2O_5 P	MnO Mn	CO_2 C	S*	H_2O^+ H	O	Σ (Δ)
2	49.72	14.94	0.90	3.05	8.47	7.43	8.03	2.86	1.41	0.35	0.19	0.35	0.13	1.81	–	99.64
K-497	38.76 -16.8	10.75 -23.2	4.26 **405**	0.34 -88	10.1 **27**	13.3 **91**	8.79 17	2.69 0.4	2.75 **108**	1.36 **315**	0.21 **18**	1.83 **457**	0.16 **31**	3.40 100	3.3	98.70 (21)
K-588	40.40 -14.2	11.50 -18.7	2.90 **241**	1.09 -62	11.5 **44**	12.2 **74**	6.90 -9.0	4.33 **60**	2.44 **83**	1.04 **213**	0.23 **28**	2.67 **708**	0.26 **111**	1.69 -1,4	2.3	99.15 (19)
K-495	37.83 -22	9.14 -37.2	5.06 **475**	0.27 -91	11.6 **41**	14.0 **94**	7.79 -0.6	3.98 **43**	2.55 **86**	1.37 **303**	0.29 **57**	4.88 **1332**	0.14 11	1.30 -26	-0.6	100.20 (24)
K-562	38.0 -17.1	10.80 -21.6	3.16 **281**	0.39 -86	11.2 **43.6**	15.3 **123**	7.49 1.1	3.88 **47.2**	2.10 **61.7**	1.08 **234**	0.29 **65.6**	2.89 **797**	0.27 **125**	2.54 52	5.5	99.39 (24)
K-486	33.80 -30.6	8.16 -44	2.72 **209**	0.02 -99	12.0 **45**	12.9 **78**	6.87 -12	2.27 -19	1.95 **41**	1.17 **241**	0.33 **77**	14.4 **4107**	0.13 2.2	1.98 12	5.9	98.70 (29)

Table 1. Chemical compositions of moderate alkali dolerite and apodolerite metasomatites of inter-ore dykes and balanced petrogene elements in apodyke metasomatic aureoles of Kedrov deposit

Note (1) Sample 2 – moderate alkali dolerite composition; (2) S* – replaced sulphur; (3) Δ – specific gravity of transported (input and output) substance (sum of petrogene element atoms) in %wt of initial dolerite (sample 2) in standard geometric volume of 10000 Å^3 ; (4) table 1 and table 2 show total chemical silicate analyses of rocks performed in Central laboratory of State Production Association (SPA) "Zabsibgeology" and in West-Siberian Research Centre (Novokuznetsk)

Dependency of metasomatic dyke alteration to host rock alteration indicates the dyke fluid conductivity, which, according to the physical law, is possible only in thermal bodies located in relatively cold environment.Therefore, usually hot vertical dykes, as well as faults and channels, provided those paths where hot solutions rose from generation areas with the melts in time and these generated dyke melts have no time to cool, so cold dykes could not have accumulated the hot solutions. Gold and phemophile element anomalies, including contrast anomalies, indicate the fact of metal-bearing fluids. The inherited potassium-sulphur-carbon dioxide profile of interdyke apodolerite metasomatism, phemophile element anomalies, determining the petrochemical diversity of basalt melts, and gold near-ore metasomatites-beresites near deep fault framings (table 2) proved the fact that apodolerite and near-ore metasomatites formed (and ores) as a result of the behavior of one and the same metal-bearing fluids, the sources of which were moderate alkali basalt magma pockets.

Mineral zone, subzone	Output-input value of element atoms in % to number of their atoms in parent rocks in standard geometric volume of 10000 Å^3													Δ
	Si	Al	K	Na	S*	**C_o**	**Ca**	**Mg**	**Fe^{2+}**	**Fe^{3+}**	**Ti**	**P**	**Mn**	
Irokin deposit Granite migmatite melting, AR (3)														
Vy(5)	0	0	-10	-10	+	20	20	0	0	10	10	50	-60	1.2
Vi(6)	0	0	-10	0	+	220	70	30	30	70	20	110	0	3.1
Ch (9)	-10	10	-40	40	0	500	70	60	0	60	-10	210	-50	6.9
A (8)	0	0	-20	-10	+	**870**	**10**	**60**	**20**	**50**	**70**	**250**	**10**	4.5
Rz (7)	-10	10	20	-90	+	**2400**	**200**	**220**	**100**	**230**	**250**	**650**	**30**	18.8
Kedrov deposit Carboniferous feldspar-quartzose shales of Kedrov suite, PR_2 Metasiltsandstone (1)														
A (1)	-17	4.9	**248**	-34	+	**1905**	**33**	**1053**	**282**	**340**	**82**	**300**	**374**	18.0
Rz (1)	-39	8.8	**445**	-93	+	**6913**	**880**	**1781**	**447**	**125**	**73**	**672**	**347**	43.0

Table 2. Petrogene element balance in zonal host rock metasomatic aureoles of mesothermal gold deposits in Southern Siberia

Note: 1) Mineral zones and subzones of near-ore metasomatic aureoles: Vy, Vi – subzones of moderate and intensive frontal zone alterations, C, Ch, A and Rz – carbon, chlorite, albite, back zones, respectively; 2) S^* – sulphide sulphur, C_o – oxidized carbon (carbonate), + –
S input , if its content in parent rock is lower detection limit assay; 3) in brackets – number of samples, which are involved to calculate the average; 4) Δ – specific gravity of transported (input and output) substance as percentage of substance mass of parent rock in standard geometric volume 10000 $Å^3$

The deficiency of biotite and hornblende in near-ore beresites resulted in the substitution crystallization of the only potassium carrier in fluids by low-temperature sericite governed by extra heating of filtering fluids in hot dykes and cooling progressively as the fluids enter the mineralization area. Recrystallization of high-temperature biotite within inter-dyke metasomatism in succession to low-temperature metasomatic minerals (chlorite, sericite, carbonate and others) is correlated to increasing crystallization temperature of early quartz generation in succession to each ore-mineral complexes within ore bodies, comparable to the formation temperature of late generated quartz of each preceding complex. Above-mentioned facts in combination with alternating melt and metal-bearing fluid injections focus on the pulsing regime of hydrothermal systems, observed within today's volcanic activity areas.

4. Conclusion

The stated facts and data indicating the composition-genetic mineralization homogeneity in crystallized and black shale substances [4], form a conformed assemblage, which, in its turn, could be a basis for the geologic- compositional-genetic model of gold within antidromic-granite-diorite of fluid-ore-magmatic complexes. The described assemblage extends beyond the existing hypotheses (granitogene, metamorphogene, polygene). Therefore, fact-based evidence for the above-mentioned depends on the possible supporting facts of mineralization convergence, i.e. ore-producing potential of granite magmatism and/or regional metamorphism heating. However, this is practically impossible.

References

[1] Kucherenko I V 2014 Rekonstrukcija geologicheskoj istorii metallov kak sredstvo ocenki donorskogo potenciala ul'trametamorficheskih kompleksov i chernyh slancev v obrazovanii gidrotermal'nyh mestorozhdenij zolota *Zoloto i tehnologii* **1(23)** 104–115
[2] Zhmodik S M, Mironov A G and Zhmodik A S 2008 Zolotokoncentrirujushhie sistemy ofiolitovyh pojasov (na primere Sajano-Bajkalo-Mujskogo pojasa) (Novosibirsk: GeO) p 304
[3] Large R R, Maslennicov V V, Francois R, Danyushevsky and Zhaoshan Chang 2007 Multistage Sedimentary and Metamorphic Origin of Pyrite and Gold in the Giant Sukhoy Log Deposit, Lena Gold Province, Russia *Economic Geology* **102**(7) 1233–1267
[4] Kucherenko I V 2007 Petrologic-Geochemic Evidence of Geologic-Genetic Uniformity of Gold Hydrothermal Deposits Formed in Black-Shale and Non-Shale Substratum *Bulletin of the Tomsk Polytechnic University* **311**(1) 24–33
[5] Spiridonov A M, Zorina L D and Kitaev N A 2006 Zolotonosnye rudno-magmaticheskie sistemy Zabajkal'ja (Novosibirsk: GeO) p 291
[6] Rafailovich M S 2009 Krupnye zolotorudnye shtokverki Central'noj Azii, associirujushhie s granitnymi plutonami *Rudy i metally* **3** 43–50
[7] Kucherenko I V and Gavrilov R Y 2011 Femophilic Elements in Wall-Rock Metasomatites and in Ores of Mesothermal Gold Deposits – Newsletter of Mantle Deep *International Journal of Applied and Fundamental Research* **1** 37–43
[8] Tauson L V 1977 Geohimicheskie tipy i potencial'naja rudonosnost' granitoidov (Moskva: Nauka) 280

[9] Mustard R, Ulrich Th, Kamenetsky VS, Mernagh T 2006 Gold and Metal Enrichment in Natural Granitic Melts During Fractional Crystallization *Geology* **34**(2) 85–88
[10] Yang HM, Lentz DR 2005 Chemical Composition of Rock-Forming Minerals in Gold-Related Granitoid Intrusion South-Western New Brunswick, Canada: Implications for Crystallization Conditions, Volatile Exsolution and Fluorine-Chlorine Activity *Contribution Mineral Petroleum* **150** 287–305
[11] Gusev AI 2012 Petrologija zolotogenerirujushhego magmatizma (Moskva: RAE) p 160
[12] Hart CJR 2006 Reduced Intrusion-Related Gold System *Economic Geology* **101**(7) 1415–1427

IOP Conf. Series: Earth and Environmental Science **24** (2015) 012025 doi:10.1088/1755-1315/24/1/012025

Noble metals in rocks and ores of Maysko–Lebed ore field (Mountain Shoriya)

A Pshenichkin[1], T Timkin[2] and Yu Oskina[3]
Department of Geology and Mineral Exploration
National Research Tomsk Polytechnic University, Russia
[1]paya@tpu.ru, [2]timkin@tpu.ru

Abstract. First the authors determined platinum and palladium in the ores and rocks of Maysk–Lebed ore deposit via stripping voltammetry. Based on research data the increased platinoid (platinum group elements) content was identified both in the source host rocks and in metasomatically altered ones in ores.

1. Introduction

Currently, integrated study on platinum group elements in gold -ore and other deposits is considered to be of significant importance, due to two major aspects: on the one hand, conventional noble metal deposit depletion, while, on the other hand, the majority of different genetic deposits embrace high platinoid concentration (including even commercial reserves) [1].

The authors performed an integrated study on Pt, Pd and Au concentrations in the rocks and ores of Maysk–Lebed ore field via stripping voltammetry within the Gold–Platinum Research and Education Innovation Center, Tomsk Polytechnic Unversity. The preliminary study results showed a high concentrations of Pt and Pd not only in source host rocks but also in near–ore metasomatites and ores.

The source material included rock and ore samples collected during the theme-based research (2004-2008), detailed mineral and geochemistry mapping of available mine openings, and drill hole cores from drill core samples from Maysk–Lebed ore deposit. The study group included the employees of the Department of Geology and Minerals Exploration, Tomsk Polytechnic University in collaboration with personnel of Tetis-T LLC (Novokuznetsk, Russia) [2].

2. Geological Setting

Maysk–Lebedev gold-ore field is located in North–East Gorny Altai of North Altai gold bearing zone within the near north-south strike intersection of Kuztensk Alatau and near EW -Western Sayan within Altay–Kuznetsk–Western–Sayan volcano–plutonic belt. Such a location could be explained by the continuous ore field formation history involving a succession of tectonic–magmatic impulses and hydrothermal activity [3, 4, 5].

The Maysk–Lebed ore field development embraced Maysk, Magalaksk and Chanysh intrusive massif of the Sardinsk complex. Two rock groups were identified in the massif: gabbro, gabbro-diorites and granodiorites, quartz diorites. This massif formed during two stages: initial-gabbrodiorite and final -granitoid. Granodiorties are regional throughout the inner massif, while quartz diorites are in endomorphosed massif zone. As a result of hydrothermal–metasomatic transformations of rocks, skarns, propylites, beresites with veinlet silicification and argillizated rocks developed.

The ore field area has a block structure. Such blocks are usually isolated by tectonic wedges and characterized by thrust-faults and strike–slip faults. The principal ore controlling structure embraces N-S strike small-scale faults and anticline fold hinges. The combination of such structures is considered to be the most favorable ones for mineralization. N-W faults are predominately post-ores and do not significantly influence the ore localization itself.

Two heterochronous geological and industrial types of ore grade mineralization have been identified in the Maysk–Lebed ore field: contact–metasomatic deposits of gold–magnetite–sulfide ores in Caledonian aposkarn propylites (Maysk area) and mineralized zones of Hercynian gold–sulfide–quartz vein–disseminated ores accompanied by the beresitization of host rocks (Pravoberezhny and Semenov areas).

Maysk gold–skarn ore bodies are localized in the eastern exocontact of Maysk granodiorite–diorite intrusive massif within thick skarnified and quartz–epidote–chlorite and quartz–sercite–carbonate metasomatite zone. The basic ore mineralization includes pyrite, magnetite, chalcopyrite and gold, as well as rare minerals – pyrrhotine and cobaltite.

Pravoberezhny and Semenov ore bodies are localized in the Lower Cambrian igneous rocks (Sardinsk suite) and include gold–sulfide–quartz lenses of 0.1–0.7 m thickness, accompanied by sulfidized beresites of 10–15 m thickness. Ore mineralization includes pyrite, arsenopyrite, chalcopyrite, galena, tetrahedrite and gold.

Hierarchical three–level ore–metasomatic zoning was defined for the Maysk–Lebed ore field [4, 6]: ore field, ore deposit and ore body. Identified cross–section zoning in all three levels is characterized as concentric convergent and three detected zones within the ore–bearing structures. In the central zone, near–ore metasomatites with gold–magnetite–sulfide and gold–magnetite–quartz mineralization are well-developed. The intermediate zone is characterized by propylites with pyrite–arsenopyrite mineralization. The external zone, showed dispersed pyrite mineralization. Cross–section zoning is governed by the structural conditions of localized ore mineralization and reflects the differentiated character of geological structure permeability. Axial ore field zoning was identified on the cross–section zoning background which was revealed in the ore deposition temperature decrease from south (Maysk area) to the north progressively in a direction away from the ore–generating intrusive bodies (Pravoberezhy and Semenov areas). Within the ore field the zoning itself reveals the regular changing of high–temperature metasomatites (skarns) to low–temperature metasomatites (beresites, argillizated rocks), regular assay of gold reduction (from 600 to 980 %), increasing Ag/Au ratio (from 1:1 to 10:1) in ores and alteration of typomorphic pyrite and metasomatite mineral properties.

3. Research Methods

The high–sensitive and economical method – stripping voltammetry (SV) – was applied in analyzing rocks and ores of the Maysk–Lebed ore field to assay the Pd, Pt and Au content. The evaluation of the platinum metal and gold content via SV method has been applied by the personnel of the Gold–Platinum Research and Education Innovation Center, Tomsk Polytechnic Unversity [7,8].

The SV method eleminates the numerous interfering base (trace) components in a sample by selecting that platinoid electrodeposition potential where the base components electroconcentration would be excluded. Therefore, the total saline background system with such elements as Na, K, Cr, Pb, Mn and others does not interfer in the identification of platinoids. However, such elements as Cu, Fe, Se, Te that oxidize under positive potentials may interfere in the SV identification of these platinoids. At the same time, a mutual influence of noble elements is observed. To identify platinum group metals and gold by the SV method, analysis outlines are designed to provide either the isolation of base (trace) components or the precipitation of single platinum group metals.

The isolation and identification of Pd, Pt and Au involve a multi–stage process. Initially, "dry" oxidation of samples is performed at 750 °C for 45 minutes. Then, the annealed samples are treated by a concentrated acid mixture ("moist" oxidation) for further deposition of some metals in solution.

To isolate palladium (II) from the solutions after the deposition of analyzed ore sample in solution, dimethylglyoxime Pd^{2+} extraction complex with chloroform is applied; to isolate gold (III) chloride Au^{3+} extraction complex with diethyl ether is applied. Identification of platinum (IV) in samples was performed after sample matrix isolation via deposition method with bromate hydrolysis.

Measurement technique for gold and palladium weight concentration is based on stripping voltammetry analysis of sample solution after its preliminary treatment. Gold and palladium electrolytic deposits settles on the surface of a graphite electrode from chloride gold (III) and palladium (II) solutions.

Identification of platinum (II and IV) in solutions by stripping voltammetry analysis is possible only with electro-negative metal settled on graphite electrode surface and further selective electro-oxidation of alloy [7,8]. Indium is used as an electro-negative metal.

TA–4 voltampermetry analyzer (produced by TomAnalit LLC Tomsk, Russia) was applied in the research. All measurements were performed in three–electrode cell with main graphite low–pressure polyethylene impregnated electrode. Saturated chloride–silver electrode was used as reference and auxiliary electrode, respectively. Deaeration of solutions was not performed.

Stripping voltammetry analysis with graphite electrode identifies the platinum group elements and gold within an interval of $10^{-8} - 10^{-2}$ wt % on weight of 1–5 grams. Element identification accuracy was evaluated by the correlation of reference sample results to parallel line sample analysis (the latter by atom absorption spectroscopy).

4. Research Results and Discussion

The stripping analysis voltammetry was used to assay platinum, palladium and gold in samples of 123 host rock, near–ore metasomatites and ores of Maysk and Pravoberezhny deposits. First determined rather high platinum and palladium content both in unaltered host rocks (tuffs, andesites) and in skarns, near–ore metasomatites and ores (table 1).

Table 1 Gold, platinum and palladium content in rocks and ores of Maysk–Lebed ore field

Rocks and ores	n	Gold, g/t			Platinum, mg/t			Palladium, mg/t		
		from	to	X	from	to	X	from	to	X
Maysk area										
Andesite tuff	3.0	0.033	0.20	0.081	1.0	7.3	3.9	1.0	14.0	6.1
Scarnified andesite tuff	12.0	0.11	8.30	2.05	6.5	50.5	26.7	1.0	45.0	10.7
Limonitized andesite tuff	19.0	0.40	9.40	1.19	1.9	50.5	23.4	1.0	48.0	8.7
Andesite	4.0	0.23	1.10	0.69	4.4	21.2	10.8	6.0	33.0	19.0
Scarnified andesite	10	0.60	4.70	2.48	5.9	94.4	31.8	!.0	4.8	2.3
Limonitized andesite	8.0	0.72	12.0	4.45	1.2	47.6	16.2	7.2	32.0	15.5
Garnet–epidote skarn	4.0	0.15	0.99	0.57	5.5	30.8	12.2	8.1	75.0	26.7
Ore	13.0	0.20	22.2	5.92	2.1	55.5	20.2	1.0	86.0	30.0
Propylite	6.0	0.011	6.10	2.10	1.0	50.0	15.2	1.0	12.5	4.1
Magnetite	3	0.02	0.12	0.07	2.1	8.1	4.0	1.0	9.3	3.6
Pravoberezhny area										
Andesite tuff	6.0	1.00	2.60	1.75	5.7	42.9	16.2	1.1	61.0	17.4
Scarnified andesite tuff	2.0	0.17	0.23	0.20	21.1	23.4	22.3	7.9	9.5	8.7
Limonitized andesite tuff	3.0	0.23	0.84	0.49	2.8	21.1	10.6	9.5	66.0	31.7
Andesite	5.0	0.10	4.40	1.40	5.7	32.8	18.2	1.8	32.0	13.2
Limonitized andesite	3.0	0.51	1.50	0.95	3.0	60.1	28.8	1.0	67.0	23.3
Beresite	7.0	0.31	3.00	1.00	4.2	58.3	22.5	1.0	240.0	49.4
Ore	8.0	0.23	5.30	2.46	3.2	50.1	19.3	1.0	130.0	27.0
Ironstone	4.0	1.00	7.00	2.90	9.9	34.5	19.7	2.0	93.0	42.4
Calcite veinlet in beresites	1.0	–	–	0.072	–	–	20.8	–	–	0.41
Quartzite	1.0	–	–	0.091	–	–	6.7	–	–	38.0

NB: n – number of analyses; X – average value; analyses performed in Gold–Platinum Research and Education Innovation Center, TPU (by analytical chemists: E Gorchakov – Au; Yu Oskina – Pd; E Ustinova – Pt).

Platinum Platinum content scatter in Maysk area is from 1.0 mg/t (andesite tuffs, propylites) to 94.4 mg/t - scarnified andesites. The latter (scarnified andesites) has highest average platinum content of 31.8 mg/t. Normal average platinum concentrations were defined in the ores (20.0 mg/t), limonitized (23.4 mg/t) and scarnified (26.7 mg/t) tuffs. Platinum content is half as high in unaltered andesites (10.8 mg/t) and in garnet–epidote barren skarns (12.2 mg/t). Lowest average platinum concentration was found in unaltered andesite tuffs (3.4 mg/t) and in magnetites (4.0 mg/t).

Rather normal platinum concentrations -16.2–22.3 mg/t with content scatter from 2.8 (limonitized tuffs) to 60.1 mg/t (limonitized andesites) have been observed in the rocks and ores of the Pravoberezhny area, whereas the highest average concentrations were also observed (28.8 mg/t). Limonitized tuffs are less platinum concentrated – 10.8 mg/t.

Palladium Palladium concentration varies from 1.0 mg/t (in most rocks and ores) to 75.0 and 86.0 mg/t (in garnet–epidote skarns and ores) in the Maysk area; whereas in the latter, the highest and normal average palladium concentration (26.7 and 30.0 mg/t) was observed. The highest and normal average palladium concentration was defined in unaltered (19.0 mg/t) and limonitized (15.5 mg/t) andesites. Ther lowest average platinum concentration was observed in scarnified andesites, magnetites and propylite (2.8, 3.6 and 4.1 mg/t, respectively).

A higher palladium concentration was observed in most rocks and ores of the Pravoberezhny area in comparison to those in the Maysk area where content scatter is from 1.0 mg/t (beresites, ores, ironstone) to 130.0 and 240.0 mg/t (ore quartz and beresites). The highest average palladium concentration has been observed in beresites (49.4 mg/t), ironstones (42.4 mg/t) and a bit lower in limonitized tuffs (31.7 mg/t) and in ore quartz (27.0 mg/t). The lowest average concentration was observed in scarnified tuffs (8.7 mg/t) and unaltered andesites (13.2 mg/t). In a single sample from a, The lowest palladium concentration (0.41 mg/t.) was observed in one veinlet calcite sample of beresites for this ore field.

A platinum and palladium concentration decrease in dependence with depth of scarnified andesites and propylites was also observed in the Maysk area.

Gold. Gold concentratrion in the rocks and ores of the Maysk area vary from 0.011 g/t (propylites) to 22.0 g/t (ores), whereas the highest average gold concentration was observed in the ores (5.92 g/t) and limonitized andesites (4.45 g/t). Normal average gold concentration could be observed in propylites (2.1 g/t), tuffs (2.05 g/t) and scarnified andesites (2.48 g/t). The lowest average gold concentration was observed in unaltered andesite tuffs (0.081 g/t) and magnetites (0.07 g/t). An average metal content increase was observed in andesite (0.69) – scarnified andesite (2.48) – limonitized andesite (4.45 g/t). (table 1).

The gold concentration is a bit lower in most rocks and ores of the Pravoberezhny area in comparison to those in the Maysk (table 1), whereas the content scatter is even less – from 0.1 g/t (andesite) to 7.0 g/t (ironstone). The highest average gold concentration has been observed in the ores and ironstones (2.46 and 2.9 g/t). Normal average gold concentration was found in andesite tuffs (1.75), andesites (1.4), beresites (1.0) and limonitized andesites (0.95 g/t). Scarnified and limonitized tuffs contain low–grade gold (0.2 and 0.49 g/t, respectively). Single gold concentrations were observed in quartzites and post-ore veinlet calcilte in the beresites of the Pravoberezhny area - lowest concentration (0.091 and 0.072 g/t, respectively).

5. Conclusion

1. Highest platinum and palladium concentrations have been found in the ores of Maysk–Lebed deposit and normal average concentrations in the Maysk and Pravoberezhny areas.
2. During skarning and hypergene formation (limonitizing) of rocks and ores, metal re-distribution and even their increase, in some cases, was observed.

3. Gold concentration level in the rocks and ores of the Maysk area is higher than of those in the Pravoberezhny area.

Acknowledgement
The investigations were supported by National Research Tomsk Polytechnic University, project No.: 114-2014
The research was performed with the financial support from Government Project "Nauka", No. 1.1312.2014.

References

[1] Korobeynikov A F 2004 Platinometallnye mestorozhdeniya mira *Kompleksnye zoloto-redkometallno-platinoidnye mestorozhdeniya* **3** 236
[2] Timkin T V 2011 Localization Criteria of Gold Mineralization in Maisk-Lebedsky Ore Field (Mountain Shoria) *Vestnik of Irkutsk State Technical University* 58–64
[3] Sanin V N 2009 Model zolotorudnykh proyavleny v strukturakh Maysko-Lebedskoy ploschhadi. Respublika Altaj *Rudy I metally* **4** 42-54
[4] Timkin T V 2012 Endogennaya rudno-metasomaticheskaya zonalnost Maysko-Lebedskogo zolotorudnogo polya *Izvestija Tomskogo politehnicheskogo universiteta* **320** 56-63
[5] Voroshilov V G, Savinova O V, Ananjev Y S and Abramova R N 2014 Anomaly Geochemical Fields in Siberian Hydrothermal Gold Deposits // *IOP Conference Series: Earth and Environmental Science* **21** 1 1–6
[6] Gavrilov R Y, Khmelev M S and Abramova R N 2014 Volumetric geochemical model of Chortovo Koryto gold-ore deposit (Patom range) *IOP Conference Series: Earth and Environmental Science* **21** 1 1-6
[7] Oskina Yu A, Pakrieva E G, Ustinova E M and Gorchakov E V 2014 Voltammetric and Stripping Voltammetric Determination of Platinum, Palladium, and Rhenium in Mineral Raw Materials *Industrial Laboratory* **7** 14–18.
[8] Oskina Y, Pakrieva E, Ustinova E and Kryazhov A 2014 Decomposition and Preconcentration Methods for the Determination of Pt, Pd Re in Minerals Raw Materials *Advanced Materials Research – Swizerland: Trans Tech Publications* **1040** 278-281.

IOP Conf. Series: Earth and Environmental Science **24** (2015) 012026 doi:10.1088/1755-1315/24/1/012026

Problem analysis of geotechnical well drilling in complex environment

A K Kasenov[1], M T Biletskiy[2], B T Ratov[3] and T V Korotchenko[4]

[123] Kazakh National Technical University after K.I. Satpaev, Almaty, the Republic of Kazakhstan
E-mail:Ratov69@mail.ru[3]
[4] National Research Tomsk Polytechnic University, Tomsk, Russia
E-mail: tatyana1003@mail.ru[4]

Abstract. The article examines primary causes of problems occurring during the drilling of geotechnical wells (injection, production and monitoring wells) for in-situ leaching to extract uranium in South Kazakhstan. Such a drilling problem as hole caving which is basically caused by various chemical and physical factors (hydraulic, mechanical, etc.) has been thoroughly investigated. The analysis of packing causes has revealed that this problem usually occurs because of insufficient amount of drilling mud being associated with small cross section downward flow and relatively large cross section upward flow. This is explained by the fact that when spear bores are used to drill clay rocks, cutting size is usually rather big and there is a risk for clay particles to coagulate.

1. Introduction

Based on the analysis of geological environment as well as the methods and technology used to drill both exploration and geotechnical wells (injection, production and monitoring wells) at uranium sites owned by "Volkovgeologiya" JSC, the most common drilling problems caused by various factors have been revealed. The present article examines the most serious drilling problems – hole caving and packing.

Hole cave-in and hole collapse can be caused by any borehole instabilities stemming from in-situ rock stress. Therefore, in order to select an appropriate drilling technology involving well design and development of preventive measures and adequate emergency response plans, it is essential to evaluate mechanical behavior of subsurface rocks.

2. Materials and methods

In accordance with the rock classification based on the mechanical behavior modes [1], the rock formations which comprise the geological cross-section of Khorasan-2 and South Karamurun uranium deposits are classified as poorly stable or unstable and presented by the following rock types: weak unconsolidated rocks, plastic, soft and loose rocks related to the corresponding drillability categories from 1 to 1Y. As is shown in figures 1 and 2, hole caves are formed while drilling sand and clay formations.

Apart from the above-mentioned, the rock stability can be estimated by cavernosity ratio proposed by K.F. Paus [1]:

IOP Conf. Series: Earth and Environmental Science **24** (2015) 012026 doi:10.1088/1755-1315/24/1/012026

$$K = \frac{V_a}{V_c} = \frac{d_a^2}{d_c^2}, \tag{1}$$

where, K- cavernosity ratio;
V_a and V_c – actual and calculated hole volumes;
d_a and d_c – hole diameter and the diameter of applied cutting tool.

The rocks are considered stable, if K=1, temporary stable rocks if 1< K < 3 and unstable if K > 3. When K < 1, forces acting within the formation push the wall of the hole inward, for example while drilling swelling clays.

While drilling the sand formations in Khorasan-2 uranium deposit, the calculated diameter is 132 mm and actual one is 230 mm (figure 1). Under these conditions, cavernosity ratio is assumed to be 3.04, which means that these rocks are referred to as unstable.

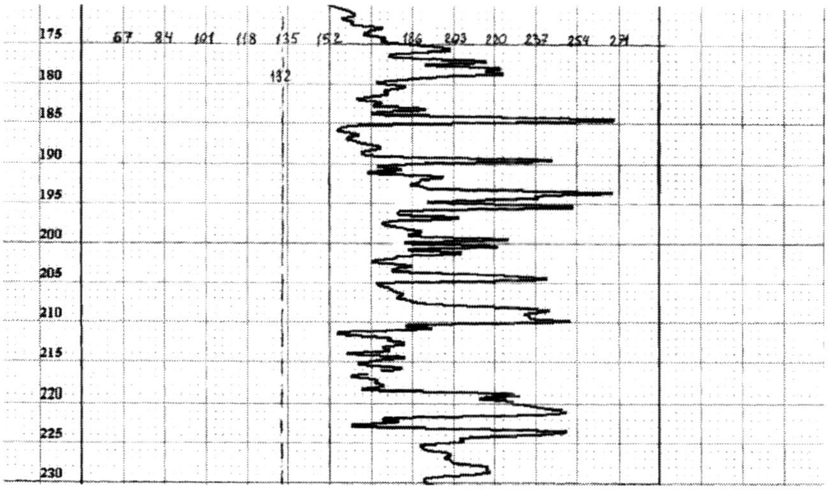

Figure 1. Caliper logs of wells № 5-6-17-18

While drilling the sand formations in South Karamurun uranium deposit (figure 2), the diameter of applied cutting tool is 161 mm and the actual diameter is up to 311 mm. In this case, cavernosity ratio is assumed to be 3.74, which characterizes the rock formation being drilled as very unstable.

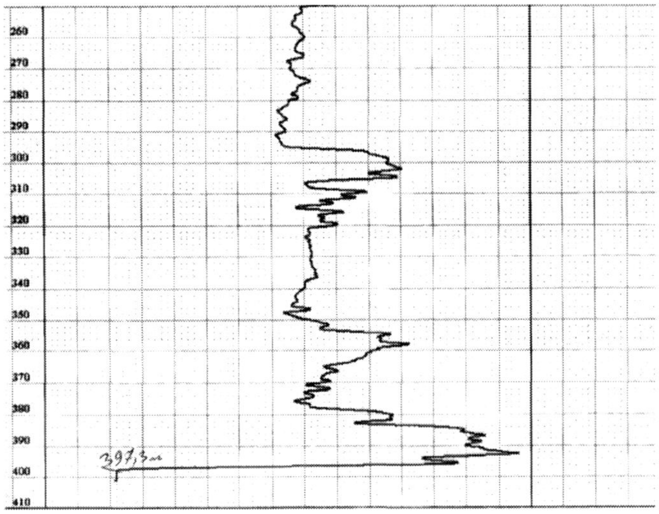

Figure 2. Caliper log fragment of well № 3-109B, North Karamurun uranium deposit

It is worth noting that dry heavy clays are considered to be stable rock. That air-rotary drilling is primarily used for drilling in such formations proves the above-mentioned fact [2]. Under these conditions, hole cave is formed due to different reasons which will be further discussed.

As previously mentioned, in-situ overburden stress directly affects hole cave formation, which in its turn may lead to borehole collapse. Because of the in-situ overburden stress, soft loose, unconsolidated, fractured and broken rocks push the hole wall outward. The value of overburden stress is defined by the following formula [3]:

$$P_{гд} = g \cdot \rho_{гп} \cdot H, \tag{2}$$

where P – overburden stress value, Pa;

$\rho_{гп}$ - rock density, t/m^3;

H - formation depth, m;

g - gravity acceleration, m/s^2.

Here, it is essential to consider horizontal stress value which is obviously dependent on overburden stress.

The value of horizontal stress is determined by [3]:

$$P_б = \frac{\mu}{1-\mu} P_{гд}, \tag{3}$$

where $P_б$ - horizontal stress value, Pa;

μ - Poisson ratio (for clays $\mu = 0.2 - 0.4$).

Besides, mud hydrostatic forces are also used to prevent formation collapse. The value of hydrostatic head can be calculated by the following formula [3]:

$$P_{гс} = g \cdot \rho_ж \cdot H, \tag{4}$$

where P – mud hydrostatic head value, Pa;

$\rho_ж$ - drilling mud weight, t/m^3;

H – the depth where hydrostatic head is measured, m;

g - gravity acceleration, m/s^2.

Generally, borehole fails when horizontal stress exceeds the tensile strength of the rock and hydrostatic mud head, which is expressed as follows:

$$P_б \succ \sigma_T + P_{гс} \tag{5}$$

where, σ_T - rock tensile strength, Pa.

Thus, in order to ensure borehole stability, it is required to observe the following conditions:

$$P_г \cdot \frac{\mu}{1-\mu} \leq \sigma_T + P_{cm} , \tag{6}$$

The above-mentioned formula fully describes hole cave development and represents almost all factors which influence this process, i.e. overburden stress value, rock mechanical properties and hydrostatic head of mud.

Mud-circulating drilling can also alter rock properties through the interaction with fluid, which in its turn influences borehole stability parameters. Under these conditions, the stability of the borehole is directly dependent on the moisture content.

Drilling mud/clay rock interaction occurs through adsorbing fluid filtrate. Depending on the moisture content, i.e. swelling potential, some authors [2] distinguish the following types of rocks: slightly swelling rock, medium swelling rock and strongly swelling rock.

When the rock moisture content is low, the boreholes are rather stable and there are no serious drilling problems.

While drilling medium swelling clay, the drilling mud/clay interaction causes the clay to swell, weaken, and eventually fall into the wellbore, which can cause tight hole conditions, however, without cave development. If the borehole diameter is small, the drilling mud/clay interaction can lead to total collapse and loss of the hole.

Drilling of strongly swelling clays presents the greatest engineering challenge in terms of cave development. In this case, pressure decreases the strength of clay by destroying cement bond between the clay platelets (clay dispersion). The clay becomes ductile and is pushed by the overburden stress to the wellbore developing the caves.

To illustrate the above-mentioned borehole stability parameters, the example of cave development while drilling wells №3-109b is provided. In accordance with the caliper log shown in figure 2 caves basically develop while drilling through clays and siltstone. The largest cave was found at the depth of 328-460 m, with cave diameter being 311 mm.

Let us define the overburden stress value by the dependence (2) if the conditions are as follows: well depth - 382 m, clay density - 2500 kg/m^3 and gravity acceleration - 9.81 m/s^2. It can be expressed as follows:

$P_г = 9{,}81 * 2500 * 382 = 9368550$ Pa = 9,4 MPa.

Then, it is necessary to define the value of horizontal stress by the following dependence (3):

$$P_б = 9{,}4 \cdot \frac{0{,}25}{1-0{,}25} = \frac{0{,}98}{0{,}75} = 3{,}14 MPa .$$

The hydrostatic head of the mud column is determined by the dependence (4), with the density being 1030 kg/m^3. At this depth, hydrostatic head is as follows:

$P_{гс} = 9{,}81 * 1030 * 382 = 3859843$ Pa = 3,9 MPa.

Thus, with the clay yield point being 2-14MPa and hydrostatic head being 3.9 MPa, borehole walls should have been stable as the sum of all the factors mentioned above exceeds the horizontal stress by 3,1 MPa. This conclusion is especially relevant for dry clays; however, even in case when clay strength equals zero, borehole walls should have remained stable as hydrostatic head exceeds the horizontal stress. Therefore, these caves developed due to chemical reactions caused by the exposure of the rock to the drilling mud, which altered physico-mechanical properties of rocks.

When humidified, the strength of clays characterized by high porosity significantly decreases as they are the most water-saturated rocks. Increasing water content weakens clay rocks. It is explained by the alteration of rock physico-mechanical properties, precisely, the dissolution of grain-to-grain cement bond.

In addition to the above-mentioned causes, the development of the caves of large diameter while drilling geotechnical well №3-109b can be explained by other reasons as well. In accordance with the selected drilling technology, hard rock formations were drilled by PDC bits being 132 mm in diameter, while soft rock formations were drilled by spear borers and roller bits 161 mm in diameter.

Figure 3 illustrates the principle of large cave development while drilling the above-mentioned well.

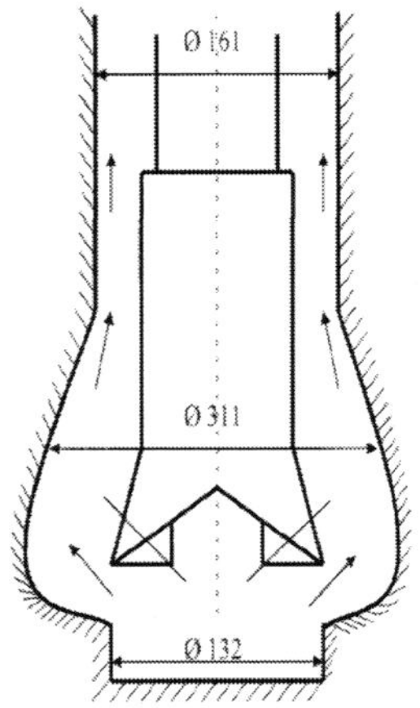

Figure 3. Cave development principle

In our opinion, due to the fact that there is great difference in drilling rate for soft rocks and hard rocks (soft rocks – 15-20 m/hr; hard rocks -0.20-0.30m/hr), the penetration rate drastically decreases when a drill bit penetrates into hard rock formation, which in its turn increases mudflow intensity near the drill bit. This leads to hole washout problems and large cave formation. The development of small caves while drilling the overlying soft formations can be explained by the presence of hard streaks where the drilling rate also decreases which results in washout enlargement due to by excessive bit jet velocity.

3. Results and discussion

Based on the above-mentioned facts, it is possible to conclude that while drilling geotechnical wells at uranium sites in South Kazakhstan, hole cave development is caused by the mud filtrate invasion in clay rocks (clays, siltstone, etc.) and excessive bit jet velocity.

Packing is another typical drilling problem which is related to borehole instability and poor hole cleaning. As mentioned above, packing leads to excessive torque and drag on drill string, as well as difficulties in logging, particularly in running logging tools into the borehole, and cementing. All this increases mud and cementing costs. For example, the injection well № 3-109b in North Karamurun uranium deposit was drilled according to the following procedure: spear borers of 161 mm in diameter were used up to the depth of 460 m, then the rock formations were drilled by PDC bits being 132 mm in diameter up to the target depth – 550 m. When well logging was being performed, logging tool got stuck at the depth of 397.3 m. To find out the reason, caliper log was examined.

4. Conclusion

The caliper log analysis has revealed that a large cave occurs at the depth of 378 m-460 m, being 311 mm in diameter and 80 m in thickness.

It is obvious that such a cave negatively influences hole cleaning performance due to the sharp decrease in drilling mud return velocity, which in its turn results in cutting accumulation. When the pump is stopped, cuttings fall back into the well bore causing pack-off in tight section of the hole (change in diameter from 161 mm to 132 mm) and, thus, preventing a logging tool from running into the wellbore.

Besides, in order to prevent clay rock dispersion, chemical reagent coagulator K-ION has been introduced into the drilling mud. This reagent forces the cutting particles to coagulate forming clay balls or packing. A further core drilling of packing has revealed that it mostly consists of clay rocks and is of plastic character.

It should be also mentioned that suspending clay platelets, i.e. packing, usually accumulate on bit surface (above the cutting tool or in place of drill pipe connections), which leads to hole drag with corresponding negative effects: stuck pipe and drill string failures, increased power consumption, etc.

Thus, packing results in significant operational problems, most of them being related to well design and drilling. The liquidation of these drilling problems leads substantial financial losses. As it became apparent, packing is basically caused by hole caves and insufficient amount of drilling mud which ensures the required velocity of mud returns in caves.

References

[1] Vasiliev V I, Ponomarev P P, Blinov G A and et al. 1983 *Otraslevaya metodika po razrabotke tekhnologii bureniya na tverdye poleznye iskopaemye* (Leningrad: VITR) p 130
[2] Gorodnov V D 1984 *Fiziko-khimicheskie metody preduprezhdeniya oslozhneniy v burenii* (Moscow: Nedra) p 229
[3] Sereda N G, Soloviev E M 1974 *Burenie neftyanykh i gazovykh skvazhin* (Moscow: Nedra) p 451
[4] Ivachev L M 1989 *Promyvka i tamponirovanie geologorazvedochnykh skvazhin* (Moscow: Nedra)
[5] *Elementarnyy uchebnik fiziki* (Moscow: Nauka)

IOP Conf. Series: Earth and Environmental Science **24** (2015) 012027 doi:10.1088/1755-1315/24/1/012027

Evaluation of plain river channel deformation in the absence of observation data

O G Savichev, M V Reshetko, I A Matveenko and Ye V Ivanova

Tomsk Polytechnic University, 30 Lenin Avenue, Tomsk, 634050, Russia

E-mail: osavichev@mail.ru, mrechetko@tpu.ru

Abstract. Evaluation and long-term forecast techniques for plain river channel deformation has been developed in the absence of observation data. Their testing was performed by the example of medium rivers (with catchment area from 2000 to 50000 km^2) in taiga zone of Western Siberia (the Ob river basin on the section of its midstream). The technique is based on determination of flow parameters, at which the maximum river bed deformations are observed. Standard data of hydrometric observations obtained at the state hydrological network are used for calculation.

1. Introduction

For design of water facilities, bridge and pipeline river crossing one needs to evaluate the value of horizontal and vertical river bed deformations [1–4]. Usually this problem is solved on the basis of river bed maps and schemes constructed in different years with the interval of not less than 5 years [3, 4]. The essence of the technique is comparison of cross-sections obtained as a result of channel survey in different years. Among its disadvantages are: high labour consumption of engineering-hydrometeorological survey and complexity or impossibility to obtain calculated parameters in a short time specified for design and survey by a customer; the technique is not applicable without reliable river bed survey made in in different years with the interval of not less than 5 years; besides, even if time different river bed survey available, the comparison process of several views and schemes presents some additional errors and is not enough formalized.

In the absence of observation data there is a technique of line construction of possible vertical river bed erosion under the condition of deformations due to river bed ridge transformations [3, 4]. The essence of the technique consists in evaluation of vertical deformations in terms of relationships between microform bed sizes and flow depth [5]. Its disadvantages: it is applicable in the condition of ridgy sediment movement and does not permit reliable forecast for vertical deformations of a peat bed or a bed with insignificant load transport in the form of ridges; one needs measurements of flow parameters and bed shapes in the free channel period as most of small streams can freeze through in winter.

Limited possible horizontal river bed deformations (the profile length of limited scour $L_{\lim}$) without observation data can be roughly defined according to [4] in the following way: 1) the type of bed process is stated on the researched river section and the width of flood plain flow is calculated $B_{FP,x}$; 2) using reference data [4] for a stated bed process the ratio of channel-forming belt width $B_{CT,a}$ and flood plain flow width $B_{FP,a}$ is taken, as well as mean square deviation of the ratio $\sigma_{CT,a}$; 3) the profile length of limited scour is taken as the product of flood plain flow width on the researched river section

$B_{FP,x}$ by the value ($B_{CT,a}/B_{FP,a} + \sigma_{CT,a}$) set for a stated bed process type. The technique allows reliable data on limited possible horizontal deformations, but without time reference (that limits the possibility of long-term forecast).

It is also known another forecast technique for river bed horizontal deformation in the absence of observation data based on the data on bed deformation of analogue river [3, 4]. It is based on data transition obtained by the techniques mentioned above for analogue river to the studied river taking into account forecast errors. The technique disadvantages: it is slightly suitable for hydrologically poorly studied areas of Siberia, where studied analogue rivers are few and presented mainly by large rivers with catchment area more than 50000 km^2; taking this fact into account it is particularly difficult or impossible to match an analogue for an uninvestigated small river with catchment area less than 2000 km^2 (river classification according to [6]).

At present, horizontal bed deformation determination without observation data is mostly performed at qualitative level taking into account bed type and/or bed process, or by means of empirical formulas based, as a rule, on criterion relationships of scour diameter, flow depth, water surface slope and other parameters [1, 7–9].

The general disadvantages of existing techniques relevant for the river bed forecast deformation are as follows: 1) most of existing methods need time different bed survey with the interval not less than 5 years on studied river or analogue river; 2) in the design of construction projects on hard access areas of Siberian taiga, tundra forest, and tundra zones the bed survey data in different years (particularly with the interval of less than 5 years) are usually absent or do not meet the requirements in accuracy specified in [3]; 3) labour consumption of engineering-hydrometeorological surveys, complexity or impossibility of obtaining calculation parameter in short terms defined for surveys and design; 4) high degree of inaccuracy in measurement and calculation of parameters used for indirect evaluation and forecast of planned (horizontal) river bed deformations; 5) significant deviations of calculated bed deformation values from measured values, if a technique is used in other region in the conditions different from those considered in development of this or that technique (for example, in case of muskeg banks).

Keeping these in mind the authors have performed the research in bed deformation of plain rivers to develop technique for obtaining their quantitative characteristics in the absence or minimum of observation material. Technique testing was made in terms of data obtained at state observation network of the Russian Federal Service in Hydrometeorology and Environmental Monitoring (Roshydromet) and the regional network of JSCo «Tomskgeomonitoring» (Tomsk) on the rivers of taiga zone in the basin of the Ob River (Western Siberia).

2. Description of the studied area

The area involved corresponds to the administrative borders of Tomsk Oblast. It is located in the Central and South-Eastern parts of West Siberian Plain and characterized by predominance of two differently directed groups of exogenous geologic processes – swamp and bed processes. The first group deals with accumulation of organic matter at catchment, decrease in river load flow and definite increase in ground elevations due to peat deposit formation at the stage of oligotrophic domed bogs, whereas the second – on the contrary, high relief roughness and increase in solid flow. The mentioned above processes are closely interconnected over the period of all Holocene, water flow was directed in accordance with different hypothesis: 1) The Arctic Ocean between open tongues of Ural and Siberian glaciers, or in the absence of land glaciation, or between different time Ural and Siberian glaciers, or through glacier (including under glacier); 2) in Aral-Caspian basin through Turgai depression; 3) in blind West-Siberian basin (sea) [10].

Being uninvolved in discussion we would like to note that, anyway, one of the reasons for West-Siberian Plain bog formation (that reaches 30 % of the territory and more [11–13]) could be continuous transformations of river beds in easily eroded soils in the absence of tree vegetation and excessive moisturizing of catchments as a result of deglaciation, precipitation, and difficulty in water flow from the area involved. In such conditions existence of numerous dead channels became a catalyst of area bog formation in some cases, moreover, eutrophication of inundated reservoirs is going on now. The analysis of engineering survey materials permitted us to make conclusion on the fact that bed process contributed to bog formation on a regional scale that, thereafter, resulted in some bed deformation stabilization, at least, of small rivers. The next stage of bog formation process became development of secondary river network of bog streams and reservoirs in the condition of excessive or moderate moistening and deteriorating conditions of water runoff, in some cases stream bog beds form at watersheds of disappeared synantetic rivers [14, 15].

Virtually, the Ob River within Tomsk Oblast borders is characterized by prevailing development of bed braidedness, in the central and northern part – by incomplete meandering and flood plain braidedness. The downstream of the Tom River is distinguished by incomplete meandering, bed and flood plain meandering, in some places limited meandering. On the Chulym River and its tributaries within Tomsk Oblast the free and incomplete meandering, bed braidedness prevails. The other medium rivers of the area involved are characterized by incomplete meandering. The rest types of bed process in terms of the State Hydrological Institute classification [3] are generally absent or conditioned by the local sections of some rivers [14]. The general characteristic of horizontal deformation of some largest regional rivers obtained in terms of observation data by JSCo «Tomskgeomonitoring» within 1996–2005 is presented in table 1.

Table 1. Characteristic of river bank erosion of Tomsk Oblast [14], m/year.

Settlement	River	Eroding velocity	
		mean	maximum
Kolpashevo	Ob	4.4	15.0
Krivosheino	–//–	0.7	20.0
Zyraynskoye	Chulym	4.9	30.0
Komsomol’sk	–//–	10.8	22.0
Podgornoye	Chaya	1.6	30.0
Moryakovka	Tom	4.4	20.5

3. Results and discussion

The program of standard hydrometric observations at the state monitoring network over the territory of the Russian Federation (at flow points) includes everyday measurement of water levels (twice and more) and measurement of stream velocity, stream depth and width, as a rule, not less than once a month. In some cases hydrometric operations have been performed since the 1930’s, more often – since the 1950-1960’s. Owing to the operations there are sufficiently long rows of observations of water levels, stream width, its mean and maximum depth.

Based on the results of the data analysis an approach to bed deformation evaluation (both horizontal and vertical) is introduced, it involves determination of flow parameters, at which maximum bed deformation is observed. The approach consists in selection of observation data corresponding to the fixed water level (in Baltic elevation system) set with some constant increment from the observed (or calculated) minimum to the relevant maximum. The example of data selection for fixed water levels (200 and 400 cm from conventional reference plane) is shown in figure 1.

In case when just such a value of water level is absent in data, it is to be found by interpolation between the next following dates. In the resultant selection there has to be an even number of selected levels corresponding both to river water rise and fall. For each data selection, at fixed water level the minimal and maximal values of the sought quantity (width, mean and maximum stream depth) is defined and its maximum oscillating amplitude (the difference between maximum and minimum) is calculated. The maximum amplitude value for the whole studied range of level water changes is a characteristic of vertical $D_{max}(h_{max})$ (in case of maximum stream depth h_{max}) or scheduled $D_{max}(B)$ (for stream width B) of river bed deformation. It should be noted that channel control is usually fixed at the least deformed and straightest river section. Taking this into account the resultant values of river bed deformation can be considered as maximums for the most stable sections that well corresponds to the project tasks directed to searching for limited values of hydrological characteristics and bridge and pipeline crossing construction and other facilities in similar regions.

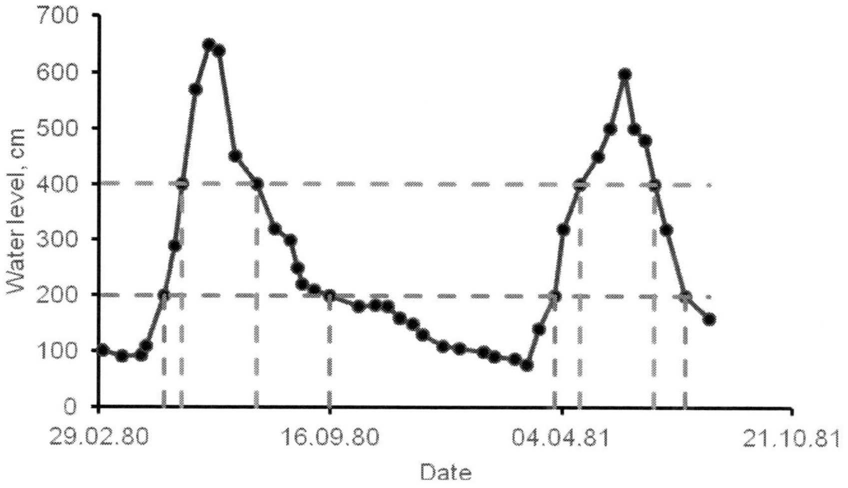

Figure 1. The scheme of data selection corresponding to fixed water levels (at 200 cm from conventional reference plane B_{max} = 30 м, B_{min} = 26 m, at 400 cm B_{max} = 55 m, B_{min} = 49 m, hence, $D_{max}(B)$= 6 m).

The testing results of the approach described concerning the data on medium rivers in taiga zone of West Siberia are presented in table 2. The analysis showed that, firstly, vertical deformations change in the range of 0.39–1.27 m, but scheduled ones – in the range of 4.55–16.01 m [16]. It corresponds to both the results of engineering survey on the area involved and (in case of vertical deformations) calculations of bottom deformations conditioned by bed shape transformations [16].

Table 2. Characteristic of water runoff and river bed deformations of plain tributaries of the Ob River in taiga zone.

River – settlement	Catchment area, km^2	Mean water discharge, m^3/s	$D_{max}(B)$, m	$D_{max}(h_{max})$, m
The River Chaya – Podgornoye	25000	78.7	9.98	1.27
The Parabel' River – Novikovo	17900	71.4	11.58	0.84
The Kenga River – Tsentral'ny	7440	24.5	13.28	0.39
The Vasyugan River – Sredniy Vasyugan	31700	162	8.22	0.89
The Vasyugan River – Maysk	3730	15.1	4.55	0.96
The Tym River – Vanzhil' – Kynak	10100	80.9	16.01	0.89

Note: characteristic of water runoff is given according to [14]

Secondly, between the indicated characteristics and difference of maximal and mean values of stream velocity and stream depth the statistically significant relationships were revealed. The physical sense of such relationships consists in consideration of: 1) solid flow as a magnitude proportional to flow energy losses at interaction with banks and bottom; 2) river bed deformation equation as a relationship between deformation in one of the directions (vertical or horizontal) and solid flow. The general view of the relationship:

$$D_x = k_1 \cdot (v_{max} - v_a) \cdot d^{k_2}, \tag{1}$$

or

$$D_x = k_3 \cdot (h_{max} - h_a) \cdot d^{k_2}, \quad (2)$$

where D_x – the characteristic of river bed deformation in x-direction (vertical or horizontal); v_{max} and v_a – maximal and mean stream velocities; h_{max} and h_a – maximal and mean stream depths; d – the typical load diameter; k_1, k_2, k_3 – empirical coefficients.

Thirdly, values of maximal change amplitude of stream width $D_{max}(B)$ and maximal depth $D_{max}(h_{max})$ are on the whole comparable with corresponding evaluations of river bed deformations obtained by river bed survey and techniques based on calculation of river bed shape parameters and criteria of river bed stability [16].

In fact, evaluation of limited possible vertical bed scour Z_{lim} and forecast of maximum possible scheduled bed deformations $\Delta B(T)$ of hydrologically uninvestigated plain rivers can be carried out using the formulas:

$$Z_{lim} = Z_{min} - D_{max}(h_{max}) - \delta_h, \quad (3)$$

$$\Delta B(T) = T \cdot (D_{max}(B) + \delta_B), \quad (4)$$

where $\Delta B(T)$ – the forecast of scheduled river bed deformation in control point within the period T, m/period in years; δ_B and δ_h – errors in measurement of flow width and depth, m. Values of $D_{max}(h_{max})$ and $D_{max}(B)$ are either taken in terms of data analysis of measured analogue river water discharges (for example, shown in Table 3) after preliminary grounding its choice, or calculated by empirical relationships using data on maximum and mean depths.

4. Conclusions

In the course of the given research the evaluation technique of river bed deformations for plain rivers is proposed and its testing in terms of medium rivers data in taiga zone of Western Siberia (the Ob River basin) is performed. The given technique allows the evaluation of vertical and horizontal river bed deformations of hydrologically uninvestigated plain rivers with permissible certainty. Undoubtedly, such evaluation cannot replace completely the results of relevant observations, but their application in other environmental conditions would require additional justification. Nevertheless, their application for this or that uninvestigated river in taiga zone, in Siberia in particularly, could efficiently contribute to the analysis of several layout options for the designed facilities crossing a river and, consequently, sufficient cost reduction of engineering survey.

References

[1] Chang H H 2008 *Fluvial Processes in River Engineering* (Malabar, Florida: Krieger publishing company) p 432

[2] PNIIIS Gosstroya SSSR 1987 *Rekomendatsii po otsenke i prognozu razmyva beregov ravninnyhk rek i vodokhranilishch dlya stroitel'stva* (Moscow: Stroyizdat) p 72

[3] *VSN 163-83 Vedomstvennye stroitel'nye normy. Uchet deformatsiy rechnykh rusel i beregov vodoyemov v zone perekhodov magistral'nykh trubuprovodov (neftegazoprovodov)* 1985 (Moscow: Goskomgidromet) p 142

[4] GGI 2009 *Uchet ruslovogo protsessa na uchastkakh podvodnykh perekhodov truboprovodov cherez reki. Standard organizatsii. STO GU GGI 08.29-2009* (Saint-Petersburg: Nestor-Istoriya) p 184

[5] Alexeevskiy N I 2006 *Gidrofizika* (Moscow: Academia) p 176

[6] Gidrologiya sushi Terminy i opredeleniya GOST 19179-73 1988 (Moscow: Izdatil'stvo standartov) p 65

[7] Leopold L B and Maddock T 1953 *The hydraulic geometry of stream channels and some physiographic implications* (United States Government Printing Office, Washington) p 57

[8] Antropovsky V I 1972 Issledovaniya morfologii i gidravliki rechnykh rusel, poim i vodoyemov dlya nuzhd stroitel'nogo proyektirovaniya *Trudy Gosudarstvennogo gidrologicheskogo instituta* **190** 5

[9] Berkovich K. M., Zlotina Л B 2003 Raschet stabil'nosti rechnykh rusel v usloviyakh antropogennoy nagruzki *Geography and Natural Resources* **2** 117
[10] Maloletko A M 2008 *Evolutsiya rechnykh sistem Zapadnoy Sibiri v mezozoye i kaynozoye* – (Tomsk: TP Univesity Press) p 288
[11] Neishtadt M I 1977 Vozniknoveniye i skorost' razvitiya protsessa zabolachivaniya *Nauchnye predposylki osvoyeniya bolot Zapadnoy Sibiri* (Moscow: Nauka) pp 39–48
[12] Bazanov V A, Berezin A E, Savichev O G., Skugarev A A 2009 The phytoindication method for mapping peatlands in the taiga zone of the West-Siberian Plain *International Journal of Environmental Studies vol 66* **4** 473
[13] Eckstein Y, Savichev O G, Pasechnik E Y 2014 Is climate change affecting ground water chemistry in the largest boreal wetland? *Geological Society of America. Annual Meeting in Vancouver, British Columbia vol 46* **6** 67
[14] Savichev O. G. 2010 *Vodnye resursy Tomskoy Oblasti* (Tomsk: TP Univesity Press) p 248
[15] Savichev O. G., Paromov S V 2014 Hydrological aspects of bog frmation in taiga zone of Western Siberia *Bulletin of the Tomsk Polytechnic University* V 324 **1** 154
[16] Savihev O G, Reshet'ko M V 2012 Estimating sediment yields and riverbed deformations of unexplored lowland rivers in the Western Siberia taiga zone *Engineering Surveys* **1** 52

IOP Conf. Series: Earth and Environmental Science **24** (2015) 012028 doi:10.1088/1755-1315/24/1/012028

Facies and sedimentation model of iron-ore sequence in Bakchar deposit

M Rudmin[1], A Mazurov[2] and A Ruban[3]
[1 2 3] Department of Geology and Mineral Exploration, Institute of Natural Resources, National Research Tomsk Polytechnic University, Tomsk, Russia

E-mail:[1]rudminma@tpu.ru, [2]akm@tpu.ru

Abstract. This article describes the investigation in the behavior and structure features of the iron-ore sequence in Bakchar ore deposit. Five facies characterizing the typical sedimentation environments were identified based on the analysis of typomorphic attributes of ore-bearing and barren deposits. The facies where oolitic ores are localized have been determined. A conceptual sedimentation model of the Bakchar ore mineralization has been proposed. Two facies- littoral argillo-arenaceous ferriferous sediments and mobile shallow alluvium oolitics were proposed as prospecting indicators of oolite hydrogoethite ore deposits.

1. Introduction

The Bakchar ore deposit is located in the south–eastern West Siberian Iron Ore Basin (figure 1), i.e. 200 km NW of Tomsk. Iron occurences in the West Siberian Iron Ore Basin have been traced within an area of 150 km wide and extenting approximately 2000 km (from the Turukhan and Bolshaya Kheta River basin in the NE and to the Tobol river upstream in the SW). The ore-bearing sediments of the Bakchar ore cluster include typical sediments of coastal-marine facies: gravelites, oolite ores, sandstones, siltstones and clays. The iron ores are characteristic of goethite–hydrogoethite and chloride–hydrogoethite oolite composition and are confined to three horizons: Narym, Kolpashevo and Bakchar [1–2].

The first findings on the facies environments in iron ore sequences of the Bakchar ore occurence and the West Siberian Basin were published by A Kondakov, I Nikolaeva and N Belous in 1964-1967 [3, 4]. The point is that facies differentiation of Bakchar ore-hosting sequences has not been furthered. The problem of ore formation has been addressed in several conference proceedings, however, positive findings have not been discussed [5, 6]. In this case, the conclusions about the typical Bakchar ore formation environments stated in the 1960s of the last century are the only and major ones available. Without doubt, ore sequence sediments are commonly considered to be coastal-marine ones. The prospecting-evaluating survey data of 2006-2014 provided up-dated information on the geological structure aspects of Bakchar area itself. The analysis of data and its correlation with modern understanding of sedimentary iron-ore deposit formation conditions (V Kholodov, R Nedoumov, Ye Golubovskaya et al. [7, 8]) provides the possible identification of facial features of Bakchar ore-hosting sequence.

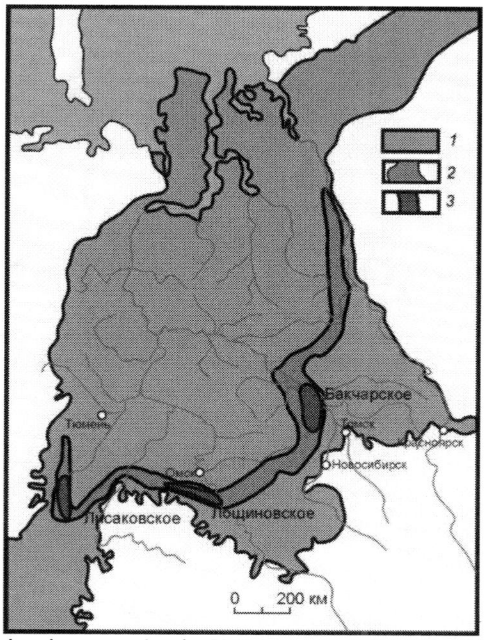

Figure 1. Map of West Siberian iron-ore basin [2, 9]: 1 – areal limits of Cretaceous–Paleogene sea; 2 – zone of coastal-marine iron-ore sediments; 3 – areas of the largest iron-ore clusters.

2. Factual Material and Research Methods

The facies analysis data source included graphics of the 1960s, as well as physical and geological maps plotted on the results of prospecting-evaluating survey data of 2006–2008 and 2006-2014 in Bakchar ore deposits and occurrences. Therefore, integrated exploration embraced geological plans, cross-sections, borehole logging and core samples. To investigate sample substance composition the optical and electron microscopy methods were applied.

3. Facial Features of Bakchar Ore-Hosting Sequences

Based on investigation results of the Bakchar ore occurences, the following ore-hosting sequence facies (according to Logvinenko classification, 1980 [10]) were identified: sand beach sediments in aggradation terraces and bars; argillic-ferrous silts in isolated stagnant depressions; oolite sands in mobile shallows – barriers and bars; sand–argillic–ferrous sediments in coastal bay zones; silt clays in the shallow marine shelf. It is should be noted that the climate and sea-floor relief are important factors infacies differentiation. N Strakhov [11] stated that "... oolite ores are most commonly found the gulf, bay and island sea areas involving complex shoreline contours..."; " such ores usually accumulated in distinctly contoured submarine pits (potholes)..."(based on the investigation of oolite iron ores in Lotharingian and Kerch basins, as well as Ayat deposit). The lithological feature of the ore sequence correlated with the established concepts indicated the fact that these sediments had formed under humid climate.

Facies of sand beach sediments in aggradation terraces and bars

Facies of sand beach sediments in aggradation terraces and plains (sea depth of up to several meters) are light–gray (often greenish) fine-grained sands, silts and poorly cemented fine–grained gray sandstones. The amount of ore oolites in these facies sediments rarely exceeds 10%.

These facies sediments formed in the ore occurence area within Ipatovskaya suite top as a result of regression having formed the Narym ore horizon. The facies is exposed in the south-eastern part of the ore occurrence (figure 2–A). The facies on the map is of a NW trending offshore bar, pinching out in the central part of the ore occurrence. This bar was enclosed by a depression in which there was a sharp transition of predominating argillic–ferrous silt facies in isolated stagnant depressions.

Facies of argillic-ferrous silts in isolated stagnant depressions
Facies of argillic-ferrous silts in isolated stagnant depressions is related to coastal–marine facies with sea depth of up to 10 meters. Under the conditions of relatively calm wave dynamics and aleuropelitic material denudation, gray thin clay layers with oolite inclusions formed.

The clays of this facies were abundant throughout the ore occurence area within Ipatovskaya suite top (figure 2–A) isolated by a sand bar, which, in its turn, was washed out after subsequent transgression. Therefore, due to indistinct unconformity these sediments are superposed by sediments of shallow marine shelf silt clay facies. In the W and NW trendingof the Ipatovsk suite top the facies of clay–iron silt facies grade to facies of sand–argillic–ferrous sediments in coastal bay zones.

Facies of sand–argillic–ferrous sediments in coastal bay zones
Facies of sand–argillic–ferrous sediments in coastal bay zones (sea depth between 10 and 40 m) includes the following lithogenetic types: cemented hydrogoethite–chlorite oolite ores, well–cemented hydrogoethite oolite ores, argillic chlorite and hydromicaceous cemented sandstones and greenish–gray aleurolites and clays. As a result of erosion caused by regression, weak (up to 2 m) gravelite lenses are observed. Sediment continuity is observed towards the shoreline.Gray, greenish sandstones change to oolite ores with argillic, chlorite-argillic and sometimes hydrogoethite cements, then to well-cemented siderite-chlorite and siderite cement at the bay bottom. Such a sequence could be explained by the regular succession of autigenous minerals in marine sedimentary iron ores in humid climate [11]. The sediments of this facies can be divided into two subfacies: aleurolite–clay–iron sediments on the bay slope and sand–iron sediments on the bay bottom.

(1) Subfacies of aleurolite–clay–iron sediments on the bay slope includes sandy aleurolites, sandstones with argillic and chlorite–argillic cement, oolite hydrogoethite-chlorite ores with argillic, argillic-chlorite and chlorite cement. The sediments of this facies formed on the bay slope where the environment reduction potential gradually increased progressively to depth, resulting in the formation of corresponding authigenic minerals (glauconite, montmorillonite, nontronite, leptochlorite, iron sulfides- pyrite framboids. Hydrogoethite-chlorite grains are similar to irregular–shaped spheric oolites, peas with inner homogeneous structure of numerous micron-sized terrigenous debris, as a result of relatively constant and unidirectional geochemical and hydrodynamic mode. Conventionally, ore accumulated at times of pH changing from 5.0 to 6.5.

Aleurolite–argillic–ferrous sediments on the bay slope extend throughout the ore sequence. The sediments of this subfacies are 65–70% of the ore occurence area in the Ipatovsk suite top (figure 2–A); 25–30 % in the Gankinskaya suite base (figure 2–B); 80–85% in the Gankinsk suite top (figure 2–C); and 10% in the Lulinvorskaya suite base (figure 2–D).

During the formation of the Ipatovsk suite top (70–80 million years ago), these sediments were located in the central part of the ore occurrence (figure 2–A), whereas, unstable underwater elevations were located in the southern and northern parts. Sandstones and oolite ores formed an inclined dipping elevation trending W-E towards the bay floor where the sediments of sandy–ferrous subfacies were extensive. A successive facies substitution of argillic cement sandstones to silt sandstones with chlorite–argillic cement, then, oolite ores with argillic–chloride and chlorite cement was observed from the central elevation areas to the bay bottom.

During the formation of the Gankinsk suite bottom as a result of the pre-Kolpashevo transgression within ore occurrence area, the shallow–water shelf environment with a NW trending underwater bar with aleurolite–argillic–ferrous sediments and ores with glauconite–chlorite cement was predominate (figure 2–B).

Then, during the final stages of the Kolpashev horizon formation in the Gankinsk suite top, the sediments of this subfacies embrace maximum distribution (figure 2–D). Within this area two depressions in the north and south were observed, where sediments of the sandy–ferrous subfacies were predominate on the bay bottom.

During the formation of the Lulinvorsk suite (50 million years ago), the sediments on the bay slope had subordinate occurrence. They occupied a narrow zone (figure 2–C) where shallow water bars separated oolite sands and shallow water shelf - aleuritic clays.

(2) Subfacies of sand–iron sediments on the bay bottom is mainly characterized by hydrogoethite oolite ores with chlorite–siderite, argillic–siderite and siderite cement. Such oolites usually have a concentrical inner zone structure and show a significant disintegration. The cleavage of oolites indicates the fact of their transportation by wave currents and parallel oxidation of chlorite minerals. The ores formed on the bay bottom silt sediment at times of pH changing from 6 to 7. As a result of reduction potential increase during diagenesis, sphalerite microspherolites, as well as chlorite and carbonate minerals (e. g. siderite) formed and cemented by transported oxidized oolites.These sediments have been lithified due to profound diagenetic changes, which, in its turn, furthered the development of the sea bottom environment.

These facies sediments extensively accumulated in the top of Ipatovsk (figure 2–A) and Gankinsk (figure 2–C) suites throughout the ore occurrence area and occupied 15–20% of this area, in the Lulinvorsk suite basement (figure 2–D) -not more than 2%. During ore formation these sediments were the result of facial sediment replacement in associated facies. The sediments on the bay slope, as a rule, successively grade to bottom sediments.

Facies of oolite sands in mobile shallows – barriers and bars

Facies of oolite sands in mobile shallows – barriers and bars (sea depth -10–20 m) is characterized by loose hydrogoethite oolite ores, poorly- cemented ores with argillic cement, fine–grained dark gray sands. Hydrogoethite oolites have a contrasting concentrical zoned inner structure which could be the result of oscillating geochemical (pH change within the range of 4–5) and hydrodynamic formation environment conditions. The sediments formed in shallow bar conditions. Due to wave action argillic fractions were transported from the upper bar zones, resulting in sediment succession. Consequently, the bars have the following structure: upper zone is composed of loose poorly-sorted hydrogoethite oolite ores gradually grading to fine–grained sands with argillic cement or poorly cemented ores in the middle or lower zones.

These facies sediments extensively accumulated in the Lulinvorsk suite bottom throughout the ore occurrence area and occupied 15–20% of this area. These sediments formed two shallow bars in the western and eastern areas of the ore occurence, which, in its turn, are associated with loose ore-bodies.

Facies of silt clays in the shallow marine shelf

This facies belongs to the group of neritic facies (sea depth of over 40 m) and marks the transition to shallow marine environment. The sediments are gray (often greenish) parallel–layered clays and aleurolites. The clays comprise montmorillonite, nontronite, chlorites and other terrigenous minerals. During the ore sequence formation, the silty clay sediment facies extensicely accumulated in the Gankinsk (figure 2–B) and Lulinvorsk (figure 2–C) suite basements and occupied about 70–75% of the ore occurrence area. Greenish–gray clays of this facies, as a rule, superpose iron–hosting sediments, indicating the closing ore sequence formation.

The sandy–argillic–ferrous sediments of coastal bay zones gradually grade to silty clays of the shallow shelf with depth increase. Observed cross–sections of such transitions showed that this was the result of transgression and normally have sharp boundaries indicating the formation of thin gravelite lenses after the rewashing of underlying sediments.

4. Formation Model of Bakchar Ore–Hosting Sequence

The analysis of obtained data on the structure and facies features of the Bakchar ore sequence indicated the fact that the above- described environments have co-occured from the very beginning of the formation itself and changed during the transgressive–regressive cycles. Some authors have forwarded [4, 12] such a theory that well–cemented and loose ores are the products of oxidation and alteration of leptochlorides with different cements. However, it should be noted that such oxidation processes occurred during the ore formation but were predominate in the formation of hydrogeothite ores which could be proved by their geochemical differentiation [13]. Therefore, a conceptual sedimentary model is proposed to explain the formation of the Bakchar ore occurence (figure 2).

Initially, (Ipatovsk suite top (figure 2–A)) there were two silty depressions within the ore occurence area where material was transported from the land and then transported by bottom wave currents resulting in the formation of a slightly north-north-west elevation, composed of aleurolite–argillic-ferrous sediments. Hydrogoethite ores with argillic–chlorite cement were formed on the elevation slope, while hydrogoethite ores with siderite cement formed on the bay bottom. In the south-east area, as a result of sand material deposition, south-west offshore bar was formed partitioning off the stagnant depression where gray and dark gray layered clays and aleurolites were deposited.

Further, during the post-Narym wave action transgression [2] (Gankinsk suite basement) resulted in the formation of an underwater bar (figure 2–B) extending from SW to NE, where shallow–water silty clays were formed on both sides. This bar was composed of oxidized hydrogoethite–chlorite ores of early generation and glauconite sandstones.

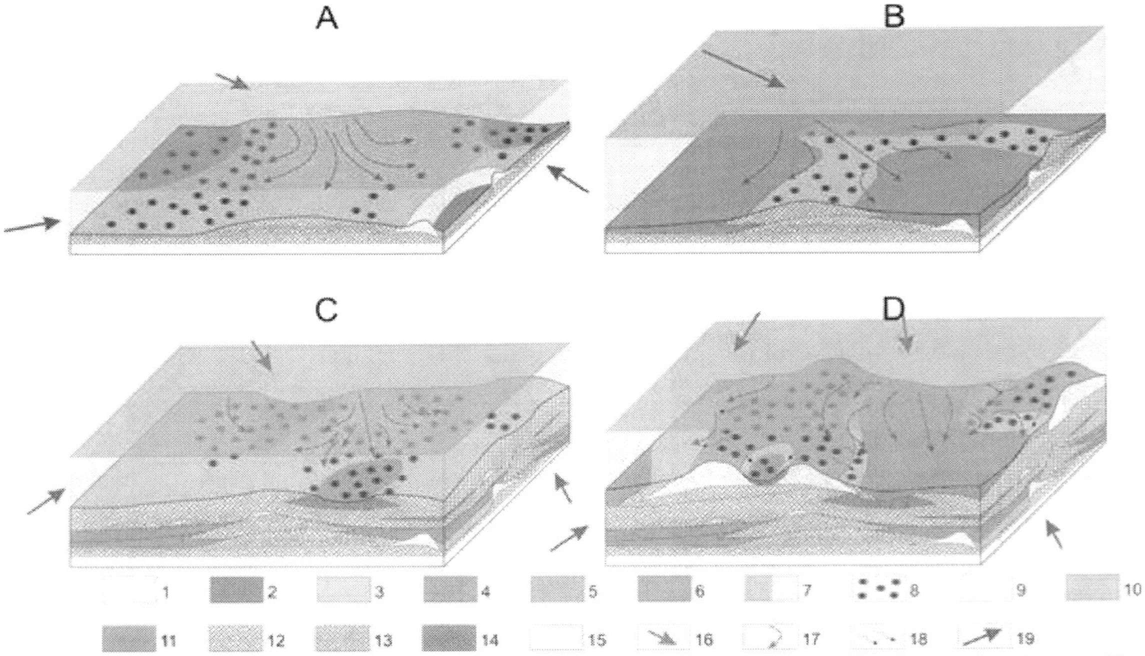

Figure 2. Conceptual sedimentation model of Bakchar iron ore occurence. A – Ipatovsk suite top; B – Gankinsk suite basement; C – Gankinsk suite top; D – Lulinvorsk suite basement; 1 –facies of

sand beach sediments in aggradation terraces and bars; 2 – facies of argillic-ferrous silts in isolated stagnant depressions; facies of sand–argillic–ferrous sediments in coastal bay zones: 3 – subfacies of aleurolite–clay–iron sediments on the bay slope; 4 – subfacies of sand–iron sediments on the bay bottom; 5 -facies of oolite sands in mobile shallows – barriers and bars; 6 - facies of silt clays in the shallow marine shelf; 7 -diagrammatic sea level; 8 - oolite ore formation area (scheme view); 9 - gray sands, silts and poorly-cemented sandstones; 10 - aleurolites and clays of argillic-ferrous silt stagnant depression facies; 11– aleurolites and clays of silt clay shallow marine shelf facies; 12 - cemented barren sandstones; 13 - cemented hydrogoethite–leptochlorite ores; 14 - closely–cemented hydrogoethite ores; 15 - loose hydrogoethite ores; 16 - sea movement direction; 17 - supposed directions of bottom waves; 18 - direction of material denudation in the bottom; 19 - direction of material input.

As a result of successive regressions (Gankinsk suite top), two silt depressions (figure 2–C) formed within the N-S ore occurence areas, where the transported ore material was found. Subfacies of sandy–ferrous sediments predominated in such depressions where ores with siderite cement were mainly formed. Chlorite–goethite oolite ores and sandstones deposited on the bottom slope.

After the following transgression, the sea advanced from the north, "heading" towards the southern bay depression, resulting in the formation of two underwater bars in consequence of dubottom wave action in the shallow zone (figure 2–D). The oolites transported from the bar sediments formed ores with siderite cement in the bay bottom, which, in its turn, graded to shallow-water clays.

After a significant post–Bakchar transgression [2], aleurite clay sediments overlapped these ore deposits after which ore genesis did not proceed.

5. Conclusion

The proposed sedimentation model does not fully exclude the existing problem, i.e. the sources of ore matter (for example, ancient crust weathering or catagen- connate waters in oil and gas bearing basins). Nevertheless, the analysis of the obtained data showed that the facies factor is predominating in determining the shape and location of ore substance. The singled out facies in the ore occurrence area only rarely develop lateral succession from the central part of the bay (silt depressions) to the coastal beach zone, which could be explained by the frequent transgression–regression cycles.

Facies of sand–argillic–ferrous sediments in coastal bay zones and facies of oolite sands in mobile shallows could be considered as a prospecting tool for oolite hydrogoethite deposits in different areas of the West–Siberian iron–ore basin.

Acknowledgements

The investigations were supported by National Research Tomsk Polytechnic University, project No.: 114-2014.

References

[1] Babin A 1969 Bakchar iron ore deposit (geology, occurrence characteristics and genesis of iron ores) *Ph.D. Dissertation*

[2] West Siberian iron ore basin 1964 *SO RAN SSS*

[3] Kondakov A 1967 Facialno-geokhimicheskie osobennosti i mehanizm oolitoobrazovaniya osadochnyh zheleznyh rud Bakcharskogo mestorozhdeniya *Ph.D. Dissertation*

[4] Nikolaeva I 1967 Bakchar oolitic iron ore deposit *SO AN SSSR* **129**

[5] Karepina K 2013 Litologo-facialnye osobennosti Bakcharskogo uzla na osnove interpretacii geofizicheskikh dannykh *Problemy geologii i osvoenija nedr: trudy XVII Mezhdunarodnogo simpoziuma imeni akademika M.A. Usova studentov i molodyh uchonyh, posvyashhennogo 150-*

letiyu so dnya rozhdeniya akademika V.A. Obrucheva i 130-letiyu akademika M. A. Usova, osnovateley Sibirskoy gorno-geologicheskoy shkoly **1** 188-190

[6] Gorjuhin E 1994 O nekotorykh geohimicheskikh kriteriyakh ustanovleniya obstanovki formirovaniya zhelezorudnykh basseynov (na primere Zapadnoy Sibiri) *Problemy Geologii Sibiri* **2** 36.

[7] Kholodov V, Nedumov R and Golubovskaya E 2012 Facies types of sedimentary iron ore deposits and their geochemical features: Communication 1. Facies groups of sedimentary ores, their lithology, and genesis *J. Lithology and Mineral Resources* **6** 447-472

[8] Kholodov V, Nedumov R and Golubovskaya E 2013 Facies types of sedimentary iron ore deposits and their geochemical features: Communication 2. Problems of the geochemistry of phanerozoic iron ores *J. Lithology and Mineral Resources.* **1** 14–47

[9] Mazurov A, Bojarko G, Emeshev V and Komarov A 2006 Prospects of Bakchar ore field development *J. Ores and metals* **2** 64–70

[10] Logvinenko N 1980 Morskaya geologiya. *Leningrad: Nedra* 343

[11] Strahov N 1963 Tipy litogeneza i ikh evolyutsiya v istorii Zemli *Moscow: Gosgeoltekhizdat* 535

[12] Assochakova E and Konovalenko S 2007 Geohimicheskie osobennosti zheleznyh rud Bakcharskogo mestorozhdeniya (Zapadnaya Sibir) *Vestnik Tomskogo gosudarstvennogo universiteta. Nauki o Zemle* **305** 219-222

[13] Rudmin M, Mazurov A and Bolsunovskaya L 2014 Mineral and elemental composition features of "Loose" oolitic ores in Bakchar iron ore cluster (Tomsk oblast) *J. IOP Conference Series: Earth and Environmental Science* **21**

IOP Conf. Series: Earth and Environmental Science **24** (2015) 012029 doi:10.1088/1755-1315/24/1/012029

Liability of the supervisor under petroleum drilling contract

A Sizov[1], K Tretyakov[2], G Boyarko[3] and I Shenderova[4]

[1,2]Department of Natural Resources Economics, Institute of Natural Resources, State Educational Institution of Higher Professional Training, National Research Tomsk Polytechnic University, Tomsk, Russia
[3] Department of Natural Resources Economics, Institute of Natural Resources, State Educational Institution of Higher Professional Training, National Research Tomsk Polytechnic University, Tomsk, Russia
[4] Department of Foreign Languages, National Research Tomsk Polytechnic University, Tomsk, Russia

E-mail: [1]Sizov_aleksey@rambler.ru, [2]Magg_1@mail.ru, [3]Gub@tpu.ru, [4]inna-shenderova@yandex.ru

Abstract. Petroleum industry supervision is not properly regulated by current legislation. This results in the lack of common vision to the provision of such kind of services, the uncertainty of the subject under contract as well as the degree of supervisor's liability for improper duty performance. The article discusses the main issues of supervisor services contract. Considering supervising specifications, place of supervisor in the system of customer-contractor relations during work realization is marked. The authors highlight the degree of supervisor organization liability and conceptually new approach to its definition.

1. Introduction
The need to improve the quality of oil and gas objects construction in the 1990s, at the dawn of Russian Federation market economy, spontaneously formed a new service market - supervising. Supervising as a phenomenon of petroleum industry is reflected in the hiring of engineers and engineering organizations to realize supervision over contractor services. At that, the relationships between the customer and the supervisor are regulated by the contract.

2. Research Methods
System approach. Supervising of well construction is studied from the standpoint of its varieties. The authors also conduct the research of supervising services in whole as a collective set.
Method of comparison. Petroleum supervising is studied in comparison with the current methods of well construction monitoring.

3. Subject of supervisor service provision under the contract
Today, supervising as a category has consolidated in the petroleum industry as self-existing economic-productive phenomena, covering a variety of activities. At this, supervising service according to its definition is common industry "slang". This is explained by the fact that during 20 years of supervising existence there was no sufficient legal and regulatory framework to manage these relations. There is one intradepartmental document of the Federal Agency for Subsoil Use - Order of 18th April 2006 "About organization of technical-technological and methodological instruction of oil and gas geological exploration, carried out at the expense of the Federal Budget," which regulates certain functions of supervisors as representatives of the client organization and their basic liabilities. However, such notion as well construction and workover supervising is not protected by current law and there is no list of supervisor services.

Supervising in petroleum engineering, as productive-economic phenomenon, has established its own system of production and economic relations. This reflects in the fact that supervisors could be on the one hand the experts working at the customer's enterprise and performing the control and supervision over the contractor. In this case, the liability of such specialists is primarily determined by Employment Regulations. On the other hand, specially attracted independent engineers or engineering organizations could realize supervisor functions. In this situation, the liability of such experts will be determined by the rules of the Civil Code of Russian Federation and contract.

Existing law does not directly regulate this kind of contracts, but at the same time allows making contracts for the provision of supervisor services based on the principle of contractual freedom [1], thus the terms of the contract will be determined at the discretion of the parties, except the cases when the contract terms are regulated by law or other legislation. [2]

The basis of the contractual relationship between the construction customer, the contractor and the supervisor is a provision of the Civil Code of Russian Federation, according to which the customer in order to realize the control and supervision over the construction work as well as to make decisions, can sign a contract for the provision of services with the relevant engineer (engineering company) without contractor's agreement. In this case, the construction contract regulates the liability of the engineer (engineering organization). [3]

The implementation of Russian Federation legislation forms a system of contractual relations including: construction contract signed between the customer and the construction contractor; contract for the provision of supervisor services signed between the customer and the construction supervisor. It should also be mentioned that there is no direct relationship between the contractor and the supervisor, their relations are only regulated by liabilities of construction contract.

The structure of supervising service is complex. Whereas there is an alliance of civil matters (two independent relations) - implementation of contract work and supervisor services in one productive-economic phenomenon, where the realization of construction works by the contractor will become a subject of supervisor's control. Realization of contract for the provision of supervisor services is impossible without contract relations. For the construction customer, the contract for the provision of supervisor services will be an independent subsidiary contract interrelated with construction contract.

Subject of the contract for the provision of supervisor services, based on the current practice of contractual relations could be defined according to the existing forms of organization supervisor services as: technical and technological supervision and construction management of oil and gas wells. Each of these forms has its own characteristics [4].

4. The uniqueness of the supervising contract subject

Supervisor service as a form of technical-technological activity has its own uniqueness and fundamental differences from the other types of production processes monitoring. Supervisor in accordance with the civil contract provides independent supervision over the construction technology, realizes final and intermediate control over the quality of contractor's work.

The main difference between supervisor's control and field supervision lies in competences. Supervisor is not the designer of project documentation, while the field supervision could be provided only by the developer of the project or by the representative of the design organization. By definition, field supervision is one of the types of services considering the project documentation developers (individuals and legal entities) supervision over the compliance of the construction works to the construction documents. Liabilities of a person realizing field supervision are as follows: monitoring the construction works compliance to the documentation and requirements of construction regulations; control over the quality and work production technology considering the strength, stability, durability of the construction and process of engineering equipment installation [5]. Supervising control is expressed in constant supervision of construction technology as a process of construction carried out by the contractor and other involved individuals. This activity is the principal sphere of supervisor's responsibility. At that, supervisor may also realize field supervision, considering customer needs, established by contract relations.

Supervisor control also differs from the construction control realized by the project developer. Project developer construction control is:

- to check that the contractor has all necessary documents (certificates in certain cases) confirming the quality of the applied materials, tools and equipment, results of entry material control and laboratory testing;

- to monitor the rules of warehousing and storage of the applied materials, tools and equipment; if violations occur a representative of technical supervision could prohibit the usage of improperly stored materials;
- to control the compliance of the executed works to the operating control requirements;
- to control correctness of the contractor's documentation, including the geodetic schemes fairness of random element accuracy control placement assessment;
- to eliminate the deficiency in project documents revealed during construction works, to return incorrect project documentation, to control and receive documented approval of the revised documentation, to give documentation to the contractor;
- to control the execution of State supervisory body and local authority requirement;
- to notify State supervisory body about all the emergency conditions at the construction site;
- to control the compliance of construction volumes and time to the contract terms and the schedule of construction;
- to evaluate (together with the contractor) the relevance of executed work, constructions, areas of utilities, signing of bilateral acts confirming the relevance; to monitor the contractor over the realization of requirements concerning prohibition of work execution before signing the act;
- to realize final evaluation (together with the contractor) of construction object in order to define its local legislation and project documentation relevance [6].

Supervising control can cover both individual and all project developer liabilities towards construction control established by independent contractual relations. Supervisor fulfilling field supervision at the construction object, represents the customer, protects its interests, while the construction supervision is organized on the basis of internal customer service for realization of production and operational control over the contractor's work. Internal supervisor is not always an independent specialist. There is a dependency from the authorities, targeted to key performance indicators. This can lead to the concealing of process deviations and irregularities committed in the course of construction.

Thus, due to the character of construction works, as well as the features established by legislation concerning supervision of work realization at the construction object, supervising can be considered as an alliance of all existing individual control elements that may be significantly expanded on a contractual basis due to the customer's needs.

5. Contract liability of the supervisor

The question, concerning the degree of service company liability in the system of customer-contractor relationship, arises on the basis of the supervisor services contract subject. At that the liabilities of the supervisor should be clearly noted in the contract and cover the specificity of the main production risks.
The following risks could be referred to the main customer's risks:
1) Project risks: non-compliance of contractors work to the cost estimating documents;
2) Presence of falsified and counterfeit materials nonconforming inventory project documentation;
3) Non-compliance of contractor's production technology to regulatory and legal framework;
4) Information risks: delay in notification, concealing of violations described above, lack of relevant information concerning the progress of work;
5) Failure to comply the rules of labor safety as well as the environment protection at the customer's site.

The specificity of above mentioned risks lies in the fact that the processes of well production (construction, workover) are of probabilistic nature and random character. Omissions could be sometimes detected only after a considerable period of time. Liability of contractor towards the customer is strictly regulated by the Civil Code of Russian Federation. The Contractor bears full liability in case any deficiencies or construction defects occur and ensures work realization in accordance with the requirements of design documentation, within 5 years of the warranty period or other specified terms of the contract between the contractor and the customer. [7]

The current tendency of contractual relations is the design of contract attachments in the form of work quality violation list and contractor's liabilities for such violations. At that, it is obligatory to attach the justification of decreasing the quality of work, namely the approval of the document confirming the fact of quality decrease. For example, in case the well fails to get in to the target area considering the tolerance range of the permissible instrumental error up to 10 m, the contractor could burden 50% of the well cost. Geophysical data will be the justification. The additional term to contractor's organization in the context of fines excess of more than 20% of the total well construction cost, could be the liability considering redrilling of wells at the contractor's expense.

Late detection of deviations associated with the work quality decrease is the basic difficulty in determining the degree of supervisor organization liability directed to the prevention of such defects. As it has been previously noted, Russian Federation legislation is lack of regulations concerning liabilities of supervisor organizations towards the customer and the contractor, so these aspects should be clearly defined in the contractual relations, as the degree of supervisor liability is the main factor motivating the proper execution of supervisor duties.

The degree of liability directly depends from the supervisor duties, that is the liability of service company at various forms of supervising will be different. Thus the main purpose of supervisor organization in the form of technical and technological control is the detection of deviations and defects that influence the quality of construction realized for the customer. At that the supervisor is not a co-executor of the contract and therefore bears secondary liability for the result of this construction. In other words, the supervisor will only be liable for the delay or undetermined violations of technology or other factors affecting further performance of the construction contract. The main method of providing liability under supervisor services contract is penalty. In this case the customer is not obliged to prove the amount of losses, despite the degree of supervisor organization fault. Customer's financial damages will be covered by the contracting organization [8]. Normally the amount of fine is 0.1% of the contract value for the improper performance of obligations. Fines do not free the supervisor from the recovery of financial damages in case supervisor's fault is proved.

In the case when supervisor is attracted as a construction manager, supervisor becomes the subject and a direct participant of the construction. That is, the supervisor becomes the executor under the contract. Supervisor's actions are regulated by liability rules of the contractor towards the customer. The liabilities become equal to that of the contractor, at that the supervisor is not liable to choose the contractors and subcontractors this is the prerogative of the customer. It is necessary to set the financial liability of supervisor organization for the construction result, since the whole process of construction should be under supervisor control under equal liabilities of general contractor and supervisor. The fine is a method ensuring the quality of supervisor organization duties performance. However, the fine could not always cover customer losses caused by faulty contract work performance. In this case, the terms of financial damage cover must be established in the supervisor services contract (in accordance with Civil Code, Article № 394).

The presence of such terms in the contract means full supervisor's financial liability for the result the work. A difficult question arises - how to determine the degree of supervisor organization fault for undetected violations, which resulted in significant losses for the customer. For example, geophysical data indicate that the contractor performed faulty run-in-hole operations, at that the inability of further well operation is proved, but the supervisor failed in noticing that. Such serious violation of construction technology leads to redrilling of wells at the expense of the accused party. The degree of supervisor's fault from the financial point of view is difficult to determine, as both parties are guilty - the contractor for procedural violations and the supervisor for being not attentive.

Therefore the following economic criteria in determining the degree of supervisor organization liability is offered, considering the establishment of supervisor organization liability towards the customer losses relating the construction contract value to the value of supervisor services contract. At such approach, the supervisor financial liability could be expressed in the range of up to 10% from the total cost of well construction, depending on the specificity of supervisor services. The penalties for the customer will not exceed the cost of supervisor services contract. The contractor is proposed to bear the rest of uncovered losses. Thus the contractor and the supervisor will bear equal liability towards the customer. However, this approach could be of negative consequences in terms of market relations development for the attraction of supervisor services. Revealing of such cases, could lead to the unprofitability of supervisor organizations, since not all organizations could bear such liability. Therefore the amount of liability could be adjusted by establishing the decreasing coefficient related to the priority of works performed by the contractor and the complexity of faults and deviations detection during the construction.

6. Conclusion

Lack of regulatory acts in Russian Federation, results in the uncertainty concerning the supervising role in the process of oil and gas wells construction and workover. The uniqueness and the fundamental difference between supervisor services lies in combination of existing oil and gas wells quality construction and workover control in Russian Federation. At that, the range of supervisor organization liabilities in implementing service management and construction control, could be significantly extended, based on the specifics of work performance and customer needs. Liabilities of supervisor organizations toward the customer are established by contractual relations. The main difficulty that arises is to determine the degree of supervisor organization liability for the performance of work under the contract. In various forms of supervising the degree of responsibility will be different. At realizing of technical and technological supervision the method of providing liability under supervisor services contract is penalty, yet the customer is not obliged to prove the amount of losses for undetected defects and failures in production technology. In case when supervisor is attracted as a construction manager, supervisor becomes the subject and a direct participant of the construction. This means that supervisor bears equal liability together with the contractor. At that an additional liability of supervisor toward the customer is offered – recovery of financial damages not covered by defective construction fine. The economic criteria in determining the degree of supervisor organization liability is offered, namely the establishing of supervisor organization liability toward customer losses relating the construction contract value to the value of supervisor services contract.

References

[1] Article 1. Civil Code of Russian Federation
[2] Article 421. Civil Code of Russian Federation
[3] Article 749. Civil Code of Russian Federation
[4] Sizov A V, Boyarko G Y and Shenderova I V 2014 Development of well construction and workover supervising in Russian Federation [Electronic resources] *IOP Conference Series: Earth and Environmental Science* Vol. **21** 1-5 Mode of access: http://iopscience.iop.org/1755-1315/21/1
[5] SNiP (Construction norms and regulations) «Construction management» approved by Gosstroy (Russian State Committee for Construction), 10th June 1999, №44
[6] SNiP (Construction norms and regulations) 12-01-2004 «Construction management» approved by Gosstroy (Russian State Committee for Construction), 19th April 2004, № 70
[7] Article 756. Civil Code of Russian Federation
[8] Shirenkova E A 2010 Legal aspects of supervising services *Drilling and Oil* **9** 52-55.

IOP Conf. Series: Earth and Environmental Science **24** (2015) 012030 doi:10.1088/1755-1315/24/1/012030

Modeling of tunneling-induced ground surface movement

L A Strokova

Tomsk Polytechnic University, 30 Lenin Avenue, Tomsk, 634050, Russia,

E-mail: sla@tpu.ru

Abstract. Surface subsidence control during tunnelling in urban areas is required to minimize any disturbance to nearby buildings and services. The main focus of previous studies concerning surface subsidence was basically on empirical solutions derived from field research. The number of analytical studies which have been carried out is not enough because they do not consider complex ground conditions; hence, it is required to model the effect of surface subsidence caused by tunnelling based on comprehensive analytical solution combined with numerical modelling. This paper presents the results of tunneling-induced surface settlement modelling based on finite element method. The effect of the overconsolidation ratio of soils expressed in terms of the coefficient of earth pressure at rest (K_0) on surface subsidence caused by tunnelling is investigated. It is demonstrated that value of K_0 is an essential factor affecting accuracy of consolidation settlement estimate in overconsolidated clays.

1. Introduction

Ground surface settlement caused by tunnelling in soft ground is one of the basic issues in all aspects of tunnel design. Surface subsidence can be caused by several factors such as ground loss at the tunnel face, behind the tail of the shield and through the tunnel support or linings. Other factors include the consolidation of the soil due to reduction of ground water level. Besides surface settlement, tunneling also produces lateral deformation of the ground and longitudinal movement of the ground ahead of the tunnel face. To control the effects of deformation various field studies and experiences are carried out, empirical and numeric methods are developed.

The semi-empirical approach described in many references [8] is applied to estimate the size and shape of the settlement trough at a “greenfield site”. These solutions are based on field observations. The equations used are based on the assumption that the settlement profile above a single tunnel is of normal probability or Gaussian form. To define the surface settlement trough the following parameters are used: volume loss V_s, point of inflexion i and maximum surface settlement S_{max} the settlement profile. The Gaussian curve is used at all levels in the ground above the tunnel. For the combined effect of multiple tunnels, the movements induced by each tunnel are simply added.

Numerical methods have become much more common in the analysis of excavations in urban environments. Such a numerical technique as FEM analyzes heterogeneous ground layers by means of more sophisticated constitutive models, involving time dependent effects, initial and boundary conditions which are similar to the actual field ones. They are particularly effective for the study of tunnels excavated in grounds that can be modelled as continuous media, with due account of non-linear behaviours, as well as complex staging and geometrical conditions.

It is important to select constitutive models for the soil capable of reproducing field behaviour. Linear elastic soil models usually give trough widths that are too wide. Most of the numerical simulations of tunnel excavation are performed with linear-elastic perfectly-plastic models without the influence of the nonlinear ground behaviour. Stiffness is the same for loading, unloading and reloading. Models based on hardening plasticity involving kinematic hardening are appropriate, as they can model the nonlinear elasticity at small strains, the effect of stress history and cyclic loading [9]. One of the great benefits of numerical analysis is that it enables to incorporate adjacent influences of surface structures or tunnels. It is also possible to model the effects of compensation grouting to protect surface structures during tunnelling projects. Numerical modelling is often carried out to get back-analysis of experimental data, to verify the input parameters that had been assumed in the planning and design stages.

As it is known, accurate determination of the in situ stress is an important issue in the numerical simulation of many geotechnical problems. The determination of lateral earth pressures is absolutely necessary at engineering structure design and application stages. Particularly, the knowledge of lateral earth pressure is required in tunnel, retaining structure and deep foundation design. Tunneling in soft ground always leads to stress rearrangements and deformations in the surrounding soil. The deformations can be detected by the evaluation of measurement data available from previous tunnel project sites – the empirical prediction – or by numerical calculations. Vertical stresses can be easily determined on the basis of depths, densities and groundwater data, but to calculate horizontal stresses is more complicated task. They usually have to be estimated using empirical values of K_0, the coefficient of earth pressure 'at rest' under conditions of zero lateral strain. The coefficient K_0 is the ratio of in situ horizontal effective stress to the vertical effective stress. K_0 is determined by field or laboratory tests. The coefficient of earth pressure at rest (K_0) changes depending on the relative density, stress history, over-consolidation ratio, plasticity index and similar soil properties. The over consolidated soil is a normal soil which has suffered a bigger load in the ancient geological time than nowadays.

According to the theory of elasticity, $K_0=\nu/(1-\nu)$, where parameter ν – Poisson ratio, taken often according to recommendations. When applying the method of final elements with using of elasto-plastic models, the calculations should be based on the laboratory and computing experiments data concerning model input parameters.

2. Experimental design

To provide a tool to improve the state-of-the-practice of controlling ground movements associated with tunnelling, this paper presents a numerical procedure for tunnelling that effectively and efficiently updates design predictions of surface settlements during construction. Several authors have studied the influence of K_0 in numerical tunnel analysis on surface settlements, e.g. Gunn (1993), Addenbrooke (1997), Franzius (2005), and stated that the settlements trough becomes wider and less deep with increasing K_0-value. In the present studies the non-linear elastoplastic Hardening Soil Model was applied to investigate the influence of K_0 on surface settlement and find parameters K_0 and OCR for modelling deformations around tunnels in Munich.

The city of Munich forms the core of a fast growing urban region in the Alpine foreland of Bavaria, in southern Germany. The subsurface is composed of Neogene and Quaternary formations made up of loose alluvial, fine- to coarse-grained sediments. Today's landscape is known as the Munich gravel plain, which comprises sand terraces formed during the Pleistocene glacial periods, as well as the modern floodplain of the river Isar. Due to major unconformity and hiatus, these rather young gravel terraces overlie Neogene Molasse deposits of the Alpine foreland basin, which, in contrast, involve fine-grained fluviatile and lacustrine facies.

The subsoil in Munich has the following geological characteristics: uppermost Quaternary sediments comprise thick and extremely thick interglacial layers of various ages. Their thickness ranges from 4 to 7 m, whereas, in the zone of superimposed lower and upper terraces. it ranges from 8 to 25 m. These sediments are underlied by soft late-Tertiary sandstone known as "flinz". The layer composition varies significantly: thick fine to medium micaceous sand layers, practically impermeable clay marl layers with high variable monaxial compressibility of about 50 kN/m^2 for plastic lime-free clays and up to 6000 kN/m^2 for marl. The average compressibility value is about500 kN/m^2. In the multi -Tertiary layers there is often a regular sandwich system within the soil layers. The layers have a wavy profile and sudden indentations can be observed in erosion trenches.

The Munich U-Bahn began operation in 1971. The Munich U-Bahn system currently comprises eight lines, serving 96 stations, and encompassing 95 kilometres of routes. The inverse analysis is carried out using data from a tunnel U-8 in Munich. The tunnel is approximately 14 m below the ground surface, with diameter being 7 m. The settlement parameters are modelled by 15-noded triangular elements in a plane strain mesh (Plaxis, Brinkgreve et al. 2004). Deformations of the tunnel and its overburden were studied using two material models: the hardening soil model (HS) and the hardening soil model with small- strain stiffness (HSS initial stress conditions were geostatic). Inclinometer data are used as observations in the inverse analysis that calibrates the numerical model in an objective way, the engineering judgments being made during the construction of the tunnel.

In this study, effects of the coefficient earth pressure at rest and over-consolidation ratio were investigated. There are different methods to determine the preconsolidation stress, σ'p, from laboratory oedometer data. Casagrande (1936) developed the most commonly used method and it is that method that was used. The table 1 shows the results of OCR estimation based on compression tests for neogene molasse deposits.

Table 1. OCR estimation based on compression tests

Sample No.	Sampling depth, [m]	σ_z', [kPa]	Preconsolidation Stress , σ_{vm}', [kPa]	*OCR*, [-]	Calculated mark of a surface, [m]
01-60221	24,8	356,8	1850	5,2	124
02-60189	28	392	1675	4,2	107
03-60193	32,6	442,6	2000	4,5	130
04-a-60194	33,4	451,4	2075	4,6	135
04-b-60194	33,4	451,4	2375	5,3	160
05-60195	35,5	474,5	1925	4,1	121
06-60477	23,5	342,5	1200	3,5	71
07-60479	32,1	437,1	2100	4,8	139
08-60482	35,8	478,4	1975	4,1	125
09-60494	28,5	356,7	2325	6,6	166
10-62151	19,5	250,5	1650	6,6	117
16-61741	25,8	319,8	1750	5,5	119
17-62170	20	292	2400	8,22	179
19-a-60052	33,6	411,6	1850	4,5	120
19-b-60052	33,6	411,6	2075	5	139
20-60146	23	373	1875	5	125

In general, the data confirm the statements of geologists about an erosion of deposits in the Quaternary Period through the reconstruction of the alpine paleorelief.

Based on the various laboratory and field test results, many methods have been developed to determine the coefficient at earth pressure at rest of different soil types. The determination of earth pressure coefficient at rest by using laboratory methods has been preferred because of the simplicity of the laboratory tests compared to field ones.

Currently, K_0-consolidation tests can be performed using an oedometer or a triaxial apparatus. Since the oedometer test has the disadvantages of unknown side friction as well as difficulties in measuring the radial stress, the triaxial apparatus more fully satisfies the K_0-condition. Recent advances in control techniques and equipment encourage the easy use of triaxial K_0 -consolidation test.

Triaxial tests have been performed to determine the model parameters and provide data for validation of the model to predict the behaviour of soil. The results of estimation of coefficient of earth pressure at rest (K_0) at K_0 consolidation for 20 samples of neogene molasse clay soil are given in fig. 1, 2. K_0 values that were obtained from empirical equations and values obtained from laboratory tests were compared.

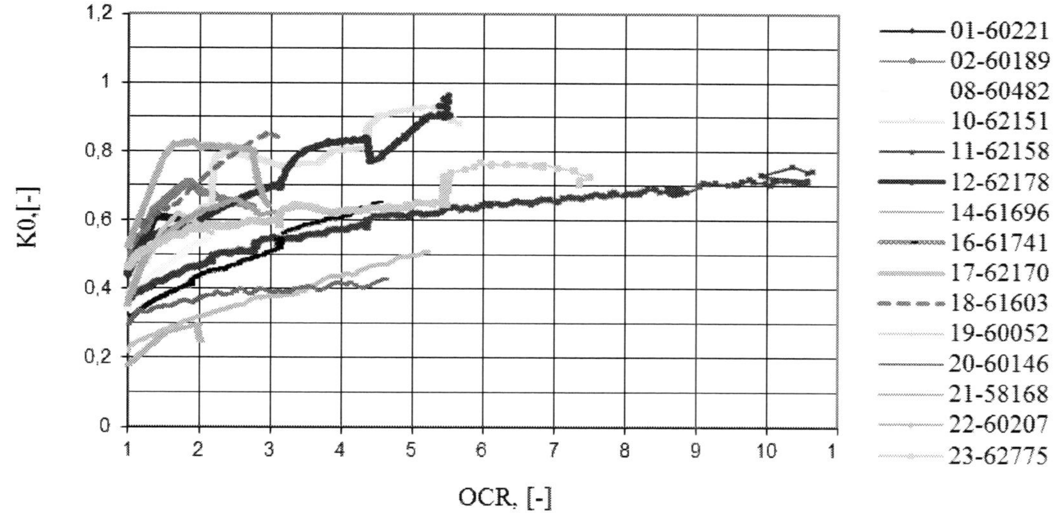

Figure 1. Variation of K_0 with OCR on unloading

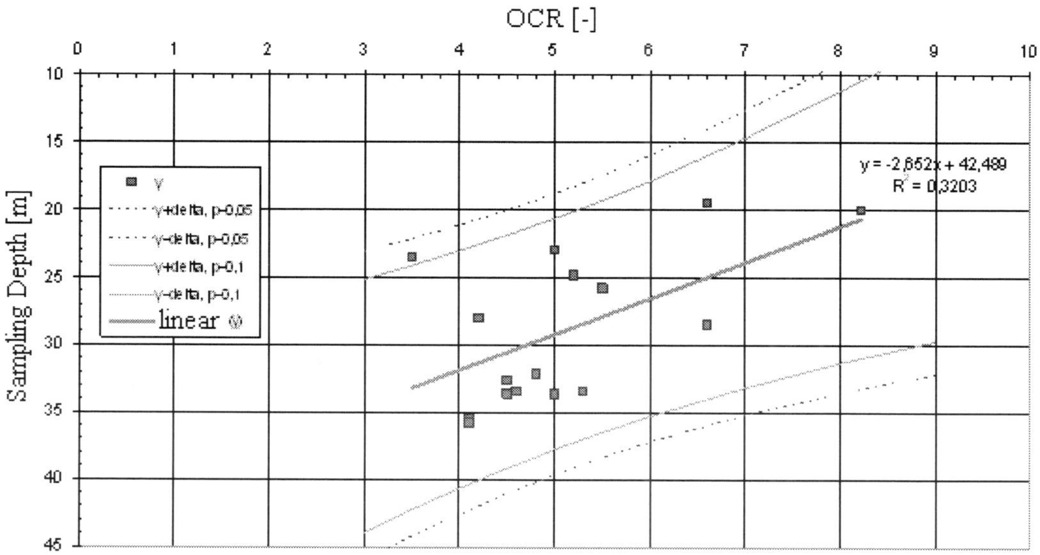

Figure 2. The confidence bands for OCR vs depth with confidence levels of 90 % and 95 %

Figure 3 shows the transverse settlement curves to demonstrate the influence of K_0 on the shape. The different surface settlement curves have been obtained for K_0=0.6, 0.8 and 1.0 respectively. A tremendous influence of K_0 on the magnitude of the maximum steady-state settlement is observed. The trough becomes shallower and wider with increasing horizontal initial stress. K_0=1 surface heave curves show that heave is primarily concentrated over the tunnel center line.

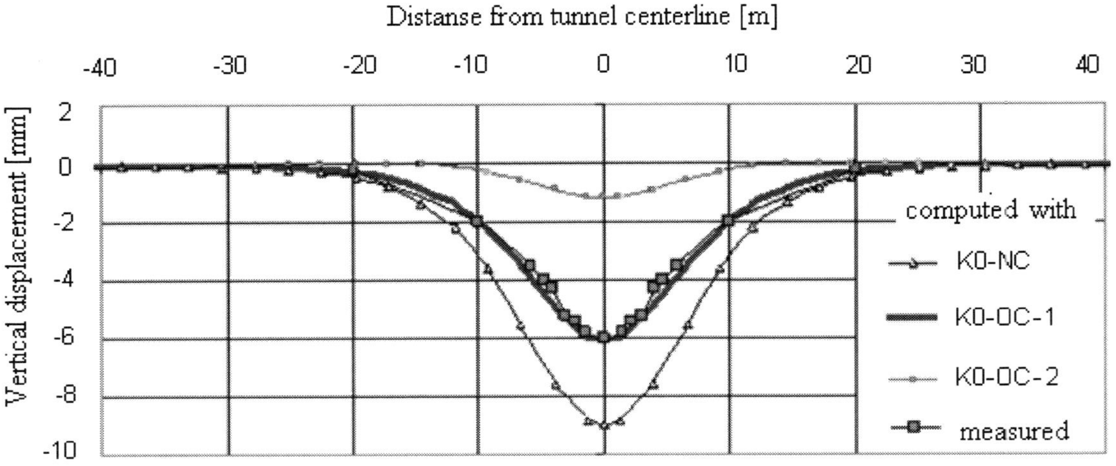

Figure 3. Influence of the K_0 value of soil on the calculated surface settlement

The construction of twin tunnels is a common requirement for underground railways and there are equations to make preliminary ground movement predictions. In practice, the tunnels will rarely be driven simultaneously and one tunnel is likely to be excavated before the other. In some cases this will give rise to an asymmetry which is not modelled by the equations.

It is generally assumed that the predicted ground movements for each tunnel can be superimposed. In relation to twin tunnels with reduced transversal distance between the axes, this assumption may be unconservative. The disturbance caused by the first tunnel drive can be simulated by assuming a greater volume loss for the second bore and superimposing the resulting ground movements.

Figure 4 shows side-by-side tunnels, the superimposed curve almost exactly matches the numerical results.

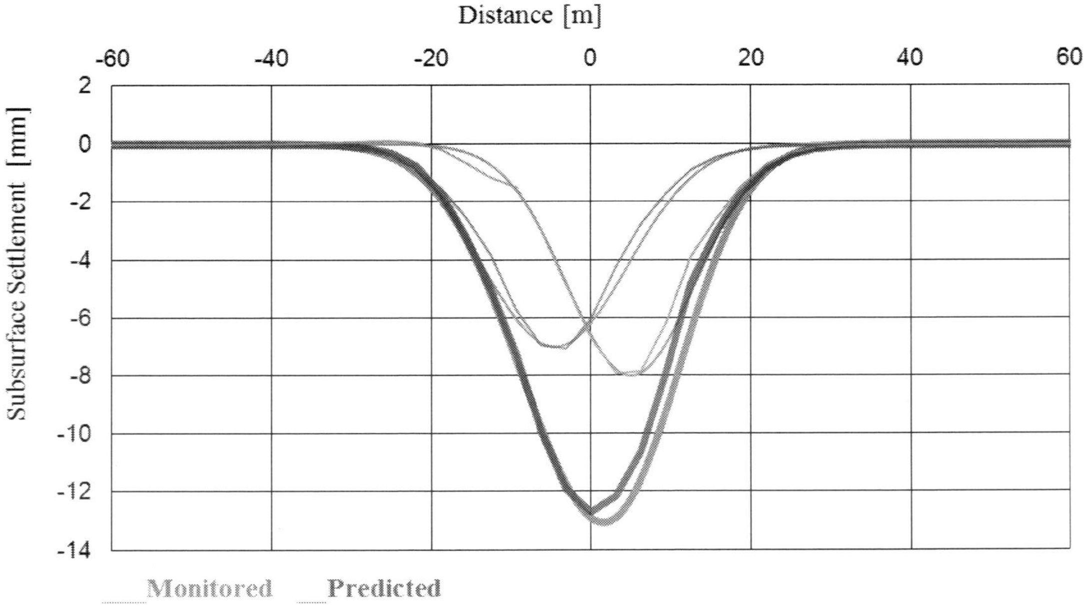

Figure 4. Predicted and observed ground settlement profile for side by side tunnels

K_0 coefficient variations are as follows: for sand 0,426; 0,6 and 0,8; for clay 0,576; 0,8 and 1,0. Vertical movements data concerning each settlement model from the PLAXIS program were taken to compare a calculated subsidence with different K_0 values and measured surface subsidence. As illustrated in fig. 2, the best coincidence with measured subsidence have settlement curve computed with K_0-OC-1 (for molasse sand – 0,6, for clay – 0,8).

Summary

The surface settlements appear to be sensitive to the K_0-value. For geotechnical calculations, K_0-value for overconsolidated soils - 0,6 (sand) and 0,8 (clay) are proposed.

1) The variation of horizontal stresses is linear. The vertical stresses are greater than pre-consolidation pressure in loading stage. Therefore, K_0 values are considered as constant for vertical stresses which are greater than pre-consolidation pressure.

(2) It was observed that experimental K_0 values are greater than K_0 values calculated from empirical equations for normally consolidated state. However, experimental K_0 values are lower than calculated K_0 values for over consolidated state. K_0 values calculated for low over consolidation ratio values are close to test results.

(3) In the unloading stage, horizontal stresses decreased slower than vertical stress, that's why K_0 values increased. Therefore, K_0 values are considerably affected by over consolidation ratio in unloading stage. Since over consolidation ratio increases, coefficient of earth pressure at rest also increases.

Thus, the application of the special software allows engineers not only to reduce labor input but also significantly increase the reliability of the obtained results, which in its turn could contribute to design efficiency increase.

Acknowledgments

The author is very thankful to the faculty members of Zentrum Geotechnics of Muenchen Technical University, in particular, G.Pelz who assisted to perform laboratory tests and Plaxis-calculations presented in this paper.

References

[1] Addenbrooke T, Potts D and Puzrin A 1997. The Influence of Pre-failure Soil Stiffness on the Numerical Analysis of Tunnel Construction *Geotechnique* **47**(3) 693

[2] Brinkgreve RBJ et.al., 1997 PLAXIS 2D Version 8 *Balkema* 200 p.

[3] Casagrande A 1936. The Determination of the Pre-consolidation Load and its Practical Significance *Proceedings of the 1st Intern. Soil Mechanics and Foundation Engineering Conference,* Cambridge, Mass., 22–26 June 1936. Edited by A. Casagrande. 3: 60–64.

[4] Franzius JN, Potts DM and Burland JB 2005 The Influence of Soil Anisotropy and K0 on Ground Surface Movements Resulting from Tunnel Excavation *Geotechnique* **55**(3) 189

[5] Gunn MJ 1993. The Prediction of Surface Settlement Profiles Due to Tunnelling. In Predictive soil mechanics: *Proceedings of the Worth Memorial Symposium*, London, p 304.

[6] Jâky J 1944 A Nyugalmi Nyomâs Tényezöje (The Coefficient of Earth Pressure at Rest) // Magyar Mérnok és Epitész Egylet Közlönye (*Journal for Society of Hungarian Architects and Engineers*) October 355.

[7] Stallebrass SE and Taylor RN 1997 Prediction of ground movements in overconsolidated clay *Geotechnique* **47**(2) 235

[8] Strokova LA 2010 Methods of Estimating Surface of Tunneling Effects on Existing Buildings *Soil Mechanics and Foundation Engineering* **47** (3) 92

[9] Strokova LA 2009 Numerical Model of Surface Subsidence During Subway Tunneling *Soil Mechanics and Foundation Engineering* **46** (3) 117

[10] Strokova LA 2013. Effect of the Overconsolidation Ratio of Soils in Surface Settlements Due to Tunneling *Sciences in Cold and Arid Regions*, **5** (5) 0637

IOP Conf. Series: Earth and Environmental Science **24** (2015) 012031 doi:10.1088/1755-1315/24/1/012031

Litho-facies factors of Late Jurassic productive sediments in Myldzhino gas-condensate field (Tomsk Oblast)

T Ten [1], E Panova[2] and R Abramova[3]

[1, 3] National Research Tomsk Polytechnic University,
30 Lenin Avenue, 634050, Tomsk, Russia
[2]TomskNIPIneft, 72 Mira Avenue, 634027, Tomsk, Russia

E-mail: [1] ten@tpu.ru; [2]elenapanova@tpu.ru

Abstract. Research results of the lithological composition for Upper Jurassic productive sediments in Myldzhino gas condensate field (Tomsk Oblast) have been described. System lithological studies have been based on the layer-by-layer profile description. Recently electrometrical research methods for facies analysis and litho-facial interpretation have been applied. Both lithological and field geophysics data provided problem- solving related to stripping conditions, structure and lithological trap locations.

1. Introduction
Myldzhino gas condensate field is located in Kargasok region (Tomsk Oblast) about 450 km north-west from Tomsk. Geological field profile is comprised of Paleozoic basement sediments, Mesozoic-Cenozoic mantle sediments. The sedimentation composition involved Jurassic, Cretaceous, Paleogene, Neogene and Quaternary sediments. Commercial oil-gas – bearing capacity has been identified in Jurassic and Lower Cretaceous sediments.

2. Geological setting
Upper Vasyugan subsuite J_1 horizon is traditionally divided into undercoal, intercoal and overcoal sequences embracing productive arenaceous layers [1, 2]. The research target involved undercoal sediment thickness which was comprised of two sections. The lower section with variable thickness and lithological composition, indexed as J_1^4, is characterized by two sequence types – aleurite-arenaceous and clay-aleurites. The upper section included mainly thick arenaceous member overlayered (in most studied profiles) by a 1m thick coaly layer and indexed as Lower J_1^3 N.

The horizon sediments formed in various facies coastal –marine environments of the sedimentary basin that resulted in the lithological composition variation. Numerous studies have defined that the undercoal sequence rock formation occurred under conditions of prevailing marine regression and areal rugged topography [3].

3. Results and discussion
To identify the specific distribution features of various lithological rock compositions within the field, pinching out nature and facies variation analysis of sediments, a series of thematic maps were plotted, including lithological composition maps, lithologic-facies maps, etc. The lithological composition of

cyclite J_1^4 is shown on the map characterizing the distribution of diverse rock composition, whereas the sediments comprise arenaceous, aleurolitic and clayey rocks (figure 1).

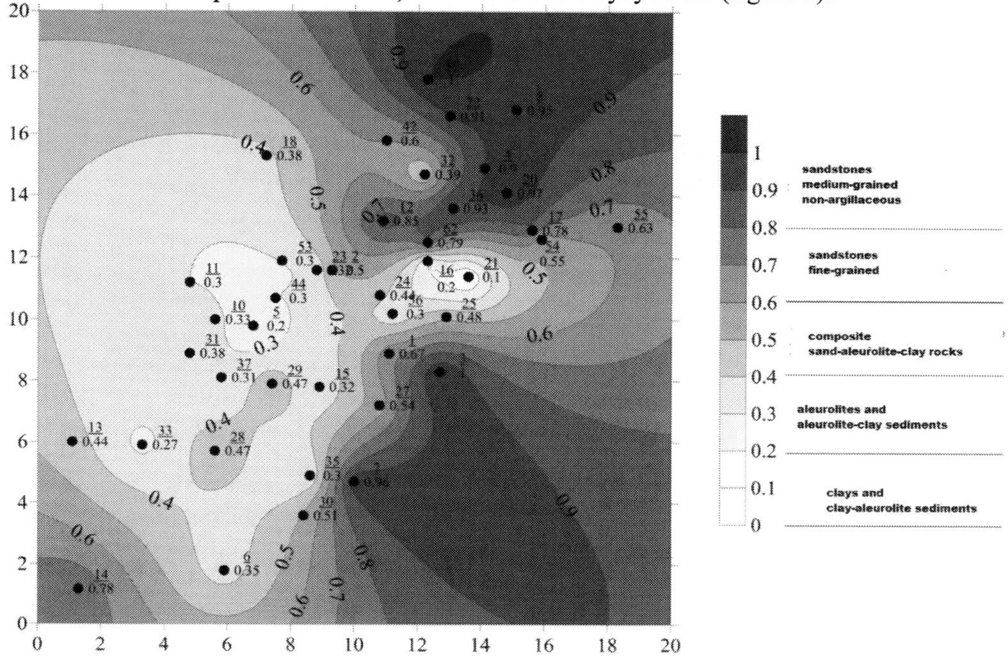

Figure 1. Cyclite J_1^4 lithological composition map

According to well logging and core data two J_1^4 sequence types were identified. The first sequence is predominately characterized by arenaceous sediments, which developed in the NE (wells № 38, 22, 42, 8, 4, 36, 12, 62), SE (wells № 1, 3, 27, 7, 30) and SW (well №14) areas. The second J_1^4 sequence type is well- developed and characterized by clayey-aleurite interbedding. The gross sand ratio of J_1^4 within the distribution zone varies from 5.9 to 78.9.

The facies map of investigated stratigraphic units was plotted on the basis of facies analysis data reflecting the distribution of stratigraphic time-sediment span types and the genetic interpretation of their accumulation environment.

Although the formation of cyclite J_1^4 is correlated to the beginning of the sedimentation cycle regression, the inherited morphology features of Nizhnevasyugan sedimentation basin bottom could be observed.

In this case, the bottom relief is more uplifted towards the east which is governed by arenaceous, aleurite-arenaceous accumulations and aleurite sediments which are mostly typical for uplifting-sloping facies. Arenaceous-aleurite-clayey and clayey-aleurite facies sediments are widely distributed throughout the studied area.

The integrated analysis of both textural-structural features and electrometric cross-section characteristics of studied sediments made it possible to identify different hydrodynamic regime zones in the sedimentation basin and corresponding facies. The distribution zones of arenaceous sediments are confined to areas of highly intensive and intensive (I and II levels) hydrodynamic activity (figure 2). These zones are located in the outermost SW (well №14), SE (wells №7, 3) and NE (wells №38, 22, 8, 4, 36, 62) areas. Maximal SP curve deviation is typical for anomaly apex, reflecting the dynamic activity of water environment in the final sedimentation stage of this zone.

Arenaceous-aleurite rock distribution zones which formed during the medium (III) hydrodynamic sedimentation level are confined to sloping uplift areas and are localized (wells № 42, 1, 27, 13).

Environment dynamic regime for these zones is less intensive than in the areas of II regime that resulted in a smaller DP curve amplitude deviation in case of formal shape resemblance.

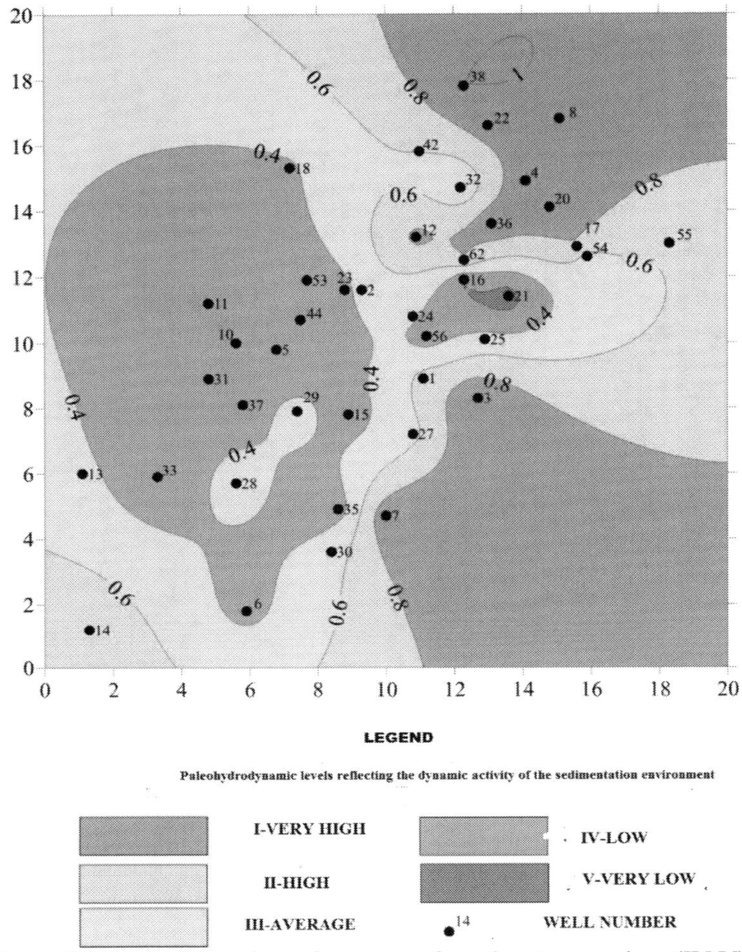

Figure 2. Cyclite J_1^4 lithologo-facial map

The fourth sedimentation environment regime is characterized by low hydrodynamic activity showing extensive distribution. SP curve, having explicit regressive character, indicates even smaller amplitude deviation. Clayey-aleurite sediments accumulated within this zone, apparently in the near-crestal uplift section. In the setting of overall marine basin regression this is characteristic for uplifted overgrowth areas and the decrease of the hydrodynamic environment activity. The localized distribution involves alternating hydrodynamic regime zones from low to very low (IV-V) and these accumulation zones are basically argillaceous sediments (wells № 33, 53, 44, 23, 15, 35).

J_1^4 sandstones, confined to the upper top of the horizon, are mainly light-grey, grey, sometimes brownish and yellowish, homogeneous, cross-bedded or lenticular-stratified due to the granulometric composition contrast and deposition of coaly-micaceous material and siderite microconcretions with coal lenses [4]. These sandstones have the following specific features: crude oil and condensate odour and gas bubbling. The marine worm burrows and traces (well №14) and glauconite inclusions in the sandstones (wells №22, 8, 62) are indicators of shallow-marine sedimentation. Arenaceous, aleurolite and argillaceous rocks have been identified in Lower J_1^3 cross section (figure 3).

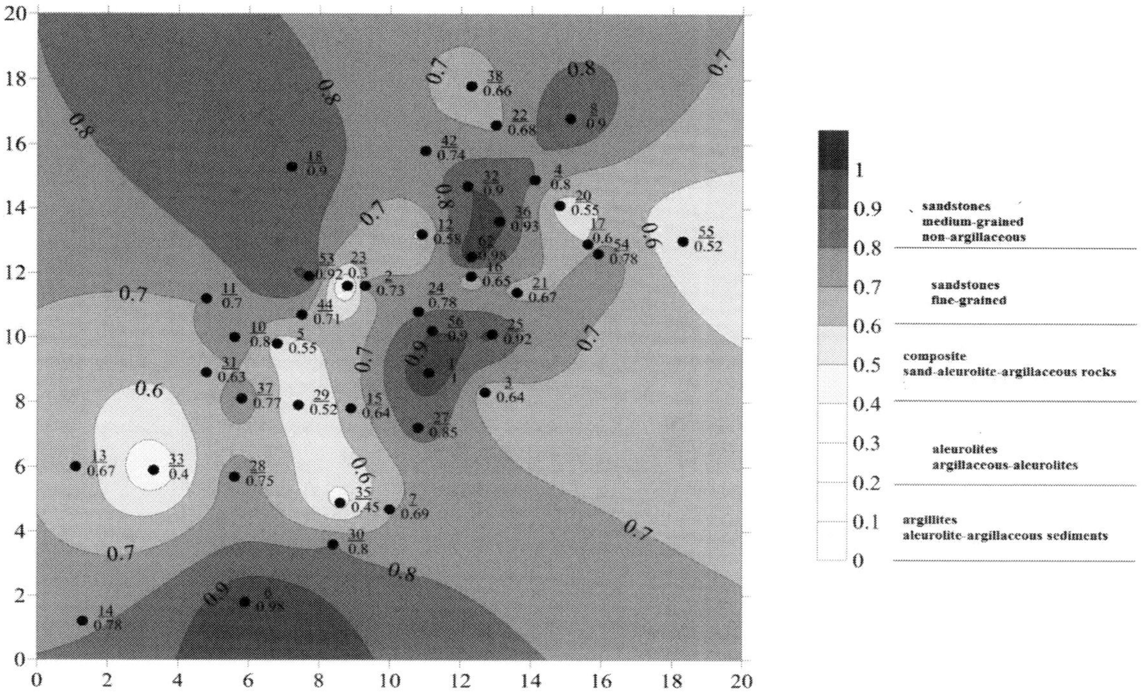

Figure 3. Cyclite Lower J_1^3 lithological composition map.

Throughout the major territory (wells № 38, 42, 17, 16, 21, 3, 11, 44, 37, 15, 7, 13, 14, 28, 31, 2) the horizon includes fine-grained sandstones (values α_{dp} from 0.8 to 0.6), fringing coarse-medium grained sandstone proveniance within the remaining territory.

Composite arenaceous-aleurite development sections (values α_{dp} from 0.6 to 0.4) are localized and can be found within the fine-grained sandstone development zones in the west (area near well № 33), in the center (wells № 5, 35, 29, 12) and in the east (wells № 55, 20).

Sandstones in Lower J_1^3 layer are light-grey, grey, brownish-grey, medium-loose cemented and rarely coherent. Structure is homogenous or stratified. Bedding is horizontal, cross, wave-like and lenticular due to granulometric composition contrast, clayey-siltstone material and coal interlayers and lenses, deposition of coaly-micaceous material and siderite microconcretions. Bioturbation structures are observed (figure 4).

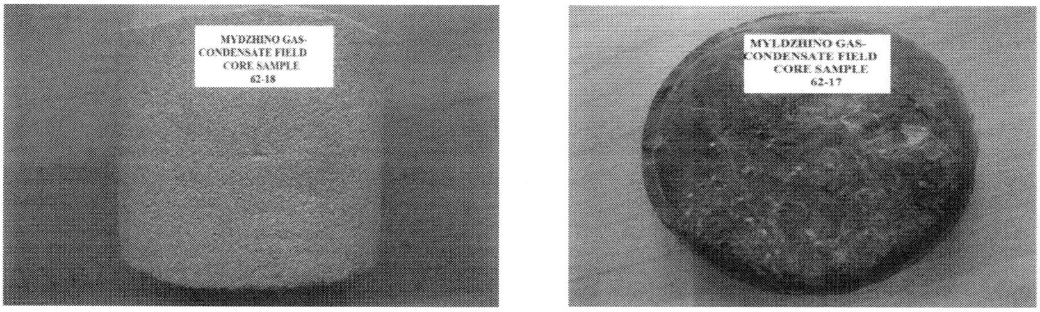

Figure 4. Core samples from well № 62

Four facies zones with different hydrodynamic regimes within sedimentation basin and facies types were identified on the basis of integrated investigation results (figure 5).

The first (I) sedimentation regime exhibited highly intensive hydrodynamic activity typical for active wave zones, where the sandstones are of medium/ fine-grained texture, well- sorted and showing a homogenous structure. The rocks formed under conditions of uplift-sloping facies conditions could be related to some extent to submarine seafloor uplifting. These sediments are characterized by highly and high intensive hydrodynamic level during the whole formation period of Lower J_1^3.

Intensive hydrodynamic activity (II) zones are widely distributed throughout the studied area. The cross-sections showed increased sand content in the upper Lower J_1^3 horizon.

Medium hydrodynamic activity (III) zones in sedimentation environment are found in the western and eastern Myldzhino areas (wells № 17, 54, 55, 33, 37, 29). SP curves showed marked regressive orientation of granularity alteration. These zone formations are confined to the lower relief areas; most likely, these are lagoons being filled up with argillaceous-aleurite material.

Low hydrodynamic activity (IV) zone has a limited distribution (well № 23). The cross-sections include clayey-aleurite material, moreover, SP curves showed marked regressive orientation of granularity alteration.

Identification of zones with similar hydrodynamic regimes, their spatial distribution and relationships proved the fact of the formation conditions of LowerJ_1^3. It was determined that Lower J_1^3 sediments accumulated under conditions of continuous Oxfordian marine regression, resulting in sea basin shallowing under severe topography of the shallow sea basin itself during intensive tectonic activity.

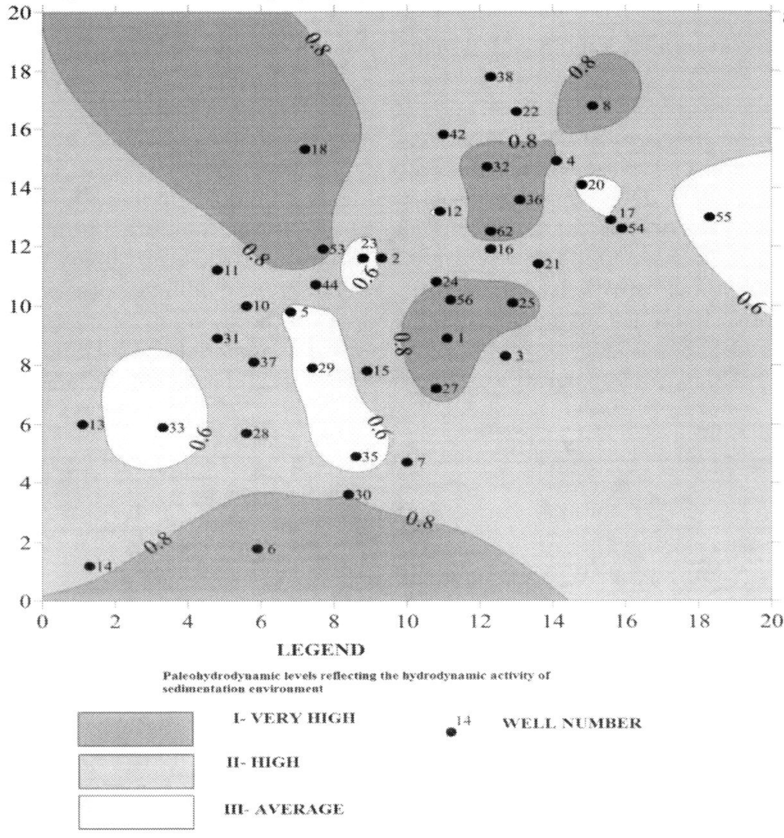

Figure 5. Lower J_1^3 lithologo-facial map

At the same time corrasion processes intensified within the clastic source areas, which, in its turn, resulted in the increased amount of transported clastics to the sedimentation basin. Non-uniform uplifting resulted in intensive hydrodynamic activity within submarine arches and bars. Such sedimentation conditions furthered the accumulation of arenaceous, aleurite-arenaceous material, its improved sorting and immense thickness increase.

4. Conclusion

Thus, integrated processing of geologic and geophysical data, structural -textural and biofacies analyses provided high accuracy identification of productive sediment formation conditions, their variability in time and space, which, in its turn, could improve geological exploration efficiency.

References

[1] Belozerov V, Danenberg E and Ogarkov A 1980 *Perspektivy neftegazonosnosti yugo-vostoka Zapadnoy Sibiri* 92–100
[2] Belozerov V, Brylina N and Danenberg E 1991 *Teoreticheskiye I regionalnye problemy geologii nefti i gaza* 171–180
[3] Belozerov V, Brylina N and Danenberg E 1984 *Problemy geologii I neftegazonosnosty verknepaleozoyskykh i mezozoyskikh otlozheniy Sibiri* 11-23
[4] Yezhova A, Ten T, Polumogina E, Serdyuk Z, Royenko N and Chekantsev V 1998 *Osobennosti stroyeniya produktivnykh yurskikh otlozheniy Myldzhinskogo gazocondensatnoneftyanogo mestorozhdeniya* 161

IOP Conf. Series: Earth and Environmental Science **24** (2015) 012032 doi:10.1088/1755-1315/24/1/012032

Designing economic and legal mechanism of land management in oil and gas companies

M R Tsibulnikova[1,2], O V Pogharnitskaya[3] and A B Strelnikova[4]
[1,2,3,4] National Research Tomsk Polytechnic University
[2] National Research Tomsk State University

e-mail: [1]tsibulnikova2011@yandex.ru, [3]pov@tpu.ru, [4]annas24@yandex.ru

Abstract. The article deals with the problem of economic and legal relationship in the sphere of land management provided by Russian government. The gas pipeline construction serves as an example to analyze the problems connected with leasing of both federal and privately owned lands. Comparative analysis of costs made by Gazprom to lease the lands at the stage of construction has been conducted. It has been concluded that the government should regulate relationships within the land sector to harmonize the interests of the Federation and private landowners.

1. Introduction

Sustainable land-use management is a significant socio-economic factor, which influences the development of all industries.

Today, deficiencies in legal framework inevitably lead to the situation when economic and legal mechanism of land management is based on economic interests of legal entities involved in the use of lands.

The government is concerned about sustainable and efficient use of all lands to meet the interests of national economy. As a result, the government aims to act on the principle of optimality which implies that every decision on land use made under the given circumstances is the best one among all possible. The difficulty is to delimit federal lands within the land fund of the country [1].

Today, the major part of land fund is federally owned, i.e. owned by the Russian Federation (745.4 million ha). A comparatively small part of lands is owned by federal subjects of the Russian Federation and municipal structures (12.7 million ha). The rest part is used by economic entities of private sector (133.1 million ha).

2. Materials and Methods

As an example of economic and legal mechanism intended to develop the system of land management, Gazprom activities within Eastern Gas Program have been analyzed (Development Program for an integrated gas production, transportation and supply system in Eastern Siberia and the Far East, taking into account potential gas exports to China and other Asia-Pacific countries) [2].

One of the program's priority is to design and enhance gas transmission system (GTS) "Sakhalin-Khabarovsk-Vladivostok".

Besides the free data of Gazprom, informational resources from The National Report on Characteristics and Use of Lands within Explored Areas and the data of Federal State Statistics Service were used.

3. Results and Discussion

The gas pipeline stretches for more than 1,800 kilometers. With 14 compressor stations being operated, the system will annually transport 30 billion cubic meters of Sakhalin gas [3].

The system includes "Komsomolsk-na-Amure" gas pipeline, 472 kilometers long, which is currently being operated.

Gas pipeline "Sakhalin-Khabarovsk-Vladivostok" runs through the territories of Sakhalin Oblast, Khabarovsky Krai and Primorsky Krai. The length of the gas pipeline is 133.9 km in Sakhalin, 777.4 km in Khabarovsky Krai and 438,7 km in Primorsky Krai. Within the investment project, land resources of various land cover including forests and woodlands were used, which caused significant land leasing costs for Gazprom. By January 01, 2012 the total land fund of Primorsky Krai, Khabarovsky Krai and Sakhalin Oblast was 103940.7 thousand ha. The information on lands owned by federal subjects of the Russian Federation, which the gas pipeline stretches through, is represented in table 1 [4, 5, 6].

Table 1. Land areas within federal subjects of the Russian Federation, thousand ha

Land use / total	Primorsky Krai	Khabarovsky Krai	Sakhalin oblast
Agricultural land	1 872.4	375.0	168.8
Residential land	245.2	421.2	86.1
Industrial lands and lands for other purposes	386.2	268.8	324.6
Lands of specially protected sites	841.8	1646.2	124.1
Forest land	12 140.6	73 707.2	6 959.4
Water	323.3	961.4	46.8
Reserve lands	657.8	1 383.5	1 000.3
Total	16 467.3	78 763.3	8 710.1

The structure and dynamics of land use by the constituent entities of the Russian Federation having been analyzed, one can make a conclusion that forest land occupies the largest territory and is predominant in percentage. It makes up from 74% in Primorsky Krai to 80% in Sakhalin Oblast and 93.6% in Khabarovsky Krai. Residential land accounts for 11% in Primorsky Krai, 1% in Sakhalin oblast and 0.5% in Khabarovsky Krai.

Investment project "Gas transmission system "Sakhalin-Khabarovsk-Vladivostok" being implemented, Gazprom bore significant costs to lease the lands located in Khabarovsky Krai and Primorsky Krai, as well as in Sakhalin Oblast.

The costs of the initial stage, when the first pipeline section, 1350 km in length, was built, were 260 372.6 thousand rubles. The distribution of costs paid by Gazprom within the period from 26.10.2009 to 31.12.2011 is represented in table 2.

Table 2. Actual costs paid by Gazprom within constituent entities of the Russian Federation

Costituent entity of the Russian Federation	Costs, thousand rubles	Proportion of costs, %	Number of contracts	Proportion in total number of contracts, %

Primorsky Krai	42 552. 3	16	153	63
Khabarovsky Krai	212 623. 8	82	85	35
Sakhalin oblast	5 196. 5	2	5	2
Total	260 372. 6	100	243	100

At the first stage of the project implementation 243 lease contracts were signed, 63% of which were made in Primorsky Krai. It is noteworthy that costs in different federal subject are not directly proportional to the number of contracts signed. The greatest amount of money Gazprom spent to lease lands in Khabarovsky Krai, while the number of contracts made there is half as many as that in Primorsky Krai. To interpret this fact, particularities of land leasing in these constituent entities should be analyzed.

Forest land leasing is a specific type of land leasing. To calculate the rent, one should take into account different criteria, such as total area, including gross forest land, gross forest cover, characteristics of forest vegetation (intended purpose of use, compartment number, predominant tree species, timber resources, forest resources data). Another particularity of forest land leasing is the term of leasing which can vary from 10 to 49 years. All forest land is controlled by Federal Forestry Agency of the Russian Federation. In accordance with Forest Code of the Russian Federation, the federal subjects are empowered to manage the forest land. Executive power is exercised by Forestry Department of Primorsky Krai, Forestry Administration of the Government of Khabarovsky Krai and Department of Forestry and Hunting of Sakhalin Oblast in Primorsky Krai, Khabarovsky Krai and Sakhalin Oblast respectively.

The distribution of costs paid by Gazprom to lease forest lands at the first stage of set-up complex construction is represented in table 3.

Table 3. Actual costs paid by Gazprom for forest land leasing

Contractor	Area, km	Area length, km	Actual cost, thousand RUB	Proportion in total amount of actual costs, %	Average cost per 1 km, thousand RUB
Department of Forestry and Hunting of Sakhalin Oblast	0.0 – 133.9	133.9	2 354. 9	3	17.6
Forestry Administration of the Government of Khabarovsky Krai	133.9 – 911.3	777.4	68 573. 3	77	88.2
Forestry Department of Primorsky Krai	911.3 – 1350.0	438.7	18 123. 2	20	41.3
Total		1 350	89 051. 4	100	49.0

The largest sum was spent to lease forest lands in Khabarovsky Krai, which is connected with the length of the gas pipeline within the territory of this region. The least amount of costs was spent on lands of Sakhalin Oblast, which is only 3% of total sum. Moreover, average cost spent on construction of 1 km of the gas pipeline in Khabarovsky Krai is twice as high as that in Primorsky Krai and five times higher than that in Sakhalin Oblast.

The rental rate is influenced by different factors, in particular – characteristics of the forest land. The most significant ones are the rental rate for an area unit set in the region, intended purpose of land use, types of areas within the forest land. The rental rate for an area unit of a forest compartment owned by the Federation within the territories mentioned above varies insignificantly (see table 4). It is important that there are no hardwood species and cedars in Sakhalin Oblast because of its geographical position.

Table 4. Rental rate for an area unit of a forest compartment owned by the Federation, in rubles

Forest plant species	Primorsky Krai	Khabarovsky Krai	Sakhalin Oblast
Coniferous trees (except for cedar)	2 978.88	2 931.8	2 970.32
Cedar	3 233.54	3 173.62	–
Hardwood trees	5 598.24	5 638.9	–
Softwood trees	2 871.88	2 835.5	2 865.46

Calculating the rent, one should take into account not only the rental rate for the area unit, but also the correction factor which depends on the intended purpose and varies from 2 (for forests of commercial use) to 6 (for protected lands). The correction factor is also influenced by timber species, the state of the forest fund etc. For example, in Khabarovsky Krai the rented area includes high percentage of lands of rich vegetation cover, so, to calculate the rent, correction factor 4 is applied, whereas in Primorsky Krai correction factors 2 and 3 are used in the majority of cases.

Besides, about 35% of lands leased in Khabarovsky Krai are not included into the forest fund as they are owned by individuals or legal entities. The costs for land leasing in Primorsky Krai, in their turn, are insignificant and represent only 0.2% of total costs.

If the land is privately owned, procedure, terms and dates of rent payment are defined in the rental contract. The particularity of such contract is not only the price of land use, but also indemnity for losses caused by land use, loss of opportunity and costs for land recovery (biological recultivation) which are included into the rent. The area is delimitated with the cadastral passport and annual lease cost is set by the landowner but not less than the tax on land.

Calculation of losses caused by using land to construct the objects under the implementation of investment project "Gas Pipeline "Sakhalin-Khabarovsk-Vladivostok" was based on the assumption that the lands were used for agricultural purposes. In civil legislation the loss of opportunity means the profit which the owner would get under the ordinary condition of civil circulation. Reimbursement of outlays including indemnity for losses caused by land use, loss of opportunity and costs for land recovery (biological recultivation) is made once a year with a lump-sum payment.

In our work we have analyzed some examples of calculating the rent for private lands. According to particular calculations of the losses of agricultural industry, as well as indemnity for losses including loss of opportunity, the average rent in Khabarovsky Krai is 35,000 RUB per hectare of herbaceous and mixed rangelands. This results in significant economic costs, as well as administrative expenses caused by registration of land rights in privately owned plots. Today, the number of lands owned by individuals or legal entities is constantly increasing and the rent they set is often overpriced, as well as the value of indemnity for losses including loss of opportunity. It is noteworthy that the rental rate over the period of gas pipeline construction varies from three kopecks to 1,200 rubles per square meter. However, in accordance with the laws of the Russian Federation, the rent for federal and municipal lands where gas pipeline objects are located is regulated: its maximum value is set by the government and can not be changed for the purpose of restraining fares.

It is clear that the rent for private lands which gas pipelines go through should be set equal to that for public lands. To regulate the relationships with some of the owners, seizure of lands for federal needs should be implemented.

To reach this goal, the amendments to the laws of the Russian Federation should be made. It is necessary to widen the basis and simplify the procedure of grabbing lands used by natural monopolies entities to operate, reconstruct or repair the objects from private owner for federal and municipal needs. It is important to provide the opportunity to seize the lands at the expenses of the natural monopoly entity so that it gets the right of property to the land.

The suggestions mentioned above might be considered within the follow-up work on draft federal law № 304493-5 "About amendments to the laws of the Russian Federation to enhance the procedure of land seizure", which was approved within the first reading and is being prepared for the second one.

It is important that all the expenses made for land rights registration influence the prices for works and services which are included into the final price for the product. This will inevitably increase the rate of inflation in modern economy, which is under the influence of activities conducted by natural monopoly entities.

4. Conclusion

Servitudes might be a solution for the problem. In this case, there is no need for delimitation of the area, decisions on transferring a plot from one category into another, multiple registrations of land transfers, expensive cadastral works to delimitate the lands which take in different sections of the pipeline.

This suggestion might be taken into account while making amendments to section 23 of Land Code of the Russian Federation, which is about the rights and obligations of land use (servitude).

The experience of foreign countries (USA, Great Britain, Australia, Canada, Germany, Sweden) demonstrates that the servitude is one of the fundamental land rights which allows service and transporting lines to be located: gas, water and other pipelines, tie-lines and power transmission lines. The servitude provides easy access to land resources and causes quick enhancement of the facilities.

Thus, construction of lines might circumscribe the rights of the landowner and probably cause some annoyance but does not prevent him from using the land for a definite purpose. If the line is located under the ground or constructed on poles or in other way, which does not terminate the use of land for a definite purpose, there is no need to seize or buy the land.

References

[1] Anisimov A P, Ryzhenkov AYa and Chernomorets AYe 2011 *Zemelnoye Pravo Rossii* 410
[2] Eastern Gas Program [Electronic resource]// Gazprom: offic. website. URL: http://www.gazprom.ru, free. - tit. from the tit. screen (usage date: 01.03.2014).
[3] Gazprom Invest Vostok [Electronic resource] // Gazprom Invest Vostok: offic. website. URL: http://www.investvostok.gazprom.ru, free. – tit. from the tit. screen (usage date: 05.03.2014).
[4] National Report on Land Conditions and Use in Primorsky Krai in 2012 [Electronic resource] / Federal Service for State Registration, Cadastre and Cartography: offic. website. URL: http://www.to25.rosreestr.ru, free. – tit. from the tit. screen (usage date: 27.02.2014 г.).
[5] National Report on Land Conditions and Use in Khabarovsky Krai in 2012 [Electronic resource] / Federal Service for State Registration, Cadastre and Cartography: offic. website. URL: http://www.to27.rosreestr.ru, free. – tit. from the tit. screen (usage date: 26.03.2014 г.).
[6] National Report on Land Conditions and Use in Sakhalin Oblast in 2012 [Electronic resource] / Federal Service for State Registration, Cadastre and Cartography: offic. website. URL: http://www.to65.rosreestr.ru, free. – tit. from the tit. screen (usage date: 07.03.2014 г.).

IOP Conf. Series: Earth and Environmental Science **24** (2015) 012033 doi:10.1088/1755-1315/24/1/012033

First native silica findings in bismuth from garnet skarns of Ribny Log - 2 gold ore target in Topolninsk ore deposit (Gorny Altai)

T Cherkasova[1], T Timkin[2] and O Savinova[3]
Department of Geology and Mineral Exploration, Institute of Natural Resources, National Research Tomsk Polytechnic University, Tomsk, Russia

E-mail: [1]tomasstik@yandex.ru, [2]timkin@tpu.ru, [3]kzzz@inbox.ru

Abstract The nanomineralogic investigation results of ore minerals in metasomatites (garnet skarns) of Ribny Log- 2 gold ore in Topolninsk ore deposit (Gorny Altai) revealed the native silica impurities (Si) of 1 – 5 nm within the grains of native bismuth (Bi). Polished sections were examined by using Tescan Vega 3 scanning electron microscope (SEM) with Oxford energy-dispersive spectrometer at the Department of Geology and Mineral Exploration, Institute of Natural Resources, Tomsk Polytechnic University.

1. Introduction

The work is based on original nanomineralogic data obtained after the execution of commercial contract projects in examining substance composition of ore and wallrock metasomatite occurrences of Topolninsk ore field (Altai Territory). According to metallogeny Topolninsk ore deposit is within the Anuisk ore area of Altai-Kuznetsk gold - ore belt and is integrated constituent of Anuisk copper-gold placer cluster in North-Altai gold- ore belt [1]. Intrusive formations occupy from 30 to 35% of the total Topolninsk ore area and involve two solid massifs and numerous dikes of mottled Topolninsk gabbro-granodiorite-granitoid complex, stockworks and linear subvolcanic bodies of Kujagan rhyolite-dacite-andesite complex. The host rocks are strongly hornfelsed and skarnified within a significant deposit area of solid massif exocontacts. The Topolninsk ore deposit within the Sarasino-Insk zone is confined to Karaminsk transverse NW fault zone controlling the Topolninsk and Malo-Topolinsk intrusive bodies. In the above-mentioned solid massif contacts (dimensions of 4.5x2 and 2.5x1.5 km, respectively) there are skarn formations with different superimposed productive mineralization. The thickness of skarn lenses and bodies vary from 3 to 50 meters, extenting along strike from 150 to 800 meters; total thickness of skarned rocks in the Western exocontact of Malo-Topolninsk massif is 500-1000 meters. The gold - ore mineralization as placer gold is identified throughout skarned rock cross-section [2]. According to A.I. Gusev data [3], the productive mineralization (4th hydrothermal cycle stage) formed after the intrusion of granodiorite-porphyry and granite-porphyry dikes into Topolninsk magma-ore-metasomatic system (MOMS) and dikes of syenites, granophyr-spherulyte leucogranites and moderately alkali leucogranites (primarily hybrid rocks) into Karaminsk magma-ore-metasomatic system (MOMS). Third- stage albite-quartz-epidote assemblage and fourth-stage mineral aggregates overlie these dikes.

The samples, including native bismuth grains with pure silica impurities, are garnet skarns with specific granoblastic texture. Mineral composition: garnet~60%, calcite~20%, epidote~8%, chlorite~10%, ore mineral~2%, quartz – single grains. Besides, the secondary minerals that accompany the last productive stages of ore formation also include actinolite and ankerite.

Andradite garnets form anisotropic zoning crystals, the zoning of which is generated by varying iron content. The crystal form is idiomorphic, while sectoral twins are commonly observed. The grain sizes vary from 0.2 to 3 mm. The mineral is transparent or slightly yellow-brownish if reflected by the transmitted light through Nicol prism, and anisotropic in cross polarized light with a significant fringe. The grains are fractured and replaced by carbonates.

Calcite forms pelitomorphic and coarse-grained aggregates. The mineral is colorless and has bright pearly interference colors in cross polarized light. Another characteristic feature is twinkling.

Chlorite forms foliated and flaky aggregates of up to 0.2 mm spatially orientated towards garnet and calcite contact and localizes in garnet fractured cracks. Under conditions of one nicol, the mineral varys from greenish to yellow-colorless, while it has low abnormal interference colors of indigo-blue in cross polarized light.

Epidote is composed of small grains of up to 0.2 cm and is confined to calcite occurrence zones. The mineral is yellow-brownish to pistache-green and has bright jet alder interference color in crosspolarized light with a significant fringe, the extinction of which is not more than 5 degrees. The sulphides detected by optic microscope include pyrite, chalcopyrite, cobaltite and arsenopyrite. Nanomineral phases of late mineralization stage: native bismuth, tetradymite, tellurobismuthite, sulphides (bismuthite), sulphosalts, sulphotellurides and selenides of Bi and Pb are of key importance. In most cases, the above-mentioned bismuth minerals are satellite resources and can be found in other similar targets within Soloneshensk ore cluster and other large gold deposits [3].

2. Methodology

To investigate the fine-dispersed composition phases of ore minerals is increasingly difficult as such phases comprise complex interrelations, the details and composition of which can be interpreted only at the micron level. The bismuth mineral samples (including native bismuth), in which pure silica was identified, were from metasomatic formations (garnet skarns) of Ribny Log-2 gold ore occurrence in Topolninsk ore deposit and were studied with Tescan Vega 3 electron scanning microscope and Oxword energy-dispersive spectrometer.

Quantitatively, the predominant bismuth-containing mineral was native bismuth, which is composed of rather large grains of up to 150 nm (figure 1).Bismuth tellurides and selenides include significantly smaller grains, apparently, forming internally solids, the proof of which is the changing content of selenium and tellurium in different grains. Sulphotellurides and sulphosalts of bismuth have practically the same grain size (from 1 ….10 nm). The smallest grains can be found in bismuthite where silver impurities are identified. These grains are fixed on the fringe region of native bismuth, rimming them as acicular crystals, filling in microcracks and voids within ankerite and actinolite (figure 1). The characteristic feature of actinolite and ankerite from studied metasomatic formations indicated manganese impurities, which, in its turn, is also specific for epidote.

Native bismuth is composed of round or, sometimes, idiomorphic grains enclosed in pyrite grains or contacting these grains (figures 1 and 2). It is considered that in most cases, native bismuth is a result of aqueous fluid-saturated solution precipitation, usually hot and containing SiO_2. Pure silica inclusions in native bismuth were not observed either in studied ore occurrence, or in other geological targets. Pure silica inclusions form isometric crystals of 1 ... 7 nm (figure 2). Oxygen (O) of 3.7% and iron (Fe) of 0.5% are impurities in pure silica.

3. Results and discussion

Recomposed gold-bearing systems spatially and paragenetically interrelated with gabbro-granitoid intrusive bodies generate specific diverse gold – ore deposits, i.e. from small to super gigantic ones, including such global gold deposits as Muruntau (Uzbekistan), Sukhoy Log (Transbaikal), Nezhdaninsko (Yakutia), Olimpiadinsko (Yenisei range), Berezovsko, Vorontsovsko (Urals), Kumtor

(Kirgizia), Bakyrchik (Kazakhstan) and others [3]. Reducing medium in the fluids of ore-generating granitoids is characterized by the fact that the principal gold carriers are chlorine and sulfur complexes. Deep reduced state of melts provides the crystallization conditions for accessory minerals - menaccanite and pyrite [3].

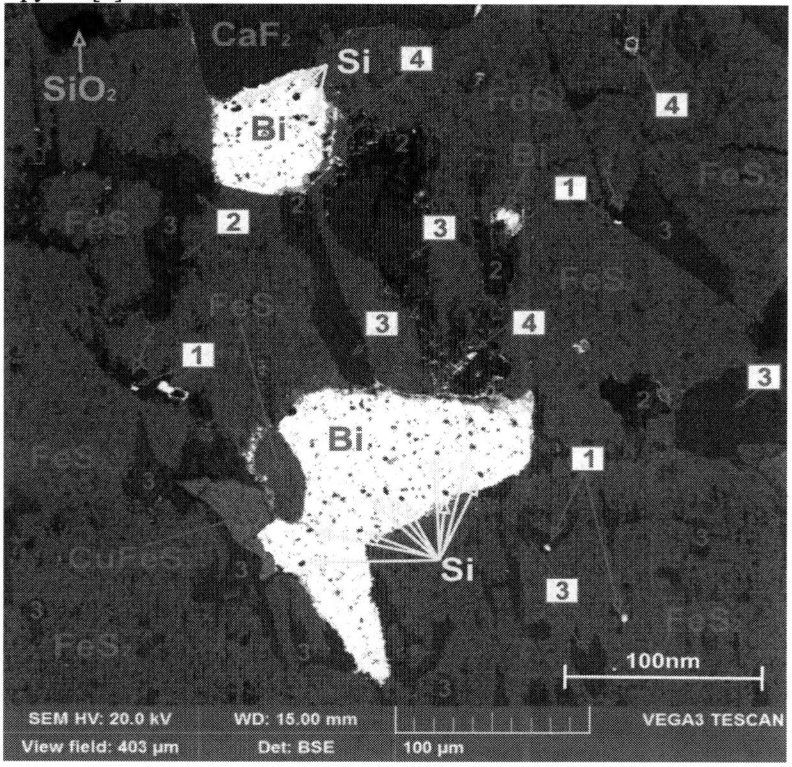

Figure 1. Electronic image of garnet skarn area containing pyrite ore mineralization, which including native bismuth grains with native silica crystals (Ribny Log - 2 gold ore occurrence in Topolninsk ore deposit): 1) bismuth selenides and tellurides; 2) actinolite including manganese impurities; 3) ankerite including manganese impurities; 4) bismuthite.

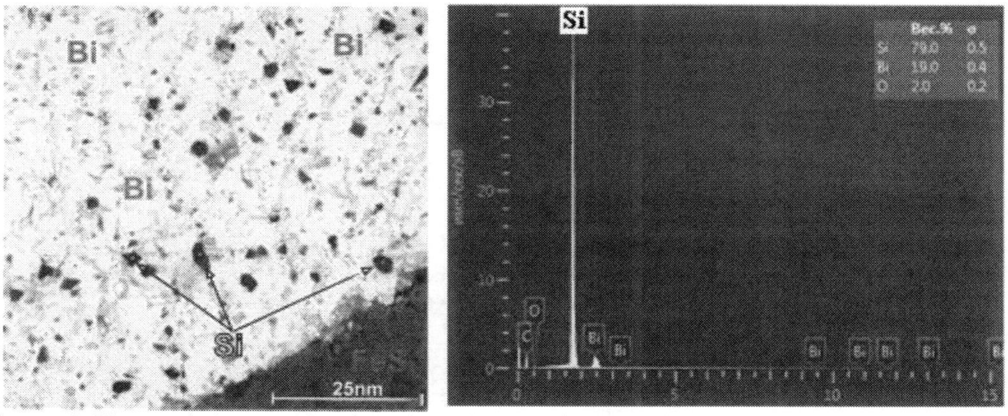

Figure 2. Electronic image of native bismuth area with impurities of pure silia crystals. Energy-dispersive spectrum of native silica.

Such minerals were continually identified within the studied targets. It has been established that sulfur as HS is present in highly-reduced gabbro granitic magmas – in this case, sulfur is more soluble in silicate melts and furthers the formation of sulfide globules, which, in its turn, isolate gold from the melt. Pyrite globular formations from metasomatic formations of Topolninsk ore deposit were found in several samples. Reduced intrusion-related gold - ore systems are frequently encountered in many countries and regions. For example, in Canada, these deposits are Dublin Gulch, Scheelite Dome, Clear Creek that are localized within the Tombstone gold belt [4-6], while in the USA; they are the reduced Fort Knox gold systems [7]. Within these gold -ore targets native bismuth and bismuth-containing sulphosalts display late paragenesis and are represented by typical alloys with Te, Pb, Sb and Au. Tellurobismuthine, tetradymite, native bismuth, boulangerite and Bi-Pb sulphosalts are abundant minerals in Fort Knox deposits [8]. Almost 40% of gold in the ores of Dublin Gulch are complex intergrowths with native bismuth [9], bismuthite, tetradymite and tellurobismuthite [4].

Reduced gold systems can also be found in eastern Australia: Timbarra and Kidston [10]. In Gorny Altai, there are several gold magma-ore-metasomatic systems (MOMS) of this type: Karaminsk, Baranchinsk, Bazlinsk and Choisk [3, 11]. It is assumed that natural reduced systems are mechanism carriers in gold deposition by chlorine and hydrosulfide complexes of magmagene fluids to ore concentration areas. In most cases, it is the fluid-saturated environment of ore-generating formation systems that generate reduced native phases of metals and their alloys [12]. The reduced magma-ore-metasomatic systems include the largest (in terms of metal reserves) global gold deposits. They are characterized by ore "complexity", where gold is accompanied by W, Bi, Te, Pt and Pd [13].

4. Conclusion

Natural reduced magma-ore-metasomatic systems that are sources of noble metal deposits and ore occurrences could apparently be a source of pure silica. Only a few cases of elemental silica findings have been observed which were formed in extremely reducing conditions. Pure silica was observed in the intergrowths with moissanite from Yakutia kimberlites [14,15], in terrestrial iron from serpentinites [16], in the intergrowths with Fe_2Si from Lower Cambrian carbonaceous sedimentation [17], as well as in associations with native gold (Au) [18] and natural phase Au_2Al [19]. Moreover, pure silica was found as microscopic crystals in fumaroles of Tolbachik and Kudryavy volcanoes (Kamchatka) in association with other native metals (Al, Ti, Fe and Pt), formed during the high activity of halogens [20]. Pure silica inclusions in fulgurites associated with iron silicides were also investigated [21]. Pure silica associated with iron silicides was found twice in the lunar regolith 61501, 22 sample from Apollo 16 [22] and in iron silicides of lunar meteorite DHOFAR 280 [23]. The above-mentioned findings of pure silica indicate its specific formation conditions within an increased fluid - saturated system.

Acknowledgement

The investigations were supported by National Research Tomsk Polytechnic University, project No. 114-2014

References

[1] Kudachin V V 1998 Mineralno-syryevaia baza Respubliki Altai: sostoianie i perspektivy razvitiya *Materialy regionalnogo soveshchaniya* – Gorno-Altaysk GAGU 120
[2] Bedarev N P and Gusev A I 1998 Geologiya i zolotonosnost Topolninskogo rudnogo polya Gornogo Altaya *Rudy i metally* **2** 100-109
[3] Gusev A I 2003 Metallogeniia zolota Gornogo Altaya i yuzhnoi chasti Gornoy Shorii *STT Press* 350
[4] Hitchins A C and Orssich C N 1995 The Eagle zone gold-tungsten sheetedvein porphyry deposit and related mineralization, Dublin Gulch,Yukon Territory *Canadian Institute of Mining and Metallurgy* **Special Volume 46** 6 803-10
[5] Mair J L, Hart C J R and Stephens J 2006 Deformation history of the northwestern Selwyn

Basin, Yukon, Canada: Implications for orogen evolution and mid-Cretaceous magmatism *Geological Society of America Bulletin* **118** 2 304-23

[6] Marsh E E, Goldfarb R J, Hart C J R and Johnson C A 2003 Geology and geochemistry of the Clear Creek intrusion related gold occurrences, Tintina gold province, Yukon, Canada *Canadian Journal of Earth Sciences* **40** 5 681-99

[7] Mustard R 2001 Granite-hosted gold mineralization at Timbarra, northern New South Wales *Mineralium Deposita* **36** 4 542- 62

[8] McCoy D, Newberry R J, Layer P W et al 1997 Plutonic-related gold deposirs of inerior Alaska *Economic Geology* **Monograph 9** 191-241

[9] Maloof T L, Baker T and Thompson J F H 2001 The Dublin Gulch intrusion hosted deposit, Tombstone Plutonic Suite, Yukon Territory, Canada *Mineralium Deposita* **36** 4 583-593

[10] Cole A, Wilkinson J J, Halls C and Serenko T J 2000 Geological characteristics, teconicsetting, and preliminary interpretations of the Jilau gold-quartz vein deposit, Tajikistan *Minearlium Deposita* **35** 5 600-618

[11] Gusev A I and Gusev N I 2011 Petrologo-geohimicheskie kriterii i flyuidny rezhim gigantskikh magmo-rudno-metasomaticheskikh zolotorudnykh sistem *Sovremennye naukoyomkie tekhnologii.* **4** 12-16

[12] Cherkasova T and Korotchenko T 2014 Noble-metal mineralization in olivine clinopyroxenite within idgimski gabbro-peridotite-pyroxenite complex (West Sayan Mountains), *IOP Conf. Series: Earth and Environmental Science* **21**

[13] Korobeynikov A F, Gusev A I and Krasova A S 2012 Vosstanovlennye intruzivno-gidrotermalno metasomaticheskie zolotorudnye sistemy *Izvestiia TPU. Geologiya i poleznye iskopaemye* **321** 16-22

[14] Marshintsev V K, Shchelchkova S G, Zolnikov G V and Voskrensenskaya V B 1967 Novye dannye o muassanite iz kimberlitov Yakutii *Geologiya i geofizika* **12** 22-32

[15] Pankov V Yu and Spetsius Z V 1989 Vklyucheniya silitsidov zheleza i samorodnogo kremniya v muassonite iz kimberlitovoy trubki "Sytykanskaya" *Docl. AN SSSR* **305** 3 704-08

[16] Bird J M and Weathers M S 1975 Josephinite: specimens from the earth's core *Earth Planet Sci. Lett.* **28** 1 51-64

[17] Novoselova L N and Bagdasarov E A 1979 Novye dannye o silitsidakh zheleza *ZVMO* **106** 3 326-333.

[18] Novgorodova M I, Boronikhin V A, Generalov M E and Kramer Kh O 1989 O samorodnom kremniy v assotsiatsii s samorodnym zolotom i drugimi metallami *Docl. AN SSSR* **309** 5 1182 - 85

[19] Sazonov A M, Zvyagina E A, Leontyev S I et al 2008 Assotciatsii mikro i nanorazmernykh obosobleny blagorodnometallnogo kompleksa v rudakh *Zhurnal Sibirskogo Federalnogo universiteta. Seriya Tekhnika i tekhnologii.* **1** 17-32

[20] Korzhinsky M A, Tkachenko S I, Bulgakov R F and Shmulovich K I 1996 Sostavy kondensatov i samorodnye metally v sublimatakh vysokotemperaturnykh gazovykh struy vulkana Kudriavy (ostrov Iturup, Kurilskie ostrova) *Geohimiya* **12** 1175-82

[21] Essene E J and Fisher D C 1986 Lightning strike fusion: extreme reduction and metal silicate liquid immiscibility *Science* V. **234** 189-93

[22] Spicuzza M J, Valley J W, Fournelle J et al. 2011 Native silicon and Fe silicides from the Apollo 16 lunar regolith: extreme reduction, metal silicate immiscibility, and shock melting *Lunar Planet Sci. 42nd.* **2231**

[23] Nazarov M A, Demidova S I, O. Anosova M O, Kostitsyn Yu A, Ntaflos T and Brandshtetter F 2012 Samorodny kremny i silitcidy zheleza v lunnom meteorite DHOFAR 280 *Petrologiya* **20** 6 560–73

IOP Conf. Series: Earth and Environmental Science **24** (2015) 012034 doi:10.1088/1755-1315/24/1/012034

Time cycle calculation procedure for the special crew during the mining mobile machine complex operation

V Shmurygin[1], V Lukyanov[2] and A Maslovsky[3]
[1, 2, 3] Department of Oil and Gas Storage and Transportation, Institute of Natural Resources, National Research Tomsk Polytechnic University Tomsk, Russia

Email: [1]shvladsv@tpu.ru, [2]lukyanov@tpu.ru, [3]Maslovsky AN@ mail.ru

Abstract. The relevance of the research is specified by the necessity to optimize the delft mobile tunneling equipment operation. Target of the research is tunneling time cycle justification for the special crew during the mining mobile machine complex operation. Methods of the research included the consideration of operation organization schemes in the drifting face and effective use of the mobile equipment during mine exploratory working operations. Time cycle calculation procedures for major processes have been considered. This has been done for the special crew during the mobile machine complex operations for several working faces and various organization schemes.

1. Introduction

Underground mine exploratory workings in the mineral deposits are often driven with the lateral derivation (the adit shaft or entry with the crosscuts and insets) including, as a rule, more or less, extensive main opening and one or several drifts which are usually shortwall driven (the insets or crosscuts).

Mine exploratory working operation is a complex technological process that includes the series of production processes (PP): drilling, charging and blasting, ventilation, shaft mucking and dirt haul, timbering, rail laying and many other operations interconnected and logically performed. So, to implement all these above mentioned processes there should be necessary machinery and equipment.

2. Cycle calculation

Workings drive operation is planned to have such distribution of work in the drifting cycle when tunneling machinery and equipment make maximum use.

Mine exploratory workings operation may be organized by one (out of three) scheme depending on mining and geological conditions and technological level. These are the following:

1 – sequential implementation of the main production processes in the drifting cycle for one working face without synchronism;

2 – concurrent execution of the main production processes in the drifting cycle for one working face with synchronism;

3 – combined execution of the main production processes in the drifting cycle for several working faces.

Let's consider the sequential scheme of the operations in the working face. Stall equipment of the complex (a loading machine and rail-mounted jumbo) is always in the working face (or at the face spur track) before the drivage completion of the relevant working, i.e. the equipment is operated by the special drift crew who sequentially implements all cycle stages. The cycle time is the following:

$$T_ц = T_{Цо} - K_{зв} \cdot T_{звп}, \tag{1}$$

$$T_{Цо} = T_у + T_б + T_{звп}, \tag{2}$$

$$T_{звп} = T_{зв} + T_{пр}, \; K_{зв} = \Delta\, T_{Цо} / T_{звп},$$

where $T_{Цо}$ – the cycle duration during the sequential implementation of the main production processes by the special drift crew (when the loading and drilling machines are always in the working face), h; $T_у$, $T_б$, $T_{зв}$, $T_{пр}$ – the time for shaft mucking, blast-holes drilling, charging and blasting, ventilation of the working face, h; $K_{зв}$ – coefficient showing the possibility of synchronism between the charging\blasting \ventilation process and a shift break.

In particular case, when the cycle begins with the underground shaft mucking, the average time for the cycle time reduction is the following [1]

$$\Delta T_{Цо} = \frac{T_{Цо} - T_{см}}{n_ц}, \text{if } n_ц\, T_{Цо} - \min\{T_{звп}, T_{МСП}\} \leq T_{см} \tag{3}$$

$$\Delta T_{Цо} = T_{Цо} - n'_ц T_{см}, \text{if } T_{Цо} - \min\{T_{звп}, T_{МСП}\} \leq n'_ц T_{см} \tag{4}$$

$$0 - \text{if not}, \; n_ц = \left]\frac{T_{см}}{T_{Цо}}\right[+ 1, \text{ if } T_{Цо} \leq T_{см},$$

$$n'_ц = \left]\frac{T_{Цо}}{T_{см}}\right[, \; T_{Цо} \geq T_{см}\,;\; T_{МСП} = 24 / n_{см} - T_{см} \geq 0,$$

where $T_{см}$, $T_{МСП}$ – the shift duration and the shift break duration, respectively, h; $T\Delta_{Цо}$ – the average time for the cycle reduction (the cycle begins with the underground water mucking), h; $n_ц$ – the number of the cycles during the shift; $n_{см}$ – the number of the shifts during the working face operation per day. Bracketed expression $]x[$ means the integer part.

A failure of the inequality (3) or (4) indicates that we can neglect the cycle reduction time at the expense of charging\ blasting \ventilation process partially synchronized to the shift break. So, $T_ц = T_{Цо}$. The equipment relocation is unreasonable when the maximum distance exceeds the value. So, the maximum distance $L_{зmax}$ between the working faces should not exceed the smallest value.

$$L_{зmax} \leq \min\,\{V_{nepj}\,(0{,}5\; T_{Цо} - T_j)\}, \tag{5}$$

where T_j – the process duration , h; V_{nepj} – the average transportation velocity of the equipment along the workings, m\h.

If $L_з \leq L_{зmax}$, then $T_Ц = T_{Цо}$, if $L_з > L_{зmax}$, then $T_Ц > T_{Цо}$.

If according to the equation (5) we have

$$L_{зmax} \leq 0, \text{ i.e. } 0{,}5\; T_{Цо} \leq T_{jm}\; (T_Ц > T_{Цо}),$$

where T_{jm} – time for the longest cycle process, h;

$L_{зmax}$ – the maximum distance between the working faces of the built workings, m; $L_з$ – the average distance between the working faces of the built workings during their drivage process, m.

In this case the complex should better be used in one working face. Figure 1 shows the work flow chart for the special drift crew during the simultaneous drivage of two workings by the mobile machine complex of the wheeled and rail-mounted equipment which was used in “Sevvostgeologiya” organizations. Let’s substitute the appropriate initial data for the equation (5) related to the drivage pattern in figure 1 ($T_у$ = 2,3 h; $T_б$ 2,4 h; $T_{звп}$ 1,3 h; $T_{Цо}$ = 6,0 h). Thus, the maximum distance between the working faces of the built workings by this complex should not exceed the following:

– for the rail-mounted jumbo (with the average travelling velocity along the workings)

$$V_{nep} = (2{,}27 / 6{,}5)^{1/3} \cdot 3{,}6 = 2{,}5 \text{ km/h}) – 1500 \text{ m};$$

– for the loading machine (with the average travelling velocity)

$$V_{nep} = 1 \cdot 3{,}6 = 3{,}6 \text{ km/h}) – 2500 \text{ m}.$$

Finally, the maximum distance between the attended working faces for this machine complex is equivalent to the minimum value – 1500 m.

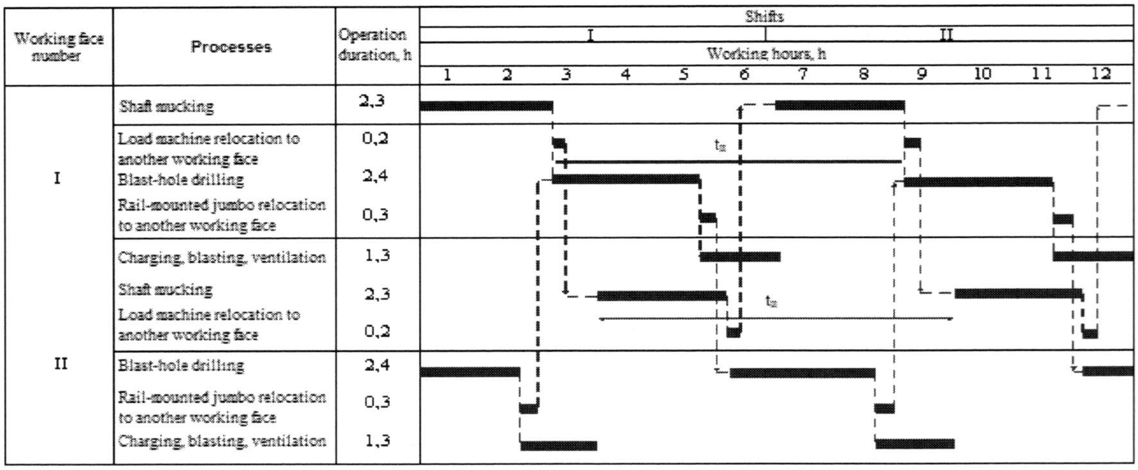

Figure 1. The work flow chart.

One tunnelling equipment set for two working faces.

3. Conclusion

To identify the average travelling velocity along all the workings, the equations for determining the longest tunneling cycle time (shaft mucking, blast-hole drilling) for each machine have been provided. The maximum distance between the attended working faces has been identified.

References

[1] Pankratov A V 1990 Prospects to use mobile equipment for underground mine exploratory survey in the Russia North-East. // Problems and prospects of mining in the USSR North-East: Scientific seminar research records **Part 2** *Yakutsk Mining Institute. Siberian Branch of the USSR Academy of Sciences* 203 – 211

[2] Shmurygin V A, Pankratov A V and Lukyanov V G 2013 Calculation procedure of the drivage pace for mine exploratory workings made by a drivage complex during single-multiple operations *Tomsk, TPU Izvestia* **V 323 № 1** 200 – 207

IOP Conf. Series: Earth and Environmental Science **24** (2015) 012035 doi:10.1088/1755-1315/24/1/012035

Understanding thin beds within the Bazhenov formation using seismic modeling

A I Bashirov[1], V B Belozerov[2], N V Bukhanov[3] and A V Evdokimov[4]

[1]RN-KrasnodarNeftegas, Krasnodar, Russia
[2,3]Tomsk Polytechnic University, Tomsk, Russia
[4]SWEPCO, Tomsk, Russia

E-mail: [3]Bukhanovnv@hw.tpu.ru

Abstract. The Bazhenov formation (Bazhen) is a nonconventional reservoir with a very high resource potential. According to different estimates, the resources of the Bazhenov formation equal 23-160 billion bbl. Currently, only about 35 million bbl have been produced from Bazhen, which is a rather low production rate in comparison with the resource potential. The reason is the insufficient number of detailed surveys on the formation and the absence of efficient production technologies. The analog of Bazhen is the Bakken formation located in North America. This formation is being actively developed due to the presence of carbonated and sand layers in Middle Bakken. Its development strategy is horizontal well drilling and multi-stage fracturing. The thickness of the carbonated and sand layers in Bazhen are about 0.5-3 m and they cannot be identified by seismic surveys, like in Bakken, as their thickness is smaller than the seismic resolution itself. The distribution of the carbonated layer within the analyzed license areas was estimated by 2D forward seismic modeling. Fractures in well K-1 were determined as planetary fractures. The paleomagnetic investigation determined the nature of these fractures.

1. Introduction

The license blocks of interest are located in Kargasok Region of Tomsk Oblast in West Siberia. There are six exploration wells within the license areas, of which three wells have provided corresponding core data. The analyzed core data and well logs of well K-1 in the Kyleevskaya area suggest the presence of two thin carbonated beds with vertical fractures in the Bazhenov formation. Well testing indicated the presence of oil in this interval. The main challenge is to evaluate the distribution of these thin beds in the Bazhenov formation and to understand the nature of fractures that can be critical for the development of this oil shale.

The carbonated layers cannot be identified by seismic profiles due to the fact that their thickness is smaller than the seismic resolution itself. In this case there are two factors that should be considered. Firstly, the influence of the carbonated bed on the seismic record is interpreted through modeling and could be used as a qualitative indicator of lateral thin bed distribution. This indicator would determine those promising zones of carbonated bed distribution. Secondly, the thickness of these beds should be determined. As the influence of the carbonated bed on the seismic record is defined, the attribute that reflects this influence could be found. This attribute could be used as a quantitative indicator. Further, an attribute map will be plotted to indicate analyzed bed distribution zones, and then a carbonated bed thickness map plotted on the basis of this attribute map.

2. Bazhenov oil shale

The detailed surveys on the Bazhenov formation began in the 1960s. It was studied by many researchers: F.G. Gurarii, I.V. Goncharov, Yu.N. Karogodin, A.E. Kontorovich, I.I. Nesterov, V.S. Surkov, etc. Bazhen is located in the West Siberian petroleum province within a territory of 1 million km^2, at the depths from 600 m to 3,500-3,800 m. Bazhen is a source rock in which the processes of oil generation from organic material (kerogen) have not been completed. The formation composition is very different, where the main organic components are liquid hydrocarbons and kerogen and non-organic – clay minerals, silica and carbonates. Today, about 70 fields with oil reserves in the Bazhenov formation have been discovered, however, they are underdeveloped due to the existing inadequate studies and lack of production technologies [1].

There are four types of kerogen divided by the composition proportion of hydrogen, carbon and oxygen. Fields that have type 1 kerogen are represented by oil, type 2 – wet gas with condensate, type 3 and 4 – dry gas. Kerogen of Bazhen refers to type 1.

The Bazhen resource potential is very high, which has been proven by the estimates of different companies. According to their estimates, the Bazhenov formation resources range from 23 to 160 billion bbl, while only about 35 million bbl have been produced from Bazhen [2]. This is very low due to the fact that the development technology of the Bazhen formation is non-efficient.

Now many researchers state that the commercial oil content of Bazhen is connected with the presence of thin compact rocks, and, in this case, the main challenge of geophysicists and geologists is to detect their lateral distribution [1, 2].

More and more attention is focused on such a point as fracturing [3]. Many researchers state that the fractures in Bazhen are not interrelated with tectonic processes [3, 4]. Trofimuk A.A. claims that fracturing (especially, vertical fractures) in Bazhen should be investigated in detail [3].

2.1. *Worldwide analogs*

There are several analog of Bazhen in the world: Green River, Bakken, Eagle Ford, Wolfcamp, Marcellus, Barnett, etc. The Bakken formation is the target identification as this formation is being developed very actively. The Bakken formation is a part of the Williston Basin in North America. It is located within a territory of 520,000 m^2. There are three main litho-stratigraphic units: Upper Bakken, Middle Bakken and Lower Bakken, where the highest production rate is within the Middle Bakken.

Production methods used in Bakken is horizontal well drilling and multi-stage fracturing. The Bakken hydrocarbon production is possible due to the presence of carbonated and sand layers in the Middle Bakken. Originally, these layers had poor reservoir properties, but multi-stage fracturing improved these properties. Long horizontal wells provide a wide drainage area. Kerogen in Bakken refers to type 1 and, partially, to type 2.

Due to the similarities between Bazhen and Bakken formations and the successful development activities in Bakken, many researchers proposed developing Bazhen taking into consideration Bakken development experience, i.e. horizontal well drilling and multi-stage fracturing [1, 2]. However, the main difference is reservoir thicknesses and their distribution. In the Bakken formation, reservoir rocks have thickness of 40 m and even more and are located between thin organic-rich rocks. In Bazhen reservoir rocks are about 0.5-3 m thick and are located within several stratigraphic layers.

If the 40-meter section of the Bakken reservoir could be determined by seismic methods, then, in the case of the thin reservoir layers of Bazhen formation, their thickness is beyond the range of resolution (less than one-fourth of the wavelength) and cannot be determined. Therefore, Bakken development strategy is not applicable in the Bazhen formation. First and foremost, it is essential to provide additional data in order to understand how to detect these thin Bazhen reservoir beds.

3. Seismic working modeling

As stated above if the layer thickness is less than the seismic resolution, information on the existing layer is encoded in the composite reflection amplitude [5]. Field & research service company LM Gochioco Inc. provides information on how thin layer thickness can be estimated by the response effect. The described research involved 2D forward seismic models to determine thin carbonated layers. Based on the data of layer velocities and formation thickness, the linear interpolation of formation properties between reference sections was computerized. Modeled seismic traces were produced as a consequence of convolution between acoustic velocities and synthetic wavelet.

The first step of modeling is to design litho-acoustic models of the wells, including well logging interpretation to determine lithology. The following formation velocity values were used in calculations: coal – 2,500 m/sec, Bazhen bituminous argillite – 3,000-3,300 m/sec, Georgiev formation argillite – 3,300 m/sec, argillite – 4,000 m/sec, carbonated siltstone – 5,000 m/sec, sandstone – 4,400 m/sec, carbonated sandstone – 5,000 m/sec. These values were used to define the rock acoustic properties that directly influence the wave pattern generation and in correlation with the petrophysical core data analysis of rocks with different lithological characteristics from Tomsk Oblast.

This modeling process will be shown on the example of well K-1, which, in its turn, embraces 2 carbonated layers. The litho-acoustic well model of the studied interval is represented in figure 1.

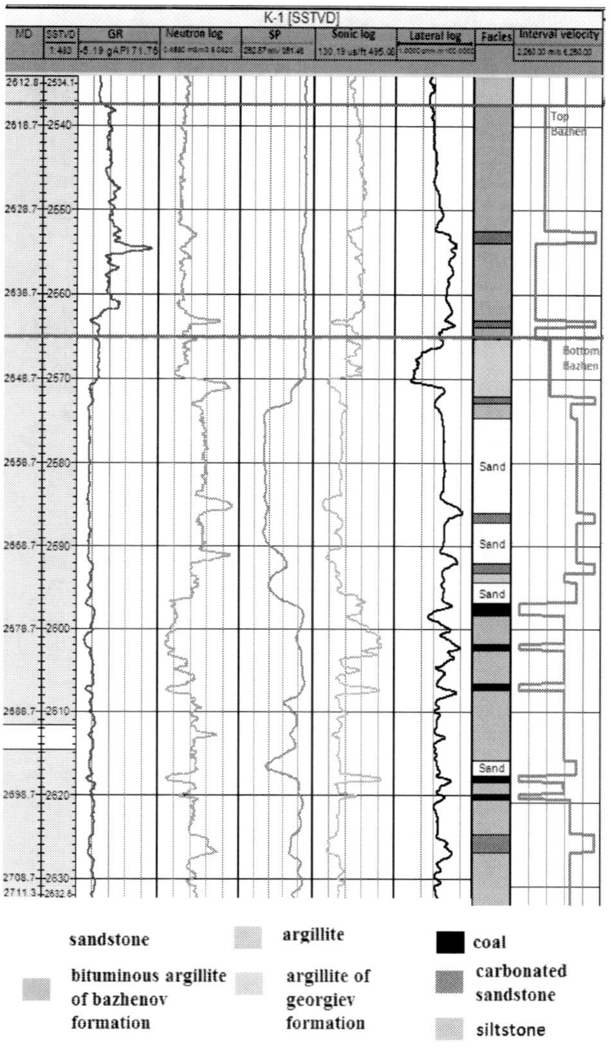

Figure 1. Litho-acoustic model of well K-1

The studied interval is Bazhen formation. The influence of the Bazhen carbonated layer on the generation of reflections in seismic traces will be determined. This influence can be expressed in reflections that correspond to the Bazhenov formation, the Bazhen bottom or in reflections referring to the neighboring layers. Therefore, modeling was not limited to the Bazhen boundaries. The layers that are 50 m higher than Bazhen and 150 m lower were also included in the modeling.

To predict the influence of the carbonated layer on the wave pattern, 2D forward modeling was performed with the involvement of synthetic wells. In these wells, the different carbonated layers thicknesses were assigned: from 0 to 6 m (figure 2). It is a good demonstration of the influence of the carbonated layer on the wave pattern. The presence of this layer is characterized by the two-phase reflection record of that corresponding to reflector IIa – bottom of Bazhen.

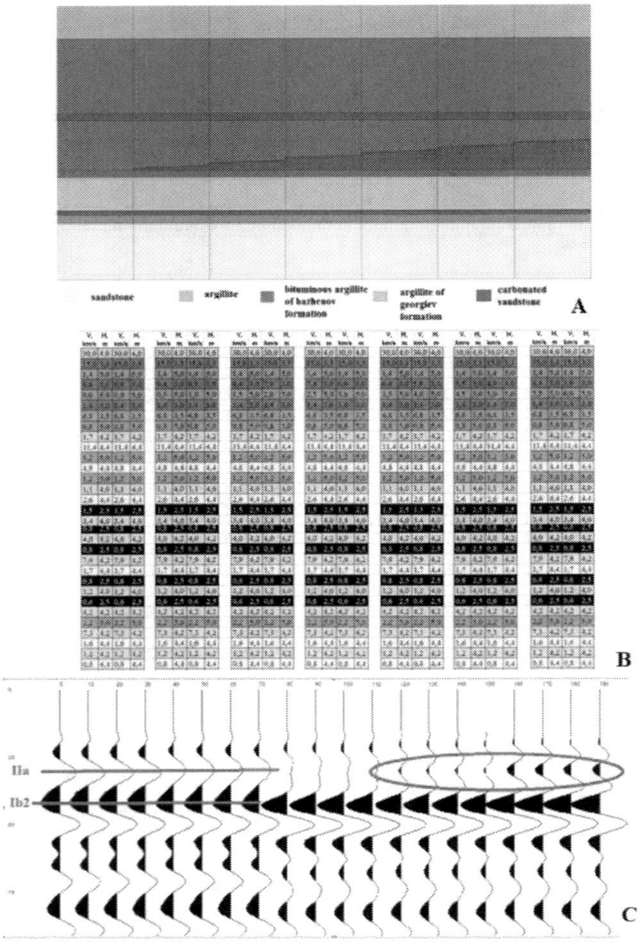

Figure 2. Determination of carbonated bed influence. A – Geological section. B – Litho-acoustic model. C – 2D seismic model

The presence of a carbonated layer in the IIa phase change from one-phase wave to two-phase wave was also proved in the modeling of other wells. However, this change can be caused by another factor. It can be the changing position of the carbonated layer, a variation in Bazhen thicknesses or coal pinching out under Bazhen. In this case, all modeled situations should be checked to determine whether they caused the phase transformation or not and then be analyzed.

Apart from one-phase to two-phase change of the IIa wave, another factor can be identified. After the increase in the carbonated bed thickness, the travel time between reflector Ib2 and the negative amplitude above reflector IIa also increases (figure 3).

Consequently, the isochron thickness attribute defined as the time difference between two horizons can be used as a quantitative indicator. This attribute was calculated in the zones of carbonated bed distribution. Then, the attribute map was plotted in order to design a carbonated layer thickness map. In well K-1, the thickness of the two carbonated beds is 2.2 m. The value of the attribute in well K-1 is 26.3. A thickness map was plotted based on this data. The average thickness of the carbonated bed is 2 m.

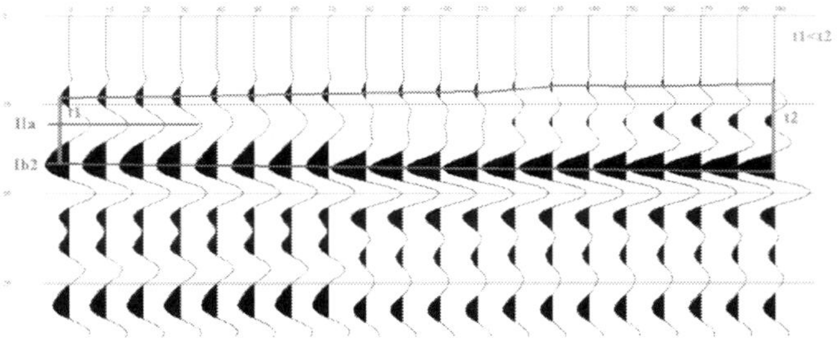

Figure 3. Quantitative indicator of the carbonated bed

4. Planetary fracturing

As already mentioned above, fractures in well K-1 are vertical and located only in one layer. Factors suggest that these fractures are planetary. The generation of planetary fracturing is interrelated with tension conversion in the faulting resulting in caused depositions, which, in its turn, is a consequence of Earth rotation activity. Planetary fracturing was studied for Russian and North American plates by Parker (1942), Shtatsky (1945), Permyakov (1946), Novikova (1952), Rac (1970), Dolitsky (1978) and et al.

These fractures have the following features:

- located both in folded and horizontal layers.
- form orthogonal systems.
- not connected with tectonic structures.
- more pronounced in compacted rocks where they are perpendicular to bedding and located only within one bed.
- relationship between fracture spacing and layer thickness.

According to the studies conducted by K.I. Mikulenko, the Jurassic and Neocomian rock fractures in the south-east West-Siberian petroleum play are divided into three orthogonal systems with the following azimuths: 335 and 65, 295 and 25, 270 and 0 degrees [7]. They correspond to the planetary fracturing described by Schultz. The Mikulenko findings are also matched with the directions of global rotational field tensions according to V.D. Dolitsky (figure 4).

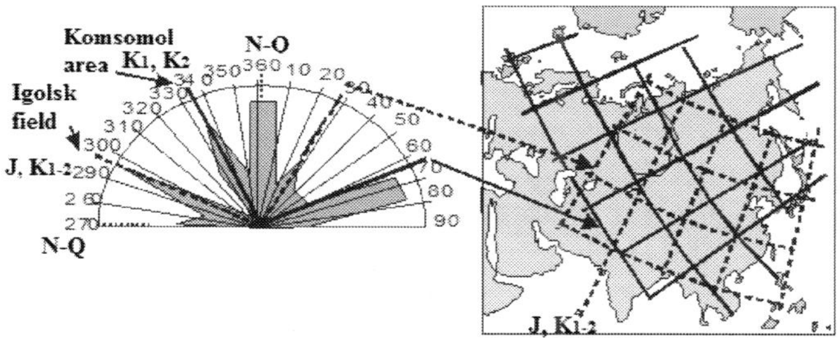

Figure 4. Correlation of research findings by K.I. Mikulenko, S.S. Schultz and V.D. Dolitsky (K.I. Mikulenko, G.G. Shemin, 1971 and V.D. Dolitsky, 1978)

To confirm which fractures in the carbonated layer in well K-1 are planetary, a paleomagnetic survey was conducted to identify the direction of the fractures. The direction was compared with the survey results obtained by Mikulenko and Shultz.

The paleomagnetic survey included the following steps:

1. Preparation of samples. Two core samples were from the interval of the carbonated layer. In these samples, an arbitrary coordinate system (in this sample) was introduced into the next magnetic measurements.
2. Magnetic measurements were performed by the MAL-036 astatic magnetometer at a TPU laboratory.
3. The samples were demagnetized to balance the external magnetic field, through Aparin thermal oven. The thermo-demagnetizing step was 25° C, while the final heating temperature was 125°C, the diapason viscous component of natural permanent magnetization is often destroyed at this temperature.
4. The obtained experiment results were used in the calculation of the vector differences to estimate the direction of viscous magnetism caused by the present geomagnetic field and equal to it. This data was used to determine the core dimensional orientation.

The paleomagnetic surveys showed that the fracture azimuths are 344° and 346° (figure 5). These values are similar to the fracture azimuths in the Lower Cretaceous deposits, i.e. 335° (figure 4). As

Bazhenov formation is related to the Late Jurassic and the formation top - to the Early Cretaceous, this could explain the similarities of the fracture azimuths in the investigated carbonated layer and the Lower Cretaceous deposits, according to Mikulenko. It should be noted that measurement errors in paleomagnetic studies may be up to 8 °.

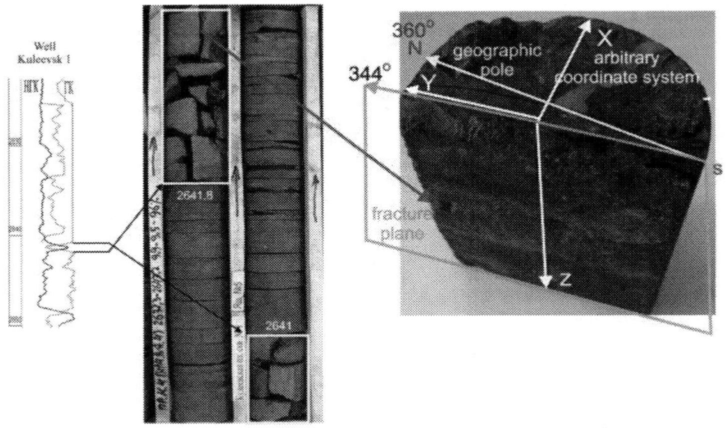

Figure 5. Paleomagnetic survey result

The vertical fracture orientation, their relation to specific lithological characteristics and the similarities of the fracture direction to the directions of planetary fractures (according to K.I. Mikulenko and S.S. Schultz) could explain the fact that the fractures in the Bazhen carbonated bed are planetary. Based on the calculated value of an average carbonated layer thickness and relationship of Rac, the fracture spacing in this bed could be determined. Therefore, according to this dependence, fracture spacing for a 2-meter bed is 2.9 m.

It should be noted that according to the well testing results, well K-1 produces water and oil. The oil chemical properties are identical to Bazhenov oil properties. Such planetary fractures may influence oil production within an interval. The fracture direction ensures the possible identification of productive zones based on seismic surveys or further design of EOR techniques.

5. Conclusion

1. It was proved that the smaller thicknesses the thin beds are, then the seismic resolution can be detected with 2D advanced seismic modeling. The zones of bed distribution can be determined by a qualitative indicator, while the quantitative indicator could estimate the bed thickness;
2. Based on seismic record data the influence of a thin carbonated layer was determined. The zones of its distribution were also defined. The map of promising carbonated bed thicknesses was plotted on the basis of the isochron thickness attribute. The average thickness within the investigated license areas is 2 m. The sign of a carbonate layer can be traced based on a 2D seismic forward model;
3. Fractures in carbonated layers were identified as planetary fractures. The paleomagnetic study showed the direction of these fractures. Based on the stated relationship between fracture spacing and bed thickness, fracture spacing for the layer under consideration was determined. For a 2-meter bed, fracture spacing is 2.9 m. Fracture spacing decreases to bed thickness decrease. For 0.8-meter carbonated bed in well K-1, fracture spacing is 1.3 m;
4. All studies in proposed license areas were considered to be conventional reservoirs. The research results proved the prospects of nonconventional reservoir within the territory of such license areas;
5. This approach based on the visual analysis of the wave pattern could be proposed for thin layer determination.

References

[1] Vygon G, Rubtsov A 2013 Nonconventional Oil: Will Bazhen Become Second Bakken? *Moscow School of Management*
[2] Alekseev D 2013 The Bazhenov Formation: In Search of Big Shale Oil in Upper Salym *Rogtec Magazine*
[3] Trofimuk A A, Karogodin Yu N 2008 Bazhenov Formation – Unique Natural Oil Reservoir *Neftegaz.ru*
[4] Gurari F G, Weitz E Ya, Melenevski V N 1988 Conditions of Formation and Methods of Exploration of Oil Fields in Argillites of the Bazhenov Formation *SNIIGGiMS*
[5] Gochioco L M 1991 Tuning Effect and Interference Reflections from Thin Beds and Coal Seams *Geophysics Vol. 56 No.8*
[6] Rac M V, Chernyshev S N 1970 *Fracturing and Properties of Fractured Rocks* (Moscow: Nedra Publishing House)
[7] Mikulenko K I, Shemin G G 1971 Fracturing of Jurassic and Lower Cretaceous rocks of Ob-Irtysh Interfluve *Geophysics and Mineral Stock*
[8] Dolitsky V D 1978 *Reconstructions of Tectonic Structures* (Moscow: Nedra Publishing House)

IOP Conf. Series: Earth and Environmental Science **24** (2015) 012036 doi:10.1088/1755-1315/24/1/012036

Petroleum Generation Potential of Bituminous Mudstones in Tomsk Region Bazhen Suite (Western Siberia)

V B Belozerov[1], V E Baranov[2] and A Y Dmitriev[3]

[1,2,3] Tomsk Polytechnic University, Tomsk 634034, Russia

E-mail: [2]BaranovVE@hw.tpu.ru

Abstract. The article reviews the prognosis for Bazhen suite HC generation potential in order to select primary commercially exploitable areas. A map of generation potential density for Bazhen suite passive reserves has been plotted. It is based on published papers on calculation of organic carbon concentration using the data of exploration and prospect well gamma-ray logging. The deliverables are compared with geochemical core examination results. Lithological heterogeneity of deposits underlying the Bazhen suite is systemized in order to estimate the possible decrease of the suite generation potential due to hydrocarbon migration to underlying deposits. Three types of cross-sections are distinguished – predominately sandy, alternating silt-argillous and argillous. An estimate of the proportion of Bazhen suite generation potential required for the discovered hydrocarbon deposit formation within oil-gathering fields of the first (sandy) type thoroughly explored by deep-hole drilling was carried out. Both quantitative (types 1, 3) and qualitative (type 2) estimations of decrease of the initial generation potential density are carried out for the above-mentioned types. Taking into account the specific character of shale oil development, a further work is required in order to design for each type its individual set of shale strata operating conditions.

1. Introduction

Combustible shale development as a promising source of HC resources has been attracting a lot of attention lately. Promising prospects of shale oil within the territory of Western Siberia are mostly based on bituminous shales of Bazhen suite the industrial oil-bearing capacity of which has been proved in a number of areas. Numerous scientific articles and production reports are devoted to studying the prospects of the suite petroleum bearing capacity. Field development programs using modern oil production technologies have been implemented.

Taking into account the necessity of drilling a considerable number of production wells for combustible shale development, it is advisable to perform such operations in deposits where the productive horizons are located stratigraphically lower than shale strata and have high production capacity for recoverable reserves. The South-Eastern part of Western Siberia is an ideal place for such innovations. Virtually all fields discovered in Tomsk Oblast are situated under a layer of Bazhen bituminous clay suite. There are no commercial HC reserves in Bazhen suite deposits found within this region, but tests have shown oil inflow in a number of areas and oil occurrence in well cores have been identified.

As for the question of shale oil exploitation efficiency, two main approaches can be singled out. They are related to Bazhen suite geographical resource distribution and its geological exploitation

IOP Conf. Series: Earth and Environmental Science **24** (2015) 012036 doi:10.1088/1755-1315/24/1/012036

background based on prognosis for various genesis reservoir rocks composing the suite [1, 3, 6]. The present paper deals with the former approach.

2. Estimation Method of Bazhen Suite Generation Potential Density

Multiple articles have shown a direct correlation between the Bazhen suite natural radioactivity and the organic carbon it contains [5], which allows using gamma-ray logging and standard electric logging data from wells drilled within the area to estimate petroleum resources distribution in the shale structure.

From this point of view, the work of Kontorovich V.A. is the most interesting for the Tomsk area [4]. His article is based on the established correlation between organic carbon content in the rock and well log data, and plotted maps of Bazhen suite content C_{org} and thickness according to the data collected from hundreds of exploration and prospect wells drilled within this area. The shown maps enable to estimate cumulative volume of organic carbon and the character of its distribution over the area, while research of suite generation potential [2] allows estimating hydrocarbon raw stock volume.

The conducted research showed that if the rock contains 10-15% organic carbon, the coefficient of liquid hydrocarbon generation may vary from 0.088 to 0.18. The generation coefficient characterizes the amount of hydrocarbons in $1m^3$ of source rock containing one ton of organic matter at present catagenesis stage. In this connection, the preliminary estimation of the generation density of hydrocarbon passive reserves in Bazhen suite can be defined as a product of the following factors – suite thickness, organic carbon percentage in rock, rock specific gravity and generation coefficient within the examined area (figure 1).

The maps presented in this article can be used to identify suite thickness and C_{org} content [4]. The values of other factors such as specific gravity (2.3 g/cm^3) and average generation coefficient within the examined area (0.1) were adopted according to the data in the article [2]. Analyzing the resulting map, it should be noticed that there are substantial territorial differences in the distribution of the calculated generation potential density.

It is more significant within negative structures of the first order – Ust-Tymsky, Nyurolsky and Koltogorsky oil and gas areas (OGA) where these values vary from 300-400 to 1000 thousand tons/km^2. The considered parameter is lower for two large positive structures Nijnevartovsky, Kaimysovsky, Srednevasyugansky and Parabelsky oil and gas areas. It is estimated at 100-500 thousand tons/km^2. Pudinsky and Aleksandrovsky oil and gas areas can be considered as exceptions , because the local heightened anomalies of hydrocarbon generation density can be noted there, due to rift zones (Pudinsky OGA) and granite massifs (Aleksandrovsky OGA). As a whole, a general decrease of hydrocarbon generation density is noted in the eastern direction which is conditioned by facial substitution of Bazhen bituminous mudstones for Maryanovsky suite mudstones.

The plotted map is in excellent correlation with laboratory research data on calculation of Bazhen suite hydrocarbon generation density that were accomplished in the Laboratory of Geochemistry and Crude Oils at OAO TomskNIPIneft by I.V. Goncharov. As from figure 1, the distribution of hydrocarbon generation density within a specific license block is similar to those maps plotted by different methods, while geochemical research carried out in step-out wells Zapadno-Ambarskaya 6 and Yujno-Nazinskaya 232 shows data that are close or correspond to the prospects of the parametric map.

Estimating the prospects of the Bazhen suite oil and gas content, it should be noted that part of its generation potential was depleted during the formation of hydrocarbon deposits situated in sandstones of Vasyuganskaya and Naunakskaya suites underlying the shale formation.

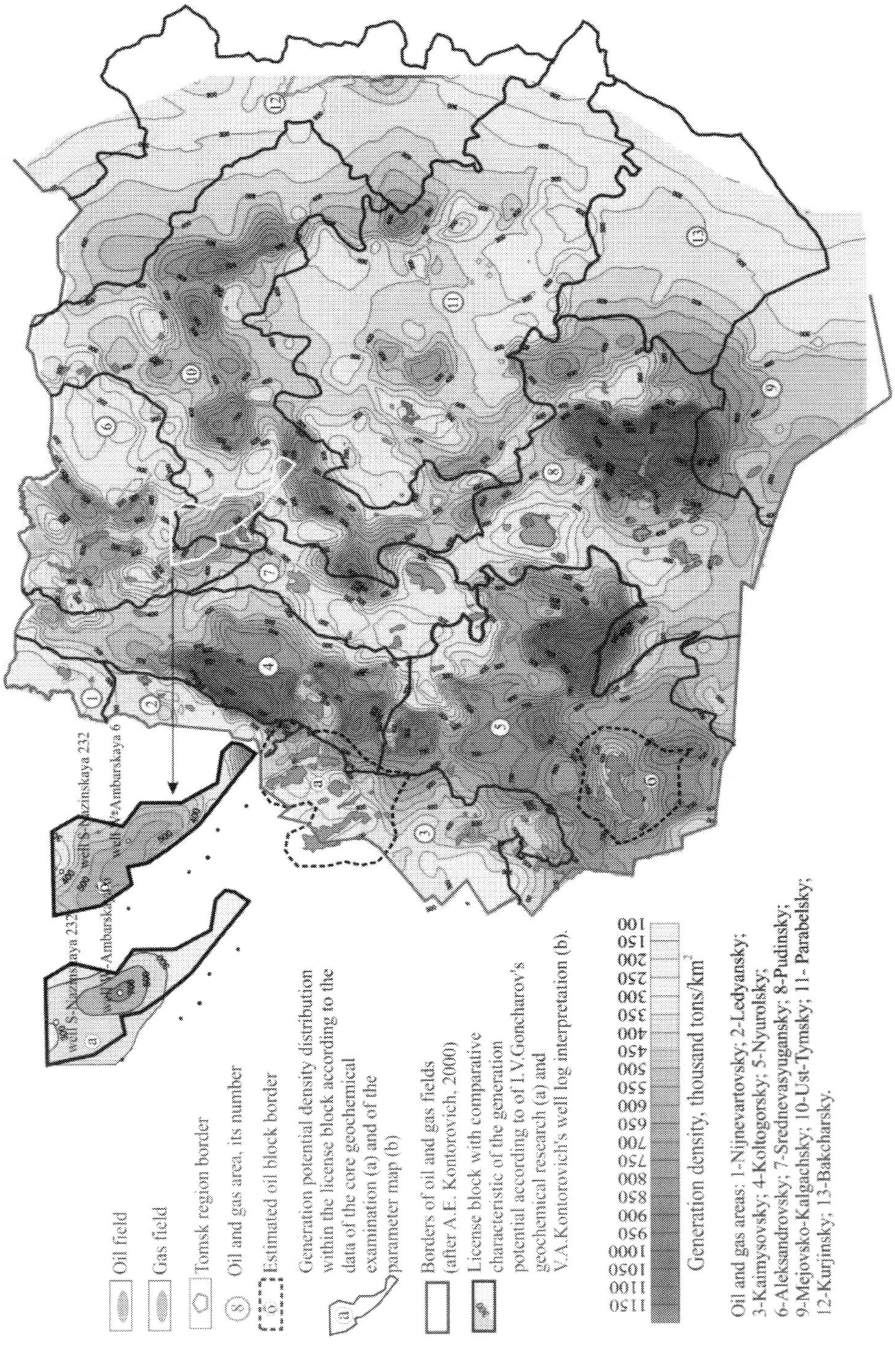

Figure 1. Map of HC generation potential density for Bazhen suite of Tomsk region area.

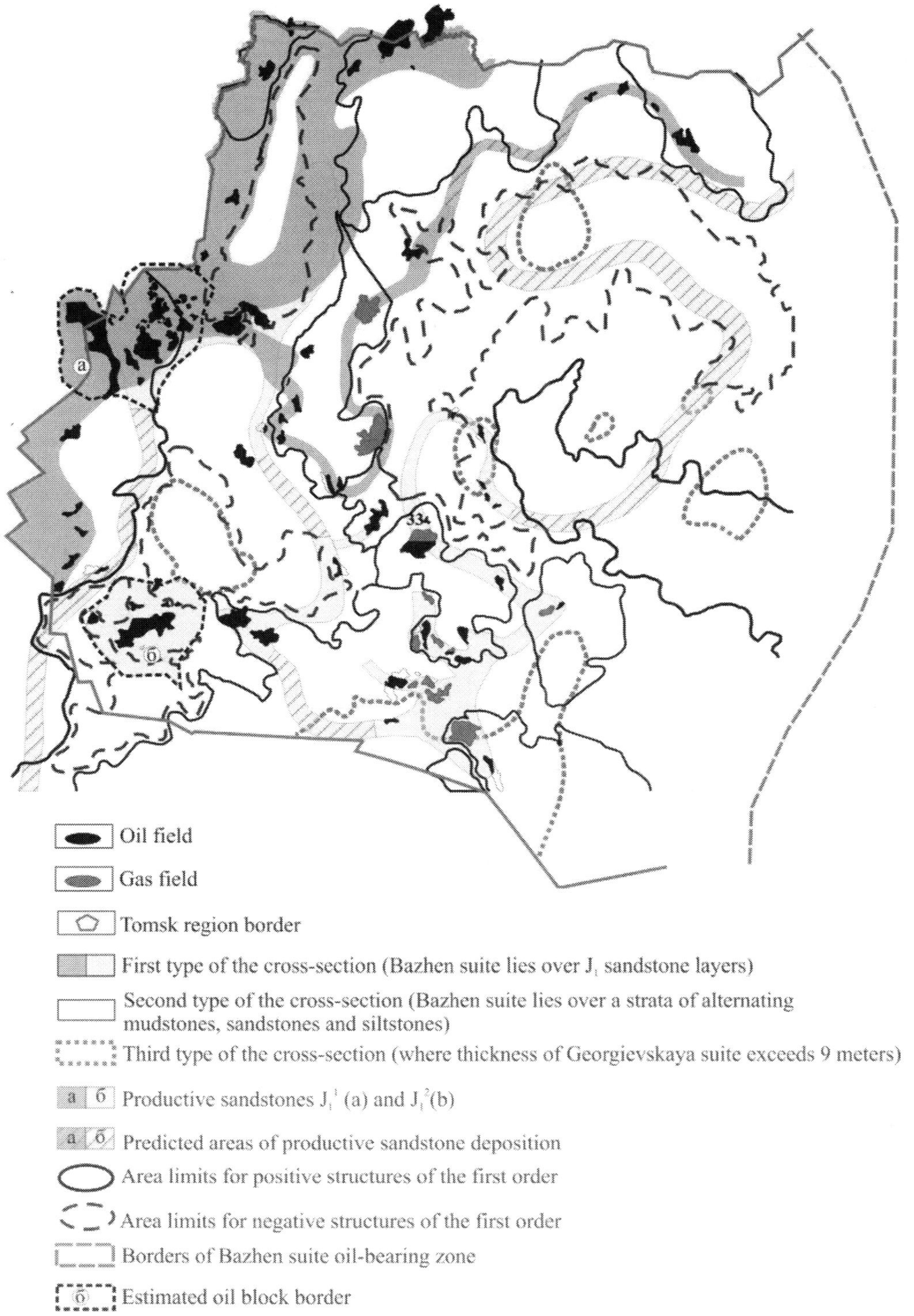

Figure 2. Bazhen suite zoning according to the type of lithological heterogeneity of underlying deposits.

Considering the scale of hydrocarbon migration, it is necessary to pay attention to the relation between the discovered deposits in Vasyuganskaya and Naunakskaya suites and lithological heterogeneity of rock directly underlying the Bazhen suite. Three types of cross-sections can be distinguished according to this characteristic. In the first type, Bazhen suite lies alone or jointly with 2-8 meters thick underlying Georgievskaya suite mudstones on top of J_1 sandstone reservoirs of Vasyuganskaya suite. Locally developed layers J_1^1, J_1^2 and layer J_1^{3-4} with a large areal extent (fig. 2) can be distinguished downright within the horizon. The second type is characterized by alteration of mudstones, low permeable sandstones and siltstones. The third type can be found in areas where the thickness of Georgievsky mudstones exceeds 9 meters, which, according to the study [2], eliminates the possibility of HC migration to underlying reservoirs. The presence of a considerably thick cluster of Kulomzinskaya suite mudstones overlying bituminous shales, significantly limits HC migration to sandstone layers overlying the deposits.

It is possible to give a preliminary estimate of generation potential decrease in the first type of the cross-section by calculating the generation potential density required for oil deposits formation within oil-gathering fields explored by seismic methods and deep-hole drilling.

There are two blocks in the considered area. The first block (figures 1 and 2) is situated within a positive structure of the first order corresponding to Kaimysovsky oil and gas area (OGA), where the Bazhen suite deposits are at MK_1^1-MK_1^2 catagenesis stage [7]. All anticline highs within the block under dense seismic observations were explored by deep-hole drilling. Commercial production of oil deposits in the strikes is carried out in layers J_1^1 and J_1^2 which supports the validity of proved hydrocarbon reserves estimation. Considering proved HC reserves, as a result of secondary migration from a specific oil-gathering area, the generation density of Bazhen suite can be determined for further deposit formation. Dividing the discovered prospective oil by the surface of oil-gathering block gives the required generation potential that can be estimated at 120 thousand tons/km^2.

The second block singled out within the negative structure of Nyurolsky OGA of Bazhen suite has similar structure and is at MK_1^2-MK_2 catagenesis stage. The density of generation potential of Bazhen suite required for oil deposit formation within this block is estimated at 140 thousand tons/km^2. According to the accomplished research the loss of bituminous shale PC generation potential within the cross-section of the first type can be up to 120 -140 thousand tons/km^2.

The cross-sectionof the second type is characterized by oil saturation identified from the cores retrieved from low-permeable siltstones. As a rule, there is no reservoir fluid inflow during well testing. This can be explained by hydrocarbons escaping into the low-porous siltstone and sandstone matrix during the first migration of hydrocarbons, with no lateral movement. It is difficult to estimate the loss of Bazhen suite generation potential within the cross-section of the second type. Nevertheless, the absence of oil deposits in underlying sand layers J_1^{3-4} can be indicative of stratigraphic scale limitation of hydrocarbon primary migration. As for the third type of cross-section, what has been stated above shows no loss in Bazhen suite generational potential.

3. Conclusion

The presented method of estimation of Bazhen suite generation potential density is based on the estimation of organic carbon content according to well log data (gamma-ray and standard electric logging) [4] and on defining zones for liquid hydrocarbon generation coefficient within the examined area [2].

Taking into account exploration and prospect well data, the resulting map shows considerable detalization improvement for HC generation potential density of Bazhen suite in Tomsk Oblast, and allows zoning the area according to the volumes of raw HC in bituminous formation. The comparative data analysis to specific geochemical core analysis results in previously and newly drilled wells showed good correlation, so it is possible to consider the presented map as a basis for commercial development of Bazhen suite reserves.

In order to differentiate the loss of bituminous formation generation potential resulting from hydrocarbon migration to underlying deposits, three types of deposit zones were identified: 1 –

sandstones, 2 – alternating mudstones, sandstones and siltstones, 3 – shales. These types of lithological heterogeneity of deposits underlying the Bazhen suite allow to estimate its residual generation potential at both qualitative (type 2) and quantitative (types 1, 3) levels. Taking into account the specific character of shale oil development, each type requires its own set of operating conditions.

References

[1] Gurari F G and Gurari I F 1974 Oil Deposit Formation in Bazhen Suite Shales in Western Siberia. *Oil and Gas Geol.* **5** 36 – 40.
[2] Gurari F G, Vaits E Y and Malenevsky V N 1988 Conditions of Oil Deposit Formation and Exploration Methods in Bazhen Suite Shales *Mineral Res.* 199
[3] Zubkov M Y, Pormeister Y A and Bondarenko P M 2002 Fractured Reservoir Prognosis Based on Tectonical and Physical Simulation Results for Bazhen and Abalakskaya Suites Deposits. *Oil and Gas Potential Enhancement in Khanty–Mansi Autonomous Okrug* **1** 244 – 253
[4] Kontorovich V A 2001 Generation potential of Volsky Deposits in South-East of Western Siberia. *Oil and Gas Geology* **1** 26-32
[5] Parfenova T M, Melenevsky V N and Moskvin V I 1999 Gamma-Ray Logging for Organic Matter Test in High-Carbon Sedimentary Formations (Illustrated by Bazhen Suite Example) *Oil and Gas Ind. Geology, Geophysiscs and Oilfield Dev.* **11** 29-34
[6] Khalimov E M and Melik-Pashaev V S 1980 On Exploring Production-Scale Petroleum Agglomerations in Bazhen Suite *Oil and Gas Geology* **6** 1 – 10
[7] Fomin A V 2011 Organic Matter Catagenesis and Oil-and-Gas Content in Mesozoic and Paleozoic Sediments of the Western-Siberian Megabasin *Novosibirsk: Inst. of Petroleum-Gas Geol. and Geoph. of the Siberian Branch of the RAS* 184 – 193.

IOP Conf. Series: Earth and Environmental Science **24** (2015) 012037 doi:10.1088/1755-1315/24/1/012037

Development prospects for Bazhenov formation bituminous claystones in the southeast of the West Siberian Plate (Tomsk Region)

V B Belozerov[1], V E Baranov[2] and A Y Dmitriev[3]

[1,2,3]Tomsk Polytechnic University, Tomsk 634034, Russia

E-mail: BelozerovVB@hw.tpu.ru

Abstract. The article considers the use of time-tested bituminous shale development technologies in the southeast of the West Siberian Plate (Tomsk Region). A research target is the bituminous claystones of the Bazhenov formation. The Bazhenov formation was divided into stratigraphic units, and the sequence of forming these units within the territory under study was restored on the basis of the peculiarities of the Bazhenov formation section structure with gamma-ray logging data. The joint analysis of logs (gamma-ray logging, resistivity logging, SP logging, neutron gamma-ray logging, induction logging) and core data revealed the wide development of carbonatization processes in the lower stratigraphic units of the Bazhenov formation, whereas the upper, most bituminous units do not have the signs of carbonatization. It is a favorable factor for using the existing technologies of producing hydrocarbons from analogous sedimentary deposits. Considering the degree of lithological disconnection of an oil-bearing object from adjacent reservoir formations allows making a conclusion that the application of the existing technologies for developing bituminous claystones can be implemented only in the insignificant territory of the area under study where the Bazhenov formation overlies Georgiev formation claystones being over 8 meters thick. To develop the oil and gas potential of the Bazhenov formation in other territories, there will be a need for new technologies that allow forming a fracture-cavernous matrix in a carbonate rock.

1. Introduction

Much attention has recently been paid to the development of oil shales as a promising source of hydrocarbons. The "shale oil" prospects within the West Siberian Plate are connected with the bituminous claystones of the Bazhenov formation where a commercial oil content has been proven for a range of areas. Many scientific publications and production reports are devoted to the oil-and-gas content of the formation. Its development projects with the use of new technical oil production capabilities are being introduced.

Taking into account that there is a need for drilling a significant number of production wells during the development of oil shales, it is most expedient to perform these operations at fields where an oil production interval is stratigraphically lower than the "shale formation" and is characterized by the high recovery of commercial reserves. The territory of the southeastern part of the West Siberian Plate is an ideal object for such innovations. Practically all fields discovered in Tomsk Region are located under the bituminous clays of the Bazhenov formation. No commercial hydrocarbon fields have been located in the Bazhenov formation deposits within the region under consideration, but its study has revealed oil influxes for a range of areas. Besides, there is practically ubiquitous oil staining in the core taken from the wells drilled.

There are two important issues associated with the profitability of shale oil development connected with the territorial distribution of Bazhenov formation resources and geological prerequisites to its development based on the application of corresponding technologies providing for oil production efficiency [4, 6, 11, 12]. The material presented is devoted to the second route.

2. Assessments of prospects for developing Bazhenov formation bituminous claystones

The efficiency of developing hydrocarbons in the Bazhenov formation deposits depends on understanding the geologic conditions of its establishment. Both the paleoclimate of the sedimentation basin and the specificity of the geochemical depositional environment have had a significant influence on the lithological and geophysical characteristic of the Bazhenov formation. The significant area of the sea basin, low tectonic activity of the adjacent alimentation zones, recurrent changes in the climate and the reducing environment in the upper part of the sedimentary layer have contributed to the long-term development of a bituminous argillaceous-carbonate, argillaceous-siliceous deposit that is unambiguously interpreted with radioactive, sonic and electric logging methods.

The direct correlation relationship between the value of the natural radioactivity of claystones in the Bazhenov formation and organic carbon in the rocks reasoned by the presence of zoo- and phytoplankton [5, 7] allows segmenting the deposits under consideration with gamma-ray logging in more detail [1, 3, 10].

In compliance with the analysis of the thickness and the observed differentiation of the gamma-ray logs of the Bazhenov formation deposits connected with the presence of lower-radioactivity rock interlayers, it is possible to single out four units in its composition [1]. These units reflecting the evolution of organic life in the Volgian basin are characterized by the specific level of gamma-ray logging values and are divided by one or a group of carbonate or siliceous interlayers of an insignificant thickness (0.5-2.0 m). (figure 1B, C).

The feasibility of dividing the Bazhenov formation into units on the basis of gamma-ray logging is confirmed by the studies of uranium content in core samples (figure 1B). The sampling points corresponding to the lower values of gamma-ray logging are characterized by the uranium concentration that does not exceed the values that are typical of the Vasyugan formation deposits, which testifies to significant changes in the ecosystem of this sedimentary basin.

The peculiarities of unit construction allow singling out two groups of sections (figure 1A, C). The first group (A) is characterized by the increased values of gamma-ray logging in the second and the fourth units, and the second group (B) demonstrates dropping values from upper unit 4 to lower units 1 and 3.

It is possible to determine 4 types of sections on the basis of the number of units within the groups under consideration. The first section type includes units 1-4. It is spread within the limits of the Pudinsky megalithic bank. The second type comprises units 2-4. This type covers the Nyurol depression, Parabelsky, Azharminsky megalithic banks, the eastern part of the Ust-Tym depression. The third section type comprises units 4 and 3. This type is spread in the wide territory of the Ust-Tym depression, Koltogor mega-downwarp, Kaymysov, Nizhnevartovsk arches and Aleksandrovsky megalithic bank. The fourth section type comprises unit 4. It is localized within the second type development area and is confined to the most structurally elevated areas of first-order positive structures.

One of the main peculiarities of dividing the Bazhenov formation into units is the fact that the boundaries of the units are isochronous, as they reflect the intensity of zoo- and phytoplankton development at different stages of the Volgian basin development. It allows using them as reliable chronologic markers in paleographic reconstruction for the Bazhenov formation development time.

In compliance with the territorial zoning in the development of these section types at the initial stage of the Volgian transgression, the bituminous deposits of the Bazhenov formation were formed in the southwestern part of the territory under study (first type sections). The step-by-step extension of the sea basin manifests itself in the formation of the second and third type sections. The fourth section type most likely characterizes paleo-elevations of the sea basin floor within the limits of major platform rises experiencing active consedimental growth during the Volgian transgression.

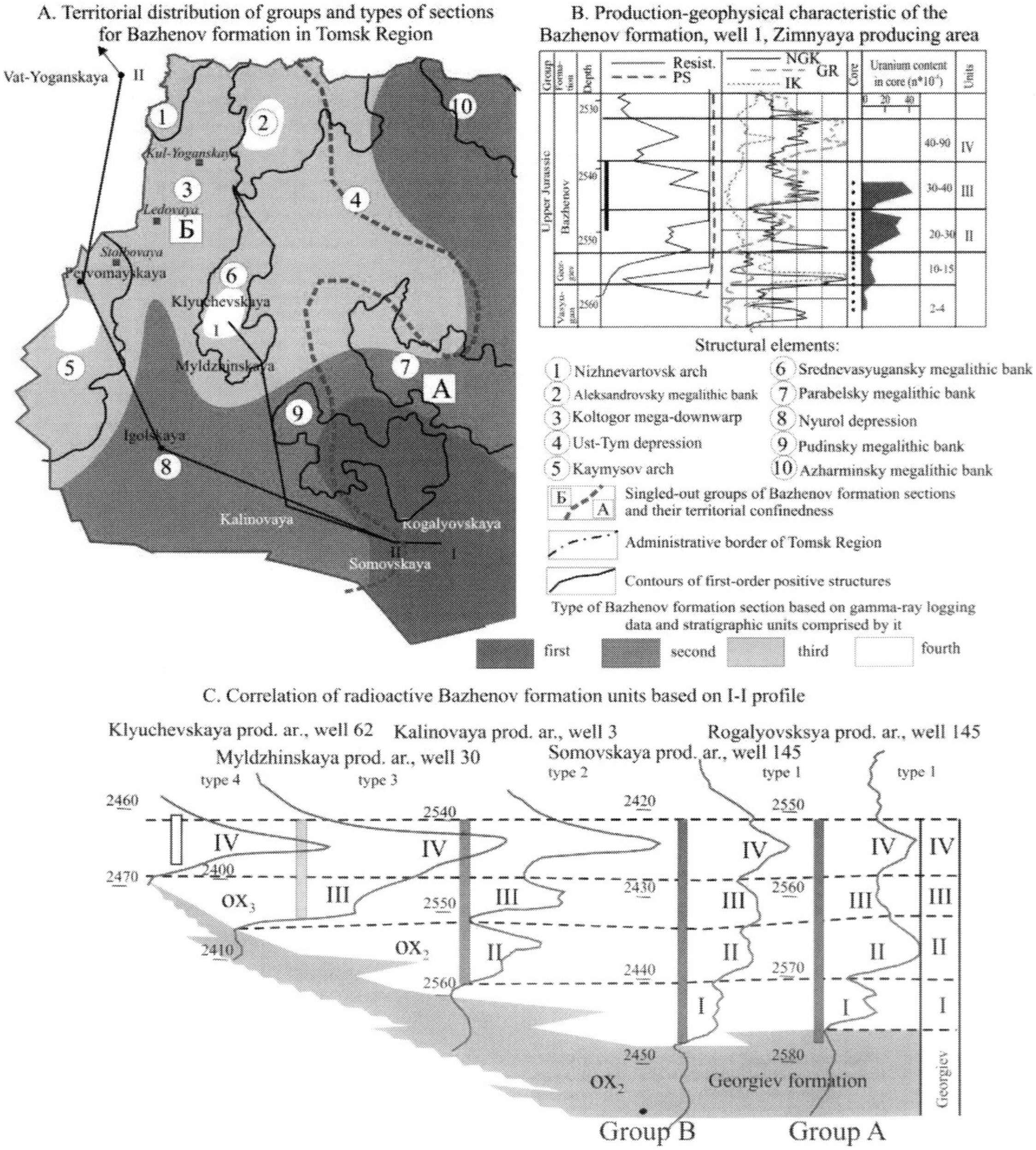

Figure 1. Segmenting (B, C) and zoning (A) the territory of the West Siberian Plate by types of Bazhenov formation section [1].

The microfaunal determinations of age for the lower Vasyugan formation deposits do not contradict the proposed Bazhenov formation development model either [2]. In compliance with figure 2, the western direction is characterized by the age slide of the Vasyugan formation top from Middle

Oxfordian (Somovskaya producing area) to Late Kimmeridgian (Vat-Yoganskaya producing area). Taking into account that there are no paleontological fragments in a good state of preservation in the Bazhenov formation territory under consideration, it allows determining the age of the first and the second units as Late Oxfordian, that of the third unit – as Kimmeridgian, and the age of the fourth unit – as Early, Middle Volgian.

The world experience shows that one of the main prerequisites to the efficient development of bituminous shales is the presence of carbonate layers in their composition [12]. Performing multistage hydraulic fracturing within the limits of these layers allows forming a conductive matrix able to drain hydrocarbons from the superincumbent and underlying beds of bituminous shales.

The presence of carbonate interlayers in the first, second and third units of all section types determined has been revealed in the territory under consideration with lithological studies and logging data [9]. These interlayers in most cases result from the replacement of radiolarites and silicites with calcite, dolomite and limestone.

As one can see from figure 3, carbonatization manifests itself as a complex of shallow-thickness interlayers. Carbonatization is absent or it is present as "traces" in the fourth unit that is characterized by the highest concentrations of organic carbon, which leads to higher gamma-ray logging values.

When considering the production-geophysical characteristic of the Bazhenov formation in terms of singling out section carbonatization zones, one should note the following:

- it is not always possible to register the presence of carbonatization in the formation section with gamma-ray logs and thermal neutron logs being the most popular types of logs for prospecting and exploration wells;
- the fourth unit having the highest content or organic carbon is represented with lower values of apparent formation resistivity (resistivity logging) vs. the third and the second units, which is connected with the absence of carbonized interlayers in it. This fact is also mentioned in work [4]. The "shift" of the apparent resistivity anomalies vs. the gamma-ray logging anomaly in the Bazhenov formation sections can be connected with carbonatization in the sections of the third, second and first units (figure 3).

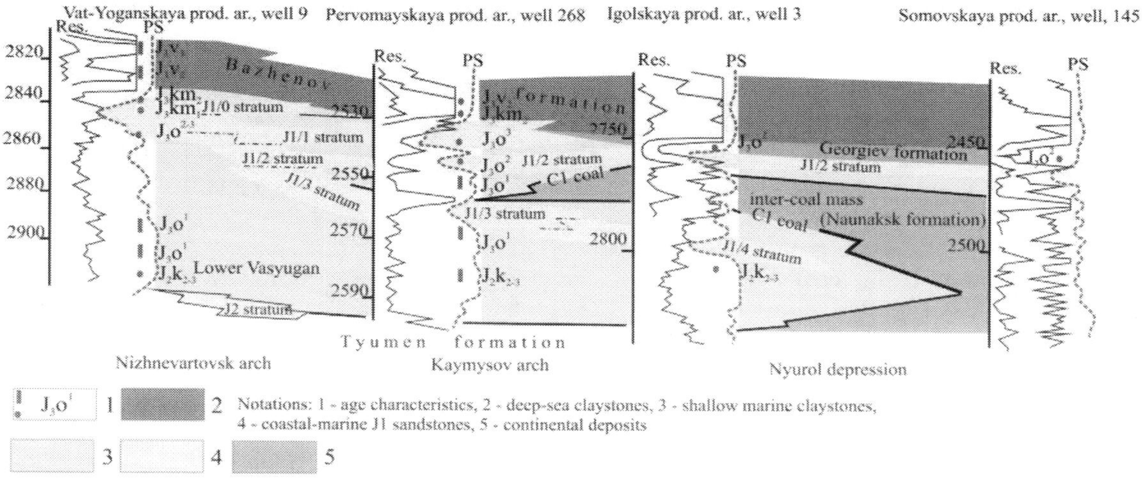

Figure 2. Lithologic-paleontologic correlation of Upper Jurassic deposits in the central and southeastern parts of the West Siberian Plate [2].

A prerequisite to the efficient development of bituminous claystones with the use of hydraulic fracturing is the presence of impermeable beds underlying and overlying carbonate deposits in the composition of the shale formation. The reason is that the need for drilling horizontal wells with subsequent multistage hydraulic fracturing through carbonate deposits requires the lithological separation of the object to be developed from the above and below lying reservoirs. Such conditions are consistently observed for the central part of the West Siberian Plate where the Bazhenov formation lies on the Abalak clay formation and is overlaid with claystones of the Kulomzin, Megion, Sortym, Akhsk and Frolovskaya formations.

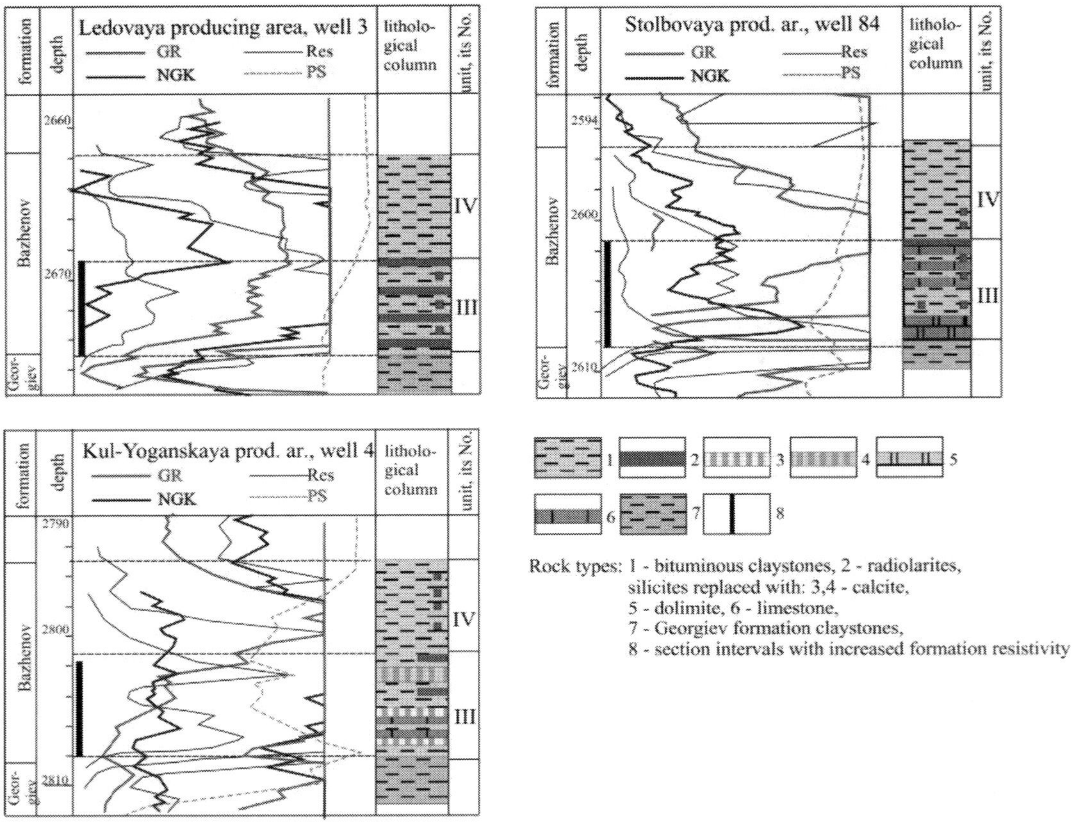

Figure 3. Main types of the Bazhenov formation section for the Kaymysov acrh and Koltogor mega-downwarp [9].

Within the limits of the territory under study, bituminous shales are deposited on the sand-clay J_1 horizon of the Vasyugan formation and on the continental deposits of the Naunaksk formation and are overlaid by Kulomzin formation claystones. The lithological separation of the Bazhenov formation from the lower deposits is possible in case of the presence of Georgiev formation claystones (figure 1 and 4).

Figure 4 presents an isopach map for the Georgiev formation based on the results of drilling prospecting and exploration wells. The map shows that the thickness of this formation varies from 0 to 6 meters in a significant territory, which is insufficient for the high-quality development of the shale formation with the existing method.

Assuming that such a method of Bazhenov formation operation is possible provided the Geogiev formation thickness is 8 meters, its area of application is limited to the central and southeastern parts of the Nyurol depression, southeastern part of the Pudinsky megalithic bank, northern and southern parts of the Ust-Tym depression.

Other territories need the development of their own methods of bituminous claystone development, excluding hydraulic fracturing for forming a fracture-cavernous matrix of a carbonate formation or restricting its penetration into underlying rocks. These methods will allow producing shale oil not only in areas with an enhanced generation potential of the Bazhenov formation, but also in the fields of Tomsk Region under development (figure 4). It will significantly enhance oil production efficiency in this region.

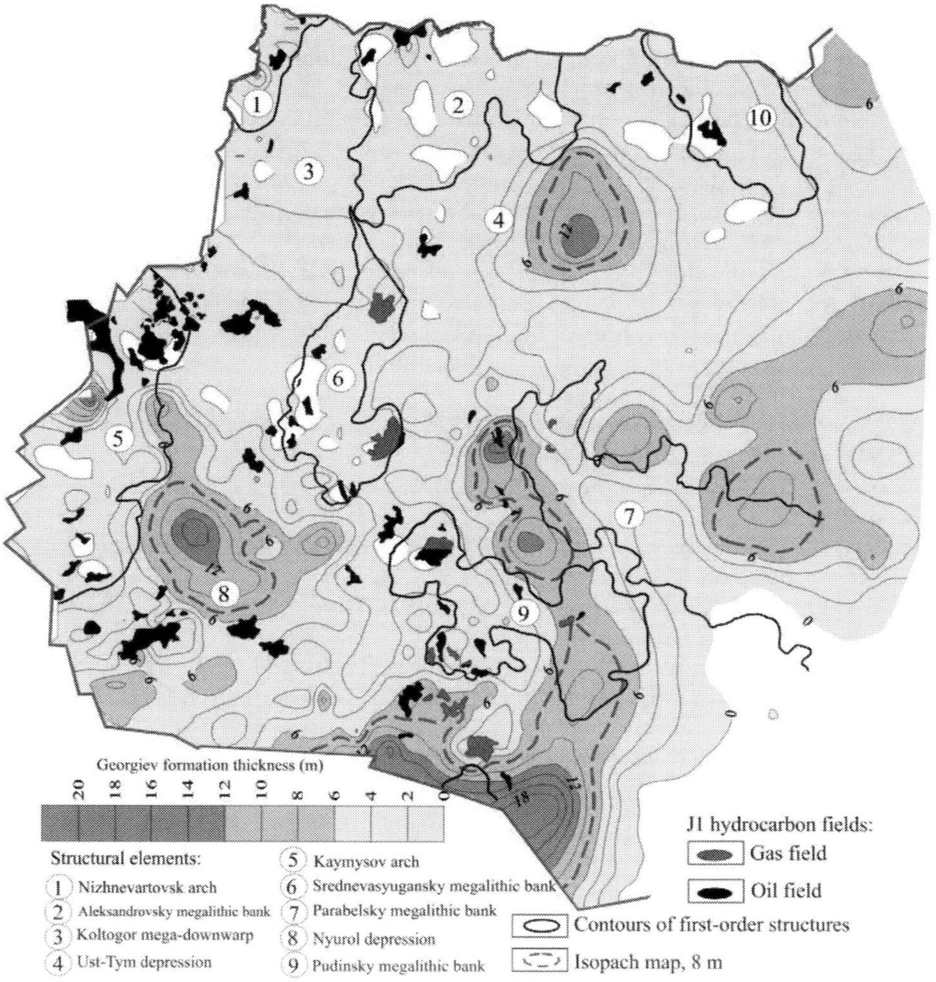

Figure 4. Thickness map for the Georgiev formation of the southeastern part of the West Siberian Plate.

3. Conclusion

This work devoted to the use of the existing bituminous shale development methods in relation to the Bazhenov formation deposits in the southeast of Western Siberia (Tomsk Region) is based on the analysis of its structure peculiarities obtained with logging techniques (gamma-ray logging, thermal neutron logging, resistivity logging, SP logging, induction logging) and analyzing core samples from prospecting and exploration wells.

The division of the formation into stratigraphic units with gamma-ray logging data and the analysis of the uranium content in the core have allowed determining the spatial sequence of their formation, whereas the existing connection between the organic carbon content and the values of gamma-ray logging has contributed to the high-quality assessment of the hydrocarbon generation potential for each of these units. The lithological studies of the Bazhenov formation rocks [9] and the analysis of the correspondence between gamma-ray logging anomalies and apparent rock resistivity testify to the active manifestation of layer-by-layer secondary carbonatization in the lower, less productive units of the Bazhenov formation. It allows using the carbonate part of the Bazhenov formation section to form a filtration fracture-cavernous matrix in it by performing multistage hydraulic fracturing with a subsequent mud-acid treatment. The purpose of this operation is draining oil from the upper bituminous unit. As there are no reliable impermeable beds in the bottom of the Bazhenov formation, using this technique is limited to insignificant territories where the underlying claystones of the Georgiev formation are over 8 meters thick, which guarantees the reliability of hydrodynamic isolation for the subjacent reservoirs of the Vasyugan (Naunaksk) formation and hydraulic fracture cracks. For a significant part of the territory under study, there is a need to develop a unique bituminous claystone development technology based on forming a filtration carbonate formation matrix without hydraulic fracturing or restricting its penetration into the underlying reservoirs of the Vasyugan and Naunaksk formations. Such technologies will allow producing shale oil not only in the areas with an enhanced generation potential of the Bazhenov formation, but also in the fields of Tomsk Region under development. It will significantly enhance oil production efficiency in this region.

References

[1] Belozerov V B, Danenberg E E and Naruta Yu S 1979 Bazhenov formation segmentation and section types *Trudy of ZapSib NIGNI* **141** 87–9
[2] Braduchan Yu V 1987 Biostratigraphic aspects of Upper Jurassic deposits *Biostrat. of Mesoz. Dep. in West. Sib.* 33–4
[3] Gaydeburova E A 1982 Types of sections of West Siberian domanikites *Sib. Doman. and their role in oil-and-gas pot. Trans. of SNIIGGiMS* 23–32
[4] Gurari F G and Gurari I F 1974 Formation of oil pools in Bazhenov formation claystones in Western Siberia *Geol. Nef. i Gaz.* **5** 36–40
[5] Zakharov V A 2006 Conditions of Volgian-Berriasian high-carbon Bazhenov formation development in Western Siberia according to paleoecology data *Evol. of biosph. and biodiver.* 552–68
[6] Zubkov M Yu, Pormeyster Ya A and Bondarenko P M 2002 Forecast of fractured reservoirs in Bazhenov and Abalak formation deposits based on tectonophysical modeling results. *Khanty-Mansi Auton. Ar. oil-and-gas pot. implem. Khanty-Mansiysk* **1** 244–53
[7] Kontorovich V A 2001 Generation potential of Volgian deposits in the southeast of Western Siberia *Geol. Nef. i Gaz.* **1** 26–32
[8] Nemota V D, Koloskov V N and Pokrovsky B G 2011 Formation processes in the carbonized reservoirs of the argillaceous-siliceous deposits of the Bazhenov formation in the west of Latitude Priob *Razv. i Okhr. Nedr* **12** 31–5
[9] Predtechenskaya E A, Krol L A, Gurari F G, Sapyanik V V, Perzio G N and Malyushko L D 2006 On the genesis of Bazhenov formation carbonates in the central and southeastern areas of the West Siberian Plate *Litosf.* **4** 131–48
[10] Fridman M Ya, Dieva E V, Muravyev P P, Glebocheva N K and Ermakova S A 2006 Experience in geotechnical study usage among geological and geophysical studies for the Bazhenov formation of the Middle Ob *NTV Karot.* **2–4** 143–5
[11] Khalimov E M and Melik-Pashaev V S 1980 On searching for commercial oil accumulation in the Bazhenov formation. *Geol. Nef. i Gaz.* **6** 1–10.
[12] Olesen N L 2010 *Bakken Oil Resource Play, Williston Basin (US). Overview and Historical Perspective.* (Continental Resources presentation)

Optimization of gas condensate Field A development on the basis of "reservoir – gathering facilities system" integrated model

E A Demidova[1] and O V Maksyutina[2]

[1,2] Tomsk Polytechnic University, Tomsk, Russia

E-mail: DemidovaEA@hw.tpu.ru

Abstract. It is known that many gas condensate fields are challenged with liquid loading and condensate banking problems. Therefore, gas production is declining with time. In this paper hydraulic fracturing treatment was considered as a method to improve the productivity of wells and consequently to exclude the factors that lead to production decline. This paper presents the analysis of gas condensate Field A development optimization with the purpose of maintaining constant gas production at the 2013 level for 8 years taking into account mentioned factors . To optimize the development of the filed, an integrated model was created. The integrated model of the field implies constructing the uniform model of the field consisting of the coupling models of the reservoir, wells and surface facilities. This model allowed optimizing each of the elements of the model separately and also taking into account the mutual influence of these elements. Using the integrated model, five development scenarios were analyzed and an optimal scenario was chosen. The NPV of this scenario equals 7,277 mln RUR, cumulative gas production – 12,160.6 mln m^3, cumulative condensate production – 1.8 mln tons.

1. Introduction

Gas-condensate Field A was discovered in 1963 and is located in the west of Tomsk Oblast. The geological cross-section of Field A is represented by sandy-clayey formations of the Mesozoic-Cenozoic sedimentary cover and Paleozoic basement formations. There are two gas-condensate deposits in this field. These deposits are located in the J_1 and J_2 formations.

The gas-condensate deposit of the J_1 formation is a layer-arch deposit. Reservoir rocks are sandstone and siltstone. A cap rock for this deposit is Georgiev and Bazhenov suites, the cumulative thickness of these formations is up to 51 m. The mean value of porosity is 16%, gas saturation – 66%, permeability – 3.5 mD, net gas thickness – 9 m. The gas-condensate deposit of the J_2 formation is a layer-arch deposit too. The deposit was penetrated by a well at the depth of 2,305-2,402 m. Reservoir rocks are medium and fine-grained sandstones covered by clay. The cap rock for the deposit is Nizhnevasyugan suite clay. The mean value of porosity is 16%, gas saturation – 66%, permeability – 3.3 mD, net gas thickness – 6.2 m.

The main development target of the Field A is the J_1 gas-condensate formation. This production horizon is divided into two productive layers, J_1^{1+2}, J_1^{3+4} and one J_1^5 interlayer. The gas water contact is located at -2,245m TVD, it is common of all the productive formations. The general reservoir and fluid properties are presented in Table 1.

Table 1. Reservoir and fluid features

Parameter	Value	
Initial reservoir pressure, atm	246	
Reservoir pressure as of 01.01.2014, atm	Area 1 152	Area 2 205
Dew point pressure, atm	246	
Reservoir temperature, ^{0}C	86	
Permeability, mD	3.5	
Gas viscosity, cP	0.018	
Condensate density , g/m^3	0.76	
Net pay thickness, m	10	
Gas compressibility factor, dimensionless	0.87	
GWC, m (TVD)	-2,245	
GIIP, MM m^3	23,908	
CIIP (geologic/recoverable), M ton	4,662/2,727	
Gas recovery factor as of 01.01.2014	0.24	
Condensate recovery factor as of 01.01.2014	0.207	

The following factors restrict well operations and limit production in Field A:

• accumulation of condensate in the wellbore at smaller gas rates relatively to the base rates (i.e. gas velocity is insufficient to lift the fluid, liquid loading).

• long periods of uncontrolled wells exploitation with reduced gas flow rates due to liquid loading.

• condensation in the reservoir due to reservoir pressure reduction below the dew point pressure resulting in condensate deposition in the near-wellbore area (condensate banking).

Reducing the impact of these factors on field development is a real opportunity to increase the productivity of the wells in the next few years.

2. Gas and condensate production technology of Field A

The gas and condensate Field A was put into production in 2002. As it has already been mentioned, the field has two deposits in the J_1 and J_2 formations. There is one exploration target and all formations are being developed simultaneously. The Field A is developed by depletion drive. Gas injection is not provided in this field. The dew point pressure equals the reservoir pressure of 246 atm. Today, the number of production wells in the field is 22 (figure 1). These wells are located within four clusters (figure 1): five wells in the first cluster, seven wells in the second cluster, six wells in the third cluster and four wells in the fourth cluster. It is planned to drill four new wells in Field A in accordance with the Field A technological development plan. Two wells in the 6th cluster are planned to be drilled at the beginning of 2017 and other wells are planned to be drilled in the 7th cluster at the beginning of 2019 (figure 1, clusters 6, 7).

Hydraulic fracturing treatments were made in wells A21, A23, A25, A27, A42, A43, A44, all of which were successful. The well experienced a 3.5-time increase in productivity on the average.

There are two development areas in Field A: Area 1 developed in clusters 1, 2, 3 and Area 2 developed in cluster 4. Area 2 was put into production only at the end of December 2012. There is no connection between these two areas. Therefore, reservoir pressures were different in 2014: 152 atm in Area 1 and 205 atm in Area 2 (01.01.2014).

The gas gathering pipelines from the clusters are interconnected and the gas condensate fluid is transported to a gas treatment unit located in Field A under its own pressure. The lengths of the gathering pipelines from the clusters to the gas treatment unit are as following:

from the first cluster – 10,435 m;

from the second cluster – 1,980 m;

from the third cluster – 7,280 m;
from the fourth cluster – 7,550 m.

The pipelines from the clusters are made of pipes with OD = 273 mm and wall thickness of14 mm (the first, second and third clusters) and with OD = 219 mm and wall thickness of 12 mm (cluster 4).

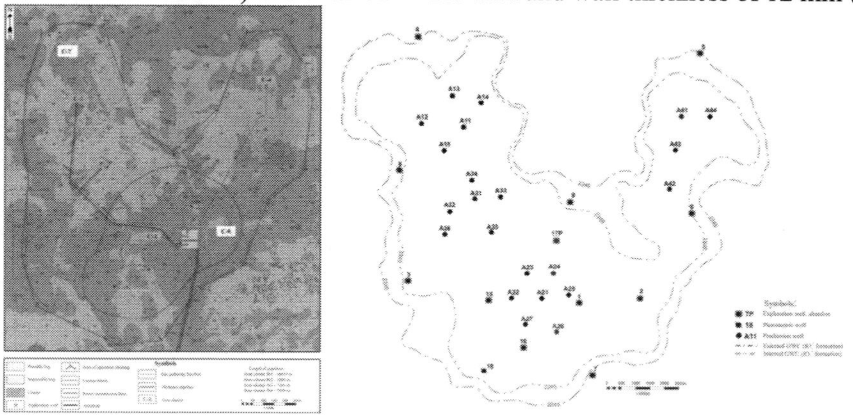

Figure 1. Cluster location layout (left) and well pattern at Field A (right)

3. Factors leading to production decline in Field A

There are two main factors leading to the production decline in Field A and they are liquid loading and condensate banking.

3.2. Liquid loading in Field A

During the development of a gas condensate field, the gas velocity should be high enough to transport liquids to the surface. It is known that gas velocity reduces with time. As a result, the gas velocity can be insufficient to transport condensate to the surface. Therefore, liquid will flow down to the bottomhole. The increase in the percentage of accumulated liquid while a well is flowing can result in a production reduction. A well can also stop if the liquids are not removed continuously [1]. To make an efficient plan and design for tackling the liquid loading problem, it is essential to predict when wells might begin to experience liquid loading [1-3].

First of all, the liquid loading velocity (critical velocity) and critical gas rate were calculated on the basis of the correlations suggested by Turner et al. "Critical velocity" is generally defined as the minimum gas velocity in a well required to move droplets of liquid upwards. In general, except for the critical velocity, the critical gas rate was also calculated. Critical gas rate is the minimum gas rate required to move liquid upwards.

The theoretical equation from Turner et al. for the critical velocity of gas V_{cg} to lift a liquid is as follows [1, 4]:

$$V_{cg} = \frac{1.593 \times E \times \sigma^{\frac{1}{4}} \times \left(\rho_l - \rho_g\right)^{\frac{1}{4}}}{\rho_g^{\frac{1}{2}}} \text{ (ft/sec)} \quad (1)$$

Where,
σ – surface tension, dynes/cm;
ρ_l – density of liquid, lbm/ft^3;
ρ_g – density of gas, lbm/ft^3.
E – correction (efficiency) factor. The value of E for Turner model equals 1.2. To convert ft/sec to m/sec, it is required to divide the above -mentioned expression by 3.28 (1 m = 3.28 ft).

From critical velocity (1), a critical gas flow rate (q_g) can be presented in the following way:

$$q_g = \frac{3.067 \times P \times V_{cg} \times A}{(T+460) \times z} \text{ (Mscf/D)} \tag{2}$$

Where,
P – surface pressure, psi;
T – surface temperature, ^{0}F;
A – tubing cross sectional area $\left(A = \frac{(\pi) \times d_{ii}^2}{4 \times 144} \right)$, ft^2;
dii – tubing ID, inches. To convert q_g (Mscf/D) to q_g (10^3 m^3/day), it is required to divide the above- mentioned expression by 35.31.

Due to these equations, it can be concluded that if the flow rate of the well is greater than the critical rate, then liquid loading will not be expected. Critical velocity equals 1.1 m/s, critical gas rate equals 26.6 10^3m^3/day for wells with 73 mm OD tubing and equals 40 10^3m^3/day for wells with 89 mm OD tubing (A12, A26, A27, A36) in Field A.

The gas velocity and gas rate of the wells are presented in figure 2. This figure also shows the critical gas velocity and the critical gas rate (straight lines). In this case, the wells A11, A12, A22, A24, A26 and A33 have a problem with liquid loading, while wells A35 and A31 have a gas rate close to the critical gas rate. There are several methods to solve this problem. For example, by using free piston; foam injection down to the bottomhole, lifting tubing with a smaller diameter, etc. [2 – 4]. All these methods are used to remove liquid from the bottomhole or increase the well productivity (to some extent). It was decided in this study to increase the productivity of the wells when liquid loading occurs in order to solve this problem. A hydraulic fracturing treatment was chosen as a method to improve production, exclude problems with liquid loading by increasing gas rate and, therefore, increasing gas velocity.

The value of erosional velocity was also estimated. To avoid the appearance of erosion in tubing, it is recommended that velocity in the tubing should be lower than erosional velocity. Erosional velocity was calculated with API RP 14E (1981, 1984) on the basis of the following empirical expression [5]:

$$v_e = \frac{C}{\sqrt{\rho_m}}, \tag{3}$$

Where,
C – empirical constant, usually between 100 and 125, it is 122 in SI units;
ρ_m – density of the fluid in the tubing, g/cm3;
v_e – erosional velocity, m/sec.

Erosional velocity changes from 10 m/sec to 13.5 m/sec in Field A (depending on the fluid density in the tubing) and on the average it is equal to 12.5 m/sec. According to figure 2, there are no wells with gas velocity higher than the erosional velocity. It is always checked due to further development prediction. No increase in gas flow velocity above the erosional velocity was detected during the years of stimulation.

3.2 Condensate banking in Field A wells

It is known that the accumulation of retrograde condensate through the reservoir and in the near-wellbore area of exploration wells differs due to changes in thermobaric conditions at the bottomhole of wells. Thus, based on the degree of condensate saturation, it is possible to separate two areas in porous media. They are the area of “static condensation” located far from a well and the area of “dynamic condensation” located near the wellbore. Condensation in the “static condensation” area is described by the process of differential condensation and depends on the pressure and composition of the initial mixture. Condensate accumulation in the “dynamic condensation” area depends on a fluid phase and mass transport of hydrocarbons. This process has the following mechanism: when there is gas passage through the point of a reservoir where pressure is lower than the dew point pressure, the condensation process occurs. In a high pressure gradient area, a dropped-out fluid (condensate) can be permanent (in case of critical saturation) or it can be moved with velocity, lower than the velocity of gas. With the passage of each new portion of gas through this point of the reservoir, retrograde condensate comes out of solution and,therefore, the process of condensate accumulation takes place here. As the result of this process, the saturation of liquid (condensate) in the near-wellbore area can exceed the average value of condensate saturation largely compared to the remaining part of the reservoir and condensate banking is formed near a wellbore [6].

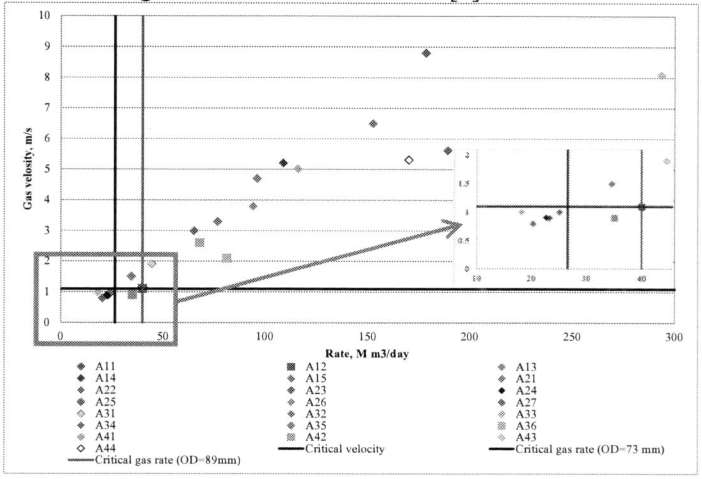

Figure 2. Gas velocity versus gas rate for wells (rhombus – wells with 73 mm OD tubing, square – wells with 89mm OD tubing)

At the pressure above the dew point, well rate is defined by permeability, net pay thickness of the reservoir and viscosity of gas. When the flowing bottomhole pressure falls below the dew point, liquid condensate builds up (condensate banking) near the wellbore area with high condensate saturation. The parameters of this condensate banking (size and saturation of condensate) depends on some additional factors, such as relative permeability of gas and condensate, pressure distribution in the reservoir and PVT properties of the system. Condensate saturation near a wellbore can reach 40-60%, therefore, there can be a several-time decrease in well productivity [7 - 9].

Thereafter, using a simulation model, the content of gas condensate in the near-wellbore area in wells without hydraulic fracturing treatment was analyzed. There are profiles of condensate saturation for one well from each cluster presented in figure 3. To achieve these results, the local refinement procedure was performed for every well left without hydraulic fracturing. From these figures, it can be concluded that there is condensate banking taking place in all the wells.

There are several technologies to improve the productivity of gas condensate wells providing bottomhole treatment with different solvents, reagents and thermal treatment to remove condensate banking [10, 11]. However, a hydraulic fracturing treatment seems to be one of the most effective methods to improve the productivity of gas condensate wells [7, 12-15]. This treatment allows increasing the contact area of a well with the reservoir and decreasing the depression on the reservoir. It leads to the more proportional development of reserves and more optimal production of gas and condensate. Condensate banking will also take place, but the size and profile of condensate saturation in the near-wellbore area will change. Condensate will drop out along the fracture and after entering the fracture it will move with gas flow to the surface [7].

4. Research techniques

The Field A development optimization was performed with the use of an integrated model built with the Avocet Integrated Modeling® software (Schlumberger). Making this model included the following steps:

1. A gathering facilities system model was created with the Pipesim® software (Schlumberger);
2. The gathering facilities system model was connected to the simulation model (input data) with Avocet Integrated modeling;
3. History matching of the integrated model as of 01.01.2014.

These steps will be described in more detail below.

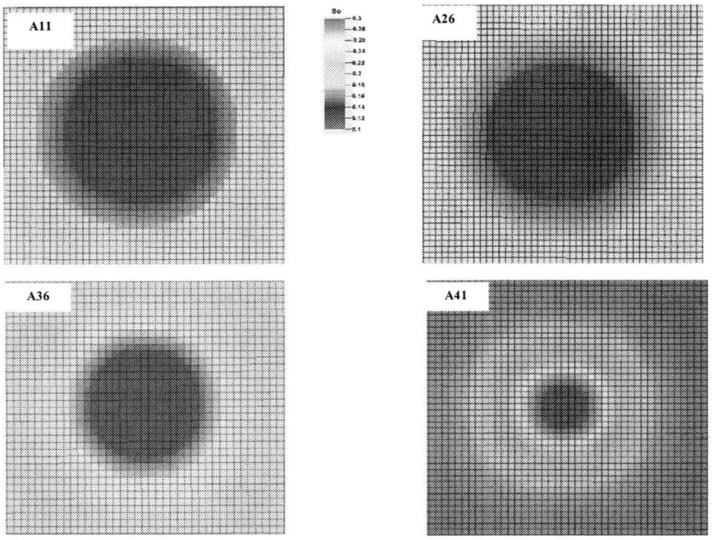

Figure 3. Condensate saturation profile for wells A11, A26, A36, A41

4.1. "Gathering facilities system" model creation

The gathering facilities system model was built with the Pipesim® software (Schlumberger). The model is presented in figure 4. For the correct description of the multiphase fluid filtration and due to the changes in the phase state of the system resulting from a reservoir pressure drop, the compositional model was used in the Pipesim software, as well as in the simulation model. The water cut of the fluid was also taken into account in the model. All perforated intervals in production wells are located above the gas-water contact, and at the beginning of 2014 the water cut was equal to 5% in Field A.

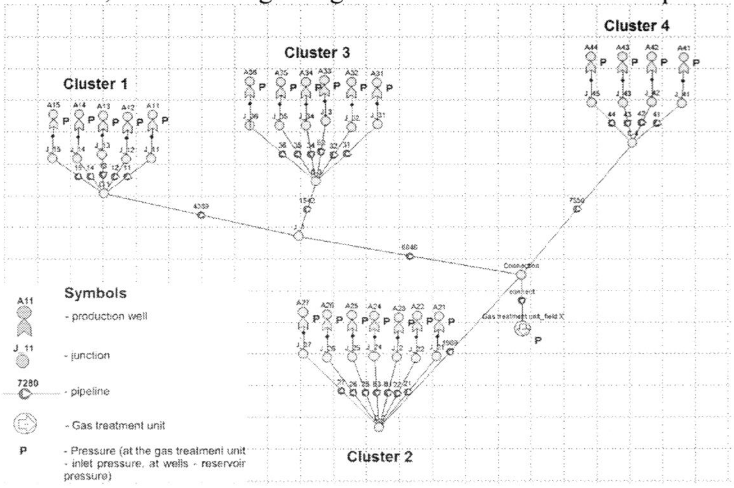

Figure 4. The gathering facilities system model of gas and condensate Field A in Pipesim® (Schlumberger)

4.2 "Reservoir – well – gathering facilities system" integrated model creation

The integrated model "reservoir – well – gathering facilities system" was built with Avocet Integrated Modeling® (Schlumberger) software. The composition simulation model was connected with the above-described gathering facilities system model. The connection point of these two models is bottomhole pressure (BHP). Therefore, all information about the reservoir is taken from the simulation model (to the bottomhole) and other information, such as construction of wells and surface facilities system were taken from the "gathering facilities system" model (above the bottomhole). A constraint between the models was the gas rate in the main phase in Field A. Bottomhole pressure was calculated as the result of balancing the simulation and gathering facilities system models. The difference in the solution of both models falls within range of 1% (set in the program). Each well in both of these models was connected to each other at the bottomhole.

As the simulation model and the gathering facilities system models are both compositional models, the integrated model is also compositional. The value and characteristics of each element in the reservoir mixture were taken into account. Each element from the simulation model was connected to a corresponding element in the gathering facilities system model. The Peng-Robinson equation of state is used in all of these models [16].

The integrated model was history-matched with the data as of 01.01.2014 and includes history matching in terms of the following parameters:

1. tubing head pressure (THP) at wells;
2. temperature and pressure distribution in wells;
3. gas and condensate rate.

5. Development optimization

5.1. Hydraulic fracturing treatment design

The design of hydraulic fracturing was created using the Unified fracture design technique described by M. Economides and P. Valco [17, 18] for three different types of proppants. It should be mentioned that hydraulic fracturing not only solved the problems mentioned above, but also allows developing the field with a lower level of depression at same flow rates.

5.2. Description of development scenarios

To obtain the optimal development plan for 8 years, different scenarios of field development were analyzed. Each of these scenarios was also assessed from the economic point of view.

Development optimization was performed in the following way: firstly, the base scenario (described below) was considered, then the results of this scenario were analyzed and economic indicators were calculated. Later, the conditions of the constant gas rate and positive NPV were checked. If both of these conditions were met, this scenario was chosen as optimal (recommended) for further optimization. If no, this scenario was optimized and the cycle was repeated until both conditions were met. Five groups of scenarios were considered.

Base scenario. This scenario is without new wells and treatment. The number of wells remains constant (22 wells). Gas and condensate cumulative production and also annual production of gas and condensate are presented in figures 5-6. From these figures, it can be seen that gas and condensate production falls continuously. After 8 years, the production level decreased to 45%. The condensate and gas recovery factors in 2022 are equal to 34.3 and 44.9, respectively.

Scenario 1. According to the Field A technological development plan, four new wells will be put into production in 2017 and 2019. Field condensate and gas cumulative production and annual production of gas and condensate are presented in figures 5-7. It can be concluded from these figures that the production of gas and condensate is also falling. An increase in production is observed in years when new wells were drilled. The condensate and gas recovery factors are equal to 0.357 and 0.47, respectively. Therefore. steps should be taken to stop the production decline.

Scenario 2. This scenario is based on hydraulic fracturing treatment in all wells at the beginning of the simulation period (January-February 2014).

It is known that production changes in one well in a gas-condensate field leading to changes in production in adjacent wells, as the gas phase is high-mobility. Modeling this scenario has brought the following results: three wells were not to become producers from the beginning of the stimulation and all wells were to show lower production rates. As it can be seen from figures 5-6, the production of gas increased up to 20% during the first year, then production trend waas falling until the year when new wells were put into production. The condensate and gas recovery factors are equal to 0.37 and 0.5, respectively.

As there are some problems with the flowing of three wells and reducing rates before the further optimization of Field A development, it was decided to analyze the influence of hydraulic fracturing for one well on adjacent wells. The task was limited to the consideration of the influence this treatment has on each cluster separately. Therefore, a hydraulic fracture treatment was modeled for every well separately from other wells on the cluster and then the influence of each treatment was analyzed. It was found out that each hydraulic fracturing treatment exerts influence on adjacent wells to some extent. Therefore it was decided to treat no more than one well per cluster at a time.

Scenario 3. In this scenario, hydraulic fracturing was consistent. It was concluded that firstly hydraulic fracturing treatment will be made in wells that have problems with liquid loading to exclude their shutdown. Then hydraulic fracturing will be performed for wells where gas rates are becoming close to the critical values and later for the remaining wells. All wells have been in production during 8 years in contrast to scenario 2.

To get the best result, different options of a hydraulic fracturing schedule were analyzed. It refers to wells with liquid loading problems. This analysis was made to exclude well shutdown. To predict the well shutdown, a restriction represented by the critical gas rate was established for each well. If the gas rate in a well fell below the critical gas rate, the well was shut down. The procedure of hydraulic fracturing will be held for three wells from different clusters according to the schedule.

When new wells are put into production (2017, 2019, as it was mentioned above), hydraulic fracturing will have already been held for all wells with the liquid loading problem. Therefore, to fulfill the objective of this project and exclude a higher increase in gas production, no hydraulic fracturing will be held in these years.

It is known that reservoir pressure declines with time. The faster reservoir pressure declines, the bigger volume of condensate remains in the reservoir. In Area 1, reservoir pressure is practically 152 atm, as it was mentioned above, as in Area 2 it has dropped to 205 atm. Gas recycling process is not provided in Field A. Therefore, the large amount of condensate starts to remain in the reservoir in the first area.

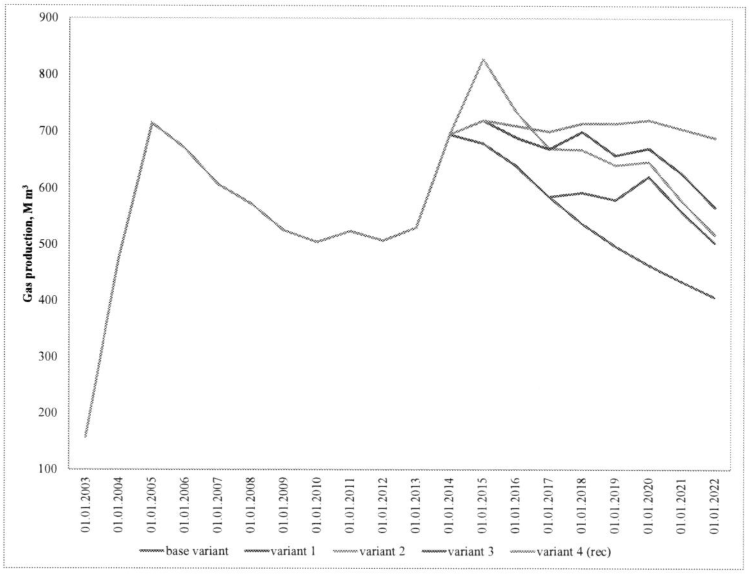

Figure 5. Annular gas production, 2002-2022

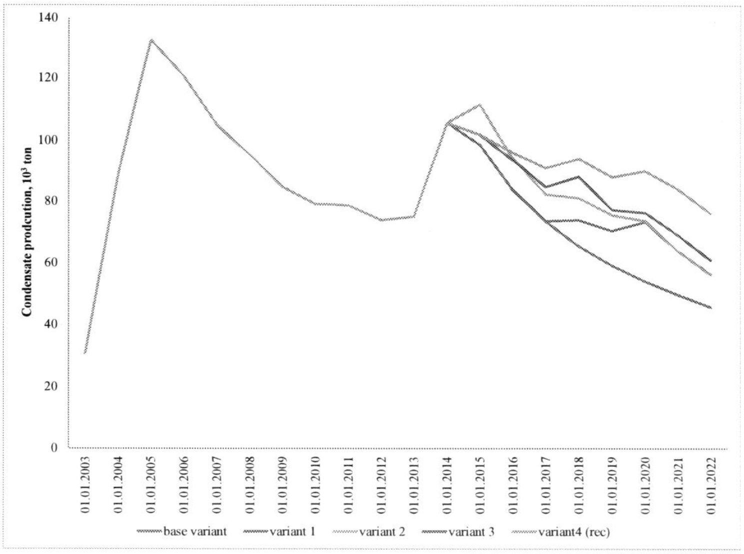

Figure 6. Annular condensate production, 2002-2022

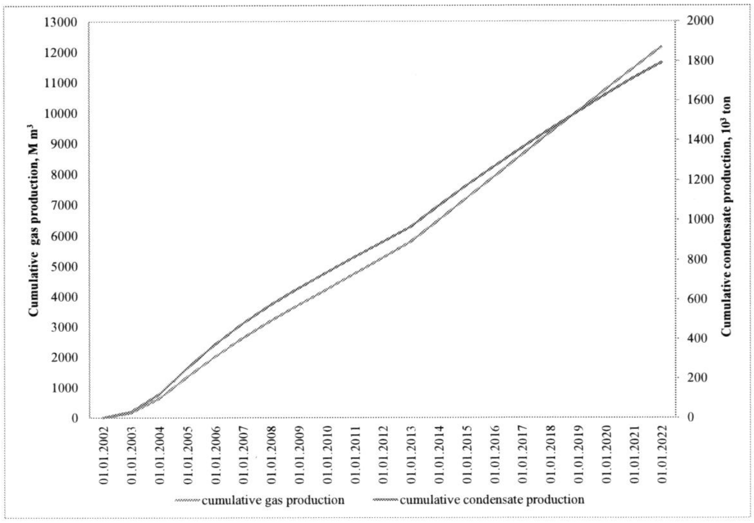

Figure 7. Cumulative gas and condensate production

In order to increase the production of condensate from the cluster, four technique described by A. Kalugin in [19] and [20] were considered. The main issue of this technique is condensate recovery maximization. It is achieved with the allocation of the gas rate in a well, namely – decreasing the gas rate in adjacent wells and increasing rates on remote wells. Condensate recovery increases due to the redistribution of pressure in the reservoir. It was decided to limit gas rates for wells A43 and A44, while wells A41 and A42 are fully opened during the first year, then during the second year of development gas rates in wells A43 and A44 will be increased (the well will be unfastened, but not fully opened) instead of wells A41 and A42 where rates will be limited and vice versa up to 2020. This year, wells A41 and A42 will be fully opened while wells A43 and A44 will be unfastened.

Three scenarios were considered. The first scenario: all wells in cluster 4 are fully opened, the second scenario: wells A43 and A44 are unfastened, the chokes of these wells stay the same as on 01.01.2014. The third (recommended) scenario is the scenario that is described above. Cumulative and annular condensate production and reservoir pressure profiles are presented in figures 8-9. According to the obtained results, it can be concluded that the third scenario is optimal for the further development of cluster 4. Despite the fact that the production of condensate in the first scenario is more than the value of 13 thousand suggested by the recommended option, reservoir pressure in the optimal scenario is 35 atm higher at the end of the stimulation period (figure 10).

The recommended scenario for cluster 4 was taken into consideration in scenario 3 of field development and in further scenarios.

As it can be seen from figure 5A, the production of gas is at the same level during the first three years and then it starts declining. The condensate and gas recovery factors are equal to 0.369 and 0.493, respectively.

During these four scenarios, the inlet pressure at the gas treatment unit is maintained at the same level (76 atm). Gas production does not correspond to the project objective.

Scenario 4. To hold gas production at a constant level, it is also possible to reduce the inlet pressure at the gas treatment unit. The minimum pressure at gas treatment is 15 atm (it is a restriction from technical point of view). This scenario is a modified scenario 3 with a reduction in pressure at the gas treatment unit with time. The profile of pressure changes is presented in Table 2.

As one can see from figure 5A, gas production stays at the same level during the whole simulation period. Also it can be pointed out that the condensate and gas recovery factors are equal to 0.384 and 0.51, respectively. It should also be noted that condensate production in this scenario is the greatest.

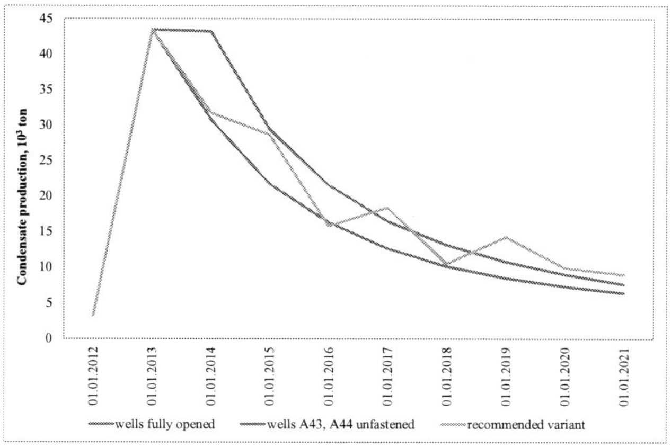

Figure 8. Annular condensate production (cluster 4)

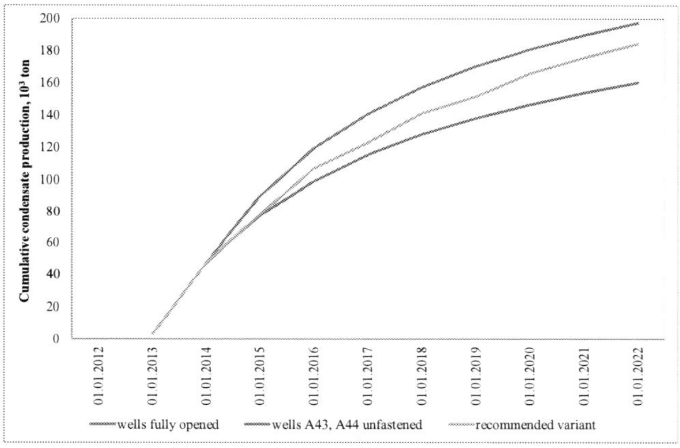

Figure 9. Cumulative condensate production (cluster 4)

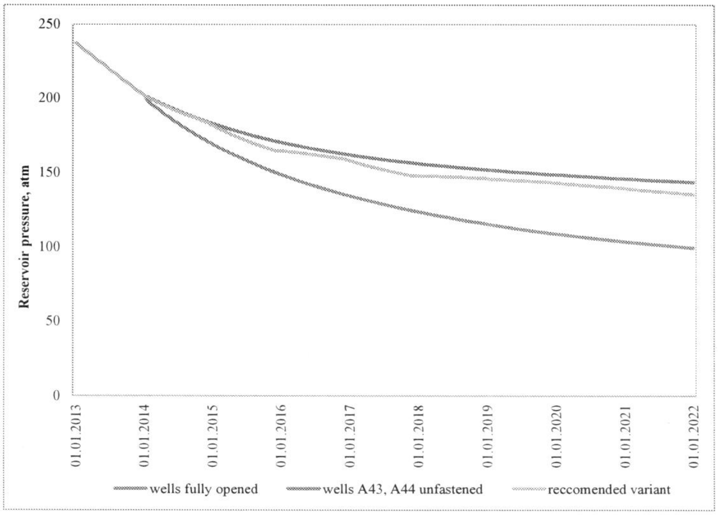

Figure 10. Reservoir pressure profile (cluster 4)

Table 2. Profile of inlet pressure at gas treatment unit

Year	Inlet pressure at gas treatment unit, atm
2014	76
2015	70
2016	64
2017	58
2018	52
2019	46
2020	40
2021	34

6. Economic factors

All these five scenarios were also considered from the economic point of view. In CAPEX, hydraulic fracturing treatments, drilling of new wells, building of a cluster base, pipeline from new clusters were taken into consideration. Also, it was concluded that acidizing treatment should be performed in a year to maintain the productivity of the wells. In OPEX, gas and condensate preparation and transportation and other production costs were taken into account. Also, the following taxes were considered: VAT, royalty. The discount rate was considered equal to 15%. All calculations were made in real terms. The condensate price equals 13,086.7 RUR/ton, the gas price is 3,463.3 RUR/103m3.

Due to the economic and simulation results, it was concluded that the optimal scenario for further development is scenario 4. It has the maximum NPV, gas and condensate production compared with other scenarios. It can also be noted that the reservoir pressure in this scenario is lower than in other scenarios as of 01.01.2022, however, this difference is not significant.

It should also be mentioned that scenario 2 is more attractive from the economic point of view because of the maximum incremental revenue in the first 2 years. However, this scenario was not chosen, as there is a problem with selling such a big volume of products.

The incremental NPV for the recommended scenario vs. the base scenario is presented in figure 11.

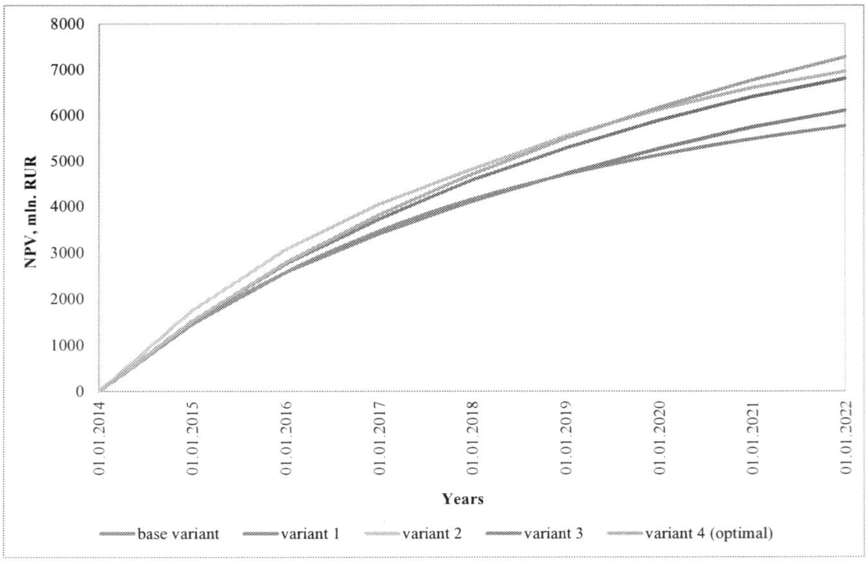

Figure 11. NVP of development scenarios of field A

Conclusions

Hydraulic fracturing treatment eliminated the liquid loading problem in the wells of Field A. Condensate banking after the hydraulic fracturing was not elimated, but, as it has already been mentioned, the size of profile and saturation changed. Therefore, condensate will be concentrated around the fracture and after entering the fracture it will flash to the surface by gas flow. Factors leading to the production decline were found. Creation an integrated model allowed optimizing Field A development for a 8-year period. Five groups of development scenarios were considered. After analyzing the results from the development and economic point of view, the recommended scenario for further development was chosen. According to this scenario, gas production in Field A was maintained at the same level (the difference being 3%) for 8 years. Cumulative gas production at the end of the optimization period equals 12,160.6 mln m^3 (Gas recovery factor – 0.51), cumulative condensate production equals 1.8 mln tons (condensate recovery factor – 0.38), the NPV of this scenario is 7,277 mln RUR.

References

[1] Yashaswini Devi Nallaparaju 2012 Prediction of Liquid Loading in Gas Wells *SPE 155356, SPE Annual Tecnical Conf. and Exh. (San Antonio Texas USA)*
[2] James F Lea and Henry V Nickens 2004 Solving Gas-Well Liquid-Loading Problems *SPE 72092*
[3] Mhunir Bayonle Alamu 2013 Gas Condensate Well Unloading Critical Rate *SPE 166350, SPE Annual Technical Conf. and Exh. (New Orleans Louisiana USA)*
[4] Kelkar Mohan 2008 *Natural gas production engineering* (Tulsa, Oklahoma) 572
[5] Economides Michael J 1994 *Petroleum Production System* ed Daniel A. Hill, Christine Ehlig-Economides (Prentice Hall PTR/ Upper Saddle River, New Jersey 07458*)* 611
[6] Jairam Kamath 2007 Deliverability of Gas-Condensate Reservoirs *Field Experiences and Prediction Techniques SPE 103433*
[7] Gritsenko A I 1997 Methods of Increasing Gas Condensate Well Productivity ed Ter-Sarkisov R M, Shandrygin A N, Podyuk V G (*Moscow: Nedra Publishing House)* 365
[8] Ignatyev A E 2011 Multiple Hydraulic Fracturing in Horizontal Wells as Method of Effective Development Gas Condensate Field of Arctic Region *SPE 1449925*
[9] Seminyakin V and Sigovatov L 2004 Features of Gas Flow in the Gas Condensate Wells *Gas industry* **7**
[10] Ter-Sarkisov R M, Shandrigin A N, Kireev C V, Vdovenko V L and Fedoseev A V 1997 Technology of Increasing Production from Gas Condensate Wells *20th World Gas Conf. (Denmark)*
[11] Settari A et al. 1996 Production of Fractured Gas Condensate Wells – A Case Study of the Smorbukk Field *The Gas Technology Conf. (Calgary, Canada)*
[12] Langedijk R A et al. 2000 Optimization of Hydraulic Fracturing in a Deep, Multilayered, Gas Condensate Reservoir *SPE Annual Technical Conf. and Exh. (Dallas, Texas)*
[13] Al-Hashmi H S and Hashmi S S 2000 Long-Term Performance of Hydraulically Fractured Layered Rich Gas Condensate Reservoir *SPE International Oil and Gas Conference and Exhibition (Beijing, China)*
[14] Murrey M.D. et al Fracture Design, Execution, and Evaluation in Retrograde Condensate: Case History of Angsi Field, Offshore Malaysia, *SPE Annual Technical Conf. and Exh. (Denver, Colorado, October 5-8, 2003)*
[15] James Lea, Henry V. Nickens and Michael Wells *Gas Well Deliquification: Solution to Gas Well Liquid Loading Problems*
[16] Duns H. Jr. and Ros N.C.J 1963 Vertical Flow of Gas and Liquid Mixtures in Wells *Proc., Sixth World Pet. Cong. (Tokyo)* 451
[17] Economides M, Valko P and Oligney R Unified Fracture Design *(London IBT Global)* 262
[18] Shirley Indriati, Xiuli Wang and Michael J. Economides Adjustment of Hydraulic Fracture Design in Gas-Condensate Wells *SPE 73751, SPE International Symp. and Exh. of Formation Damage Control (Lafayette, Louisiana, February 20-21 2002)*

[19] Kalugin A 2012 Increase in Condensate Production from Gas Condensate Fields with Depletion Drive by Development Regulating *Applied Fluid Mechanics*. **2 (14)** *ISSN 1561-9087*

[20] Kalugin A 2012 Increase in Condensate Production from Gas Condensate Field Through the Optimal Allocation of Gas Production *Exploration and development of oil and gas fields* (*ISSN 1993 – 9973)* **4 (45)**

IOP Conf. Series: Earth and Environmental Science **24** (2015) 012039 doi:10.1088/1755-1315/24/1/012039

Regularities of changes in fluid composition and properties in Vankor field pools: from light to heavy oil

I V Goncharov[1,2] and N V Oblasov[1,2]

[1]National Research Tomsk Polytechnic University,
30 Lenin Avenue, 634050, Tomsk, Russia
[2]TomskNIPIneft, 72 Mira Avenue, 634027, Tomsk, Russia

E-mail: goncharoviv@nipineft.tomsk.ru; oblasovnv@nipineft.tomsk.ru

Abstract. Oil in layers Nkh 3-4, Nkh 1, Sd 9, Yak 3-7 and vYak 2-4 of the Vankor field occurs at the depth of -2,767 to -1,357 meters at strongly different temperatures: from 62 to 26 °C. Such temperature conditions contribute to oil biodegradation processes in the pool. Therefore, oils in different pools significantly differ from each other in terms of composition and properties depending on the intensity of biodegradation. At the same time, pools might embrace both oils that have practically been not exposed to biodegradation processes and significantly biodegraded oils. The most seriously altered oils are found in vYak 2-4 layer pools. They are the heaviest and the most viscous oils among the samples under study. Many typical oil components (alkanes, alkylbenzenes, naphthalenes, phenanthrenes, dibenzothiophenes) are absent in their composition. Besides, the initial distribution of hopanes in the composition of biomarkers is altered. Apart from the molecular composition of degassed oil samples, the work also studies the effect of biodegradation on the properties and the component and isotopic composition of oils, gases and formation fluid samples.

1. Introduction

The Vankor oil and gas field is located in the northeast of Western Siberia (Russia). The field is multilayer and major. Oil and gas are accumulated predominantly in its Lower Cretaceous deposits. At the same time, the main pools are located in sand layers Nkh 3-4, Nkh 1 of the Nizhnekhetskaya formation (age – K_1 b_1-v_1), layer Sd 9 of the Sukhodudinskaya formation (age – K_1 v_1-g_1) and Yak 1-2, Yak 3-7 and vYak2-4 of the Yakovlevskaya formation (age – K_1 a_2-al_2) at depths with not high formation temperatures (25-65°C). Such temperature conditions are rather favorable for the occurrence and development of a formation microflora that consumes and generates hydrocarbon components of crude oil with its life-sustaining activity. It results in significant changes in the composition and properties of the formation fluid depending on the degree of biodegradation. Biodegradation can be the reason for forming heavy and viscous oil pools that are much harder to develop vs. usual oils. This fact has already been discussed with regard to the Vankor field. [1][2][3][4][5][6][7].

Currently, a significant amount of experimental material on the Vankor field has been accumulated, including not only standard studies of the physical and chemical properties of wellhead and downhole fluid samples, but also detailed geochemical studies (Rock-Eval pyrolysis of oil-saturated sandstones, GC/MS and IRMS studies of oils, gases and extracts from oil-saturated rocks). All this allows comprehensively characterizing the pools of the Vankor field and track changes in the fluid composition and properties at different levels. The field section demonstrates that all its pools contain the same oil generated by the Yanovstanskaya formation [2]. Therefore, this case is an excellent example of the influence of secondary oil transformation processes in a pool that are evident at the level of physical and chemical properties of oil, as well as its molecular and isotopic composition.

2. Signs of oil biodegradation in the Vankor field

The influence of the biodegradation process in reservoir conditions on the composition and properties of oil is well-known and has been described in the examples of numerous fields. Besides, biodegradation has been studied within the framework of numerous laboratory experiments. It is a well-known fact that one of the first signs of biodegradation is the reduced content and disappearance of light n-alkanes. Usually, they are C_3-C_{15} n-alkanes. Specific groups of biomarkers, such as diasteranes, triaromatic steroids and porphyrins, are considered most resistant to biodegradation among the oil components that can be identified [8][9].

Performed microbiological studies have proved the presence of hydrocarbon-oxidizing microorganisms in the microflora of waters from layers Nkh 4, Sd 9 and Yak 3-7 (table 1). Apart from hydrocarbon-oxidizing microorganisms, methane-producing bacteria have also been found in the composition of the formation microflora. By consuming the metabolism products of hydrocarbon-oxidizing microorganisms (acetates and carbon dioxide), methanogens generate methane [10][11] and are one of the main reasons for forming gas caps in pools with biodegraded oils [12][13][14], including those in the Vankor field.

Table 1. Microbiological characteristic of formation water samples from the Vankor field (the sampling is near the OWC)

Layer	Eh, mB	pH	Number of microorganisms, cells/ml			
			Heterotrophs	Denitrifiers	Methanogens	Hydrocarbon-oxidizing microorganisms
Yak 3-7	-100	7.3	2.3×10^6	45×10^3	7.5×10^3	800×10^3
Sd 9	-65	8.4	1.8×10^6	150×10^3	1.5×10^3	620×10^3
Nkh 3-4	-60	7.7	1.1×10^6	1.4×10^3	1.2×10^3	370×10^3

It is widely accepted that microorganisms in a formation live in water environment and the biodegradation process itself occurs at the OWC. However, formation microflora can also be found above the OWC level, therefore, there are assumptions about possible biodegradation along the whole depth of an oil pool [9]. It is expected that the residual water adsorbed by a mineral rock matrix is quite sufficient for the live-sustaining activity of microorganisms. Nonetheless, it should be taken into account that the population and activity of microorganisms in these conditions are inconsistent with their activity at the OWC level and is expected to significantly drop in a direction away from the OWC.

The easiness of n-alkane utilization by microorganisms vs. other oil components is often applied to assess the level of biodegradation and is the basis for specific molecular parameters, such as $Ki=(Pr+Ph)/(nC_{17}+nC_{18})$ – isoprenoid coefficient [15][16], Pr/nC_{17}, Ph/nC_{18} [8][9]. However, the efficacy of such an approach is restricted with the complete utilization of n-alkanes by microorganisms at high-level biodegradation, which often does not allow assessing the degree of microbial changes accurately. Taking into account that formation oil is a complex mixture being predominantly composed of hydrocarbons, it can be stated that changes in its composition and properties under the influence of biodegradation are to a certain extent step-by-step and consistent and depend on the biodegradation-resistance of separate component groups. Therefore, the resistance of different classes of compounds

IOP Conf. Series: Earth and Environmental Science **24** (2015) 012039 doi:10.1088/1755-1315/24/1/012039

and their generalized sequence of utilization by microorganisms [9] has become the basis of the 10-point scale of biodegradation level assessment [8] that allows ranking not only oils with not high microbial activity level, but also severely biodegraded oils. The degree of oil biodegradation in the Vankor field determined in such a way changes within a relatively wide range and corresponds to 0-7 levels.

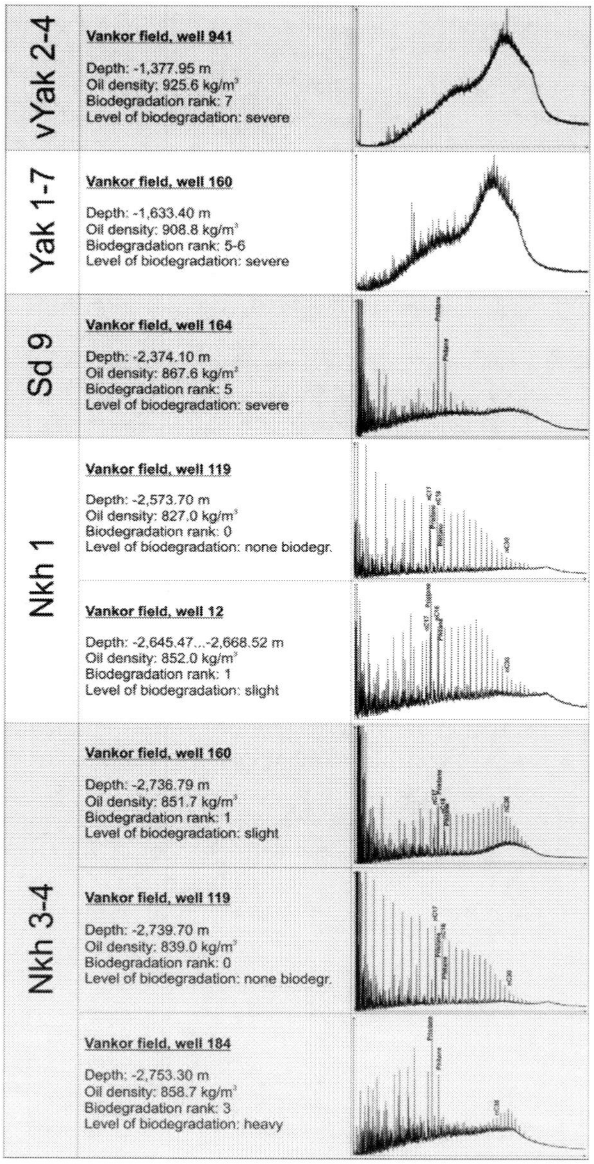

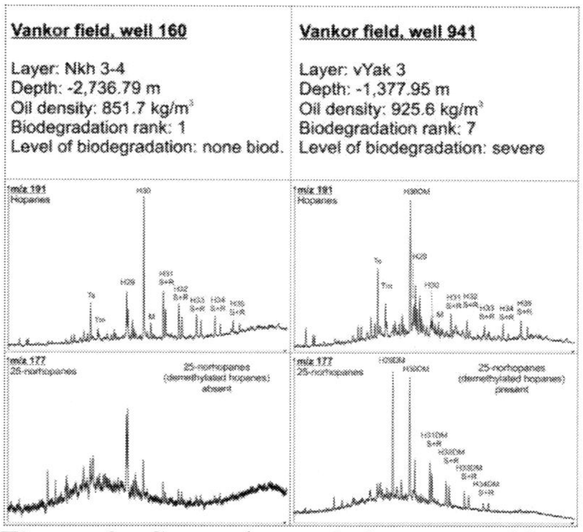

Figure 2. Comparison of the composition and distribution of hopanes and demethylated hopanes (25-norhopanes) in oils with slight and severe biodegradation.

Figure 1. Typical chromatograms of crude oils from different layers of the Vankor field (the level of biodegradation is determined with a scale [8], the level of biodegradation is assessed in compliance with [9]).

Oils from Nkh 3-4, Nkh 1, Sd 9, Yak 1-2, Yak 3-7 and vYak2-4 are significantly different in terms of the biodegradation degree. Figure1 shows typical chromatograms for oils from the pools of different formations and some characteristics of degased oil samples. At the same time, the deepest layers, Nkh 3-4 and Nkh 1, comprise both oils with heavy biodegradation (level 3) and oils that have been practically not been affected by biodegradation processes. This becomes evident from the drastically different chromatograms (figure 1) and the presence or absence of a range of n-alkanes. However, the chromatograms for samples from less deep-lying layers of the Yakovlevskaya formation (Yak and vYak) are characterized by the complete absence of normal- and isomeric-structure alkanes. In comparison with oils from other formations, the chromatograms show the prevalence of a hump or unresolved complex mixture (UCM). Nonetheless, the oils of the Yakovlevskaya formation itself reveal significant differences. The oils from its upper part (vYak) practically do not have the peaks of individual components at all, whereas, the hump itself becomes "balder" vs. other oils from the Yak layers.

As for all oils from the Vankor field under study, special attention should be paid to oils from the vYak layers that demonstrate the highest level of biodegradation. These samples lack not only the easiest to utilize components, such as n- and iso-alkanes, but also components being more biodegradation-resistant, i.e. alkyl-cyclohexanes, alkylbenzenes, naphthalene and alkyl-naphthalenes, phenanthrene and alkyl-phenanthrenes, dibenzothiophene and methyl-dibenzothiophenes. Besides, this severe biodegradation leads to the distorted distribution of hopanes (figure 2). Earlier, biodegradation of hopanes in Russia's oils from the Timan-Pechora and Volga-Ural regions, the Pre-Caucasian region and Sakhalin, as well as in Cenomanian oil pools in Western Siberia has already been mentioned in a range of publications [1][17][18] [19].

Hopanes are rather resistant to the impact of microorganisms, whereas their disappearance from the composition of oils depending on the mechanism usually complies with biodegradation of the 7th or 9th rank [8]. Biodegradation in Vankor field oils follows the demethylation route with the generation of 25-norhopanes, therefore they are revealed in the composition of vYak oils on the basis of the m/z 177 characteristic ion (figure 2), and the level of biodegradation is considered equal to 7. Besides, no significant changes have been found in the composition of steranes, which testifies the fact that the degree of biodegradation is not higher than 7.

3. Consequence of biodegradation effect on the composition and properties of Vankor field fluids

Indubitably, all changes in the composition of oils resulting from biodegradation lead to changes in their physical and chemical properties. As it can be seen from figure 3, the lightest oils with the lowest viscosity, low sulfur content, the highest production of light ends, high content of parafins and lower content of resins are found in layers Nkh 1 and Nkh 3-4. As the depth of the pools drops and the level of biodegradation resulting from utilization of the lightest ends in the oil composition grows, the composition and properties change to heavier. The least deep pools of the vYak layers where formation temperatures are about 25 °C contain oil that is characterized by the complete absence of fractions boiling off under 200 °C. The density of oil from the vYak layers, on average, makes 925.6 kg/m^3, whereas, its viscosity reaches the values of 415 cP. These oils have the highest contents of resins (up to 15 %), but they do not stand out from other samples under study in terms of the amount of asphaltenes.

As for the oils from the Yakovlevskaya formation under consideration, there is one sample from well 164 (layer Yak 3-7, sampling depth -1,594.5 m) that stands out in terms of a range of physical and chemical characteristics. In comparison with other oils from the Yakovlevskaya formation, this oil is characterized by significantly lower density and viscosity, lower sulfur content and higher content of parafins, as well as a lower biodegradation level (figure 3). Its presence cannot be explained with a crossflow or inappropriate formation tests. The presence of so strongly different oil in Yak 3-7 can be explained by the occurrence of zones having a worsened hydrodynamic connection with the main part of the pool and the OWC area in Yak 3-7.

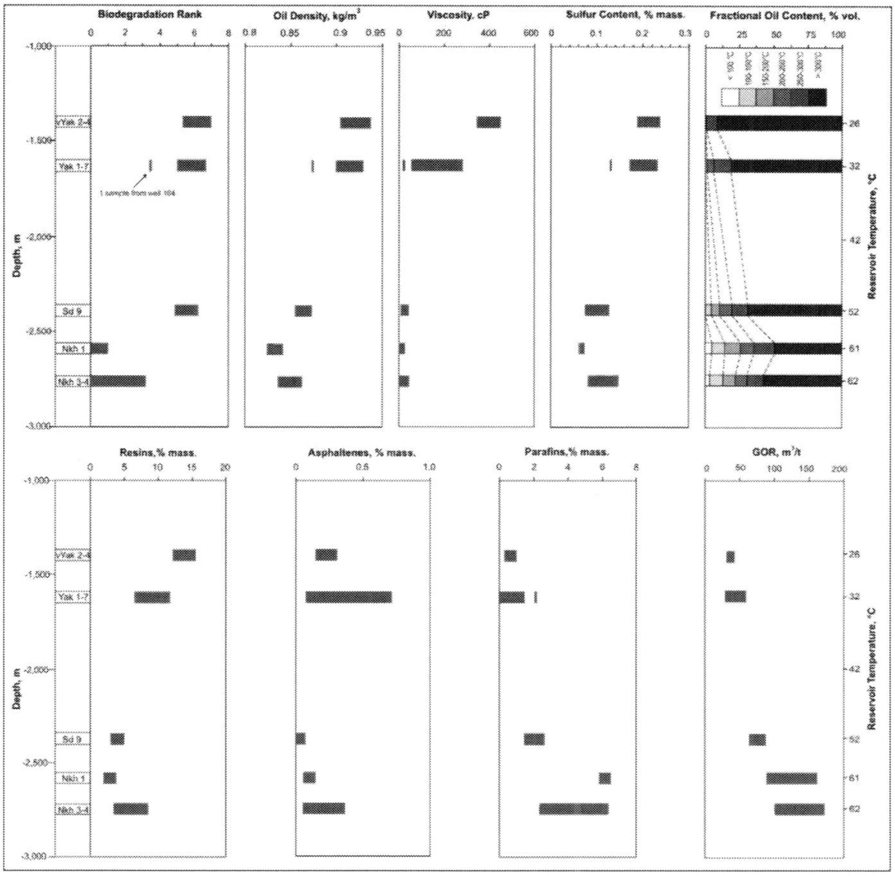

Figure 3. Physical and chemical properties of degassed oils and GOR of formation oils from different layers of the Vankor field (the biodegradation rank is determined with [8]).

In addition to the studies of physical and chemical properties and molecular composition, the following were also analyzed- the isotopic composition of oil carbon, extracts from oil-saturated rocks and fractions of saturated, aromatic hydrocarbons, resins and asphaltenes derived from them (figure 4). The figure shows that as the degree of biodegradation grows from layer Nkh 3-4 to the Yak and vYak layers, the strongest weighting of the isotopic composition of carbon occurs in saturated hydrocarbon fractions. It is quite logical and expectable, as saturated hydrocarbons are most prone to biodegradation. Taking into account the sufficiently high level of biodegradation in Yakovlevskaya formation oils vs. oils from the Nkh layers where many aromatic hydrocarbons (naphthalenes, phenanthrenes, etc.) and some heteroatomic compounds (dibenzothiophenes, etc.) are impacted, it is also possible to state that the heavier isotopic composition of the carbon in the fractions of aromatic hydrocarbons and resins is quite explainable. The differences in the isotopic composition of carbon in asphaltenes derived from the oils and extracts of various layers are insignificant. However, it is not clear enough whether this slight difference is the consequence of a biodegradation effect or it is initially implied in the composition of original fluids. It can result from both the step-by-step formation of the pools (portions of oil belonging to different generation stages) and an insignificant distortion in the process of secondary migration (influence of the oil travel path length).

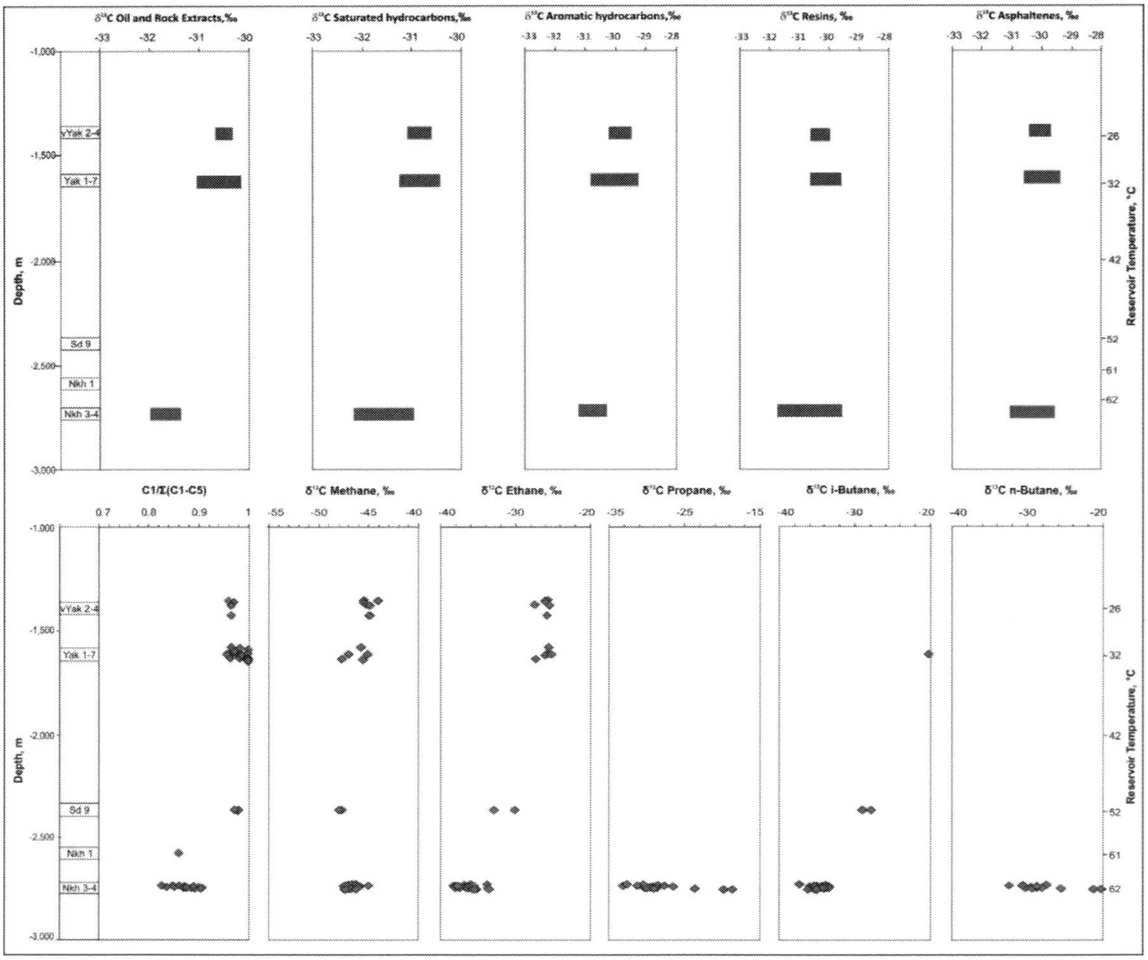

Figure 4. Isotopic composition of carbon in oils, extracts from oil-saturated rocks and their fractions, and changes in the component and isotopic composition of gases dissolved in oils from different layers of the Vankor field.

The composition of solution gas experiences significant changes too, and these changes are evident not only at the level of the component composition, but can also be detected in the isotopic composition of carbon (figure 4). The gas from the upper beds with the most significantly biodegraded oil (Sd 9, Yak 1-7, vYak 2-4) is methane-containing. Its share vs. other homologous compounds is 0.96-0.99. Unfortunately, it is impossible to determine the influence of biodegradation and methanogenesis on $\delta^{13}C$ methane values at this stage, as the Vankor field does not comprise pools that are not influenced by secondary transformations, the $\delta^{13}C$ methane values of which could be used as the reference characteristic of the initial fluid filling in the traps. Even the methane from the least biodegraded pool in Nkh 3-4 practically does not differ from the upper pools with a higher biodegradation rank in terms of its isotopic composition. Besides, there are no significant changes in the isotopic composition of methane carbon from layer to layer. The values of $\delta^{13}C$ methane – both dissolved in the oil rim and methane from the gas caps of the pools – vary within -48…-44 ‰. Such values are quite typical for both thermogenic methane and secondary microbial methane [13]. On the other hand, it is necessary to take into account the experience of geochemical studies [20] for the oil fields of the southeastern part of Western Siberia (Tomsk Oblast). Several fields were formed there due to generation of the Bazhenov formation, being the age and facies analog to the Yanovstanskaya formation, a source rock generating oil at the Vankor field [2]. The methane carbon in the non-biodegraded oil pools from these fields is isotopically lighter (-52…-57 ‰). It is quite possible that the methane of the initial fluid filling in traps in the Vankor field could have a similar isotopic carbon composition before being distorted by biodegradation and methanogenesis.

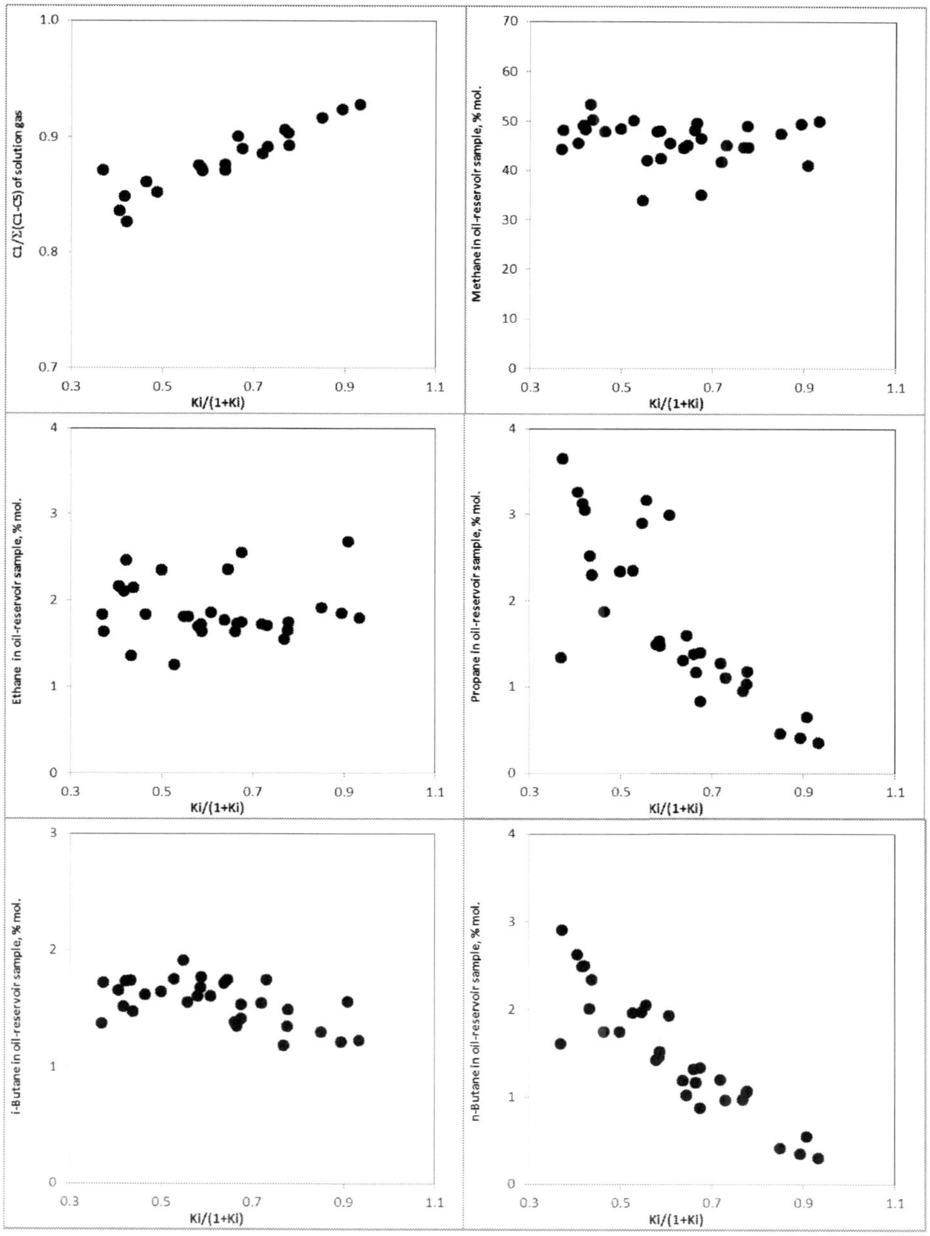

Figure 5. Relationship between separator gas composition and content of gaseous components in Nkh 3-4 downhole oil samples and isoprenoid coefficient Ki/(1+Ki).

The isotopic composition of ethane carbon consistently becomes heavier from the bottom to the top of the section: from -38 ‰ in Nkh 3-4 to -25 ‰ in the Yakovlevskaya formation layers. A more significant weighting is relevant to iso-butane: from -37 ‰ in Nkh 3-4 to -20 ‰ in Yak 3-7 (due to the low iso-butane content, this evaluation is performed only for one gas sample from the Yakovlevskaya formation). Because of the low content of propane and n-butane in the gases of heavily biodegraded fluids, isotopic assessments for these components are mainly performed only for gases from a deeper-lying layer, Nkh 3-4.

The wide dispersion of the methane content, 0.84-0.92, in gas from Nkh 3-4 and the significant range of changes in the isotopic composition of propane (-35…-18 ‰) and iso-butane carbon (-36… -20 ‰) testify the fact of active biodegradation processes in dissolved gaseous components.

Really, if the isoprenoid coefficient is used as a biodegradation criterion for oil from Nkh 3-4 and compare it with the share of solution hydrocarbon gases in formation oil (based on the results of downhole sample studies), it can be seen that there is a good correlation between the decreasing propane and n-butane content and the growing share of prystane and phytane vs. alkanes (growing isoprenoid coefficient) (figure 5). Consequently, the solution gas itself becomes leaner, which is evident from the growing $C1/\Sigma(C1+C5)$ values. In addition, there is a slight trend towards the decrease in the iso-butane content with biodegradation, which probably testifies to its insignificant biodegradation. It is noteworthy that there is a clear connection between the decreasing content of propane and n-butane in formation oil and the isotopic composition of carbon in these components (figure 6). Some researchers [9][21] have demonstrated the selective utilization of propane vs. other homologous compounds, which results in changes to the isotopic composition of carbon only in propane without affecting other components. Layer Nkh 3-4 is characterized by the synchronous consumption of both propane and n-butane by microorganisms, which is testified by the relationship of their isotopic compositions (figure 6). Besides, the relationship between the parameters of oil and gas dissolved in it (figure 7) reflecting the proportion between the content of branched and normal alkanes testifies the synchronous utilization of low-molecular and heavier compounds by microorganisms, which is somewhat out of keeping with the generally accepted viewpoint about the easier biodegradation of lighter compounds.

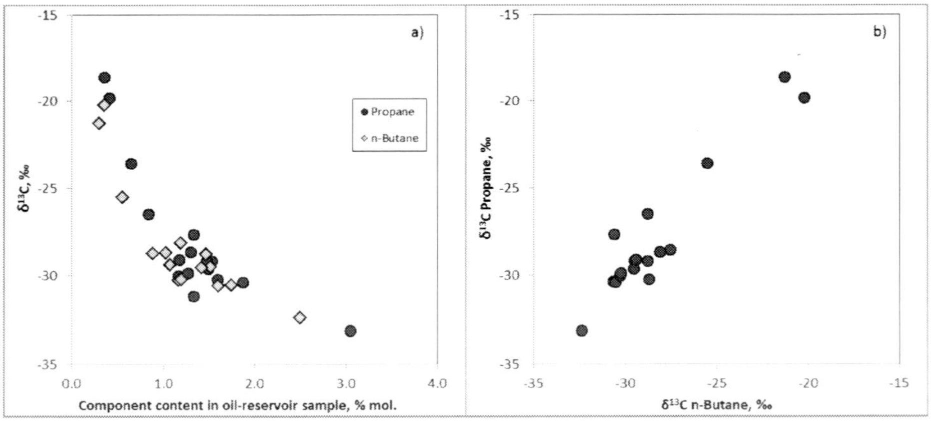

Figure 6. Relationship between isotopic composition of carbon in propane and n-butane and the share of these components in reservoir oil samples from Nkh 3-4 (a) and relationship of the isotopic composition of carbon in propane and n-butane (b).

The reason for the weighting of the isotopic composition of carbon in gaseous components with biodegradation is rather obvious. As the bonding strength of ^{12}C-^{12}C is lower vs. ^{12}C-^{13}C, the metabolism products of bacteria utilizing hydrocarbons are expected to be enriched with a lighter carbon isotope. Consequently, methanogens, ethanogens and propanogens using these products will generate isotopically light methane, ethane and propane. In other words, biogenic (newly formed) gases cannot be isotopically heavier than initial oil. On the other hand, microbiological degradation is relevant not only to liquid oil compounds, but also gases dissolved in it. In this case, molecules with ^{12}C-^{12}C bonds will also be primarily utilized in them, whereas the remaining gas will be enriched with heavy carbon. It has been found that the isotopically heaviest propane in Nkh 3-4 in a sample is from the near-OWC area (well .184). In this sample, it was detected that $\delta^{13}C$ propane is equal to -18.6 ‰, and the authors [22] have obtained the value of -2.6 ‰. Apparently, the deeper degradation of $\delta^{13}C$ propane can have a positive influence (over zero). Obviously, the speed of biochemical gas utilization drops in the following order: propane, n-butane, iso-butane, ethane. Consequently, the residual accumulation of isotopically heavy molecules in gases will grow in a reverse order: from ethane to propane.

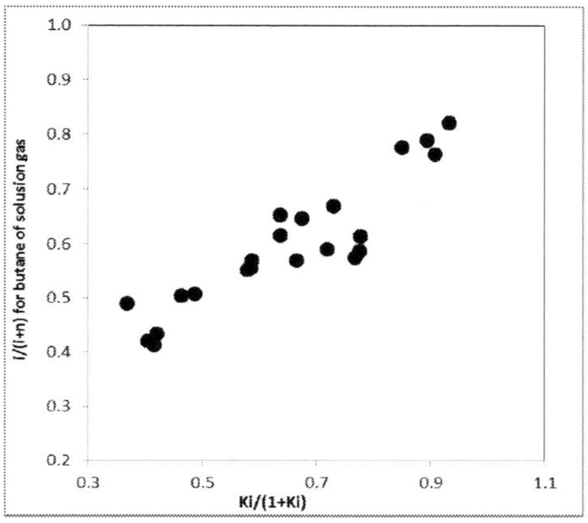

Figure 7. Relationship between i/(i+n) for butane of solution gas in Nkh 3-4 oil and isoprenoid coefficient Ki/(1+Ki).

4. Conclusion

Oil fluids from Nkh 3-4, Nkh 1, Sd 9, Yak 3-7 and vYak significantly differ from each other in terms of their physical and chemical properties, gas saturation level, molecular and isotopic composition. As oils from different layers have the same origins (generated by the organic matter of the Yanovstanskaya formation at close catagenesis stages), the differences revealed are the consequences of secondary transformations in a pool, and namely – biodegradation.

The value of oil biodegradation in the pools of different layers varies from 0 to 7 on a 10-point scale. The deepest layers, Nkh 3-4 and Nkh 1, where reservoir conditions do not contribute to high microflora activity, comprises zones where non-biodegraded and slightly biodegraded oils (the value of biodegradation making 0-3) are spread. The most biodegraded oil samples under study are those from the vYak layers (the biodegradation rank being 6-7). These samples lack not only the easiest to utilize components, such as n- and iso-alkanes, but also components being more biodegradation-resistant, i.e. alkyl-cyclohexanes, alkylbenzenes, naphthalene and alkyl-naphthalenes, phenanthrene and alkyl-phenanthrenes, dibenzothiophene and methyl-dibenzothiophenes. Besides, such severe biodegradation leads to the distorted initial distribution of hopanes.

The influence of biodegradation processes is reflected in changes to the component and isotopic composition, as well as properties of reservoir oils due to the step-by-step utilization of primarily light components from propane to nC_{18}. Thus, the evaluation of the biodegradation level based on the study of degassed oil samples and extracts from oil-saturated core allows predicting the composition and properties of reservoir oil.

Another equally important process of secondary transformation occurring in the pools is methanogenesis that resulted in the formation of gas caps in the Vankor field pools.

References

[1] Goncharov I V, Oblasov N V, Samoilenko V V, Fadeeva S V 2009 Biodegradation impact on molecular parameters by the example of oils in the Krasnodar Territory, Western Siberia and Sakhalin. *Materials of the VII International Conference OIL AND GAS CHEMISTRY* (Tomsk: Publishing House

of the Institute of Atmospheric Optics, Russian Academy of Sciences, Siberian Branch) 221-5
[2] Goncharov I V, Oblasov N V, Samoilenko V V, Krinin V A and Oshmarin R A 2011 Nature of oils of the Vankor field area *Oil and Gas Industry* **3** 12-6
[3] Goncharov I V, Oblasov N V, Samoilenko V V, Smetanin A V, and Zhurova E L 2012 Using geochemical studies for reserves calculation and developing oil-gas condensate fields *Nedropolzovanie XXI Vek* **5** 50-5
[4] Goncharov I V, Samoilenko V V, Oshmarin R Aand Zimin S V 2012 Gas-oil ration in pools with a gas cap (by the example of layer Nkh 3-4 of the Vankor field) *Gorny Zhurnal* **11** 71-3
[5] Goncharov I V, Oblasov N V, Samoilenko V V, Veklich M A and Fadeeva S V 2013 Biodegradation impact on the isotopic composition of carbon in gases by the examples of oils from the east of Western Siberia *Materials of the 20th Symposium on Isotope Geochemistry named after Academician A.P. Vinogradov (GEOKhI, November 12-14, 2013)* (Moscow: Akvarel) pp 110-113
[6] Goncharov I V, Krinin V A, Samoilenko V V, Oblasov N V and Fadeeva S V 2009 Oil biodegradation in Vankor oilfield *Abstracts of the communications presented to the 24th International Meeting on Organic Geochemistry* (Bremen: University of Bremen) Abstract No P-374 p 482
[7] Goncharov I V, Oblasov N V and Samoilenko V V 2013 Effects of biodegradation on the oil composition in Vankor oil field *Organic Geochemistry: Trends for the 21st Century: Book of Abstracts of the Communications presented to the 26th International Meeting on Organic Geochemistry* (Published on-line in: Digital.CSIC, http://digital.csic.es/) vol. 1 pp 470-471
[8] Peters K E and Moldowan J M 1993 The biomarker guide. Interpreting molecular fossils in petroleum and ancient sediments. (NJ, Englewood Cliffs: Prentice Hall) 363 p
[9] Wenger L M, Davis C L and Isaksen G H 2002 Multiple controls on petroleum biodegradation and impact on oil quality *SPE Reservoir evaluation and engineering* **5** 375-383
[10] Peters K E, Walters C C and Moldowan J.M. 2005 The biomarker guide (Cambridge, U.K.: Cambridge University Press) 1155
[11] Vandecasteele J-P 2008 Microbiology of methane and of C1 compounds *Petroleum microbiology: concepts, environmental implications* (Paris: Edition Technip) **1** 79-172
[12] Goncharov I V, Krashin D I and Spilman K A 1983 On the nature of gas in the north of Tyumen Region *Geologiya Nefti i Gaza* **3** 34-8
[13] Milkov A V 2010 Methanogenic biodegradation of petroleum in the West Siberian Basin (Russia): Significance for formation of giant Cenomanian gas pools *AAPG Bulletin* **94** 1485-1541
[14] Milkov A V 2010 Role of methane-producing biodegradation of hydrocarbons in forming giant Cenomanian lean gas pools in the north of Western Siberia *Geologiya Nefti i Gaza* **4** 55-62
[15] Petrov Al A 1984 Petroleum hydrocarbons (Moscow: Nauka) 263
[16] Goncharov I V 1987 Geochemistry of oils in Western Siberia (Moscow: Nedra) 181
[17] Petrov Al A 1994 Geochemical typification of oils *Geokhimiya* **6** 876-91
[18] Petrov Al A 1995 Biomarkers and geochemical conditions of oil formation in Russia *Neftekhimiya* **35** 25-34
[19] Peters K E, Moldowan J M, McCaffrey M A and Fago F J 1996 Selective biodegradation of extended hopanes to 25-norhopanes in petroleum reservoirs. Insights from molecular mechanics *Organic Geochemistry* **24** 765-83
[20] Goncharov I V, Korobochkina V G, Oblasov N V and Samoilenko V V 2005 Nature of hydrocarbon gases in the southeast of Western Siberia *Geochemistry International* **43** 810-15
[21] Boreham C and Edwards D 2007 Effect of source, maturity and biodegradation on the D/H ratio of Australian natural gas *Abstracts of the communications presented to the 24th International Meeting on Organic Geochemistry* (Bremen: University of Bremen) Abstract No. O84 1085-6
[22] Wang W-C, Zhang L-Y, Liu W-H, Kang Y and Ren J-H 2005 Effects of biodegradation on the carbon isotopic composition of natural gas *Geochemical Journal* **39** 301-9

IOP Conf. Series: Earth and Environmental Science **24** (2015) 012040 doi:10.1088/1755-1315/24/1/012040

MDBT estimation ratio for transformation organic matter ratio in Bazhenov Formation of Western Siberia (Tomsk Oblast, Russia)

I Goncharov[1], V Samoilenko[2], N Oblasov[3] and S Fadeeva[4]

[1] Tomsk Polytechnic University, Tomsk 634034, Russia
[2,3,4] JSC «TomskNIPIneft», Tomsk 634027, Russia

E-mail: [1]goncharoviv@nipineft.tomsk.ru

Abstract. The Bazhenov Formation is the main source rock in the West Siberian Basin, because of high organic-rich Bazhenov black shales, with a total organic carbon (TOC) content of 10-15wt.% generally. Based on Rock-Eval pyrolysis data, the initial generative potential for a significant part of West Siberian Bazhenov Formation organic matter (OM) is from 10 to 60% expended. Most West Siberian oils are in the Bazhenov Formation due to very high oil potential (initial hydrogen index = 710mgHC/gTOC) and widespread regional distribution). Vitrinite reflectance (Ro) does not correctly indicate the transformation ratio (TR) of the OM in Bazhenov black shales. However, GC-MS analysis of extracts indicated a good correlation between some molecular maturity and Rock-Eval maturity parameters. Molecular maturity parameters based on naphthalenes and phenanthrenes are not applicable for Bazhenov Formation OM. While the parameters based on 4-methyldibenzothiophene and 1-methyldibenzothiophene ratio (4MDBT/1MDBT) is applicable in estimating the hydrocarbon potential of Bazhenov Formation black shales and, consequently, provides more reliable information about marine OM TR than does Ro. The 4MDBT/1MDBT ratio is more consistent and precise than pyrolysis data. Therefore, it is more applicable in the case of scarcity of core samples throughout the whole formation.

1. Introduction

One of the main questions in petroleum exploration is the quantity of generated hydrocarbons in this or that new area. So, the following parameters should be taken into consideration: type and thickness of the source rock, TOC value, residual hydrogen index (HI), initial hydrogen index (HIo), and TR.

The first three parameters are not problematic. However, it can be difficult to determine the initial generative potential of source-rock OM, i.e. generative potential before entering the oil window (phase).

The simplest approach is to find areas, where OM is not involved in the starting stage of oil generation, i.e. areas with thermal immature source rock. However, this is practically impossible, especially in those areas where OM is already in the active oil generation phase (oil window).

One can refer to the kerogen type documented in research literature. According to the case studies on the West Siberian Basin by Lopatin et al. [1] and Peters et al. [2], the Bazhenov black shales contain marine Type II kerogen. Classical Type II kerogen is characterized by the value 627 mg HC/g TOC [3].

If HI and TR of OM are known, then HIo is calculated: while Ro calculates TR [3]. In this case, however, Ro does not always indicate the transformation degree of OM. Some problems in Ro determination are the low- quality polished samples and incorrect population selection of vitrinite. Ro determined by various laboratories commonly differs due to the above-mentioned factors [4]. Moreover, it is very difficult to measure Ro in marine kerogens, such as those in Bazhenov Formation, where vitrinite is absent. Obviously, no direct measurement of Ro for oils is possible.

Thus, this paper provides an alternative geochemical parameter based on molecular composition of extracts for Ro in the Bazhenov source rock. Similar investigations were performed for source rocks in different sedimentary basins [5-7], where molecular parameters were calibrated against Ro. Many molecular parameters are used for maturity evaluation of source rocks and oils, while (some are limited by a certain maturity level) a maturity level may limit some of them [8, 9] or even organic facies affect the application of these parameters [10]. Thus, molecular maturity parameters must be selected and calibrated for each OM type and sedimentary basin. They must indicate the hydrocarbon potential of the source rock (Rock-Eval pyrolysis data).

2. Geological setting

West Siberian Basin, one of the largest basins in the world, contains more than 3500 m Mesozoic and Cenozoic sedimentary rocks, and has a relatively simple geologic structure. There are big oil & gas pools, occurring throughout the stratigraphic section from Paleozoic to Upper Cretaceous.

Literature data [2, 11] and geochemical analyses [12-14] of oil, gas and rock samples indicate at least three source rocks in this basin. The Bazhenov Formation is the major oil source rock in Western Siberia and about 90 % of originated oil in it [1,15].

The Bazhenov Formation is located in the south, center, and north parts of the West Siberian platform, covering about 1 million km^2 in area. The base of the formation is Early Tithonian, while the top is Early Berriasian [16]. The Bazhenov Formation is a classic source rock consisting of black and brownish-black deep-marine carbonate-siliceous shales with high quality OM. TOC contents range from 5 to 20wt.% and up to 35wt.% in the central part of the West Siberian Basin [1]. The Bazhenov Formation is characterized by many anomalous parameters, including high natural radioactivity, high electrical resistance, low density, and high trace element content. The formation varies in thickness from 20-50m in the central part and up to 10-30m in peripheral part.

In a significant part of the West Siberian Basin, the Bazhenov Formation is within the oil window zone. In the central part of this basin it is no more than 25% of the initial generation potential, while more extensive conversion has occurred only in areas of major faults.

Oil pools related to OM in the Bazhenov Formation occur throughout the stratigraphic section, as types and lithology of overlying and underlying rocks are highly variable (figure 1) and hydrocarbon migration conditions from the source rock are also different. In the eastern part of the West Siberian Basin, the Bazhenov Formation is overlain by Neocomian marine shale of Kulomzin Formation (about 300m thick), providing an efficient seal. Below the Bazhenov Formation are the Georgiev and Vasyugan Formations, consisting of alternating mudstone, siltstone, coal and sandstone with good reservoir properties.

In the western part of the basin, the Georgiev and Vasyugan Formations are replaced with Abalak Formation sediments, consisting of marine shale. However, Neocomian sediments occur as regressive sandstones, the so-called "clinoform complex" in the Achimov Formation. These sandstones have good reservoir properties and, simultaneously, provide migration paths for Bazhenov hydrocarbons to Aptian-Albian shallow-marine terrigenous reservoirs.

Furthermore, in high catagenesis zones lacking favorable conditions for hydrocarbon migration, some oil pools may occur in nontraditional reservoirs within the source rock itself in the Bazhenov Formation (e.g. Salym oilfield region).

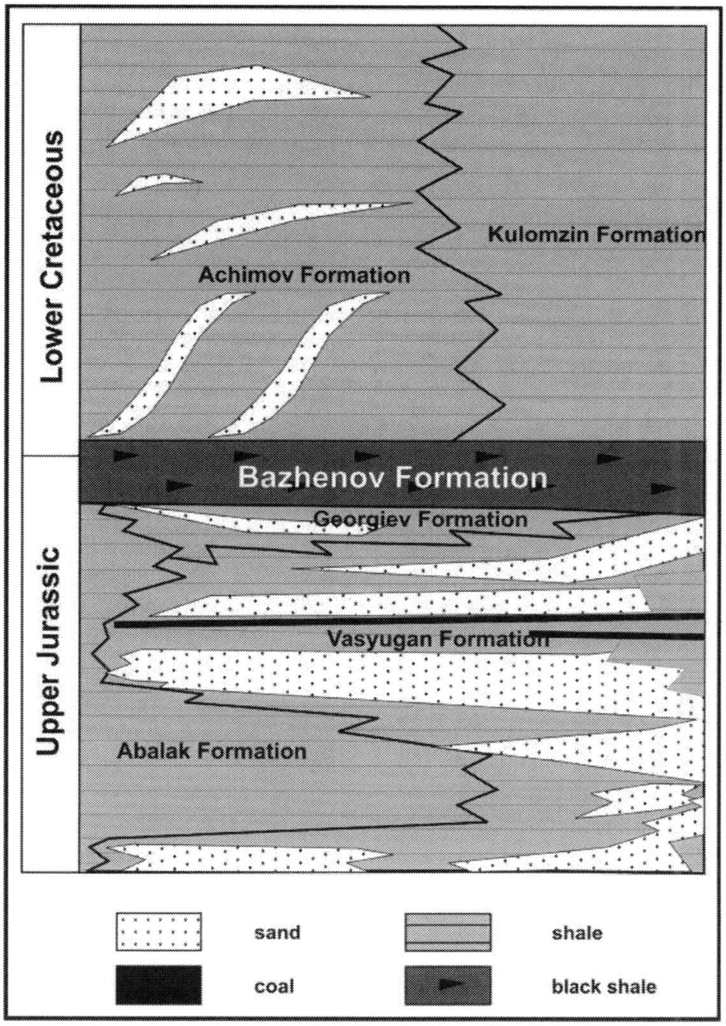

Figure 1. Generalized lithology column of Upper Jurassic and Lower Cretaceous section for the studied area.

3. Samples and methods

The above-mentioned study investigated the Bazhenov Formation black shale core samples, Upper Jurassic coals (the Vasyugan Formation), and oils related to OM in the Bazhenov Formation. More than 1500 core samples from over 100 wells in the Bazhenov Formation were investigated (figure 2). Samples were obtained from within 1-m spacing of each well throughout the Bazhenov Formation. Samples with average values from each well were used for correlation.

All crushed core samples were washed and dried beforehand and pyrolysis was carried out in Rock-Eval 6 Turbo design (with two infrared cells). The programmed temperature for pyrolysis was at 300°C (3min), 25°C/min up to 650°C, and for oxidation at 300°C (1min), 20°C/min up to 850°C (5min). Therefore, transformation ratio is given by [17]:

$$TR = \frac{1200 \cdot (HI_0 - HI)}{HI_0 \cdot (1200 - HI)} \cdot 100\%$$

Crushed rock and coal samples were subjected to chloroform solvent extraction in Soxhlet apparatus for over 1 hour. Deasphalted crude oils and source rock extracts (bitumens) separated into

IOP Conf. Series: Earth and Environmental Science **24** (2015) 012040 doi:10.1088/1755-1315/24/1/012040

saturated hydrocarbons, aromatic hydrocarbons, and resins in a silica gel column. These fractions were eluted with hexane, benzene, and benzene: methanol (1:1v/v), respectively.

GC-MS analyses of total extracts and oils, as well as, their fractions were carried out with Hewlett Packard 6890/5973 instrument in 30-m HP-1-MS column. The programmed oven temperature was at 45°C (3min) to 310°C (30min) at 3°C/min, in a helium atmosphere, where flow rate was 0.7ml/min. Their areas were calculated on the basis of biomarker peak ratios. In some publications the recognition of appropriate biomarkers and component ratio calculations from mass fragmentograms have been described [9,18].A.N. Fomin (Institute of Geology, Russian Academy of Sciences, Siberian Branch, Novosibirsk) and N.V. Lopatin (VNII Geosistem, Moscow) defined Ro.

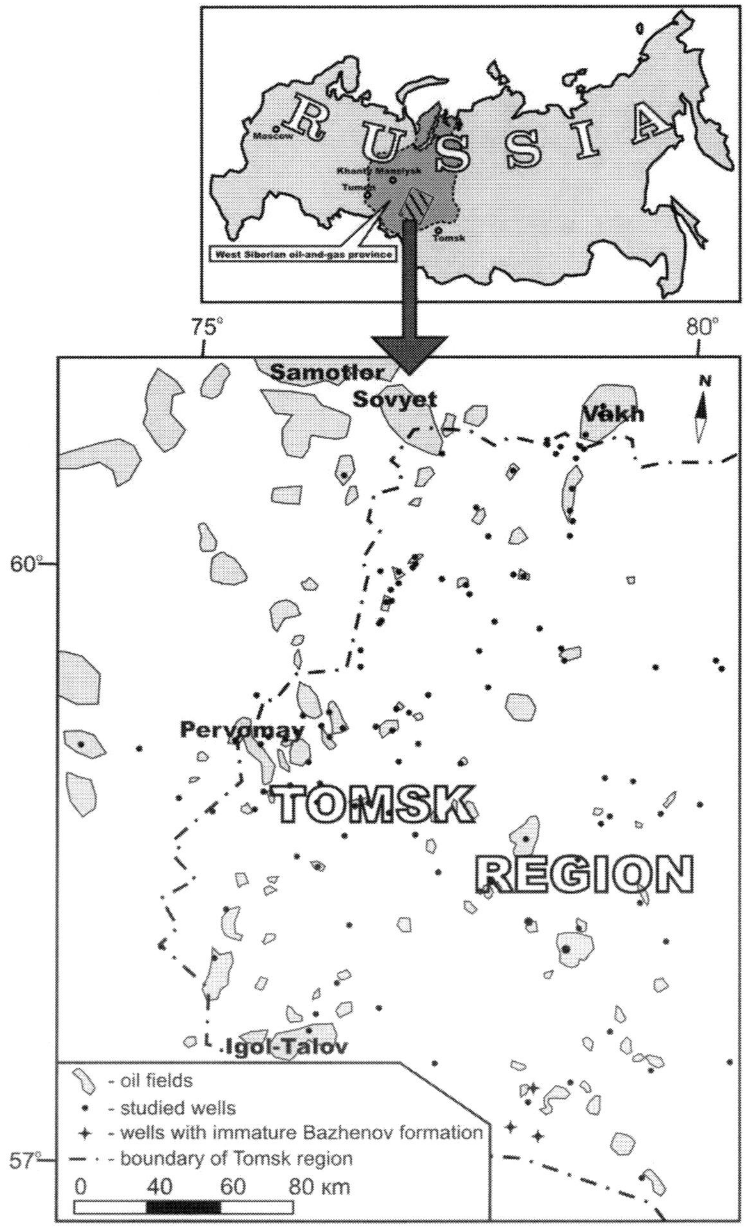

Figure 2. Map- location of the studied area.

4. Results and discussion

Rock-Eval pyrolysis data for Bazhenov Formation rock samples indicate that current HI ranges from 120 to 770mg HC/gTOC. Consequently, if it is assumed that HIo is 627mg HC/gTOC [3], then TR negative value is determined, although there exist oil fields in this region. Therefore, HI for some Bazhenov Formation rocks should be higher than for typical Type II OM.

In this case, Ro can be applied to determine TR [3], though Ro does not objectively indicate the OM thermal maturity in the Bazhenov Formation (figure 3). This is obvious from the fact that Ro precisely characterizes thermal maturity of Type III OM in coals of Upper Jurassic deposits rather than the thermal maturity of Type II OM in the Bazhenov Formation.

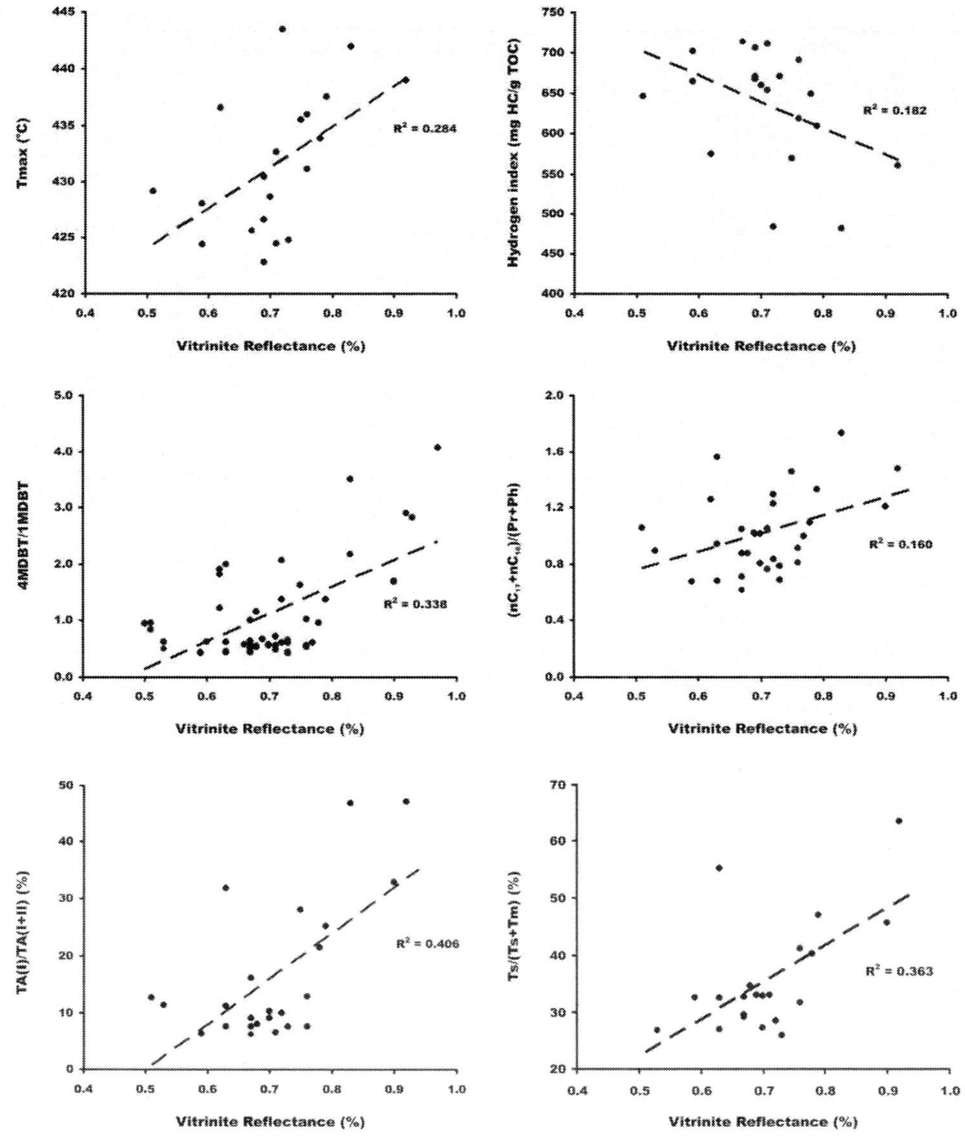

Figure 3. Rock-Eval 6 pyrolysis data for organic matter and molecular maturity parameters for Upper Jurassic Bazhenov Formation black shale extracts and Ro (%) of underlying Upper Jurassic coals.

Obtained Bazhenov Formation rock extracts were tested according to various parameters as described in Alexander et al. [19], Radke et al. [20], Peters et al. [9]. The parameters based on naphthalenes and phenanthrenes did not work out, (figure 4), probably as they indicate thermal maturity of Type III OM [6, 20-22]. Interestingly, a number of other parameters (4MDBT/1MDBT, TA (I)/TA (I+II), Ts/(Ts+Tm), (nC17+nC18)/(Pr+Ph)) increased with the increase of Ro [23]. However, the low level of thermal maturity corresponds to a significant distribution in values (figure 3). Furthermore, the relations between molecular parameters and Rock-Eval data are much more uniform with each other (figure 5,6), than with the Ro (figure 3). Two main reasons are: (1) all parameters characterize the state of OM in the source rock, and (2) the measurement error of molecular parameters is lower than that of Ro.

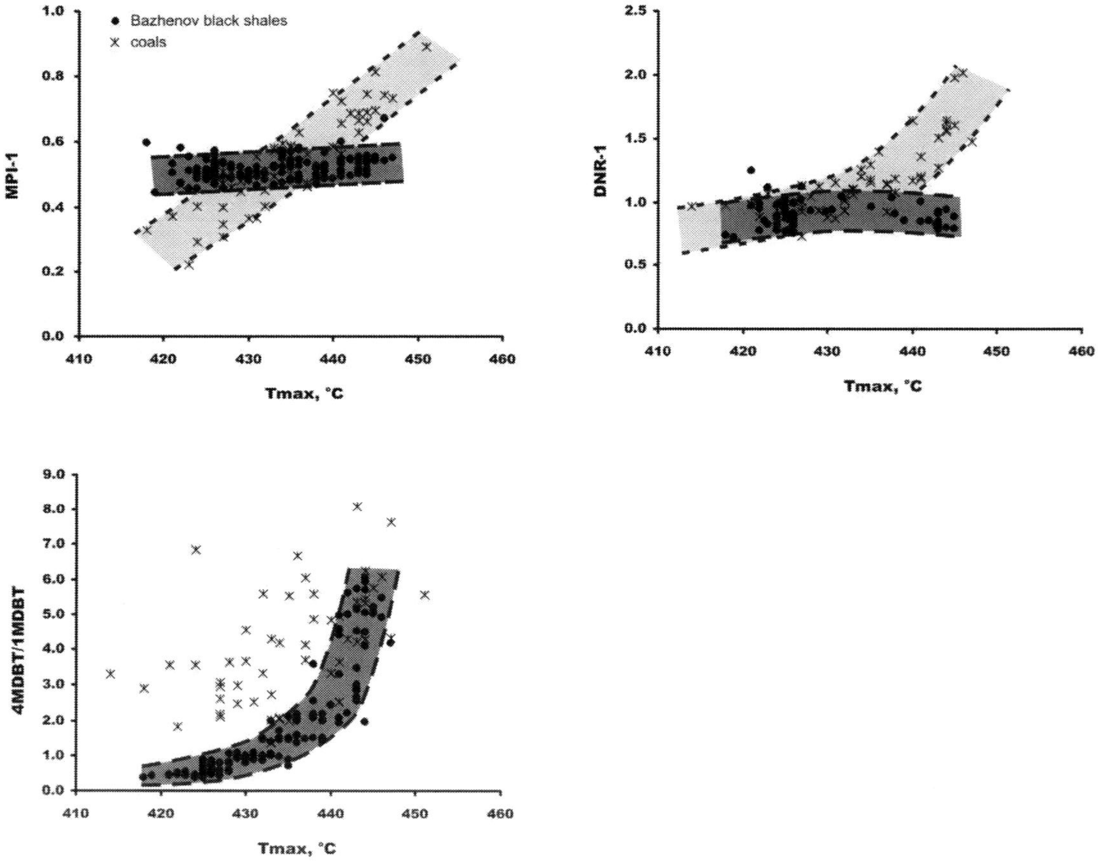

Figure 4. Molecular parameters vs. Tmax for Upper Jurassic Bazhenov Formation black shales and Jurassic coals.
Methylphenantrene Index 1 (MPI-1) = 1.5 x (2-MP + 3-MP)/((P x 0.69[a]) + 1-MP + 9-MP), using m/z 178 and 192, [a]GCMS data corrected to FID response [21];
Dimethylnaphthalene Ratio 1 (DNR-1) = (2,6-DMN + 2,7-DMN)/1,5-DMN, using m/z 141 response [19].

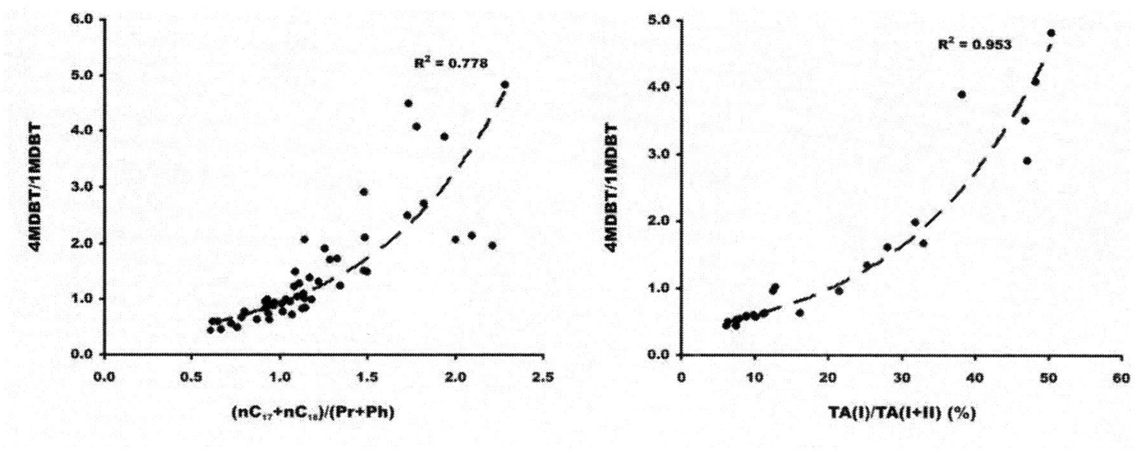

Figure 5. Relationship between molecular maturity parameters for Upper Jurassic Bazhenov Formation black shale extracts.

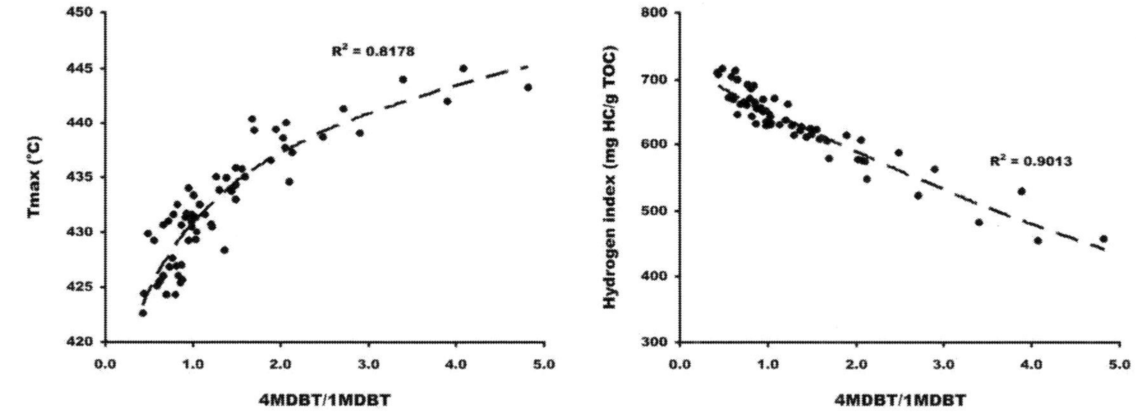

Figure 6. Rock-Eval 6 pyrolysis data vs. 4MDBT/1MDBT for Bazhenov Formation black shales.

Based on log characteristics, the Bazhenov Formation can be divided into four layers with different contents of OM, carbonates, clays and siliceous minerals [1]. In figure 7, molecular parameters change throughout the Bazhenov Formation in two wells having different thermal maturity levels. The 4MDBT/1MDBT ratio is the most stable parameter in the 10-15m thickness of both wells so, it can be stated that the maturity difference in such a small depth range is not possible. The variation limits of these parameters for all Bazhenov Formation samples in the studied area are detailed and shown in figure 8. The data show that Ro changes only two and a half times within the oil-window, while the molecular parameters change 5 or 8; while 4MDBT/1MDBT ratio changes to 55. However, at high thermal maturity levels, the biomarker content decreases, resulting in a sharp error increase. Therefore, 4MDBT/1MDBT and (nC17+nC18)/(Pr+Ph) are the most applicable in evaluating the thermal maturity level of OM in the Bazhenov Formation. So, their significant content (dibenzothiophenes and alkanes) in extracts and oil is important, but they are not trace contaminants. However, (nC17+nC18)/(Pr+Ph) ratio has a serious drawback as it partial depends on the TOC content [24].

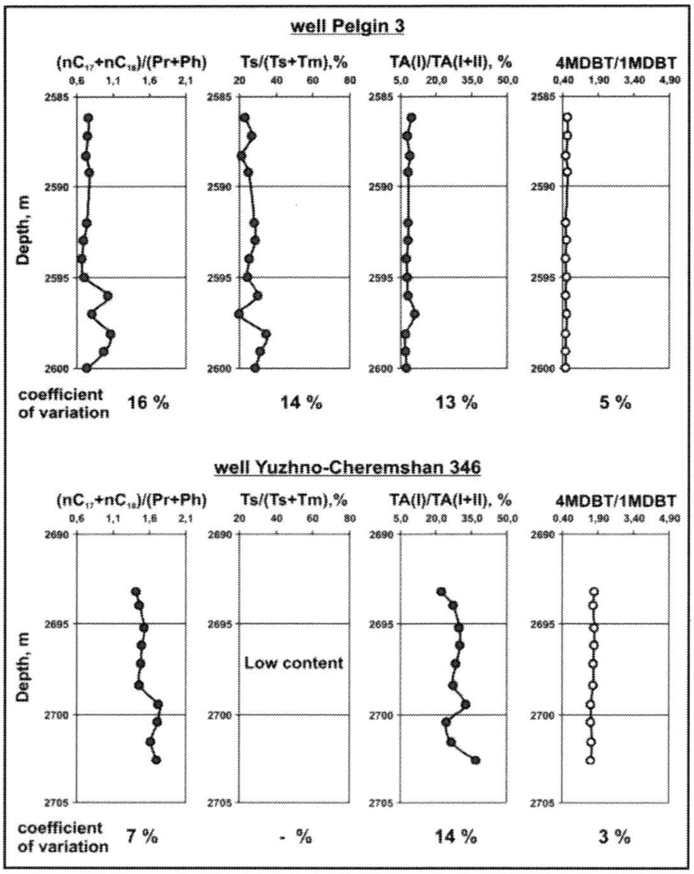

Figure 7. Variations of some maturity parameters through Bazhenov Formation in the well Pelgin-3 (low maturity) and Yuzhno-Cheremshan-346 (medium maturity)

$$\text{The coefficient of variation} = \frac{\sqrt{\Sigma(X_i - X_{av.})^2/n}}{X_{av.}} \times 100\%$$

where, X_i is the parameter value; $X_{av.}$ is the average value; n is the number of values. The range of parameters is from minimum to maximum values of the parameters in the studied area (see Figure 8).

	main part of Western Siberia	Exception: "the hot zone" - Salym oil field	*Sensitivity of parameters max/min*
TR, %	5 20 40 60	80-90	
4MDBT/1MDBT	0.45 0.90 2.1 4.9	25	*55*
$(nC_{17}+nC_{18})/(Pr+Ph)$	0.6 1.0 1.5 2.1	3.0	*5*
TA(I)/TA(I+II), %	6.0 13 27 50	Low content	*8*
Ts/(Ts+Tm),%	26 35 50 80 Low content	Low content	*2*
Ro, %	0.51 1.00	1.2	*2.5*

Figure. 8. Variation limits of molecular maturity parameters for Bazhenov Formation black shales.

The above-mentioned parameter and systematic analyses of the Bazhenov black shales from different wells identified several zones with abnormally low parameters of thermal maturity level. Many parameters based on the biomarker composition indicate that OM has already reached the oil window level (table 1). But at the same time, the large collection of West Siberian oils has no samples with 4MDBT/1MDBT values below 0.8. So, it can be concluded that in these zones OM has not only reached the oil window level but still retains its initial generative potential. Thus, the mean value of HI for these zones is 710mg HC/g TOC (table 2) and this HIo can be recommended for the Bazhenov Formation in the south-eastern part of Western Siberia. Though, it is much higher than is generally accepted for Type II OM.

Table 1. Maturity parameters of Bazhenov Formation rock extracts from the "cold zone" and some low maturity oils.

Extracts from samples of immature Bazhenov Formation black shales							
Depth, m	4MDBT/ 1MDBT	$(nC_{17}+nC_{18})/(Pr+Ph)$	TA(I)/ TA(I+II)[a], %	Ts/ (Ts+Tm)[a], %	20S/(20S+20R) $SteraneC_{29}$[a], %	$\beta\beta/(\alpha\alpha+\beta\beta)$ $SteraneC_{29}$[a], %	Mor/Hop $TerpanesC_{30}$[a]
Archin 51, Ro=0.67 %							
2611.1	0.42	0.64	6.5	24.7	27.3	46.1	0.22
2612.7	0.38	0.50	5.3	29.6	21.6	39.4	0.20
2613.9	0.49	0.58	6.0	28.9	21.7	38.7	0.19
2614.9	0.44	0.64	5.9	31.6	20.9	36.7	0.21
2616.0	0.52	0.67	6.3	31.5	18.4	38.6	0.20
Kulgin 145, Ro=0.53 %							
2622.8	0.52	0.62	6.0	30.2	20.6	40.0	0.22
2623.8	0.47	0.64	6.2	35.7	17.9	37.4	0.22
2624.7	0.50	0.61	7.7	34.6	18.4	38.6	0.20
2625.7	0.46	0.74	5.5	32.5	24.9	40.6	0.19
2626.7	0.47	0.87	7.0	32.1	19.4	40.5	0.20
2628.1	0.53	1.10	6.4	32.9	17.1	38.8	0.19
Smolyanaya 3, Ro=0.59 %							
2613,7	0,42	0,68	7,6	19,0	32,1	46,0	0,19
2614.7	0.60	0.73	7.4	42.8	30.0	42.4	0.23
2616.8	0.42	0.54	6.1	35.9	32.2	42.0	0.21
2617.8	0.48	0.53	5.7	34.7	28.2	40.8	0.20
2618.7	0.39	0.63	5.2	36.0	24.6	40.8	0.18
2619.7	0.46	0.66	6.9	33.2	26.1	40.5	0.19
2620.7	0.46	0.72	6.4	31.4	26.3	41.8	0.19
2621.7	0.38	0.72	6.0	30.2	25.0	45.6	0.21
Low maturity oils (Vasyugan Formation, Upper Jurassic)							
Well[b]	4MDBT/ 1MDBT	$(nC_{17}+nC_{18})/(Pr+Ph)$	TA(I)/ TA(I+II), %	Ts/ (Ts+Tm), %	20S/(20S+20R) SteraneC29, %	$\beta\beta/(\alpha\alpha+\beta\beta)$ SteraneC29, %	Mor/Hop $TerpanesC_{30}$
1	0.84	1.05	8.3	56.0	30.0	54.0	0.14
2	0.84	1.02	11.3	44.5	48.1	62.3	0.15
3	0.88	1.14	11.7	43.4	31.8	49.0	0.11
4	0.92	0.96	11.6	47.0	44.3	61.7	0.16
5	0.92	1.05	9.9	39.0	47.7	58.9	0.11
6	0.95	1.02	7.6	47.0	50.2	75.5	0.09
7	0.96	1.11	9.6	55.0	46.3	57.5	0.19

4MDBT/1MDBT = 4-methyldibenzothiophene/1-methyldibenzothiophene; $(nC_{17}+nC_{18})/(Pr+Ph)$ = (n-heptahexadecane+n-oktahexadecane)/(pristane+phytane); TA(I) – sum of C_{20} and C_{21} TA-steroids; TA(II) – sum of C_{26}-C_{28} (20S+20R) TA-steroids.

[a] Parameters are described in Peters at al. [9]

[b] 1 – Nizhnepervomay1; 2 – Lontyn-Yakh 70; 3 – Severo-Pervomay 2285; 4 – Zapadno-Vesenneye 241; 5 – Lontyn-Yakh 68; 6 – Uzhno-Festival 1; 7 – Kolotush 268.

Table 2. Some Rock-Eval 6 pyrolysis for wells from the "cold zone"

Depth, m	S1	S2	PI	TOC	Tmax	HI
Archin 51, Ro=0.67 %						
2611.1	4.1	49.2	0.08	7.1	423	692
2611.6	4.0	53.8	0.07	7.8	418	689
2612.7	3.4	60.3	0.05	8.4	424	720
2613.3	5.0	103.8	0.05	13.5	423	772
2613.9	3.2	68.9	0.04	10.0	422	692
2614.3	3.0	65.1	0.04	9.2	425	706
2614.9	3.4	71.7	0.05	10.3	424	697
2615.4	3.0	61.8	0.05	8.9	419	692
2616.0	3.7	84.4	0.04	11.6	422	727
Kulgin 145, Ro=0.53 %						
2622.8	2.9	60.0	0.05	8.4	425	718
2623.8	2.7	52.6	0.05	7.0	426	747
2624.7	1.6	28.5	0.05	4.1	425	692
2625.7	2.5	56.8	0.04	7.9	422	716
2626.7	3.1	75.6	0.04	10.5	421	722
2628.1	1.6	41.1	0.04	6.1	428	678
Smolyanaya 3, Ro=0.59 %						
2613.7	3.0	51.6	0.05	7.7	421	670
2614.7	3.4	52.6	0.06	7.6	425	694
2616.8	4.7	102.6	0.04	13.6	425	753
2617.8	3.8	73.3	0.05	9.7	424	756
2618.7	2.8	48.3	0.06	6.6	425	731
2619.7	2.3	37.5	0.06	5.4	425	690
2620.7	3.3	65.6	0.05	9.1	424	718
2621.7	3.1	56.1	0.05	8.4	424	672
2622.7	2.8	55.6	0.05	8.1	424	688
average	**3.1**	**59.6**	**0.05**	**8.4**	**424**	**710**
min-max	**1.6 -5.0**	**28.5-103.8**	**0.04 - 0.08**	**4.1 - 13.6**	**418 - 428**	**670 - 772**

S1 and S2 values in mg HC/g rock; PI=S1/(S1+S2) – production index; TOC – total organic carbon, wt.%; Tmax – temperature at which the maximum amount of S2 hydrocarbons is generated, ^{0}C; HI=S2/TOC – hydrogen index, mg HC/g TOC.

HIo makes it possible to determine TR and the dependent TR versus 4MDBT/1MDBT (figure 9). Figure 9 shows that 4MDBT/1MDBT determining TR gives a more precise depiction of the state of the OM than does Ro. Furthermore, TR determination by 4MDBT/1MDBT is often more simple and precise than when based only on pyrolysis results. It is difficult to find enough core samples to precisely characterize a source rock section. One or two samples are sufficient enough if 4MDBT/1MDBT changes insignificantly throughout the section and the variation coefficient are only 3-5% (figure 7). At the same time, there is a more detailed picture of thermal maturity and petroleum yield based on 4MDBT/1MDBT rather than on Ro. Due to the above-mentioned parameter the change in the petroleum yield can be established by four at a distance of 20km.

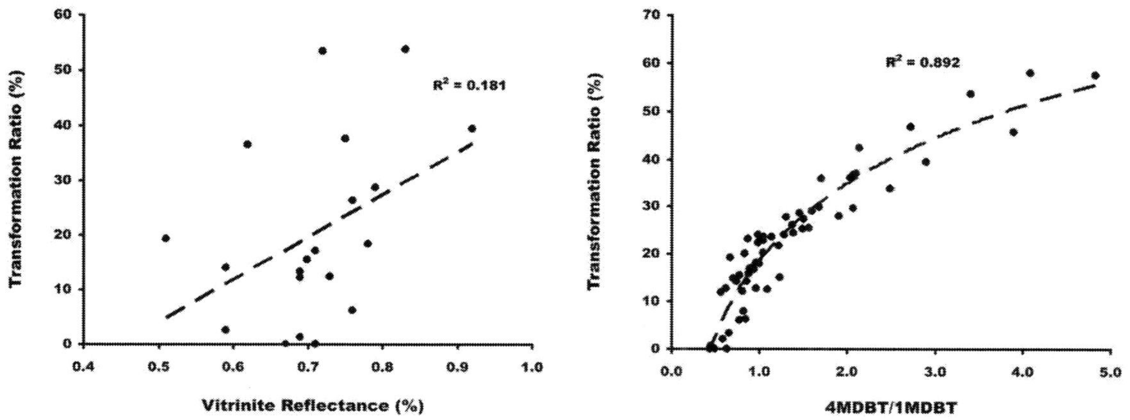

Figure 9. Transformation ratio vs. 4MDBT/1MDBT and Ro (%) for Bazhenov Formation black shales.

5. Conclusion

High sensitivity of 4MDBT/1MDBT parameter to changes in source rock thermal maturity makes it possible to evaluate areas of prospective oil and gas accumulations. Analysis of oils genetically related to the Bazhenov Formation in Western Siberia shows that 4MDBT/1MDBT value ranges from 0.80 to 18. However, oilfields with 4MDBT/1MDBT values of 0.80 – 0.90 occur very rarely and have small oil reserves. This is probable because of insufficient generated hydrocarbon amounts within catchments areas, which are necessary for larger oilfield formations. So, in the specific geological conditions of Western Siberia, 4MDBT/1MDBT values of 0.80 – 0.90 in Bazhenov Formation extracts could be considered as the minimal values of OM maturity when commercial oilfields are possible. Oils in larger oilfields in Western Siberia have 4MDBT/1MDBT ratios of 1.3 – 2.5.

References

[1] Lopatin N V, Zubairaev S L, Kos I M, Emets T P, Romanov E A and Malchikhina O V 2003 Unconventional oil accumulations in the Upper Jurassic Bazhenov Black Shale Formation, West Siberian Basin: a self-sourced reservoir system. *Journal of Petroleum Geology* **26** (2) 225-44

[2] Peters K E, Kontorovich A Eh, Huizinga B J, Moldovan J M and Lee C Y, 1994 Multiple oil families in the West Siberian Basin. *American Association of Petroleum Geologist Bulletin* **78** 893-909.

[3] Espitalié J, Deroo G and Marquis F 1985 La pyrolyse Rock-Eval et ses applications. Deuxième partie. *Rev. Inst. Fr. Pét.* **40** 563-579, 775-784

[4] Dembicki G Jr 1984 An interlaboratory comparison of source rock data *Geochimica et Cosmochimica Acta* **48** 2641-9

[5] Mackenzie A S, Leythaeuser D, Altebäumer F-J, Disko U and Rullkötter J 1988 Molecular measurements of maturity for δ Lias shales in N.W. Germany *Geochimica et Cosmochimica Acta* **52** 1145-54

[6] Radke M 1988 Application of aromatic compounds as maturity indicators in source rocks and crude oils *Marine and Petroleum Geology* **5** 224-36

[7] Santamaria-Orozco D, Horsfield B, di Primio R and Welte D H 1998 Influence of maturity on distributions of benzo- and dibenzothiophenes in Tithonian source rocks and crude oils, Sonda de Campeche, Mexico *Organic Geochemistry* **28** 423-39

[8] Mackenzie A S, Maxwell J R 1981 Assessment of thermal maturation in sedimentary rocks by molecular measurements. In: Organic Maturation Studies and Fossil Fuel Exploration ed J Brooks (London: Academic Press) 239-54

[9] Peters K E, Walters C C and Moldowan J M 2005 *The biomarker guide* (Cambridge University Press, Cambridge, UK) p 1155
[10] Ten Haven H L, de Leeuw J W, Peakman T M and Maxwell J R 1986 Anomalies in steroid and hopanoid maturity indices *Geochimica et Cosmochimica Acta* **50** 853-5
[11] Petrov A 1995 Biomarkers and geochemical conditions of occurrence of Russian oils *Petrochemistry* **35** (1) 25-34
[12] Goncharov I V 1987 *Geochemistry of oils in Western Siberia* (Moscow, Nedra) p 181
[13] Goncharov I V, Korobochkina V G, Oblasov N V and Samoilenko V V 2005 Nature of hydrocarbon gases in the southeast of Western Siberia *Geochemistry* **43** (8) 810-5
[14] Goncharov I V, Oblasov N V, Smetanin A V, Samoilenko V V and Fadeeva S V 2012 Genetic types and nature of fluid of hydrocarbon deposits south-east of Western Siberia *Oil industry* **11** 8-13
[15] Kontorovich A 1984 Geochemical methods for the quantitative evaluation of the petroleum potential of sedimentary basins *Petroleum geochemistry and basin evaluation: AAPG Memoir* ed G J Demaison and R J Murris **35** 79-109
[16] Braduchan Y, Golber A and Gurary F 1986 *Bazhenov horizon of Western Siberia (stratigraphy, paleogeography, ecosystem, presence of oil) (*Novosibirsk: Nauka) p 217
[17] Pelet R 1985. Evaluation quantitative des poduits formes lors de l'evolution geochimique de la matiere organique. *Rev. Inst. Franc. du Petrole* **40** 551-62
[18] Waples D W and Machihara T 1992 Biomarkers for geologists *American Association of Petroleum Geologist Methods in Exploration* Series **9** p. 91
[19] Alexander R, Kagi R I, Rowland S J, Sheppard P N and Chirila T V 1985 The effect of thermal maturity on distributions of dimethylnaphthalenes and thrimethylnaphthalenes in some Ancient sediments and petroleums *Geochimica et Cosmochimica* Acta **49** 385-95
[20] Radke M, Welte D H and Wilsch H 1986 Maturity parameters based on aromatic hydrocarbons: Influence of the organic matter type *Organic Geochemistry* **10** 51-63
[21] Radke M, Willsch H, Leythaeuser D and Teichmuller M 1982 Aromatic components of coal: relation of distribution pattern to rank *Geochimica et Cosmochimica* Acta **46** 1831-48
[22] Chakhmachev A V and Chakhmachev V A 1995 Aromatic sulfur compounds as maturity indices of hydrocarbon systems *Geochemistry* **11** 1656-69
[23] Goncharov I V, Samoilenko V V, Oblasov N V and Fadeeva S V 2013 Catagenesis of organic matter Bazhenov Formation rocks in the south-east of West Siberia (Tomsk region) *Oil industry* **10** 32-7
[24] Goncharov I V, Samoilenko V V, Oblasov N V and Nosova S V 2004 Molecular parameters of OM catagenesis in Bazhenov suite, Tomsk area *Oil and gas geology* **5** 53-9

IOP Conf. Series: Earth and Environmental Science **24** (2015) 012041 doi:10.1088/1755-1315/24/1/012041

Submarine fan reservoir architecture and heterogeneity influence on hard-to-recover reserves. Achimov Fm.

A Kondratyev[1,2], V Rukavishnikov[1], L Shakirzyanov[2] and K Maksyutin[1]

[1]Tomsk Polytechnic University, Tomsk, Russia
[2]Gazpromneft NNG, Noyabrsk, Russia

E-mail: rukavishnikovvs@hw.tpu.ru

Abstract. Due to the fact that simulation model calculation is the basic method used for estimating the efficiency of a development strategy, it is necessary to design geological and simulation models within which reservoir properties and heterogeneity are defined. In addition, the estimation of the influence of various kinds of geological uncertainties on reservoir properties will allow defining a more effective development strategy.

The Achimov formation of the Vingapur oil field was considered in the current study. The northern part of the field is now quite attractive for the development of this formation. The goal of this paper was the complex investigation of petrophysical properties to make a prognosis for the field and assess the effect of geologic uncertainties on production. The first step implied studying the western part of the field where core data are available, the next stage was developing an algorithm to make a prognosis for properties and the geologic and reservoir simulation models were eventually constructed to study the effect of geologic uncertainties in the northern part. As the result of the sedimentary analysis, a model of deposition was defined within which structural elements were also determined. On the basis of wireline and core data analysis, the petrophysical model of the reservoir was build where the method of Rock Types identification using specific cut-off values for wireline logs was applied for the evaluations. In addition to this, the Hydraulic Flow unit approach was employed, which allowed estimating the less extensively explored areas of the field where core had not been retrieved from. Also, this paper provides the results of the seismic attribute analysis and calculations in order to characterize uncertainty in cumulative oil production under the influence of petrophysical and geological heterogeneity.

1. Introduction

Nowadays, the dropping number of high-quality reservoirs raises the issue of developing low permeability reservoirs that were previously considered non-promising. In Western Siberia, as a rule, these types of reservoirs were deposited in a deep-marine environment and represented by turbidite flow or gravity flow deposits. Apart from low permeability, they are also characterized by a complex geologic structure, high petrophysical heterogeneity and compartmentalization. Therefore, the study and evaluation of geologic and petrophysical heterogeneity in connection with a depositional environment, the influence of the uncertainty degree on a subsurface flow are of immediate interest. This study will provide a scenario of model refinement with geological information being useful for reservoir development and making management decisions.

The Achimov formation of the Vingapur field is extensively explored from the viewpoint of sedimentology and petrophysical heterogeneity, thus the accepted modeling approach and sedimentary environment interpretation does not reflect the reservoir complexity. The novelty of this work is in the

IOP Conf. Series: Earth and Environmental Science **24** (2015) 012041 doi:10.1088/1755-1315/24/1/012041

improvement of the depositional model, the methods of reservoir prognosis, estimation of the effect of geological uncertainties on field development.

The objectives of this work are as follows:

1. Integrated study of petrophysical properties;
2. Estimation of reservoir properties;
3. Effect of geological uncertainties on production assessment

2. General information

The Vingapur field is located in Purovsky region of the Yamalo-Nenets Autonomous District and partially in Niznevartovsky region of the Khanty-Mansi Autonomous District, and is geographically located in the northern part of the West Siberian Plate [1].

The Lower Cretaceous deposits are represented by the Megionskaya group conformably underlain by Bazhenov formation argillites. The top of the Megionskaya group is represented by the BV8 formation with the biggest share of production amounts. The bottom of the group consists of predominantly shales being 10-30 m thick. The Achimov formation is above it and is highly variable laterally and vertically. Lithologically, the formation consists of fine-grained sandstones interbedded with siltstone and shale within argillites, there are also thin layers of sandstone and siltstone being 1-40 mm thick. Horizons BV8, BV7 are located above the Achimov formation. The age of the Megionskaya group is Berriasian- Valanginian (K1b-v) [13].

3. Field geologic structure

This section summarizes the data and results of the previous studies related to the geologic architecture and genesis of the Achimov formation that are relevant to this particular study. The first concept of the Achimov formation was proposed by Gurari in 1959. According to it, it is composed by laterally and vertically extensive lenticular beds consisting of sandstone and siltstone [6]. The fact that nowadays the genesis and architecture of the Achimov formation is an open question is quite important. Thus, it was assumed that sedimentation occurred here as the result of common transgression [1]. Some authors relate the Achimov formation to submerged shoals [10], where the architecture was associated with a sea bed relief and submarine currents. The clinoform Neocomian architecture was introduced by Naumov, according to which the deposits of the same age plunge in the basin-ward direction with facies variability [6]. The shelf zone represents the place of erosion and sedimentation depending on sea level variations and amount of deposits. Filling-in the fondoform part was related to the prevalence of the sedimentation velocity over the rate of sea bed sinking. According to the Naumov theory, turbidity currents and slumps were the prevailing mechanisms of sediment transport. The seismostratigraphic complexes include the undaform (shallow marine environment), the clinoform (shelf slope) and the fondoform (slope base), according to Peiton terminology [12]. The individual parts of seismic cyclite are related genetically, whereas seismic cyclite itself has a sigmoid shape. Thus, based on the regional 3D seismic surveys, it can be inferred that the area of the Vingapur Neocomian deposits has a three-stage structure with steep angles of the upper transgression complexes, typical presence of the fondoform part of a significant thickness deeply advanced in the western direction, that is a part of a lower, more regressive cycle of sedimentation [4].

Depending on the flow density, the coherency of sediments, the shale content and position relatively to the shelf slope type of process may change from a slide, slump to debris flow and turbidity current [15]. The type of this process determines the dimension and geometry of the sand bodies, while from the reservoir point of view its petrophysical properties are closely linked with the energy of the current, position in the depositional environment [15]. The fact that similar structures can be related to different types of processes brings an additional uncertainty in interpretation.

The facies distribution within the reservoir, interaction between them lies within the stratigraphic and structural framework where reservoir connectivity has a significant influence on the subsurface flow. The first step is recognizing the reservoir boundaries of various orders related to sequence boundaries, which is commonly implemented using 3D seismic surveys. [8]. Next, with the sufficient

3D seismic qulity and resolution, it is possible to reveal facies distribution within the defined structural framework according to the difference in the acoustic properties on the boundaries between facies units [11].

Petrophysical heterogeneity in turbidity reservoirs is reflected in variable permeability within several orders of magnitude for the given values of porosity, therefore it is quite important to determine major factors controlling reservoir quality in order to appropriately predict it in the model. In this study, the authors [3, 7] suggest an approach to rock type discrimination based on differences in the pore throat size, lithology, particularly in the shale volume content. In work [2], the authors used the Multi-Resolution Graph-Based Clustering method with Facimage where data points could be clustered in multi-dimensional space for log facies prediction. The log identification was based on certain cut-off values defined on the basis of the cross plot of grain density versus the absolute value of the separation between the resistivity in the washed zone and the true formation resistivity characterizing the effectiveness of mud filtrate propagation in the reservoir. It was established that the deposition processes, as well as the diagenetic alteration have an impact on porosity and permeability distribution within the reservoir. The authors also studied the statistical moments of grain size distribution versus permeability and porosity and it was concluded that permeability is a function of the skewness and standard deviation rather than grain size [5].

As a rule, facies distribution within the reservoir plays a significant role in a subsurface fluid flow. Uncertainties in this parameter connected with the lack of knowledge about distribution, connectivity and poroperm properties may cause variations in displacement efficiency, production and economic values. The example of such an uncertainty study in a turbidite reservoir is represented in work [9]. Net oil, sweep efficiency and cumulative recovery factor were chosen as the measures of uncertainty.

4. Workflow

The first part of the present study is focused on the western part of the Vingapur field description due to the fact that this area is characterized by such core data that make it possible to validate formation evaluation results. On the basis of sedimentary logs, the model of the depositional environment was constructed. After that, an attribute analysis was carried out in order to define the distribution and shape of the sand bodies on a large scale. It must be said that it was decided to perform the re-interpretation of the depositional environment due to the fact that it stated the common features of the Achimov formation that were defined in other fields. Thus, it was suggested in general that this formation represented turbidity current deposits, slump, mass flow deposits with elements of sand waves. Therefore it was decided to define the particular type of deep marine deposits in order to infer significant architectural elements from it.

Further, a petrophysical analysis was made after the sedimentary environment description. It was meant to estimate the degree of formation heterogeneity using statistical techniques. On the basis of the poroperm data and wireline response, Rock Types were defined with the use of specific cut-off values on the basis of the Hydraulic Flow Unit approach. After Rock Typing, the correlation on the basis of wireline logs was carried out in order to investigate the inter-well correlation of the formation itself and the Rock Types within it. Eventually, as the result of the first part of the study, the petrophysical and geological heterogeneity of the Achimov formation was described, controlling factors were defined and the relation with the processes of deposition was established. The authors also defined an approach making it possible to predict poroperm values based on Rock Types in areas that are not characterized by core data, but are close enough to make one suggest a similar sedimentary environment.

An attribute analysis was also performed in the northern part for two purposes: first – to validate the assumption about depositional environment similarity and second – to plot porosity and Net-to-Gross maps. It was subsequently used to define the trends of these properties. Further, on the basis of the acquired data, several geological and simulation models were constructed. The simulation results of these various models were analyzed in order to define the degree of heterogeneity distribution within the formation in terms of production and well performance.

5. Results. Sedimentary Environment

It should be noted that the successions in the wells under study have a similar character and interpretation as turbidites. It should be noted that there is a possible alternative interpretation of sedimentary processes in the area under consideration. This uncertainty in interpretation is illustrated in figure 1. Despite the similarity of structures in these two interpretations, there is a significant difference in the processes of deposition. A turbidity current implies a turbulent flow, while a debris flow suggest a laminar flow, where mudclasts typical of turbidite in the debris flow interpretation are oriented parallel to the mass flow suggesting a linear flow. Also, (C) the Bouma division may represent the evidence of the bottom current reworking of sediments [14]. The conceptual model of deposition is represented in figure 2. On the basis of core description, attribute analysis and seismic interpretation, it may be concluded that the sandstones and siltstones of the Achimov formation were deposited in the basement of the shelf slope. Despite the uncertainties in interpretation, the turbidity model of deposition where sediments were deposited in a channelized lobe system (figure 2) was selected for further studying.

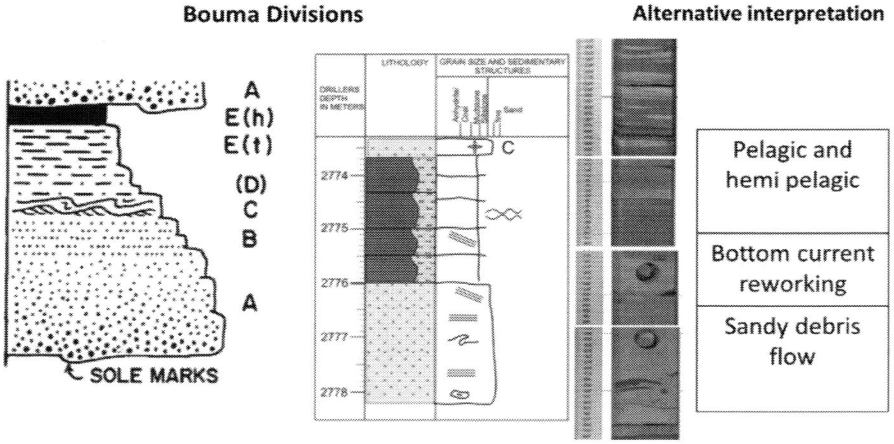

Figure 1. Interpretation [14]

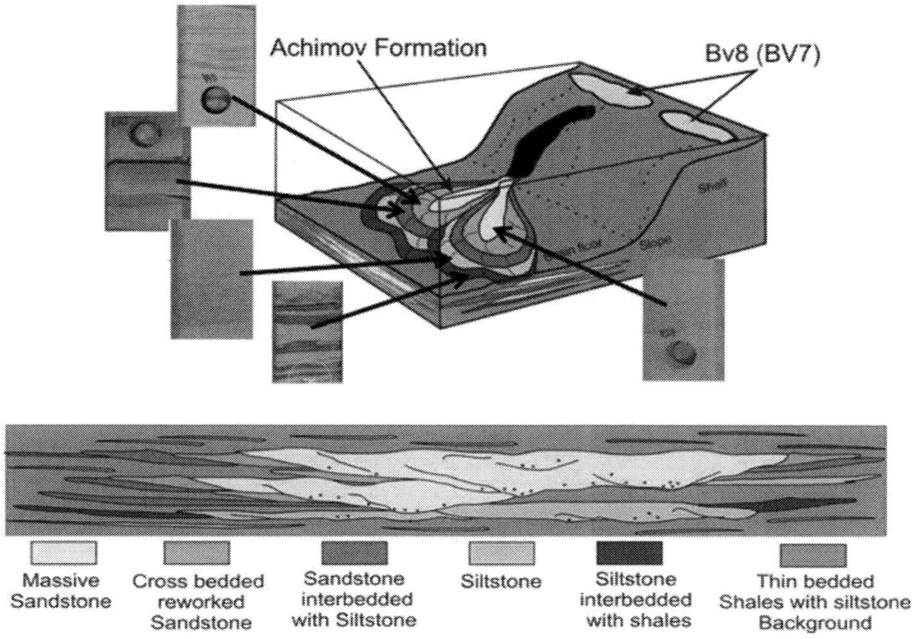

Figure 2. Sedimentary environment model

6. Attribute analysis in the western part of the field

In order to identify depositional features, the large-scale interpretation and attribute analysis of seismic data were carried out. The seismic resolution is sufficient for analyzing the Achimov formation with the average thickness of about 100 m, but it is insufficient to clearly resolve individual beds with the thickness of less than 10 m suggesting the presence of the tuning effect, as it can be seen from the time cross-sections.

From the average energy attribute calculated between the surfaces of the ACH top (the top of the Achimov formation) and B (the top of the Bazhenov formation) in a 20 ms window, the amalgamated system of individual fans represented by the ACH2 and ACH3 horizons deposited in a depression on the shelf slope can be inferred. The evidence of the shelf edge progradation can be seen in the east. Surface attributes, Variance and Envelope, were calculated in search window 15 ms below the ACH2 top in order to minimize the edge effect related to the pinching out of the formation. On this attribute maps, one can definitely see the evidence of the channel (figure 3) of sediment transport on a shelf slope that further turns into a submarine fan system.

The time section in the normal direction and in the direction being parallel to the sediment flow reveals the architecture of these deposits. The Achimov formation is pinching out in the western direction, and has a lenticular shape in the normal cross-section that coincided with the suggested depositional environment. It is a typical Neocomian clinoform structure that has a sigmoid shape and is prograding basin-wards. During the detailed examination of the time cross-sections of the impedance cube in direction being normal to the shelf edge, one can see anomalies that are related to the downcutting of the submarine fan system and the evolution of the geometry from proximal to distal parts (figure 4). The following sections of this work are focused on the petrophysical heterogeneity description of the Achimov formation in order to define the model of deposition.

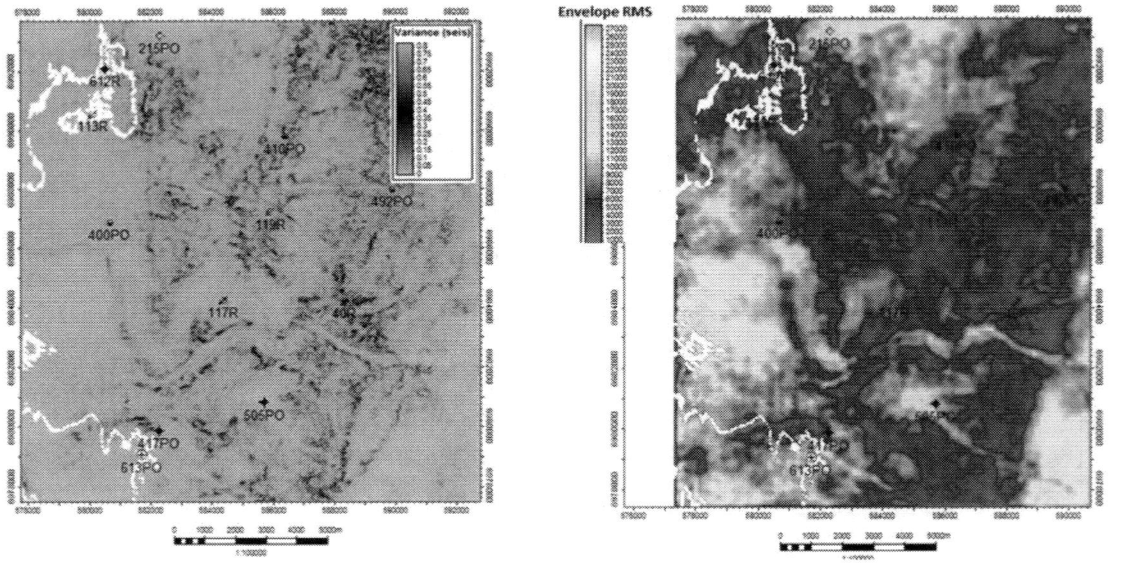

Figure 3. Variance and Envelope seismic attributes

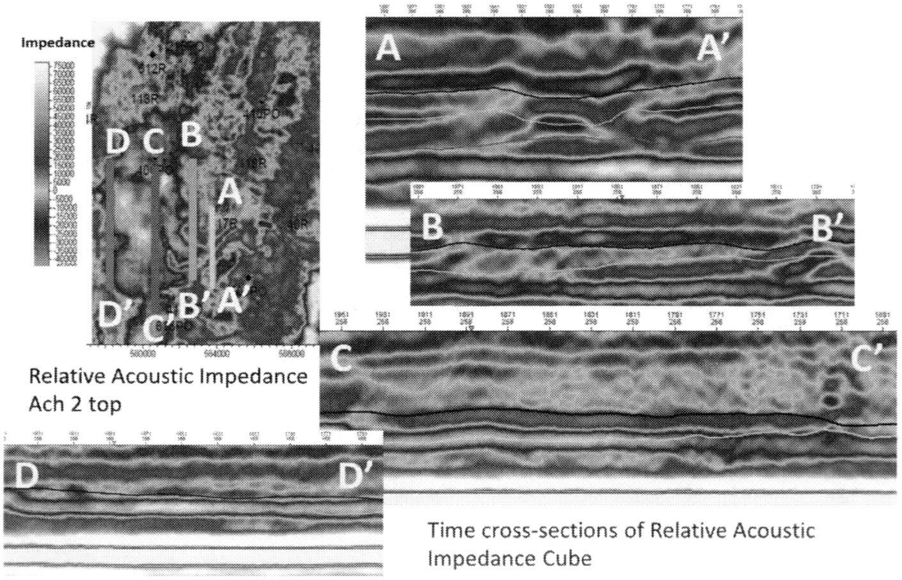

Figure 4. Time cross sections of relative acoustic impedance (Western part)

7. Petrophysical description of the reservoir and rock typing

When analyzing dependences for the Achimov formation, one should note that this formation is characterized by the high degree of permeability heterogeneity, thus, for example, with the porosity value being equal to 0.17, the value of permeability may vary within several orders of magnitude. Therefore, it is necessary to employ methods focused on defining certain dependences within the data point cloud.

With due account for the methods revised previously, it was decided to define specific lithotypes corresponding to particular lithology and including specific petrophysical information and related to some defined elements of the depositional environment. The integration of HFU and rock types allows classifying and defining rock types not only on the basis of lithology, but also from the viewpoint of the depositional environment and providing information about petrophysical properties related to pore geometry.

The first step in rock type discrimination was the definition of the boundary conditions of specific cut-off values on the basis of wireline log curves. Using the values of neutron and gamma ray tools, the authors distinguished between tight sandstones and siltstones being essentially non-reservoir rocks. Further, they identified thin interbedded shales and siltstones using the cross plot shown in figure 5a. After the discrimination of non-reservoir rock types, the cross section still comprises intervals consisting of sandstone and siltstone. To discriminate these lithology types, the authors used the cross plot of Gamma Ray versus apparent bulk density defined on the basis of a density tool response (figure 5b). The application of this type of curves is governed by the fact that these two methods are mutually independent, which in turn provides a more reliable subdivision of rocks. Speaking about the application of the other wireline logs, it should be noted that the sound wave velocity depends on density that as the result leads to the overlapping of data points complicating subdivision.

During the next step, Hydraulic Flow Units (HFU) were defined within the sandstone and siltstone intervals. In this case HFU are directly related to the defined lithology.

Next, on the basis of Gamma ray and Later log tools having quite a high definition from log suite were acquired for the Flow Zone Indicator predictor.

On the porosity versus permeability cross plot, Rock Type 1 is well-defined, while Rock Types 2 and 3 are overlapping with Rock Type 4 that is a transitional variety between sandstone and siltstone.

This function was approximated by the third-order polynomial equation using a multiple regression approach in Statistics software.

FZI_log=-22.52-1.23*BK+0.17*BK2-0.0067*BK3+11.83*GR-1.79*GR2+0.089*GR3;

Eventually were established the following cut-off values:

Rock Type 1 if (DENSITY<2.4 and GR<6.5) and (FZI_log>0.78 and FZI_log<2.17);
Rock Type2 if (DENSITY<2.4 and GR<6.5) and (FZI_log>0.46 and FZI_log<0.78);
Rock Type3 if (DENSITY<2.4 and GR<6.5) and (FZI_log>0.1 and FZI_log<0.46);
Rock Type4 if (DENSITY>2.4 and GR>6.5) and (FZI_log>0.43 and FZI_log<2);
Rock Type5 if (DENSITY>2.4 and GR>6.5) and (FZI_log>0.1and FZI_log<0.43);
Rock Type6 if GR>7.3 uR/h;
Rock Type7 if NKTD>3.73uE;

The average properties and description of the defined rock types are represented in figure 6.

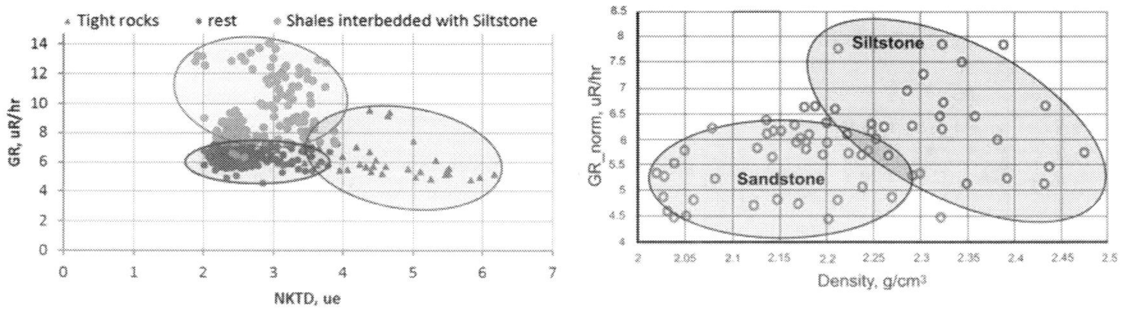

Figure 5. GR vs. Neutron tool response cross-plot (a) and GR vs. density tool response cross plot (b)

Rock Type 1

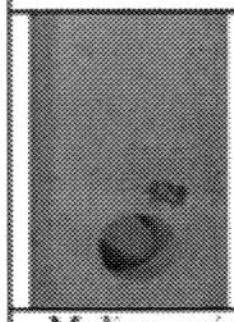

Porosity range	Permeability range	FZI range	Swi	Kv/Kh	Grain size	sorting	Bedding style
0.16-0.22 0.19	4-76 15	2.17- 0.78	0.5	0.8	0.13	1.65	Structure less

Medium grained Sandstone with shale cement chlorite and kaolinite of pore-film type, silty, flke. Quartz 46%, Fieldspar 33%, Rock fragments 13%, Mica 8%.

Rock Type 2

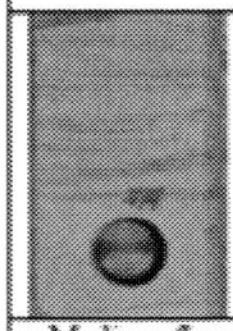

Porosity range	Permeability range	FZI range	Swi	Kv/Kh	Grain size	sorting	Bedding style
0.03-0.2 0.17	0.02-6.74 2.98	0.78- 0.47	0.58	0.72	0.11	1.49	Plannar or cross laninated

Medium-fine grained sandstone with shale cement chlorite and kaolinite of pore-film type, silty, flke. Quartz 42%, Fieldspar 31%, Rock fragments 20.5, Mica 6.5%.

Rock Type 3

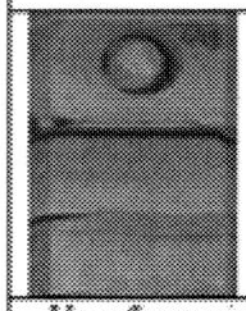

Porosity range	Permeability range	FZI range	Swi	Kv/Kh	Grain size	sorting	Bedding style
0.06-0.19 0.16	0.02-2.23 0.85	0.47- 0.01	0.75	0.48	0.11	1.46	Cross laminated lenticular bedding

Very fine grained sandstone with shale cement chlorite and kaolinite of pore-film type, silty, flke. Quartz 34%, Fieldspar 33%, Rock fragments 20.5, Mica 12.5%.

Rock Type 4

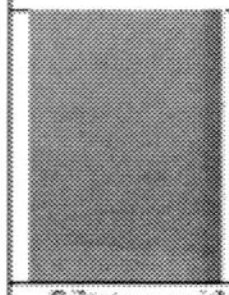

Porosity range	Permeability range	FZI range	Swi	Kv/Kh	Grain size	sorting	Bedding style
0.09-0.2 0.17	0.4-8.5 3	1-0.42	0.59	0.51	0.09	1.46	Structureless

Siltstone with shale cement chlorite and kaolinite of pore-film type, sandy, flke. Quartz 37%, Fieldspar 32%, Rock fragments 20%, Mica 11%.

Rock Type 5

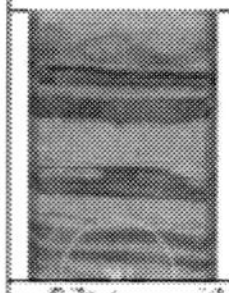

Porosity range	Permeability range	FZI range	Swi	Kv/Kh	Grain size	sorting	Bedding style
0.06-0.2 0.15	0.03-8.47 1.42	0.42-0.1	0.73	0.3	0.08	1.48	Lenticular, cross laminated

Siltstone with shale cement chlorite and kaolinite of pore-film type, sandy, flke. Quartz 35%, Fieldspar 35%, Rock fragments 17%, Mica 13%.

Figure 6. Rock type characteristics

8. Permeability heterogeneity

For the purpose of permeability heterogeneity assessment, well 492PO located in the proximal part of the submarine fan system and well 612R located in the distal zone were chosen. According to the values of Coefficients of variation and Lorenz, the formation in the area of well 612R is more heterogeneous, which can be explained by the depositional environment. Besides, well 612R is located farther from the bottom of the shelf slope, therefore, the distance from the sedimentary source is longer, corresponding to the middle-distal part of the channelized lobe complex. This difference in the position controls the degree of properties degradation in the formation.

9. Statistical analysis of grain size distribution

The next step after the sedimentary environment analysis and Rock Types identification was the analysis of statistical moments of grain size distributions in order to define parameters controlling the porosity and permeability of the formation.

It can be seen from the permeability vs. median grain size relation that there is a trend towards permeability incensement with the grain size. Besides, the correlation for Rock Type 1 is quite high while that for other Rock Types is not so robust. Permeability increases with improvement in sorting as well and, according to the standard deviation values, varies from moderate to poorly sorted. One can see from the plot that the deposits in the distal part of the submarine fan system and reworked by bottom currents and their sorting is better.

The skewness in distribution shows that the prevalence of fines is typical of all Rock Types, although there is a trend towards permeability increment when distribution is skewed to coarse grains, the relations are not robust and characterized by significant scattering. It should be noted that the major factor apart from statistical moments is the lithological composition of rocks that is directly related to the depositional environment, the energy of a sediment flow that, in turn, resemble the defined Rock Types.

10. Prognosis for the reservoir properties of the northern part

The second part of the study implied the translation of methods of properties estimation from the western part to the northern one using a defined algorithm for Rock Types identification, and petrophysical relations for Rock Types on the grounds of the fact that according to regional 3D seismic surveys, using seismic facies identified the presence of a submarine fan system in both zones being parts of the same depositional system. In addition the formation has a fixed stratigraphy position and located within the same field.

The application of seismic attributes and time sections allowed evaluating the sand body geometry and depositional environment elements. Prediction was carried out using nine basic wells where the average values of porosity and NTG were calculated in the interval of the ACH2 horizon. The typical structure of the Neocomian anticlines can be seen on the time cross-sections.

Porosity prediction was based on the calculated attributes employing multiple regression where the coefficient of correlation was R2=0.92. With a similar multiple regression tool, a function for the NTG prognosis with the coefficient of correlation, R2=0.79, was received.

The calculated porosity and NTG maps are represented in figure 7. The porosity map analysis shows that higher porosity values belong to the central zone, while the average porosity values decrease as they approach the proximal zone. The NTG values in the proximal eastern part are higher due to the proximity to the sedimentary source and significant erosion resulting from higher flow energy.

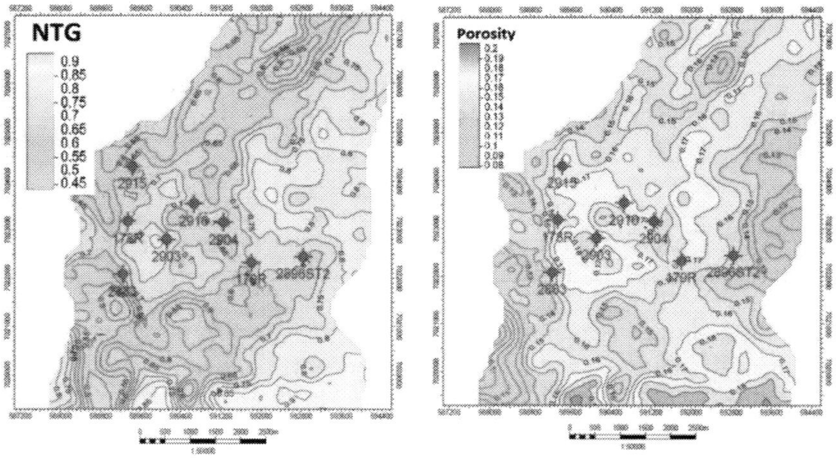

Figure 7. Property maps based on seismic attributes

11. Effect of geological uncertainty on subsurface flow estimation

Rock types within the model were distributed using the Sequential Indicator Simulation algorithm of pixel modeling. Type 1 model represents a simple structure consisting of a reservoir and non-reservoir unit being the present approach to Achimov formation modeling at V-Field. The model belonging to the second type includes Rock Types, where Rock Type 6 and 7 are non-reservoir and correspond to tight rocks and thin bedded shales and siltstone. In the model of the third type, shales and tight rocks were assigned as discrete bodies with the correlation range being equal to 100 m representing the case when carbonatization zones are locally developed whereas shales have a small length. The model belonging to the fourth type comprises shales and tight rocks with a correlation range being equal to 2,500 m. These bodies are laterally extensive, and it is also a probable scenario. The fifth model is based on the model with extensive non-reservoir rocks with changed transmissibility between Rock Types being equal to 0.1 that represents the absence of erosion between the reservoir units and can have thin layers of shale between them being a barrier to flow. The conceptual schemes of the models are presented in figure 8. Figure 9a provides the comparison of the simulation results for type 1 and type 2 models. Cumulative production in case of a simple model is higher in 70,000 m^3, which is connected with the absence of reservoir heterogeneity related to the reservoir architecture, while the prognosis based on the model comprising the full set of submarine fan depositional elements represented by Rock Types gives more pessimistic results.

The next figure 9b compares the third and the fourth models. The plot of cumulative production shows that production is higher in 40,000m^3 for the model with continuous shales and tight rocks. As a result, the presence of the significant amount of small-scale heterogeneities prevents a subsurface fluid flow that,in turn, leads to heterogeneity incensement and related to a production decrease.

The presence of laterally extensive units leads to reservoir homogenization, which results in a production increment, but due to the early breakthrough, production rates subsequently demonstrate the same trend. Figure 9d presents calculation results for the model where the coefficient of permeability anisotropy varies from 0.1 to 1 representing extreme cases. Cumulative production for the case with high Kv/Kh is less in 20,000m3, which is connected with a vertical permeability restraint. Initially, the difference in the rates is not so significant, even if water breakthrough occurs. Simultaneously, the rate of water cut for the case when Kv/Kh=1 is higher, which leads to higher production rates. Figure 9c presents the results of calculation for the fifth and the second models. Cumulative production for the case where there are erosional boundaries between the Rock Types is higher in 45,000m^3. The occurrence of such contacts may be related to the erosion of Rock Types by the Rock Type 1 that is associated with the channel axis. Thus, increasing transmissibility between the units, which, in turn, leads to higher production, occurs as the result of erosion.

The influence of transmissibility between the Rock Types on water flooding effectiveness was further analyzed. Figure 10 shows the slice of a geologic model that represents the distribution of Rock Types and oil saturation for the cases with various unit transmissibility levels. In the first case, the propagation of the displacement front within the units is more frequent than between the units. As one can see from figure 10, the displacement in Rock Type 1 corresponding to the channel axis is more effective. It is less effective in Rock Type 2, as it has poorer reservoir qualities. The worst results are for the region of Rock Type 3. It must be said that due to the transmissibility reduction between the units, the spatial organization of the units becomes more significant. In addition, erosion leads to an evenly distributed displacement front and more symmetrical saturation distribution within the element of development (figure 10).

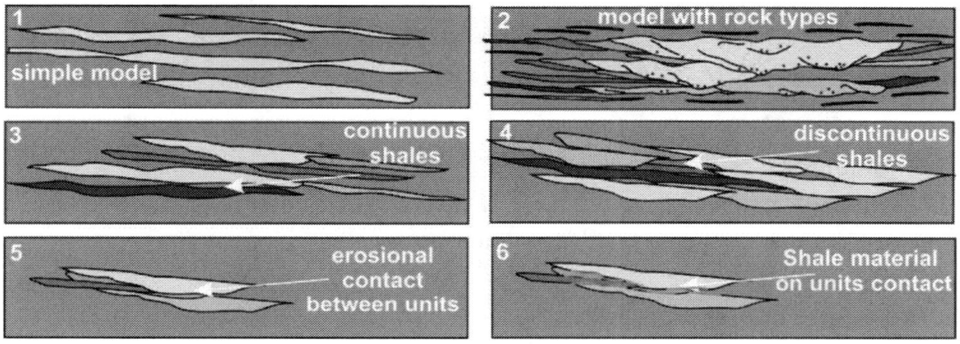

Figure 8. Types of models used for simulation

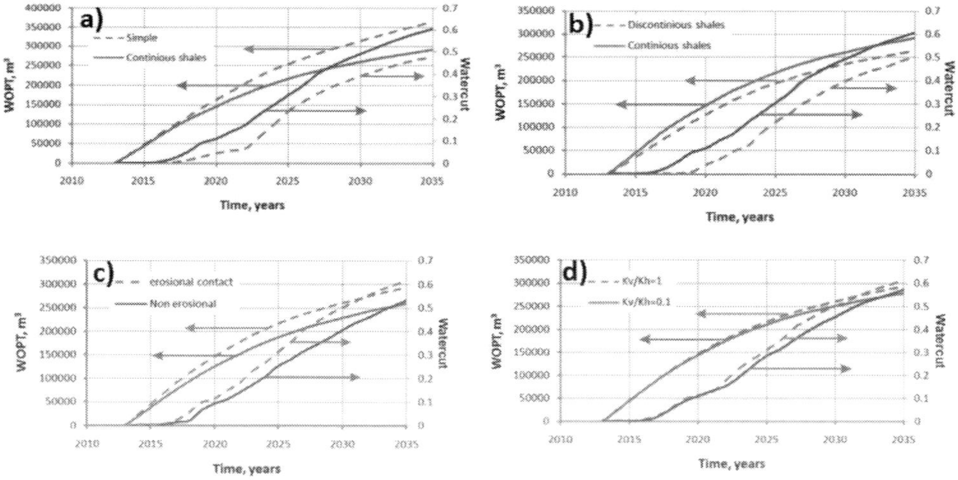

Figure 9. Simulation results

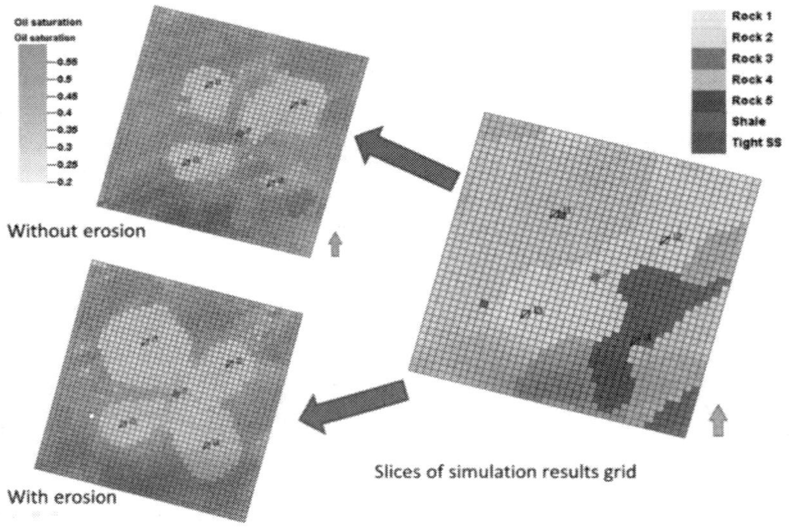

Figure 10. Slices of simulation model representing oil saturation distribution in case: (a) without erosion, (b) with erosion

12. Conclusions

The present study allows making the following conclusions:

1. The Achimov formation of the Vingapur field is a complex system of amalgamated channelized lobes where parts of this system are represented by specific Rock Types associated with structural elements: axial part represented by massive fine grained sandstone (Rock Type 1), off-axis part composed of cross-bedded very fine-grained sandstone (Rock Type 2), farther from the lobe, the axis is represented by very fine-grained sandstone interbedded with shales (Rock Type 3). The distal deposits of the lobe also include massive siltstone in the axial zone and siltstone interbedded with shales in the off-axis zone.
2. The authors applied the method of Rock Types Identification using certain cut-off criteria for the Wireline log curves and Hydraulic flow unit approach. An algorithm for properties and Rock Types prognosis within the field based on Wireline logs were only proposed on this basis.
3. According to the calculation results, permeability heterogeneity increases from the proximal to distal zones of the submarine fan system. There is a trend towards the increase in NTG from the proximal to the distal part and also higher porosity values are associated with the central zones on the basis of the attribute analysis.
4. The main factor controlling permeability is variations in lithology related to depositional environment elements.
5. The main influence on production is exerted by the complexity of reservoir representation in the model, additional details lead to a more pessimistic prognosis.
6. Reservoir heterogeneity grows in case of discrete shales and tight rocks, which prevents a subsurface fluid flow, while the reservoir is more homogeneous in the laterally extensive case, which leads to a production increase. The type of contacts between the units is also quite significant, the barriers between the units impede the inter-unit flow leading to displacement front propagation preferentially within the units.

References

[1] Braducham J V, Nesterov I I, Sokolovski A P "Stratigraphy of Meso-Cenozoic Deposits of Sredneobskaya Oil Province", *ZapSibNIGNI.*–Tyumen', p.11-57, Vol.**7**.–1968.

[2] Chen J *"Understanding Uncertainties in Deepwater Depositional Facies Modeling for Better Reservoir Conditioning"*, SPE Europec/EAGE Annual Conference, Copenhagen, Denmark, June 4-7, 2012.

[3] Davies D K et al. 1999 "*Geometry, Internal Heterogeneity and Permeability Distribution in Turbidite Reservoirs, Pliocene California*", SPE Annual Technical Conference and Exhibition, Houston, Texas, 3-6 October 1999

[4] Drozdov S S 2012 *Seismic Report Stage 1, Hydrocarbon Potential of Achimov Formation* Gazpromneft Research and Development Center

[5] Griffiths J C 1967 *Scientific Method in Analysis of Sediments* (New York: McGraw Hill Book Company)

[6] Gurari F G 1959 *Geology* and Hydrocarbon Potential of Ob-Irtishskoe Interfluve Area *SNIIGGiMS* **3** 1-13

[7] Dahlberg K E et al. 2002 *Petrofacies: Enhancing the Deepwater Reservoir Characterization Effort in West Africa* SPWLA 43 Annual Logging Symposium

[8] Mancini E A 2008 Paleogene Lowstand Systems Tract Sand Deposits of the Eastern Gulf Coastal Plain: Potential Reservoir Facies in the Offshore Northeastern Gulf of Mexico *GCAGS Transactions* **58** 669–75

[9] Milliken W J, Marjorie L and Strebelle S *The Effect of Geologic Parameters and Uncertainties on Subsurface Flow: Deepwater Depositional Systems* SPE Western Regional and Pacific Section AAPG Joint Meeting, Bakersfield, California, USA, 29 March – 2 April 2008

[10] Nezdanov A A et al. 2000 "*Geology and Hydrocarbon Potential of Achimov Formation*" Publisher Academy of Mining Sciences

[11] Nurhono A A and Kantaatmadja B P *Multidisciplinary and Integrated Methodology for Deepwater Thin Bed Reservoirs Characterization* SPE Annual Technical Conf. and Exh. (San Antonio, Texas, USA, 8-10 October 2012)

[12] Peiton C *Seismic Stratigraphy* MIR pp 846–1982

[13] *Refined Mesozoic Stratigraphic Schemes of Western Siberia* SNIIGGnMS pp 144-204

[14] Shanmugam G 2006 *Deep-Water Processes and Facies Models: Implications for Sandstone* (Elsevier Science)

[15] Walker R G and James N P 1992 *Facies Models. Response to Sea Level Change*

IOP Conf. Series: Earth and Environmental Science **24** (2015) 012042 doi:10.1088/1755-1315/24/1/012042

Existing approaches to tight rock laboratory petrophysics: a critical review

D V Konoshonkin[1] and S V Parnachev[2]

[1,2] Tomsk Polytechnic University, Tomsk, 634034, Russia

E-mail: [1]konoshonkindv@hw.tpu.ru

Abstract. A review of the existing methods for tight rock porosity, saturation, and permeability determination was performed taking into account that these methods should be applicable for Bazhenov formation evaluation. The following methods were considered: Archimedes mercury immersion; mercury displacement; caliper; helium pycnometry on crushed samples; nuclear magnetic resonance; modified retort method; modified Dean-Stark extraction; pulse decay method; and pressure decay test on crushed samples. The applicability of the pressure decay test on a crushed sample for Bazhenov formation evaluation is checked experimentally with the SMP-200 commercial permeameter. All the above listed methods were combined into five protocols for tight rock petrophysical evaluation. These protocols were analyzed and compared according to the following criteria: accuracy of the results; usage experience; time of measurements; easiness of interpretation; reliability and safety; price. The obtained results revealed that the most effective protocol is the one that includes pressure pulse on a core plug for permeability determination, He pycnometry and modified retort analysis on crushed samples for porosity and saturation determination. As there were cases when the proposed protocol was less effective vs. other protocols, a special scheme was suggested in order to choose the most effective protocol for tight rock petrophysical properties evaluation in definite conditions.

1. Introduction

The demand for oil is constantly rising in the world, whereas traditional oil resources are becoming depleted. These conditions make oil companies take interest in unconventional resource studies. According to the US Department of Energy and Russian researchers, the Bazhenov formation has the highest unconventional resource potential in Russia. Thus, the Bazhenov formation is becoming intensively studied. However, the traditional methods of core analysis are not effective for evaluating the tight rocks of the Bazhenov formation. Therefore, there is an exigency to review the existing approaches to analyzing tight rocks in order to propose the most effective one for Bazhenov formation core evaluation.

2. Lithology and petrophysical properties of the Bazhenov formation

The Bazhenov formation is a geological body composed of shale, silica and carbonate rocks in different proportions and with a varying amount of organic matter [1, 2]. The Bazhenov formation consists of tight rocks with porosity being less than 16% and permeability usually being less than 0.1 mD (table 1), although there are "sweet spots" with higher permeability.

Table 1. Average petrophysical properties of the Bazhenov formation.

Parameter	Value
Porosity	1.4 – 16 %
Permeability	10^{-6} – 10 mD (predominantly < 0.1 mD)
Total organic content (TOC)	5 – 20 %
Thickness	20 – 40 m

3. Review of published materials on tight rock laboratory testing methods

3.1. Sample preservation

There is much controversy about tight rock sample preservation. Zhou, et al. [3] claimed that a hydrocarbon-bearing shale core should be preserved and core samples are to be stored in desiccators prior to measurements. The main reason for that was the reaction of shale with atmospheric water and changes in core saturation and properties. Unfortunately, Zhou, et al. did not describe the types and properties of shale for which they recommended preservation.

On the other hand, Handwerger, et al. [4] showed that in case of tight rocks (an experiment was conducted on samples with porosity being 2.2-2.4% and permeability of $1.1 \cdot 10^{-4}$ – $1.8 \cdot 10^{-4}$ mD) the results of the lab analysis for porosity, saturation and permeability measurements made on recently drilled core are equal to the results of the lab analysis for the same core after its 2-year storage in plastic bags without any special storage protocols. Therefore, Handwerger, et al. concluded that the petrophysical properties of tight rock samples (especially important for saturation) change insignificantly due to the irreducible nature of fluid saturation in ambient conditions. It is a very important conclusion, because there are a lot of core samples from the Bazhenov formation that were drilled several years ago and have not been studied yet.

3.2. Porosity determination

The core sample of the Bazhenov formation rock may be represented with the model showed in (figure 1). Three volumes are distinguished according to that model: bulk volume (V_b), grain volume (V_g) and pore volume (V_t). Porosity is the ration of the pore volume to the total volume of the rock and may be determined with the following equation:

—— —— — (1)

Therefore, in order to determine porosity, two of three parameters (V_b, V_t, V_g) should be studied in laboratory conditions.

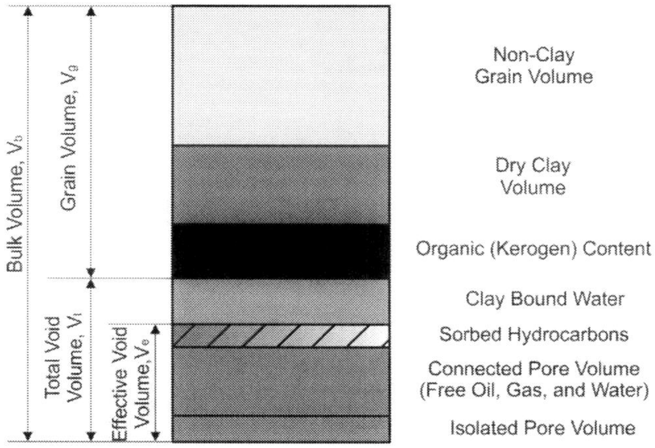

Figure 1. Model of kerogen-rich tight rock based on [5].

3.2.1. Methods of bulk volume determination

There are three main methods for bulk volume determination in tight rock samples (table 2) [6]. Two of the methods (Archimedes mercury immersion and mercury displacement) are based on mercury usage. Mercury is used because it does not penetrate into sample pores, it does not change sample saturation and does not react with sample components (due to its high surface tension and low wettability). Mercury-based methods allow performing fast and accurate bulk volume measurements but cannot be applied to samples with surface fractures and vugs.

The third (caliper) method implies the direct measurements of sample dimensions. It is very simple and fast, but the method cannot be used for samples with an irregular shape. Also, porosity results have high uncertainty if the caliper method is used with grain volume measurement to obtain porosity.

Table 2. Methods of tight rock bulk volume determination.

Method	Major advantages	Major disadvantages	Accuracy
Archimedes (buoyancy) mercury immersion	Samples can be used for subsequent tests; the method is very accurate	Trapping air around the samples; samples with a vugular surface or containing open fractures cannot be used	±0.01 cm^3 (using balance with 0.01 g accuracy)
Mercury displacement	Rapid measurements; samples are suitable for subsequent tests	Trapping air around the samples; samples with a vugular surface or containing open fractures cannot be used	±0.01 cm^3 (if the pump was calibrated and is zeroed for each sample)
Caliper	Samples can be used for subsequent tests; rapid measurements	Only for samples of an even shape (e.g. core plugs); higher level of errors for porosity derived from grain volume	±0.15 cm^3 (in case of ±0.15 mm for length and ±0.04 mm for diameter measurements)

3.2.2. Methods of grain volume determination

The grain volume of a tight rock may be determined using the double-cell helium pycnometry method on crushed samples (table 3). The method is relatively simple and gives almost the same results for the same sample (high repeatability). However, cells should be accurately calibrated, temperature fluctuations should be reduced or accounted and adsorption and the molecular sieving effect should be taken into account (adsorption effect is significant for gases other than helium, for example, methane).

Table 3. Method of tight rock grain volume determination.

Method	Major benefits	Major drawbacks	Accuracy
Helium pycnometry on crushed samples	No damage to the sample; simple and quick; high repeatability	Changes in ambient pressure and/or temperature may induce errors; adsorption and sieving effects	±0.2% of the true value (for well calibrated system)

3.2.3. Methods of pore volume determination

In order to determine the pore volume of a tight rock sample, nuclear magnetic resonance (NMR) may be used (table 4). The NMR method is based on hydrogen nuclei precession in a magnetic field. As hydrogen is present predominantly in pore fluids, nuclear magnetic resonance may be used to determine the quantity of fluids in pore space and, thus, the pore volume.

Table 4. Method of tight rock pore volume determination.

Method	Major benefits	Major drawbacks	Accuracy
NMR	Sample lithology does not affect measurements; rapid method	Indirect method; samples should be saturated with one fluid for high accuracy	±20%, but depends on measurement conditions

3.3. Saturation determination

The saturation of a tight rock sample may be determined using one of the following methods (table 5):

1) modified retort method at atmospheric pressure;
2) modified Dean-Stark method using toluene extraction;
3) magnetic resonance saturation scan.

The retort method [4] is the direct method of fluid volume determination that can be performed in a sufficiently short time (about 1 day) and can be used to separate mobile (free) and bound water and oil. However, the method destroys the sample and can give erroneous results in case of the high amount of montmorillonite, gypsum or kerogen presence.

Dean-Stark extraction [7] is thought to be an applicable method in case of a high kerogen content (nitrogen gas blanket flowing through the apparatus should be used [8]). However, Handwerger, et al. [9] showed that some water can be extracted from clay and cause water saturation exaggeration. Also, the method requires time-taking measurements (about two weeks).

Magnetic resonance measurement [10, 11, 12] is the only method that does not destroy a core plug in any way. Unfortunately, the method is indirect and there are difficulties and uncertainties in fluid response separation.

Table 5. Methods of tight rock saturation determination.

Method	Major benefits	Major drawbacks	Accuracy
Modified retort method at atmospheric pressure	Direct measurement of fluid volumes; ability to separate free water from bound water; rapid (1 day is required)	Errors for samples with montmorillonite or gypsum; errors for samples with a high kerogen content; samples cannot be used further	For water: ±5% of the measured volume For oil: ±2.5% of the measured volume
Modified Dean-Stark method	Sample material can be used for further testing; simple and requires little attention during distillation; applicable for kerogen-rich samples	Salt can precipitate inside the sample; errors for samples with montmorillonite or gypsum; oil density should be known; time-taking (a week is required for extraction and a week – for drying)	±50% of the measured volumes (in case of relatively small samples or samples with high gas saturation with residual volumes of oil and water)
Magnetic resonance saturation scan	The only method to determine saturation on a core plug that does not destroy the sample	Indirect method; difficulties with fluid response separation	Depends on differences in the NMR properties of measured fluids

3.4. Permeability determination

Permeability can be determined by means of steady state methods and unsteady state methods. However, the steady state methods are usually not used for tight rocks, because they need too much time for measurements and because of the need to measure very low flow rates [6, 13]. Therefore, unsteady state methods are predominantly used for permeability measurements in tight rocks.

There are two widely used unsteady state laboratory methods for permeability measurements in tight rocks [13] (table 6):

1) pressure pulse on a core plug;
2) pressure decay on crushed rock.

Pressure pulse on a core plug allows measuring permeability anisotropy at reservoir conditions; however, the permeability obtained can be exaggerated because of the presence of micro fractures generated during coring. Core crushing (the method of pressure decay on crushed rock) reduces the influence of micro fractures, but measurements are made at ambient temperature and pressure.

Table 6. Unsteady state methods for permeability measurements of tight rocks.

Method	Permeability range, mD	Major benefits	Major drawbacks	Accuracy
Axial flow, pulse-decay in core plugs	0.00001 – 0.1	Porosity can be determined with the same apparatus; confining and pore pressures are applied; permeability anisotropy can be measured.	Requires high pressure, leak-tight with high-quality transducers and data acquisition system – high capital cost; permeability can be affected by micro-cracks.	3% (low/no leaks, adsorption is accounted, low/no temperature fluctuations present)
Pressure-decay on crushed samples	Gas: 0.00001 – 0.01 Liquid: 0.1 – 2000	Porosity can be determined with the same apparatus; elimination of micro-cracks.	No confining pressure; low repeatability of measurements; difficulties with slip-correction.	±10% (low/no leaks, adsorption is accounted, low/no temperature fluctuations present)

4. Determination of permeability and matrix volume of the Bazhenov formation sample

The applicability of the pressure decay test on a crushed sample for Bazhenov formation evaluation is checked practically with the SMP-200 commercial permeameter.

4.1. Sample preparation

The core sample was not by any means cleaned, but was dried in a vacuum oven at the temperature of 60°C. Then it was cut and crushed using a geologic hammer and mortar.

4.2. Running of the experiment

The experiment was run using the SMP-200 permeameter. Firstly, reference cell calibration and dead volume calibration were made. After that, a leak off test was conducted. Next, one of the disks (the disk that was almost equal to the volume of the crushed sample) was removed from the sample chamber and the sample was placed into the sample chamber. Then, helium at the pressure of about 200 psi (13.61 atm) was expanded into the sample chamber. A pressure decay curve was recorded using a pressure gauge (the accuracy being 0.001 psi or $6.8\cdot10^{-5}$ atm) within 2,000 seconds. Permeability was determined by matching simulated and measured curves.

4.3. Results

The results of permeability measurements are presented in (table 7). It can be noticed that permeability for sample 1 and sample 2 differs within 10.7%.

Table 7. Results of the experiment.

Sample	Particle size, mm	Sample weight, g	Grain volume, cm^3	Grain density, g/cm^3	Permeability, mD
Sample 1	2-5	44.8193	17.772	2.522	$1.1662\cdot10^{-7}$
Sample 2	1-2	30.0360	11.813	2.543	$1.0410\cdot10^{-7}$
Difference, %	160	33	33.5	0.8	10.7

The experiment has revealed the following drawbacks of SMP-200:

1) The modeled curve has a low correlation with the measured data. This can be explained by the poor quality of the theoretical model. It should account for the deviation from the Darcy's flow due to the comparable sizes of pores and helium molecules and also it should account for temperature fluctuations.
2) The SMP-200 permeameter has insufficient thermal protection. The temperature influence on pressure measurements was determined during the leak off test. The negative values of the leak off were determined. It can be explained by a temperature growth that causes a pressure increase

in the sample chamber.

3) The possibility to choose the part of the pressure curve that is then correlated with simulated pressure decay reduces the repeatability of the results. Because of the low correlation between the theoretical and measured pressure decays, the choice of different parts of the pressure curve leads to different permeability values that can differ within 20%.

5. Protocols of tight rock analysis

5.1. Protocol description

All the above -mentioned methods can be combined into five protocols of porosity, saturation, and permeability determination in tight rocks. *The first protocol is GRI* used for the crushed rock analysis (figure 2). In this protocol, porosity is determined through bulk and grain volumes obtained by the mercury immersion method and the helium pycnometry method, respectively. Saturation is established using modified Dean-Stark extraction. Permeability is determined with pressure decay on a crushed sample.

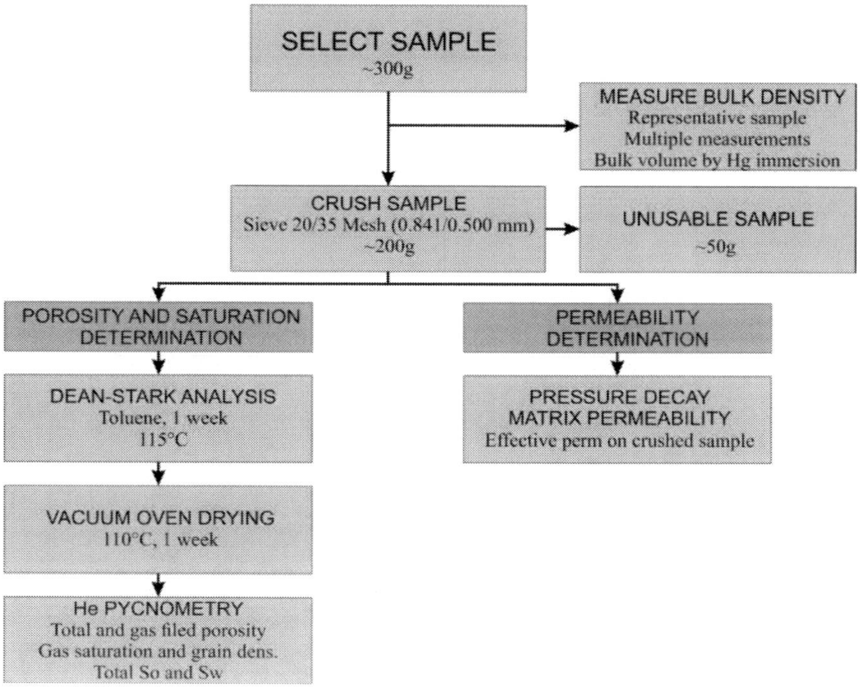

Figure 2. First protocol of porosity, saturation, and permeability determination of tight rocks (GRI).

The second protocol is used by the TerraTek Company and called Tight Rock Analysis (TRA) (figure 3). In this protocol, instead of modified Dean-Stark extraction (as in the first protocol), the modified retort analysis is used for saturation determination. Also, it is important that helium pycnometry is performed before oil and water volume determination (before core sample extraction or drying). Therefore, gas filled volume is determined in the second (TRA) protocol. It excludes mistakes that can be caused in the GRI protocol, if a core sample is not cleaned after drying and salts are precipitated to reduce pore space. As for the rest, the second protocol (TRA) is similar to the first protocol (GRI).

The third and the fourth protocols imply permeability determination by a pulse decay test on a core plug, crushing the core plug and further determination of porosity and saturation using Dean-Stark (*the third protocol*) and retort analysis (*the fourth protocol*) with helium pycnometry similar to the GRI and TRA protocols, respectively (figure 4).

The fifth protocol involves nuclear magnetic resonance method for saturation and porosity determination and a pulse test for permeability determination (figure 4). This protocol allows performing measurements without core crushing or destroying.

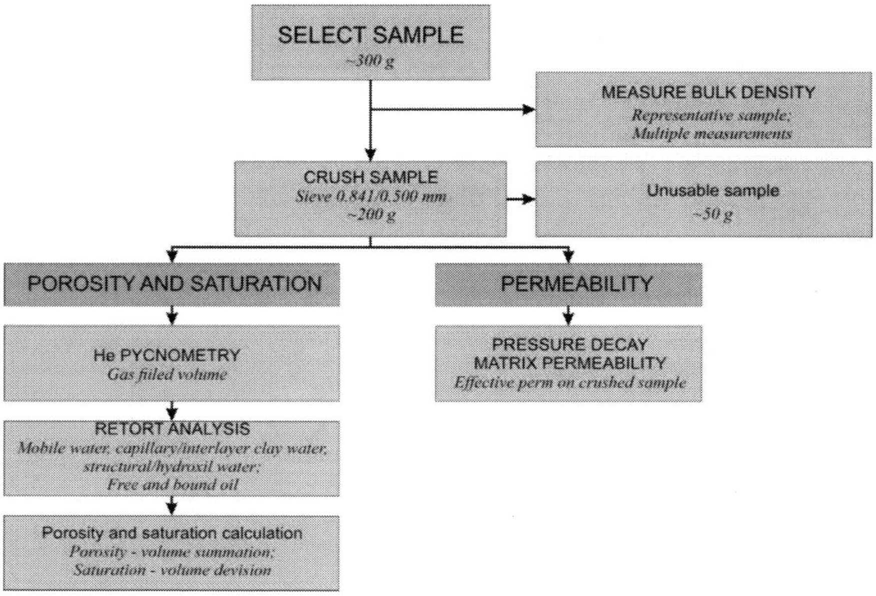

Figure 3. Second protocol of porosity, saturation, and permeability determination of tight rocks (TRA).

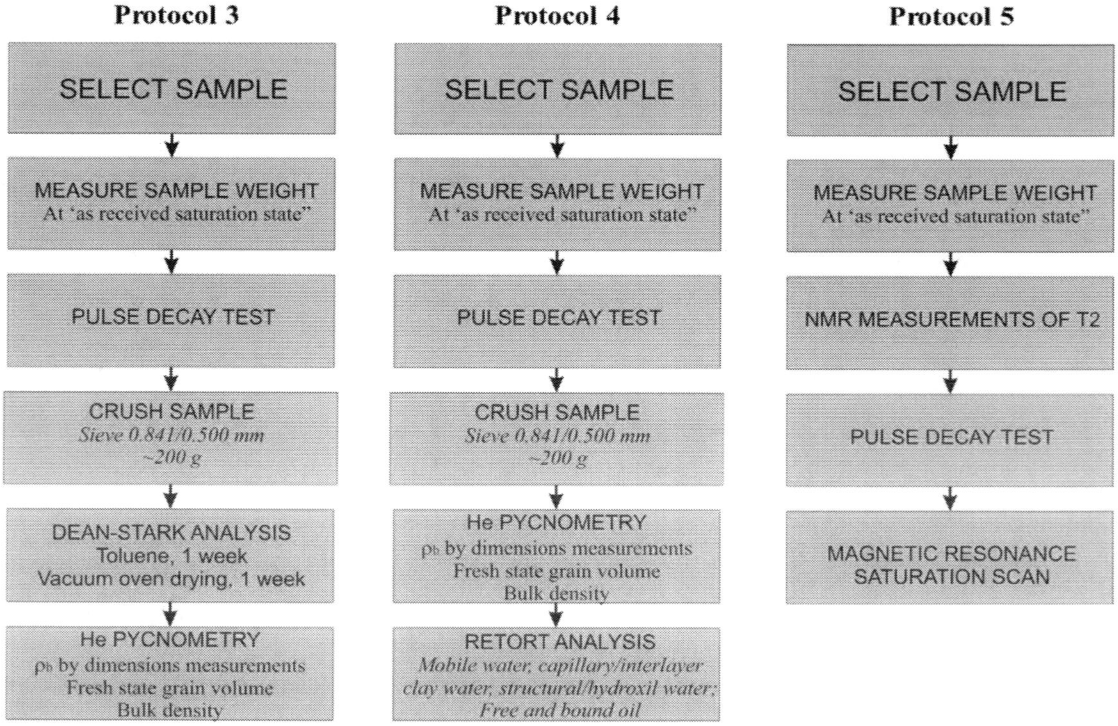

Figure 4. The third protocol (left), the fourth protocol (middle) and the fifth protocol (right) of porosity, saturation, and permeability determination of tight rocks.

5.2. Protocol selection

In order to choose the most effective protocol for tight rock analysis (and thus for the Bazhenov formation), the marketability analysis of all the five protocols was made. The main result of the analysis performed is a marketability evaluation map (table 8).

Table 8. Evaluation map for five protocols of tight rock analysis.

Criteria	Weight of criterion	Marketability				
		Protocol 1	Protocol 2	Protocol 3	**Protocol 4**	Protocol 5
Accuracy of the results	0.35	0.7	1.05	1.05	1.75	1.4
Usage experience	0.25	0.75	0.25	1.25	1	0.5
Time of measurements	0.2	0.4	0.8	0.2	0.8	1
Easiness of interpretation	0.1	0.5	0.5	0.5	0.5	0.3
Reliability and safety	0.05	0.2	0.2	0.15	0.15	0.15
Price	0.05	0.25	0.2	0.15	0.1	0.05
Total	1	2.8	3	3.3	**4.3**	3.4

The marketability of every protocol for every criterion was defined according to the expert judgment and using a five-point grading scheme: 1 – the weakest position, 5 – the strongest position. The matrixes of quantitative relations were made in order to increase grading objectiveness (tables 9, 10 and 11).

Table 9. Matrix of quantitative relations for the "usage experience" criterion for five protocols in case of porosity and saturation determination.

	Protocol 1	Protocol 2	Protocol 3	Protocol 4	Protocol 5	**Sum**	**Contribution**
Protocol 1	1	1.5	1	1.5	1.5	6.5	0.26
Protocol 2	0.5	1	0.5	1	1.5	4.5	0.18
Protocol 3	1	1.5	1	1.5	1.5	6.5	0.26
Protocol 4	0.5	1	0.5	1	1.5	4.5	0.18
Protocol 5	0.5	0.5	0.5	0.5	1	3	0.12
Total						25	1

Table 10. Matrix of quantitative relations for the "usage experience" criterion for five protocols in case of permeability determination.

	Protocol 1	Protocol 2	Protocol 3	Protocol 4	Protocol 5	**Sum**	**Contribution**
Protocol 1	1	1	0.5	0.5	0.5	3.5	0.14
Protocol 2	1	1	0.5	0.5	0.5	3.5	0.14
Protocol 3	1.5	1.5	1	1	1	6	0.24
Protocol 4	1.5	1.5	1	1	1	6	0.24
Protocol 5	1.5	1.5	1	1	1	6	0.24
Total						25	1

Table 11. Sum of contributions and grading of the protocols to the "usage experience" criterion.

	Contribution to porosity and saturation	Contribution to permeability	Sum	Grade
Protocol 1	0.26	0.14	0.4	3
Protocol 2	0.18	0.14	0.32	1
Protocol 3	0.26	0.24	0.5	5
Protocol 4	0.18	0.24	0.42	4
Protocol 5	0.12	0.24	0.36	2

In the matrixes of quantitative relations:

1) "0.5" means that the protocol in the column is inferior to the protocol in the raw;
2) "1" means that the protocol in the column is equal to the protocol in the raw;

3) "1.5" means that the protocol in the column is more effective than the protocol in the raw.

For example, in case of porosity and saturation determination, the usage experience of protocol 1 (GRI) is higher than the usage experience of protocol 2 (TRA).

The contribution was calculated with the following equation (tables 9, 10 and 11):

$$\frac{\quad}{\quad} \qquad (2)$$

Contributions to porosity, saturation and permeability were summed up and every protocol was graded according to the total contribution value. The marketability of the protocols was determined using the following equation:

$$, \qquad (3)$$

where, M – marketability of the protocol; W_i – weight of criterion; G_i – grade.

The marketability results presented in table 8 have revealed that **<u>the most effective protocol is protocol 4</u>** (modified retort analysis and helium pycnometry for porosity and saturation determination and pulse decay on a core plug for permeability determination). The high efficiency of *protocol 4* is reasoned by the high grades of this protocol in the following criteria.

5.2.1. Accuracy of the results

Protocol 4 gives the most accurate results of porosity, saturation and permeability determination. The application of the retort analysis in protocol 4 allows the direct measurements of a fluid volumes saturating core sample that reduces uncertainties caused by calculations. Also, the retort analysis was shown to give more accurate results than modified Dean-Stark extraction (protocols 1 and 3), because modified Dean-Stark extraction systematically exaggerates the water volume obtained [9]. Besides, the fourth protocol excels protocol 5, because NMR used in protocol 5 is not the direct method of porosity and saturation determination and it is prone to uncertainties caused by the longitudinal relaxation time (T_1), transverse relaxation time (T_2), and diffusion coefficient (D) overlapping for different fluids and subsequent uncertainties in saturation determination. In addition, the permeability determined with the pulse decay (protocols 3, 4 and 5) on a core plug is more accurate than the permeability determined on a crushed sample (protocols 1 and 2) in case of confining pressure sufficient for micro fractures closing. Pulse decay is made at reservoir conditions and account for the flow direction, while the pressure decay on a crushed rock is made at ambient conditions.

5.2.2. Usage experience

The usage experience of protocol 4 is only lower than the usage experience of protocol 3 (modified Dean-Stark extraction, helium pycnometry and pressure decay on a crushed rock). The high usage experience of protocol 4 is caused by numerous research projects implemented for permeability determination in tight rocks using pulse decay on a core plug. The application of pressure pulse on a core plug began in 1968 [14], while the application of pressure decay on a crushed sample began in 1993 [15]. On the other hand, protocols 2 and 4 have lower usage experience than protocols 1 and 3 for porosity and saturation determination. It is explained by the fact that the modified retort analysis for tight rock evaluation began in 2011 when Handwerger, et al. [4] proved its applicability, while modified Dean-Stark extraction has been used from 1992 [7]. The application of NMR for tight rock analysis (protocol 5) has begun only recently [10, 11, 12].

5.2.3. Time of measurements

In the time criterion, protocol 4 is only inferior to protocol 5 in terms of obtaining porosity, saturation and permeability results due to the relatively long time necessary for a modified retort analysis. On the other hand, this drawback of protocol 4 is neutralized by the possibility to perform a modified retort analysis for several samples at a time.

5.2.4. Easiness of interpretation

All the five protocols, except protocol 5, are comparable in terms of interpretation complexity. Protocol 5 includes the NMR method that implies a mathematical inversion process, creation and analysis of D-T_2 maps and, thus, implicates a highly proficient specialist and a complex interpretation process.

5.2.5. Reliability and safety

The reliability and safety of the protocols were evaluated according to the complexity of the apparatuses used. The apparatuses used for porosity and saturation determination are assumed to have the same complexity. The apparatus for permeability determination using pressure pulse includes additional tools for confining and pore pressure application in comparison to the pressure decay apparatus. Therefore, protocols 3, 4, 5 are less reliable than protocols 1, 2.

5.2.6. Price

The low grade of protocol 4 is caused by the higher price for the modified retort apparatus in comparison with the modified Dean-Stark apparatus and the higher price for the pulse decay apparatus in comparison with the pressure decay apparatus.

5.2.7. Scheme for determining the most effective protocol for tight rock petrophysical properties evaluation

There are cases when proposed protocol 4 is less effective than other protocols. Therefore, in order to choose the most effective protocol for tight rock petrophysical properties evaluation in definite conditions, scheme (figure 5) is suggested.

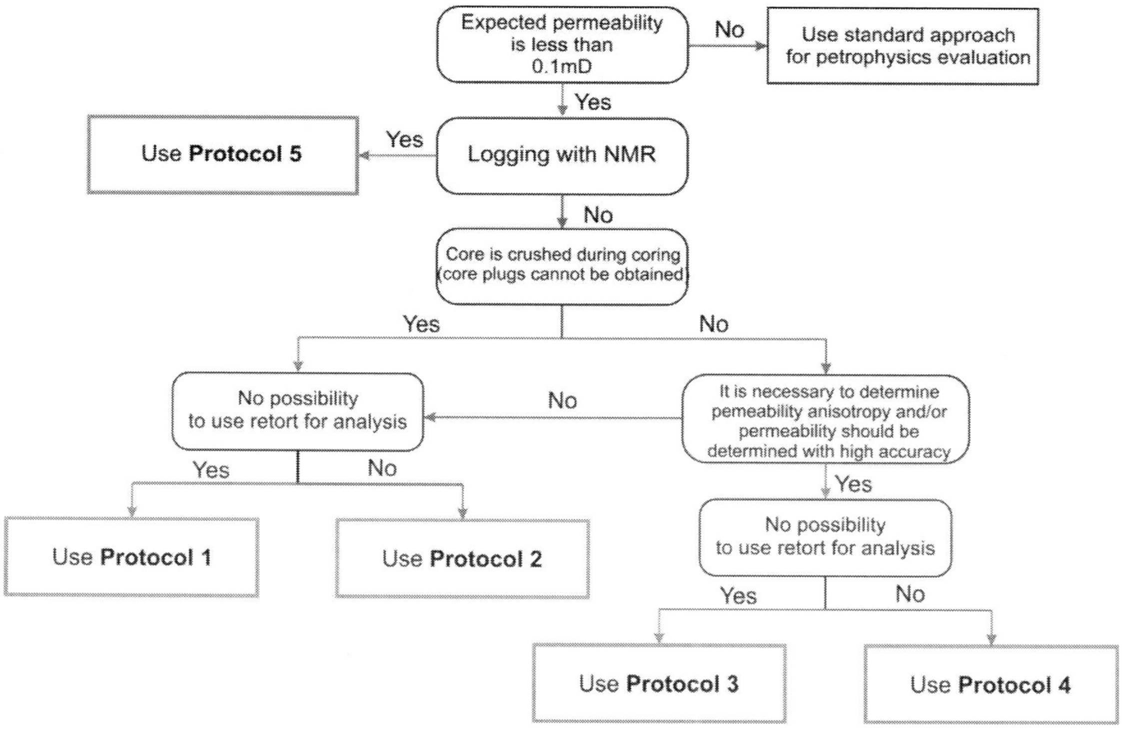

Figure 5. Scheme for determining the most effective protocol for tight rock petrophysical properties evaluation in definite conditions.

6. Conclusions

The most effective protocol for porosity, saturation, and permeability determination in the tight rocks of the Bazhenov formation is **protocol 4** that includes pressure pulse on a core plug for permeability determination; helium pycnometry and modified retort analysis on a crushed sample for porosity and saturation determination.

If NMR is used in logging or if a core is crushed during the coring process or if a retort cannot be applied for analysis, the scheme is suggested to choose the most effective protocol for tight rock petrophysical properties evaluation.

References

[1] Ulmishek G F 2003 Petroleum Geology and Resources of the West Siberian Basin, Russia *Virginia: U.S. Geological Survey*

[2] Vigodin G, Rubcov A, Klubkov S, Ezhov S, Bolova M and Kozlova D 2013 Unconventional Oil: Will Bazhen Become the Second Bakken? Moscow: Skolkovo

[3] Zhou J, Jung C M, Chenevert M E and Sharma M M 2013 Petrophysical Characterization of Organic-Rich Shales: a New Standardized Protocol *47th U.S. Rock Mechanics/Geomechanics Symp.*

[4] Handwerger D A, Suarez-Rivera R, Vaughn K I and Keller J F 2011 Improved Petrophysical Core Measurements on Tight Shale Reservoirs Using Retort and Crushed Samples SPE 147456 1-21

[5] Ambrose R J 2011 Multi-Component Sorbed-Phase Considerations for Shale Gas-in-Place Calculations SPE 141416 1-10

[6] American Petroleum Institute 1998 Recommended Practices for Core Analysis (Washington: API Publishing Services)

[7] Luffel D L and Guidry F K 1992 New Core Analysis Methods for Measuring Reservoir Rock Properties of Devonian Shale SPE 20571I184-90

[8] ResTech 1996 Development of Laboratory and Petrophysical Techniques for Evaluating Shale Reservoir" (GRI-95/0496) Houston

[9] Handwerger D A, Willberg D, Pagels M, Rowland B and Keller J F 2012 Reconciling Retort Versus Dean Stark Measurements on Tight Shales SPE 159976 1-13

[10] Minh C, Crary S F, Zielinski L, Liu C, Jones S and Jacobsen S J 2012 *2D-NMR* Applications in Unconventional Reservoirs SPE 161578 1-17

[11] Jiang T, Rylander E, Singer P M, Lewis R E and Sinclair S M 2013 Integrated Petrophysical Interpretation of Eagle Ford Shale with 1-D and 2-D Nuclear Magnetic Resonance (NMR) *Society of Petrophysicists and Well-Log Analysts* 1-21

[12] Tinni A, Odusina E O, Sulucarnain I D, Sondergeld C H and Rai C S NMR Response of Brine, Oil and Methane in Organic Rich Shales SPE-168971-MS 2014 1-9

[13] Cui X, Bustin M M and Bustin R M 2009 Measurements of Gas Permeability and Diffusivity of Tight Reservoir Rocks: Different Approaches and Their Applications *Geofluids* 208-23

[14] Brace W F, Walsh J B and Frangos W T 1968 Permeability of Granite under High Pressure *Journal of Geophysical Reserch* 2225-36

[15] Luffel D L, Hopkins C W, Schettler P D 1993 Matrix Permeability Measurement of Gas Productive Shales SPE 26633 261-70.

IOP Conf. Series: Earth and Environmental Science **24** (2015) 012043 doi:10.1088/1755-1315/24/1/012043

Oil extraction optimization case: horizontal permeability heterogeneity evaluation and application

M O Korovin[1], V P Merkulov[2] and S Z Gojkovic[3]

[1,2]Tomsk Polytechnic University, Tomsk, Russia
[3]NIS a.d. Novi Sad, Serbia

E-mail: [1]korovinmo@hw.tpu.ru

Abstract. Permeability is a parameter influenced by a direction and measurement scale that defines permeability as an anisotropic property. This parameter is very important, as it influences a fluid flow direction and velocities in the subsurface, which in turn causes problems with an uneven and unwanted water front, high early water production, losses of production and many other problems, if not addressed properly. Unfortunately, anisotropy becomes evident only when those unwanted symptoms occur. The knowledge of a magnitude and direction of anisotropy can be very important in production planning and management. Well pattern orientation, spacing, perforation program can all be optimized in the very early stages of development.

1. Introduction

Permeability is a parameter influenced by a direction and measurement scale that defines permeability as an anisotropic property. J.H. Baas et al. [1] made an observation that compares permeability measured in several different samples and in different directions. The authors concluded that permeability measured in a direction being parallel to a paleoflow in the bedding plane has the biggest value, while that measured perpendicularly to the bedding plane has the lowest one. These two extremes can be observed in the majority of cases, while the values of permeability measured in other directions fall between them [2].

The goal of this paper was to analyze data gathered and form general recommendations for an optimal production practice in an anisotropic reservoir to be proposed and, possibly, to create a potential strategy for further development.

2. Theory

Many geological methods were developed in the past to determine the direction and quantification of anisotropy and permeability alone. Conventional methods used for this purpose were: microscopic petrographic description, mineralogical analysis with infrared spectroscopy, as well as permeability and porosity correlation, conductivity and acoustic velocity measurements, etc. The authors also mention some novel techniques: permeability scans on rock surfaces, petrographic image analysis from a thin section, and density profiles obtained with a computerized tomographic (CT) scanner [3].

Wade et al. [4] wrote about particle methods that were used to determine anisotropy at the Eldfisk field in preparation for water flooding. The authors connect permeability anisotropy with horizontal stress anisotropy and use methods like: elastic strain relaxation, borehole image logs and oriented acoustic logs to determine the orientation of naturally occurring fractures in a depleted reservoir.

In the laboratory analysis of anisotropic sandstones, Auzerias et al. [3] showed that electrical and sonic anisotropies were found to be small in otherwise very anisotropic rock because in this case permeability was governed by texture and not by mineralogy.

J.H. Schon et al. used several log responses (induction, neutron and NMR) for the derivation and comparison of log-obtained permeability anisotropy with that obtained from core analysis and well tests. Very poor correlation was obtained for all the samples, which was expected, because all these responses represent scalar values, while the resulting parameter derived from them should be a vector [5].

3. K oilfield description

Field K is located in the West Siberian Plate in Tomsk Oblast. It was discovered in 1985 and the first oil was produced in 1998. The reservoir includes Jurassic sands of shallow marine and deltaic facies. Four possibly oil bearing layers were detected and are not all present in the cross section in different areal sections of the reservoir.

4. Core analysis

For the purpose of anisotropy investigation, quantification and orientation determination, several sets of data were available:

- Petrographic image analysis for the oriented core thin section of reference well 187
- Permeability analysis to liquid in principle directions

Petrographic grain orientation analysis of core samples

Being a clastic reservoir, the permeability anisotropy of field K formation was not expected to be the consequence of natural or other fracture occurrence, but the aftermath of depositional environment processes and matrix. That is why grain fabric analysis was performed on the core sample of reference well 187. Particle methods use thin section of core and microscope to determine mean grain orientation [1, 9]. Firstly, core samples were oriented to the geographical north by magnetic measurements, then they were observed under the microscope and one by one, a representative number of grains was processed. The core sample was rotated and the grains of satisfactory properties were inspected and counted. The result of this analysis is a rose (polar) diagram (figure 1). The number of grains with similar orientation is plotted onto the diagram with a predetermined step-angle, in our data it was 15°. Six rose diagrams were obtained as input data (Table 1.); they were all produced from well 187, samples from 2,762 to 2,787 m MD.

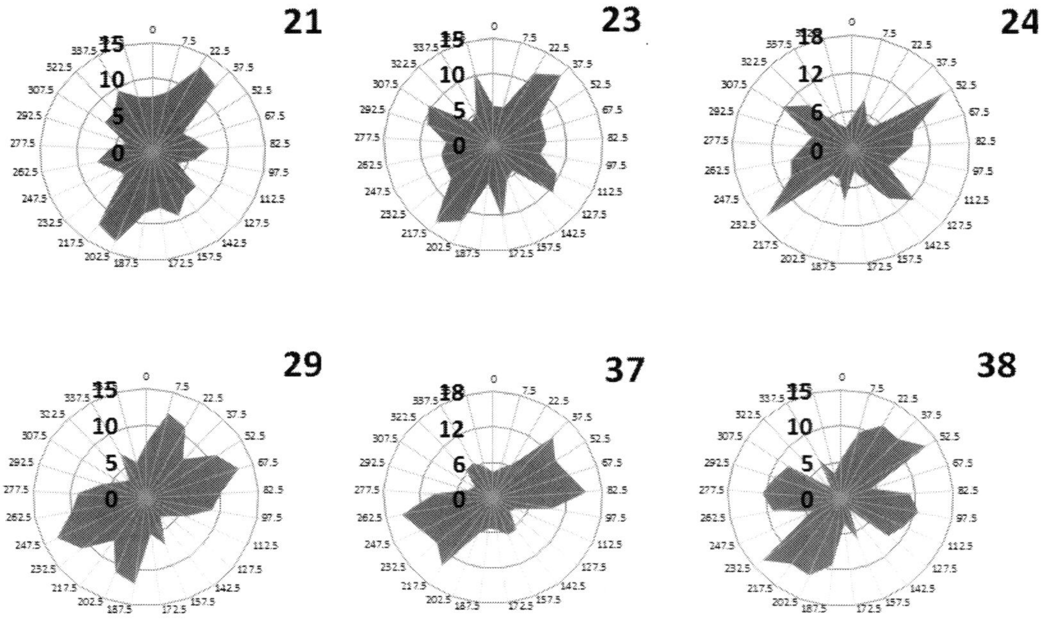

Figure 1. Rose diagrams of long particle quartz grain directions

Table 1. Permeability measured for six oriented core samples in two directions

Sample No.	*Interval, m*	*Porosity, %*	*Permeability, mD*	*Direction*	k_{Hmax}/k_{Hmin}
21	2,762-2,769	18.6	212.8	NE	2.00
			106.4	NW	
23	2,762-2,769	6.5	0.67	NE	1.14
			0.59	NW	
24	2,762-2,769	9.8	5.4	NE	1.86
			2.9	NW	
29	2,769-2,776	12.2	1.5	NE	1.15
			1.3	NW	
37	2,776-2,783	12.1	0.5	NE	2.00
			0.25	NW	
38	2,776-2,783	10.8	1.6	NE	1.30
			1.23	NW	

It is evident that the majority of grains have the SE – NW direction, it is even more obvious on the cumulative rose diagram for all samples. A value of orientation, 48° from N, was obtained graphically and adopted as a general anisotropy orientation for the entire field (figure 2).

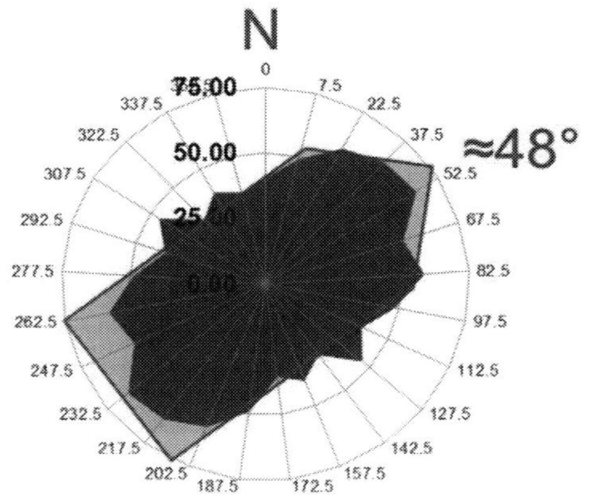

Figure 2. Cumulative rose diagram

Having the orientation of grain alignment (the main permeability anisotropy cause in field K), a measurement of maximum and minimum permeability was performed in direction parallel (NE) to grain orientation and perpendicular to it (NW). The results are presented in Table 1 where it can be seen that the uppermost part of the reservoir has high values of permeability (212 mD) that only grow smaller with depth to only few mD in the bottom part. The magnitude of anisotropy can also be noticed in the same table, which varies from 0.5 to 0.88. In proportion to the sample length and the layer that they represent, the values of k_{Hmax}/k_{Hmin} ratio were calculated for three layers that can be easily separated in the J_1^3 formation: 2 for the upper part, 1.5 for the middle and 1.3 for the lower layer. Those values were used later on in geological model simulation.

5. Results

Simulation model history matching comparison

The newly obtained data about overall horizontal anisotropy were, at this point, included into the existing geological model, simulation was performed and results were analyzed in comparison with the isotropic model.

The simulation of the isotropic model was performed and history matching was compared. The main parameter that was observed and compared was the water cut. Most of the wells showed improvement in prediction of that parameter compared to the isotropic simulation. A reference well simulation result comparison can be seen in figure 3, while changes in the relative difference (d) between the observed and predicted water cut for all wells in the investigated area can be noticed in figure 4.

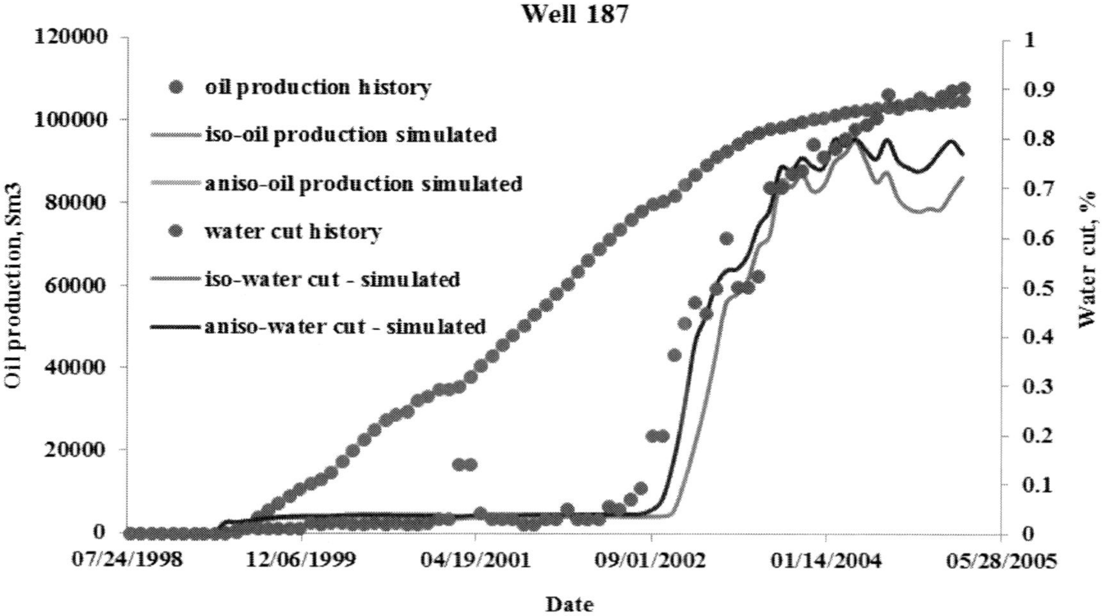

Figure 3. Comparison of isotropic and anisotropic model match

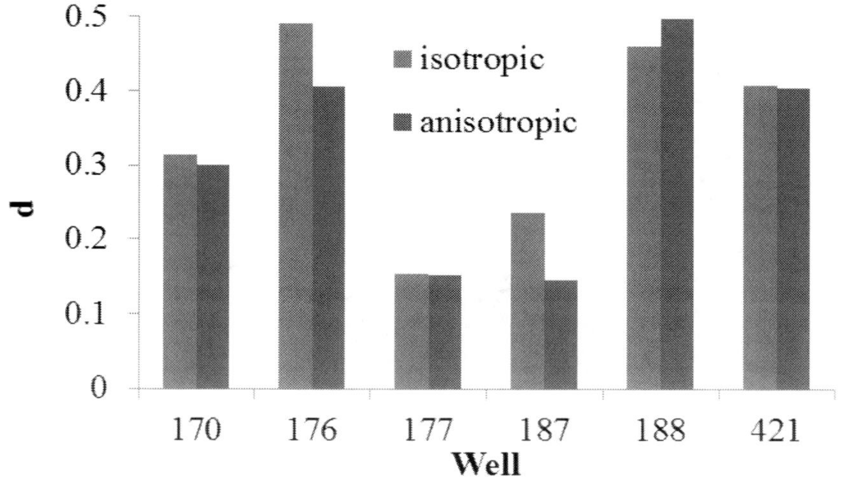

Figure 4. Parameter of match success (d) in both isotropic and anisotropic model

For comparison of the history matching success, one more parameter was created. This parameter is abbreviated with d and represents the square root of the sum of squared difference between the simulated and historical water cut normalized by the square of the historical water cut ($d=\sqrt{\frac{\sum(WC_S-WC_H)^2}{\sum WC_H^{\,2}}}$). It is evident that all the wells except well 188 show improvement in matching if anisotropy is incorporated in the model. In the case of 188, the previous artificial matching could have caused specific irregularities.

It is possible to make a conclusion that taking anisotropy into account can help and improve reservoir and well history matching. As it can be seen, this is also true in the case of field K. It could be said that the introduction of reservoir anisotropy in the beginning of the model adaptation, before individual well history matching by artificially increasing permeability values in some grid blocks, can lead to a more efficient, less time consuming and more accurate model.

Optimization and proposition of further production strategy

The information on anisotropic reservoir behaviour obtained by simulating different scenarios of a conceptual model play a role in the further optimization of water injection and secondary oil recovery. Available production data only come to 2005, so it becomes the starting year for further optimization. The base case would be production until 2010 without any changes. An optimum scenario was sought after only with the transformation of production wells into injectors for the period of 2005 to 2010. In this way, otherwise an unfavourable water front can be distorted to meet the needs. It was performed and simulated in several different variations, resulting in the optimum case where production was enhanced by obeying the general rules described and discovered in the previous sections and by converting production wells into injectors. In the optimum simulation, beginning from 2005, wells 1001, 1002, 407 and 177 had their function altered.

The results show (figure 5 and figure 6) that the simple targeted transformation of a producer into an injector can enhance production in field K up to 70,222 cm3 (around 2% of the overall oil production). The above- mentioned figure also shows enlargement in cumulative oil production for two wells, 391 and 425. The first one has undergone the cumulative oil production increase of 13.2%, while well 425 showed a growth by 8.5%. The magnitude of oil production enhancement differs from well to well in a range from unchanged to as high as 18.2% (well 379).

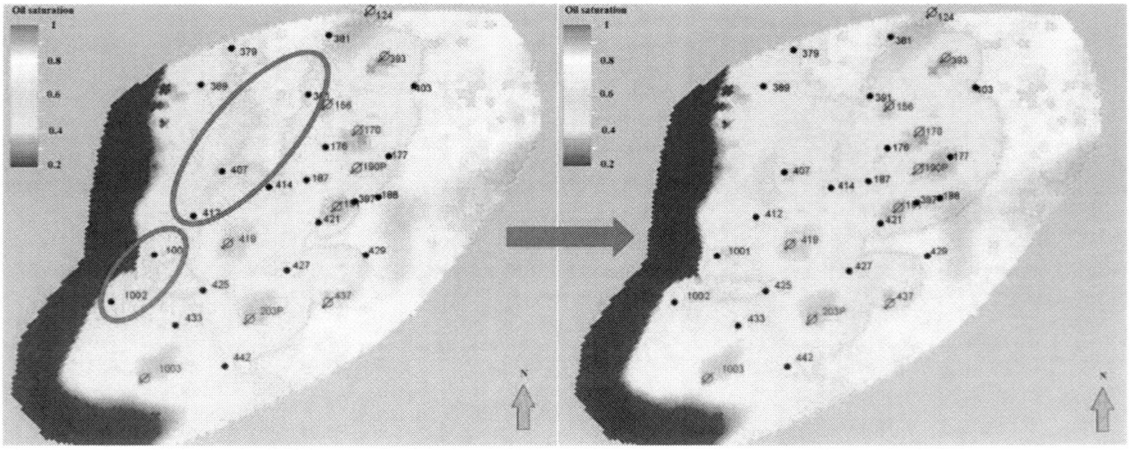

Figure 5. Oil saturation areal view for base and optimal case

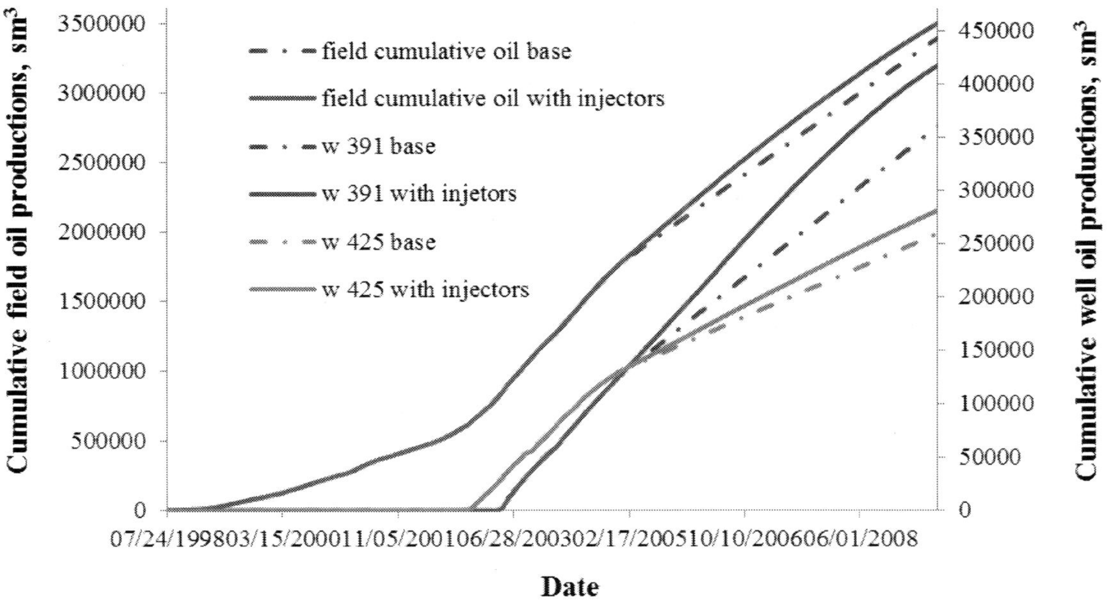

Figure 6. Diagram of cumulative field and well oil production for two cases

At this point, there were no changes in the perforation program; it can be one of the sources of extensive water injection and production enhancement. In that light, the optimization of the perforation program is proposed for further work as well as more detailed optimization of injector flow-rates.

6. Conclusion

These relatively inexpensive and quick methods can be applied during field exploration in order to gather information regarding formation properties and anisotropy. It can lead to the early understanding and prediction of possible future production problems that can arise from them. With every new well that is drilled and set of data being analyzed, a more complete picture is formed, potentially saving considerable funds in the future.

Summing up the results obtained in this study the following conclusions can be made:

- The orientation of grain alignment and the axis of the maximum permeability were determined to be 48 ° from N. This azimuth was also adopted as the direction of spatial absolute permeability spread that revealed parts of the reservoir with good properties and was confirmed with permeability map construction.
- The petrographic grain fabric analysis revealed microscopic anisotropy within a range from 0.5 to 1. This was the result of the depositional environment and, as expected, in further analysis it was proved to be present in the entire field.

References

[1] Jaco H Baas, Ernie A Hailwood, William D McCaffrey, Mike Kay and Richard Jones 2007 Directional Petrological Characterization of Deep-Marine Sandstones Using Grain Fabric and Permeability Anisotropy: Methodologies, Theory, Application and Suggestions for Integration *Earth-Science Reviews* **82** 101–42
[2] Fikri Kucuk and William E Brigham 1979 Transient Flow in Elliptical Systems *SPE* 7488 401-410

IOP Conf. Series: Earth and Environmental Science **24** (2015) 012043 doi:10.1088/1755-1315/24/1/012043

[3] Auzerias F M, Ellis D V, Luthi S M, Dussan E B V and Pinoteau B J, Schlumberger-Doll Research Laboratory Characterization of Anisotropic Rocks *SPE* 20602 751-8
[4] Wade M, Phillips Petroleum Co. Norway, Hough E V, Phillips Petroleum Co. Norway and Pedersen S H, Phillips Petroleum Co. Norway 1998 Practical Methods Employed in Determining Permeability Anisotropy for Optimization of a Planned Waterflood of the Eldfisk Field *SPE* 48961 49-63
[5] Schon J H, Georgi D T, Fanini O 2003 Imparting Directional Dependence on Log-Derived Permeability *SPE* 82058 48-56
[6] Tarek Ahmed and Paul McKinney 2005 *Advanced Reservoir Engineering* (Elsevier Inc.)
[7] Her-Yuan Chen, Dewi Triarti Hidayati and Lawrence W. Teufel A Quick Method to Diagnose Flow Anisotropy Using Pressure Interference Data *SPE* 60290
[8] Her-Yuan Chen and Lawrence W. Teufel 2003 A Quick Method to Determine Permeability-Anisotropy Orientation from interference test *SPE* 84090
[9] Korovin M O 2014 Specialized Core Analysis to Study the Anisotropy of Oil and Gas Reservoirs // *Tomsk Polytechnic University* **324** № 1: Earth Sciences 87-92

IOP Conf. Series: Earth and Environmental Science **24** (2015) 012044 doi:10.1088/1755-1315/24/1/012044

Permeability inhomogeneity accounting in terms of efficient development strategy

M O Korovin[1], V P Merkulov[2] and S Z Gojkovic[3]

[1,2]Tomsk Polytechnic University, Tomsk, Russia
[3]NIS a.d. Novi Sad, Serbia

E-mail: [1]korovinmo@hw.tpu.ru

Abstract. The main questions concerning flow and permeability anisotropy are: presence of anisotropy (is a reservoir anisotropic or not), orientation and degree of anisotropy. Answers were obtained through the analysis of grain fabric, spatial permeability spread and tracer injection data. The magnitude of permeability was determined to be within a range from 0.5 to 1 and with the azimuth of 48 from the North. Thereafter, the incorporation of new information about field K into the simulation model matched and the comparison of isotropic (the existing model) and anisotropic (the model upgraded with novel information) were performed. This process showed that the simple introduction of obtained anisotropy properties improved the history match, which was interpreted as the proof of anisotropy being present on both the macroscopic and microscopic scale. At the end, general recommendations for the optimal production practice in the anisotropic reservoir were proposed and the potential strategy for further development was suggested. This optimal strategy was created by changing several producing wells into injectors and it resulted in an overall oil production increase and individual well production increase amounting to as much as 20%.

1. Introduction

Permeability and its anisotropy influence a fluid flow direction and velocities in the subsurface, which in turn causes problems with an uneven and unwanted water front, high early water production, losses of production and many other problems [1]. Kuchuk and Brigham [2] showed that in the presence of anisotropic horizontal permeability production rate decreases with the increase in anisotropy. Unfortunately, anisotropy becomes evident only when those unwanted symptoms occur. The knowledge of the magnitude and direction of anisotropy can be very important in production planning and management. Well pattern orientation, spacing, perforation program can all be optimized at the very early stages of development.

The goal of this paper was to analyze gathered data and draw conclusions about anisotropy features. The main questions concerning flow and permeability anisotropy are: presence of anisotropy (is the reservoir anisotropic or not), orientation and degree of anisotropy. Thereafter, the incorporation of new information about field K into the simulation model matched and the comparison of isotropic (the existing model) and anisotropic (the model upgraded with novel information) were performed.

2. Theoretical justification

The following methods are performed on wells upon the start of production. The most widely used methods are well tests and inter well tracer tests.

There are several variations of well tests that can be used for this purpose, but all of them have the same principle of triggering the impulse and measuring the response. Ramey et al. [6] derived an interference method that can be used to determine anisotropy direction and magnitude. It can be performed for two observation wells, but if $\phi \mu ct$ - factor is unknown, at least three wells are needed. This formal method is very sensitive if permeability anisotropy is high and if well-anisotropy configuration is unfavorable. If this is the case (but cannot be known a priori), a new quick method was developed and can be used [7, 8, 10]. This technique uses the log-log plot of pressure change versus time/(distance)^2 responses and the direction of anisotropy can be obtained by comparing them for different observation wells. Even without the knowledge of reservoir properties ($\phi \mu ct$), a unique solution, if at all, can be obtained in the 90 ° range for two wells and in the 45 ° range for three observation wells.

On the other hand, the tracer injection technique is more favorable. It uses specially engineered substances, tracers, and after analysis can show the degree of inter-well pathways, orientation of anisotropy and velocity at which tracers are traveling and evaluate areal water breakthrough between injectors and producers. This method gives more precise information about the fluid flow in the subsurface compared to others, which only give assumptions about responses and permeability, but in combination with them the whole picture is formed.

3. Permeability analysis

For the purpose of anisotropy investigation, quantification and orientation determination, several sets of data were available:

- Interpreted logs for 29 wells
- Tracer injection test results for injection wells 156, 190P and surrounding production wells

3.1 Absolute permeability isoline distribution

Another data set considered in this case study was interpreted logs. They were used to construct absolute permeability maps. The resulting maps showed the spatial distribution of a formation with better reservoir qualities. As it has already been mentioned, $J_1^{3\text{-}4}$ can be divided into three layers with different composition and properties. It was used in order to plot the maps of the absolute permeability spatial spread. Three layers that can be separated by log response interpretation where they are very noticeable (especially in the GR and SP logs) (figure 1) are as follows:

- YU 1-3-A – The upper part of the section is characterized by well sorted medium-fine sand with permeability ranging from 10 to few hundreds of mD. It lies between the top of the reservoir and the depth of 6 m (about 30% of the reservoir thickness).
- YU 1-3-B – The middle section of the reservoir of all subtypes is characterized by massive sandstone ranging from fine to medium and being poorly sorted. The permeability of this layer ranges from one to few tens of mD. This layer begins at the depth of 6 m and ends at about 11 m from the top (covering about 25% of the J13 thickness).
- YU 1-3-C – Poorly sorted fine sand and siltstone characterize the lowest section of the reservoir in all subtypes. The permeability of this layer does not exceed one mD and this last section covers the thickness from 11 to about 20 m ($\approx$ 45% of the reservoir thickness).

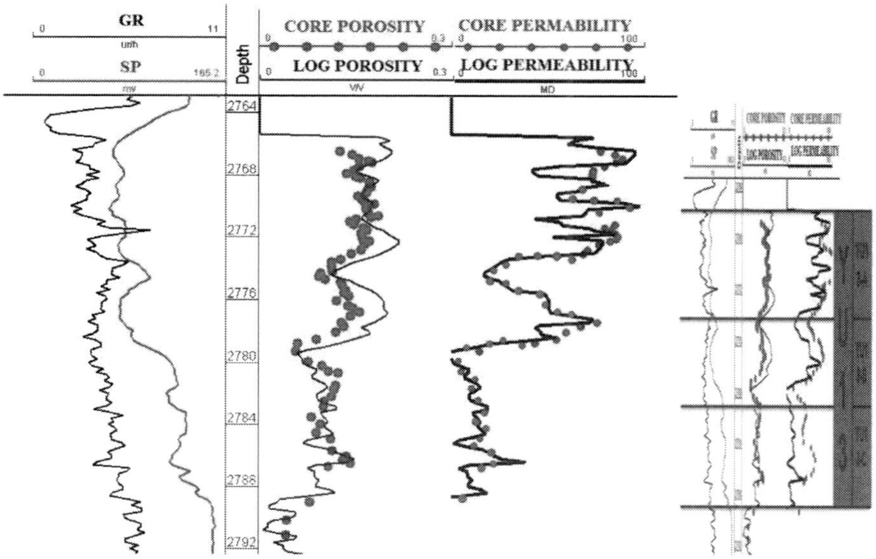

Figure 1. Separation of sub-layers in a reference well

Keeping that in mind, three sets of permeability maps were constructed (figures 2 to 4). The simple Kriging method was used and permeability data for every map and every well were obtained by averaging permeability in a particular sublayer (from YU 1-3-A to C). In recapitulation, the maps constructed for two upper layers show the evident orientation of horizontal permeability and it corresponds to the azimuth of ≈ 45 °. It is very close to the azimuth that was revealed earlier with the petrographic grain analysis (48 °). Furthermore, both those layers have an elliptical shape that stands out as a formation with better reservoir properties and can be linked to the mouth bar structure that was mentioned in the geological depositional model [9].

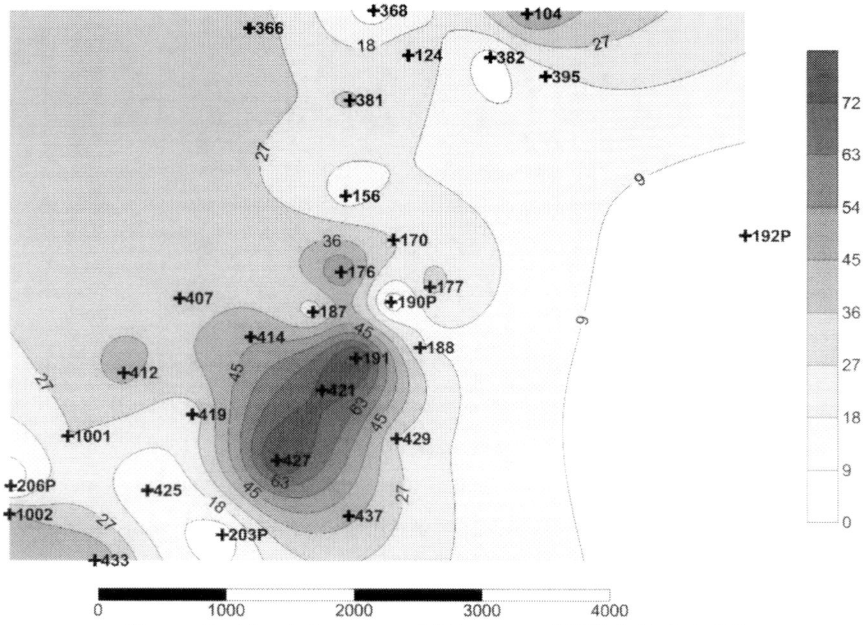

Figure 2. Absolute permeability map of YU-1-3-A (mD)

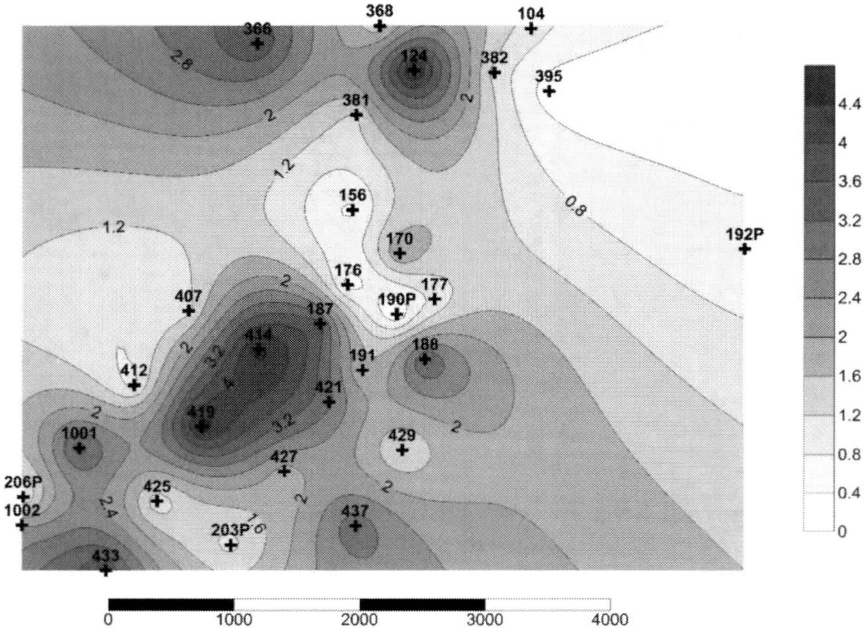

Figure 3. Absolute permeability map of YU-1-3-B (mD)

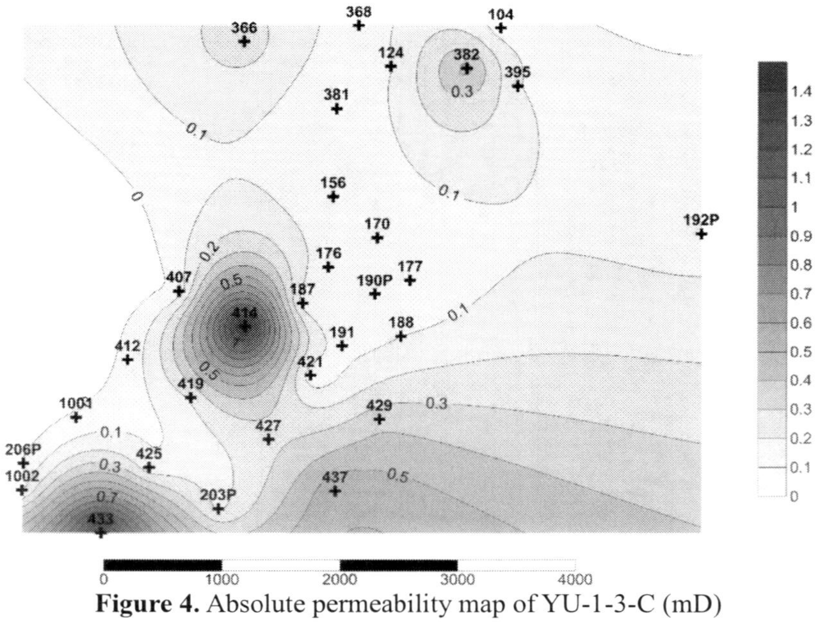

Figure 4. Absolute permeability map of YU-1-3-C (mD)

3.2 Tracer injection analysis of injection wells 156, 190P and adjacent production wells

In order to study the macroscopic reservoir characteristics of this field, an inter well tracer test was performed. The test started in September 2003 and took several months. Two injection wells were selected (156 and 190P) in which two different tracers were injected. Water injection was planned to

IOP Conf. Series: Earth and Environmental Science **24** (2015) 012044 doi:10.1088/1755-1315/24/1/012044

increase with time and the amount of tracer injected ranged from 43 to 162 m3/day for well 156, while injection for well 190P started with 500 and finished with 1111 m3/day. The samples of water produced were taken in the predetermined periods and analyzed for indicators and their concentration. The first appearances of measurable indicator concentrations were observed in 8 to 15 days after the injection.

The largest amount of tracers was recovered in connection between 156 – 170 and 190P – 170 and comprised 55.7% and 43.93%, respectively (figure 5). It can be concluded that well 170 has a major impact on filtration in the investigated part of the formation. It has the highest values of a relative filtration flow: 68.08% in the direction towards 156 and 53.55% towards 190P. It can be explained by the proximity of well 170 to injection well 156 or by the difference in flow-rates for different injection wells (up to 7 times). Also, both injection wells and the adjacent producer are surrounded with a low permeability formation from NW and SE and one more injector (393) in the north that can also influence flows from 156-170 and 156-177. This influence can also be proved by the fact that both tracers were first sampled at about the same time in well 170. It can be claimed that the maximum velocities correspond to the direction of maximum permeability obtained in the previous analysis. Many other parameters influence the water front. One of them can be a pressure profile in the reservoir, the fact that the wells can be positioned in the part of the reservoir with poor qualities and hence a small production and diffusivity factor.

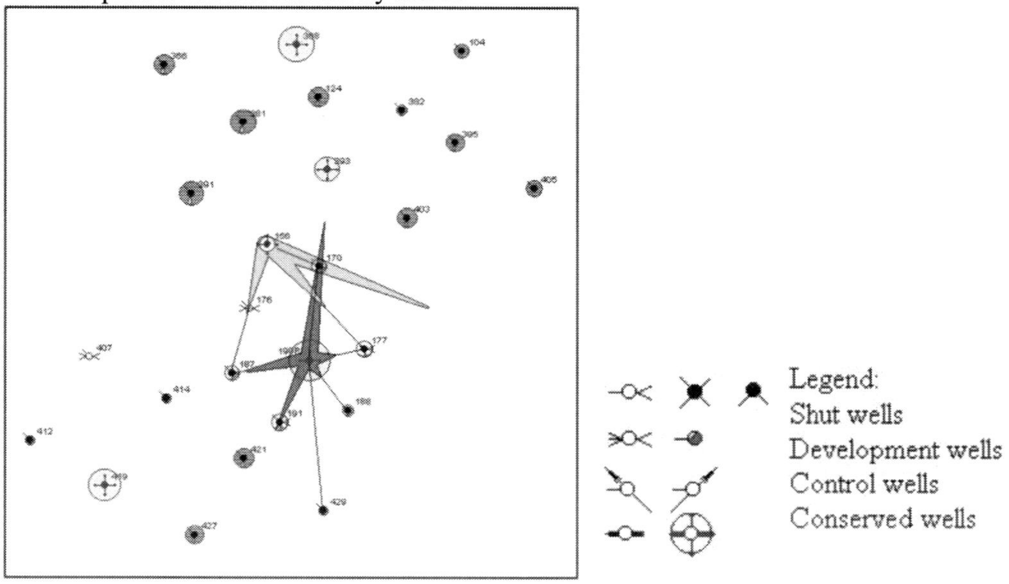

Figure 5. Tracer flow pattern

4. Results

4.1 Anisotropic model sensitivity analysis

In order to understand the anisotropic reservoir behavior, pressure distribution, water front shape, time of water breakthrough, well placement and pattern, a simple conceptual model was constructed with parameters being closest to field K. A grid with 100x100x 5 cells was formed, while anisotropy in the perpendicular direction was introduced by multiplying kx with 0.5 (the value most frequently revealed with petrophysical data). Well spacing was set to be 500 m. Relative permeability curves, PVT data, initial reservoir pressure and vertical anisotropy were taken from literature [9]. Two different sensitivity analyses were performed:

- Well grid orientation influence on water breakthrough in the anisotropic reservoir
- Comparison of five spot and different segregated well patterns in terms of oil production

IOP Conf. Series: Earth and Environmental Science **24** (2015) 012044 doi:10.1088/1755-1315/24/1/012044

In both cases, five wells were used: four production wells and one injection well. They also had the same control parameters and perforation program, only their position changed.

4.2 Well grid orientation influence on water breakthrough in the anisotropic reservoir

The idea was to rotate a well grid and to monitor water breakthrough. A rotation step was chosen to be 15 ° with a range from 0 (parallel) to 45 °. An injection well was placed in the center of the reservoir. Around it, a five-spot well pattern was made with four more production wells. Rotation was performed around the injection well.

Next, figure 6 shows that the parallel orientation of the five-spot well pattern towards the maximum permeability direction is most appropriate and has a potential to provide the most effective sweep area, while postponing water breakthrough.

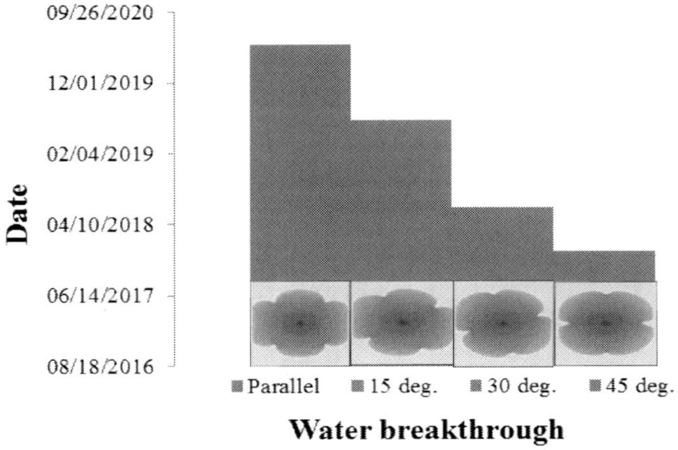

Figure 6. Comparison of different well-anisotropy orientations

4.3 Comparison of five spot and different staggered well patterns in terms of oil production from the anisotropic reservoir

The fact is that the anisotropic formation causes elliptical pressure distribution around the well, and with that, a drainage area of the same shape. On this basis, the idea that the parallel five-spot pattern that gave the best results in the previous analysis can achieve even better results if it is transformed into a staggered pattern that corresponds to the actual anisotropy was applied. The methodology was as follows: by changing the ratio of 2d/a from 1/5 to 5 (one being a parallel five-spot pattern), attention was focused on finding the ratio that, with field K anisotropy, leads to the biggest oil production.

As it can be seen in figure 7, different ratios induce different production levels; this is related to the elliptical distribution of the water front and its influence on the sweeping area. On the basis of the conclusion that production is highest within the ratio range from 0.5 to 1, one more simulation was performed for 2d/a ratio of 0.75. Finally, it resulted in the most optimal well distance.

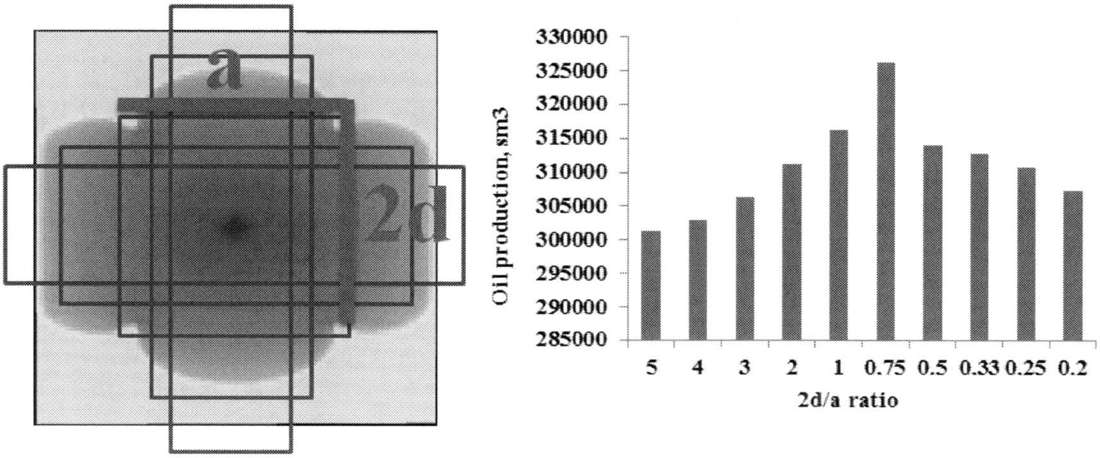

Figure 7. Schematic representation of well patterns and their results

5. Conclusion

Upon the detailed study, the data analyzed provided an insight into permeability distribution in both the orthogonal and vertical direction, microscopic and macroscopic scale, changes in that permeability and gave an answer to three important questions regarding its anisotropy. The knowledge about the presence, direction and magnitude of anisotropy was used to adjust the model that can be further used to optimize oil production by water flooding. Summing up all the results obtained with this study allows making several conclusions:

- The incorporation of anisotropy into the history matched model of the field and comparison with the isotropic one showed improvement in matching, which was interpreted as one more proof of anisotropy presence
- Several sensitivity simulation runs revealed that a staggered well pattern and parallel orientation of a well grid were most appropriate for anisotropic reservoirs development and exploitation.
- Governed by the above- mentioned principles, optimization of water-flooding program was performed by changing several producing wells into injectors. The result was a total oil production increase by 2% and a well oil production increase up to 18.2%.

References

[1] Jaco H. Baas, Ernie A. Hailwood, William D. McCaffrey, Mike Kay and Richard Jones 2007 Directional Petrological Characterization of Deep-Marine Sandstones Using Grain Fabric and Permeability Anisotropy: Methodologies, Theory, Application and Suggestions for Integration *Earth-Science Reviews* **82** 101–42

[2] Fikri Kucuk and William E. Brigham 1979 Transient Flow in Elliptical Systems *SPE* 7488 401-10

[3] Auzerais F M, Ellis D V, Luthi S M, Dussan V and Pinoteau B J 1990 Laboratory Characterization of Anisotropic Rocks *SPE* 20602 751-8

[4] Wade M, Phillips Petroleum Co. Norway, Hough E V, Phillips Petroleum Co. Norway and Pedersen S H, Phillips Petroleum Co. Norway 1998 Practical Methods Employed in Determining Permeability Anisotropy for Optimization of a Planned Waterflood of the Eldfisk Field *SPE* 48961 49-63

[5] Schon J H, Georgi D T and Fanini O 2003 Imparting Directional Dependence on Log-Derived Permeability *SPE* 82058 48-56
[6] Tarek Ahmed and Paul McKinney 2005 Advanced Reservoir Engineering (Elsevier Inc)
[7] Her-Yuan Chen, Dewi Triarti Hidayati and Lawrence W. Teufel 2000 A Quick Method to Diagnose Flow Anisotropy Using Pressure Interference Data *SPE* 60290
[8] Her-Yuan Chen and Lawrence W. Teufel 2003 A Quick Method to Determine Permeability-Anisotropy Orientation from Interference Test *SPE* 84090
[9] Danenberg E E, Belozoryev V B and Brylina N A 2006 Geological Structure and Oil/Gas Saturation of Upper Jurassic - Lower Cretaceous Sediments of Southeastern West Siberian Plate (Tomsk)
[10] Korovin M O and Merkulov V P 2014 Formation Evaluation and Well-Test Analysis for Complex Interpretation of Reservoir Permeability Distribution" *XVIII International Scientific Symposium in Honor of Academician M. A. Usov: Problems of Geology and Subsurface Development, April 7–11, 2014, Tomsk, Russia*

IOP Conf. Series: Earth and Environmental Science **24** (2015) 012045 doi:10.1088/1755-1315/24/1/012045

Application of miscible displacement for Field MTX low permeability formations

V Kuntsevich[2], S Slivkin[1] and V Belozerov[1]

[1]Tomsk Polytechnic University, Tomsk, Russia
[2]LLC Gazpromneft Science and Technology Center, Saint-Petersburg, Russia

E-mail: belozerovvb@hw.tpu.ru

Abstract. Miscible displacement is a very effective method of recovery efficiency improvement. It is widely used in the world, but this technology is not widely used in Russia. For this reason, it is necessary to study global experience and physical aspects of this EOR method. The most important factors and limitations of miscible displacement application from the geological point of view (heterogeneity) and from the physical point of view (properties of injected fluids and reservoir fluids) should be determined. The results of this analysis should be tested on the low permeability reservoir of field MTX with analytical, proxy calculation and simulation methods.

1. Field description

The geological cross section of Field MTX consists of alternating sand and shale layers confined to the Mesozoic-Cenozoic deposits. Field MTX has 9 productive layers that were formed in the same depositional environment. The oil-bearing zones are represented by sand lenses of a varying size and shape (figure 1). The main interest in the current study is applying a gas injection technology to the three upper layers (B6-21, B6-3, B6-31) where STOIIP equals 51 mln tons.

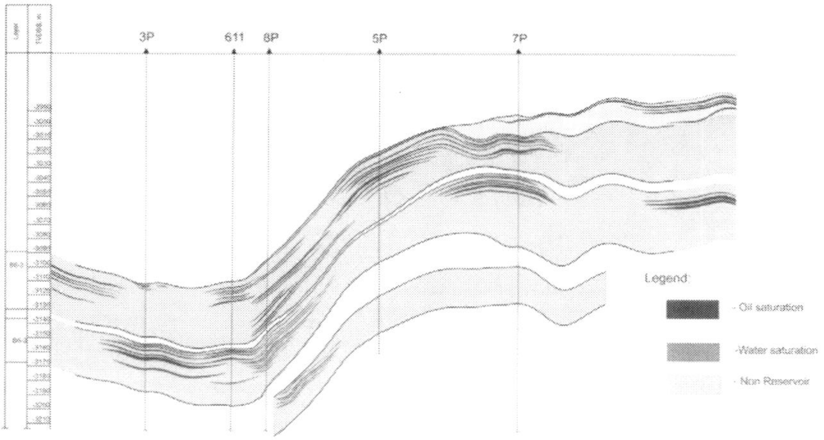

Figure 1. Cross section

These three layers of Field MTX have similar reservoir characteristics and fluid properties (table 1).

IOP Conf. Series: Earth and Environmental Science **24** (2015) 012045 doi:10.1088/1755-1315/24/1/012045

Table 1. Reservoir characteristics

Parameters	Value
Depth, m	-3,100
Trap type	layer – arch, lithologically screened
Oil-bearing area, m^2*103	133,145
Thickness of effective oil-bearing formation, m	10
Porosity, %	16
Permeability, mD	20
Oil saturation, %	58
OWC level, m	No
Reservoir temperature, ^oC	100
Reservoir pressure, MPa	32.7
Oil viscosity (at reservoir conditions), cp	0.28
Oil density (at standard condition), tons/m^3	0.834
Oil formation volume factor	1.491
Bubble point pressure, MPa	21.9
Gas – oil ratio, m^3/ton	200

According to the accepted oil classification, the oil is very light, has a low sulfur, tar and paraffin content. Field MTX is considered "very complicated" from the geological point of view. All these three layers have a common development strategy represented by:

- Seven spot pattern (500 meters between the wells)
- A pattern with horizontal wells with multi-stage hydraulic fracturing (6 stages for production wells and 3 stages for injection wells). The length of the horizontal part of the well equals 1,000 meters.

2. Screening criteria

Table 2 was constructed on the basis of information analysis and field experience [10].

Table 2. Screening criteria for miscible and immiscible gas injection

Technology/ Factors		Formation type	Permeability, mD	Heterogeneity	Reservoir and injected fluid composition and reservoir conditions		Density, kg/m^3	Viscosity, cp
					On the left side of critical tie line	On the opposite or the right side of the critical tie line		
Miscible displacement	Hydrocarbon	S & C	0.1 – 5,000		N/A	A	749 - 914	0.04 – 18,000
	CO_2	S & C	1.5 – 4,500		N/A	A	800 - 920	0-35
	Nitrogen	S & C	0.2 – 35		N/A	A	761 - 848	
	WAG	S	13 – 1,000		N/A	A	828 - 859	
Immiscible displacement	Hydrocarbon	S	40 – 1,000		A	N/A	767 - 914	0.25-4
	CO_2	S & C	30 – 1,000		A	N/A	848 - 991	0.6 – 592
	Nitrogen	S	3 – 2,800		A	N/A	761 - 957	0 – 18,000

	WAG	S & C	100 – 6,600		A	N/A	819 - 1003	0.17 – 17,000

S – Sandstone
C – Carbonate
A – Available
N/A – Not available
High dependence
Medium dependence
Low dependence

Thus, the reservoir characteristics are appropriate for all types of miscible displacement and for immiscible displacement with nitrogen. Also, there are several criteria for the reservoir fluid, including the range of densities and viscosities at which these technologies can be applied. The reservoir oil of Field MTX meets these criteria for all types of gas injection, except CO_2 immiscible displacement. It should be noted that there is no available sources of nitrogen and carbon dioxide near Field MTX. Thus, only miscible gas injection with hydrocarbon gas and miscible WAG injection with hydrocarbon gas is available for development and will be further considered. Two more conditions for the application of these technologies were considered by [6]: reservoir pressure should be higher than 20 MPa (the reservoir pressure of Field MTX equals 32.7 MPa) and oil density – smaller than 0.991 tons per m^3. The Field MTX characteristics meet both of these conditions. Thus, miscible gas injection can be used for Field MTX.

3. PVT modeling

The PVT modeling of the reservoir fluid was provided with the help of the Schlumberger PVTi software. The reservoir oil composition is presented in table 3.

Table 3. Reservoir oil composition (Field MTX)

Component	At standard conditions		Oil at reservoir conditions
	gas	oil	
Carbon dioxide	0.6	0.006	0.35
Nitrogen + other	0.36	0	0.21
C1	56.11	0.065	32.47
C2	15.73	0.42	9.27
C3	12.72	1.69	8.07
iC4	2.86	1.07	2.10
nC4	5.14	2.92	4.20
iC5	1.53	2.05	1.75
nC5	1.77	3.15	2.35
C6	3.179	1.07	2.29
C7+	0.00	87.57	36.94
MW of reservoir oil	82		
MW of C7+	166.87		

Three parametric Peng – Robinson EOS were used to match the model with the real PVT data. The matching process was provided to achieve a convergence between the real and calculated properties of the fluid (GOR, bubble point pressure, oil formation volume factor and viscosity) by the varying parameters of the C7+ fraction (it has higher uncertainty).

4. Minimum miscibility pressure

Gas miscibility grows as pressure increases. Thus, the recovery factor should increase too. It actually occurs, but there is a pressure above which the further pressure increase only entails the minimum increase in the oil recovery. The pressure at which the oil recovery aligned is the minimum miscibility pressure. There are various impurities that can modify the minimum miscibility pressure. Methane and nitrogen increase the minimum miscibility pressure, whereas H2S decreases it.

The ability of impurities to increase or decrease the minimum miscibility pressure depends on whether there is an increased dissolving ability of the gas. The dissolving ability is improved (the minimum miscibility pressure is decreased), if the gas is diluted with an impurity, critical temperature of which exceeds the critical temperature of the gas. The dissolving ability deteriorates (the minimum miscibility pressure increases), if an impurity has a critical temperature lower than the critical temperature of the gas.

There are a lot of equations trough which minimum miscibility pressure can be determined [3]:

For pure and impure CO_2:

- Orr and Silva
- Extrapolated Vapor Pressure (EVP) Method
- Yelling and Metcalfe
- Alston et al
- National Petroleum Council (NPC)
- Enick-Holder-Morsi
- Croquist

For lean gas and nitrogen:

- Firoozabadi and Aziz
- Hudgins-Liave-Chung
- Glaso

In the current work, the minimum miscibility pressure was estimated through four different methods that gave similar results.

Input data for calculation:

- reservoir oil composition presented in Table 3.
- only associated petroleum gas used for miscible displacement (Table 3) that contains 56% of the C_1 fraction.
- oil at reservoir conditions is undersaturated (Pres = 32.7 MPa, Pb = 21.9 MPa, GOR = 200 m^3/ton)
- *Firoozabadi and Aziz* [2]

$$MMP = 9{,}433 - 188(10^3)F + 1{,}430(10^2)F^2 = 4{,}378.57\ psi = 29.8\ MPa \qquad (1)$$

with

$$F = \frac{I}{M_{C_{7+}}(T-460)^{0.25}} = 0.03769 \qquad (2)$$

$$I = x_{C_2-C_5} + x_{CO_2} + x_{H_2S} = 23.99\% \qquad (3)$$

where I – concentration of intermediates it the oil phase, %; T – temperature (671.67 °R); $M_{C_{7+}}$ – molecular weight of C_7+ (166.87)

- *Hudgins-Liave-Chung* [2]

$$MMP = 5{,}568e^{R1} + 3{,}641e^{R2} = 3{,}958\ psi = 27\ MPa \qquad (4)$$

with

$$R1 = \frac{-792.06x_{C_2-C_5}}{M_{C_{7+}}T^{0.25}} = -0.345066 \qquad (5)$$

$$R2 = \frac{-2.158(10^6)C_1{}^{5.632}}{M_{C_{7+}}T^{0.25}} = -5.5347 \qquad (6)$$

where T – temperature (212 °F); C_1 – mole fraction of methane; $x_{C_2-C_5}$ – sum of the fraction of $C_2 - C_5$ in the oil phase.

- *Glaso for condensing-vaporizing drive* [1]

To calculate MMP with injection gas with MC_2-C_6 = 20, the following equation was used

$$MMP = 43.74 - 0.1752M - (32.23 - 0.127M)y_1 + \left(0.777(10^{-14})M^{5.258}e^{31980y_1M^{-1.703}}\right)(1.8T - 460) = 32\ MPa \quad (7)$$

with

$$M = 572.7 / S_{C_{7+}}^{-5.573} \quad (8)$$

where T – temperature (373.15 °K); y1 – mole fraction of methane in the injection gas; $M_{C_2-C_6}$ – sum of the fraction of $C_2 - C_6$ in the injection gas.

- *Equation based on Maklavani correlation* [13]

$$MMP = 43.664 - 4.542\alpha + 0.689\alpha^2 - 0.132\beta = 30.6\ MPa \quad (9)$$

with

$$\alpha = \frac{X_{C2-C6}^{1.72785} X_{C1}^{0.1}}{(1.82T+32)^{0.5} M_{C7+}} = 0.211472 \quad (10)$$

$$\beta = Y_{C2+}^{(1.064+0.00686M_{C2+})} = 91.65134 \quad (11)$$

where T – temperature (100 °C); C_1 – mole fraction of methane; $X_{C_2-C_6}$ – sum of $C_2 - C_5$ in the oil phase, %; X_{C_1} – mole percent of methane in the oil, %; $Y_{C_{2+}}$ – mole percent of C_2 in the injected gas, %; MC2+ – molecular weight of C_2+ in the injected gas.

All calculations are summarized in Table 4. It should be noted that the difference in MMP determination with different equations is smaller than 10%. Also, all the above calculated MMP values are smaller than the initial reservoir pressure that equals 32.7 MPa.

Table 4. MMP summary

Equation	MMP, MPa
Firoozabadi and Aziz	29.8
Hudgins-Liave-Chung	27
Glaso for condensing-vaporizing drive	32
Maklavani correlation	30.6

The pseudo ternary plot obtained with PVTi simulation shows that miscibility will occur at reservoir conditions (figure 2).

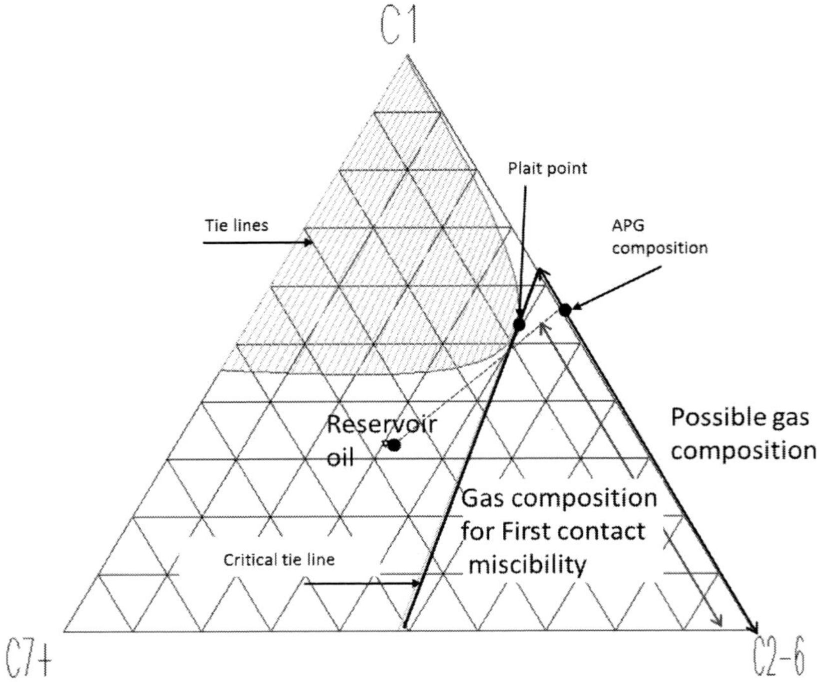

Figure 2. Pseudo ternary plot for reservoir oil at reservoir conditions (P = 32.7 MPa, T = 100°C)

It should be noted that injected gas for miscibility displacement at reservoir conditions should not contain more that 62% of the C_1 fraction.

5. Displacement efficiency coefficient

The displacement efficiency coefficient for the B6 layers of the field was obtained from the core data and equals 0.407. Stepanova G.S. provides an equation for the analytical calculation of the displacement efficiency coefficient for miscible gas injection and WAG injection with the maximum inaccuracy being equal to 9%. On the basis of her study [12], analytical calculations were provided.

Firstly, it is necessary to estimate a viscosity ratio between the reservoir fluid and gas in the transition zone.

$$\mu_{0\,min} = 0.1017 \cdot 10^{-3}F^4 - 0.02417F^3 + 2.1614F^2 - 86.887F - 6.147 \cdot 10^{-3}p^2 - 23.945p - 18.044\,{}^{p_b}/_{p} + 1646.39 + C \; = \; -43.0424 = 1 \quad (12)$$

with

$$C = -0.1243M^2 + 1.454M + 19.6 = \; -45.4145 \quad (13)$$

where F – percent of oil volume recovery till 300 °C, (65 %); M – MW of gas; P – reservoir pressure (32.7 MPa); Pb – bubble point pressure.

If $\mu_{0\,min}$ is lower than zero, it should be equated, because it characterizes the condition of full miscibility. The next step is determining the displacement efficiency coefficient in case of miscible gas injection:

$$DE_{misc.gas\,inj.} = 1.0192 - 0.2706 \cdot 10^{-2}p - 0.1573 \cdot 10^{-2}t + 0.3243 \cdot 10^{-4}k + 0.5464 \cdot 10^{-2}\mu_{0\,min} - 0.6310 \cdot 10^{-3}p\mu_{0\,min} - 0.2054 \cdot 10^{-6}(p\mu_{0\,min})^2 = 0.759 \quad (14)$$

where P – reservoir pressure (32.7 MPa); t – reservoir temperature (100 °C); k – permeability, (20 mD); $\mu_{0\,min}$ – viscosity ratio between the reservoir fluid and gas in the transition zone.

And the last step is determining the displacement efficiency coefficient in case of miscible WAG.

$$DE_{misc.gas\,inj.} = 1.3367 + 0.8898 \cdot 10^{-2}p - 0.6998 \cdot 10^{-2}t + 0.5286 \cdot 10^{-4}k + 0.8099 \cdot 10^{-2}\mu_{0\,min} - 0.4438 \cdot 10^{-3}p\mu_{0\,min} - 0.1508 \cdot 10^{-6}(p\mu_{0\,min})^2 - 0.4868 \cdot 10^{-1}\theta = 0.874 \quad (15)$$

with θ – factor that characterizes the type of impact and can vary from 1 to 6 (for miscible WAG with APG, it equals 1). Table 5 reflects calculated displacement efficiency coefficients, the sweeping efficiency coefficient for the 7-spot flood pattern (from the input data) and calculated recovery efficiency for the 7-spot pattern.

Table 5. Displacement efficiency

Technology	Displacement efficiency coefficient	Sweeping efficiency coefficient	RF, %
Waterflooding	0.407	0.789	32.1
Miscible gas injection	0.759	0.789	59.9
Miscible WAG	0.874	0.789	68.9

It should be noted that the highest recovery efficiency is achieved for miscible WAG injection and the lowest level is relevant to waterflooding.

6. Development strategy

Layer B6 at Field MTX has the following characteristics:

- Cumulative oil production equals 397*10^3 tons
- Recovery factor – 0.4%
- Current number of appraisal wells – 14
- Current number of exploitation wells – 0
- There are two accepted development strategies: waterflooding with a 7-spot pattern (500 meters between the wells); waterflooding with horizontal wells with multi-stage hydraulic fracturing (6 stages for production wells and 3 stages for injection wells). A fracture with specific parameters allows involving all the 3 layers into development with 1 well (the fracture is somewhat smaller than the interval between the top of the upper layer and the bottom of the lower layer). The length of the horizontal part of the wells equals 1,000 meters.
- Well operation conditions were determined with the following input data: production wells with ESP and BHP = 20 MPa, maximum BHP for injection wells = 50 MPa.

A three-phase black oil model was constructed. This model was converted into the Schlumberger Eclipse – 300 compositional model with the help of the PVT model for the reservoir fluid obtained with the PVTi Schlumberger software. Simulation was carried out for the sector of Field MTX. Different strategies on three different well patterns were assessed (in addition to the patterns adopted by the Oil Company, a 5-spot pattern was considered; it implied similar well density values with the 7-spot pattern). The following strategies were assessed:

- Waterflooding
- Miscible gas injection (restrictions on the injected gas volume with the quantity of APG)
- Miscible gas injection (no restrictions on quantity, the lacking volume of gas will be imported)
- WAG with alternation performed every 6 months
- WAG with alternation performed every year
- WAG with alternation performed every 2 years

The number of wells for each development strategy was as follows:

- The 7-spot pattern includes 32 production wells and 20 injection wells
- The case with horizontal wells includes 8 production wells and 6 injection wells
- The 5-spot pattern includes 30 production wells and 25 injection wells

Waterflooding was the first strategy to be considered (Figure 3).

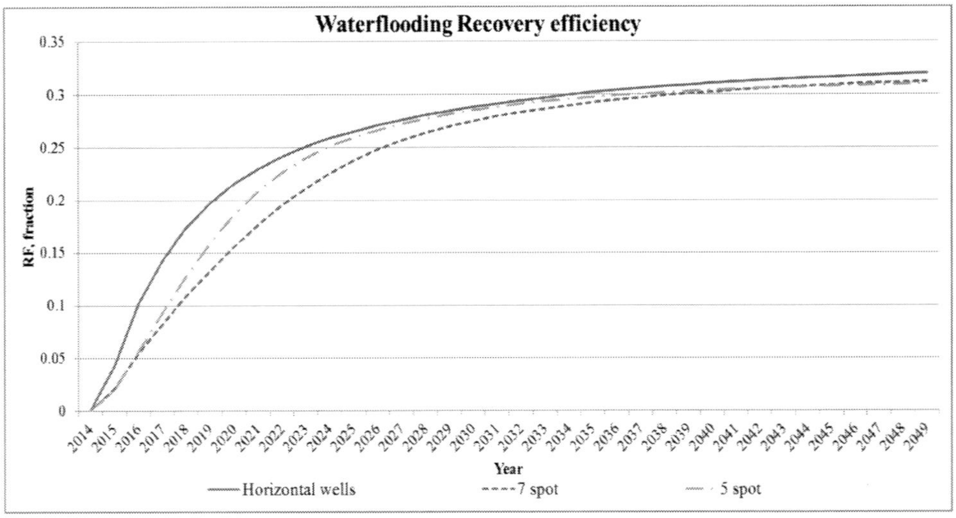

Figure 3. Waterflooding strategy

As one can see from the graph, the highest recovery factor is achieved with horizontal wells and equals 0.32. The intermediate result is achieved with a 7-spot pattern and equals 0.312 and almost the same result is achieved with a 5-spot pattern and equals 0.31. The highest recovery efficiency with horizontal wells was achieved due to the higher drainage area of the horizontal wells and better pressure support. The maximum economic efficiency (NPV) is also relevant to horizontal wells due to the 75% smaller number of wells (Table 6). The 5-spot pattern takes the second place in terms of economic efficiency due to the better pressure support vs. the 7-spot pattern and thus leads to a higher production rate.

The second strategy was reinjection of APG separated from oil (figure 4)

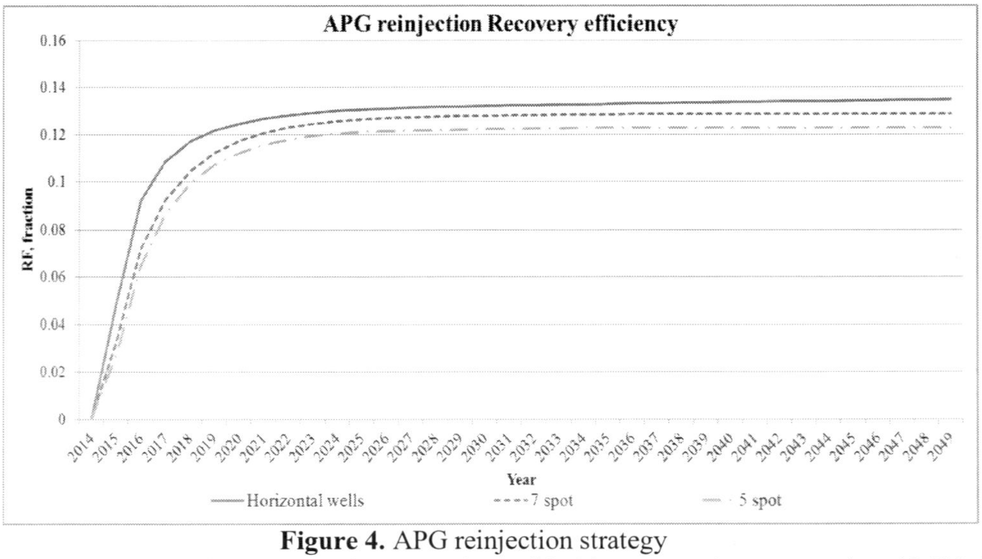

Figure 4. APG reinjection strategy

It should be noted that the recovery factor reached with this strategy is no more than 13.5% due to the poor pressure support, as the amount of associated petroleum gas is too small for good pressure maintenance (for all the patterns, NPV is negative). Thus, the next strategy to be considered was also miscible gas injection (figure 5), but without restrictions on the quantity of gas with the volume of APG separated from oil (the lacking volume of gas will be imported from another field).

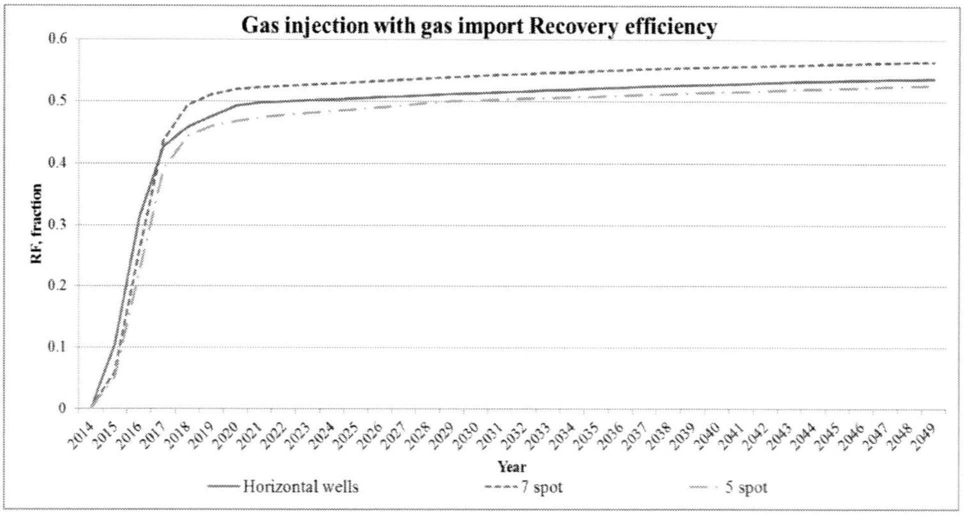

Figure 5. Gas injection (gas import) strategy

This development strategy shows a better result for all types of patterns than the previous one. Also, there are recovery factors being higher than those in case of waterflooding. The maximum recovery factor is achieved for the 7-spot pattern and equals 56.5%. The lowest recovery factor equals 52.7% for the 5-spot pattern. The highest NPV is achieved with horizontal wells due to the small number of wells. The recovery factor in this case equals 53.7%, which is smaller vs. the 7-spot pattern due to a faster gas breakthrough from injectors to producers. The lowest economic efficiency is relevant to the 5-spot pattern due to lower production rates at the early stage of field development (smaller number of producers vs. the 7-spot pattern, lower productivity index and higher quality of producers (higher CAPEX) in comparison with the horizontal wells pattern). Pressure support in all these cases is good and the average reservoir pressure does not falls below 300 bars.

The main benefits of WAG injection are as follows: better sweep control, mobility control over water and gas phases and thus displacement efficiency improvement. There is one more benefit for Field MTX: there is a less acute need for imported gas because Field MTX can be divided into several sectors with different injected agents at the same time (i.e. gas is injected in some sectors while water is injected in others). Besides, one of the main drawbacks is a more complex infrastructure in comparison with the previous strategies. The next stage was the assessment of miscible WAG injection with sensitivity to the time periods of agent injection. The first one was alternating agents every 6 months (figure 6).

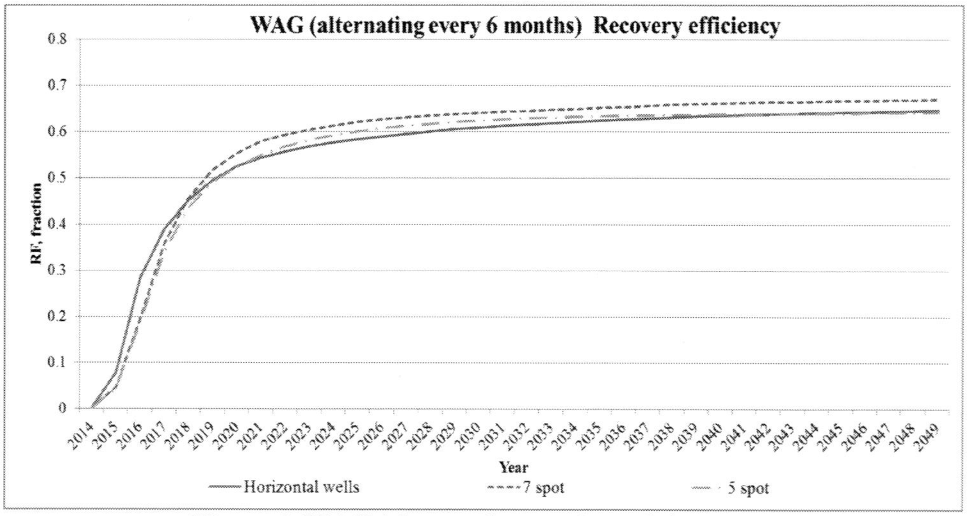

Figure 6. WAG (alternated every 6 months) strategy

The WAG injection strategy shows an increase in recovery efficiency in comparison with the miscible gas injection strategies. The highest recovery is achieved with a 7-spot pattern and equals 67%. The highest NPV is relevant to horizontal wells with a recovery factor being equal to 64.6%. The worst case in terms of NPV and recovery efficiency was the 5-spot pattern. The next strategy was alternating agents every year (figure 7).

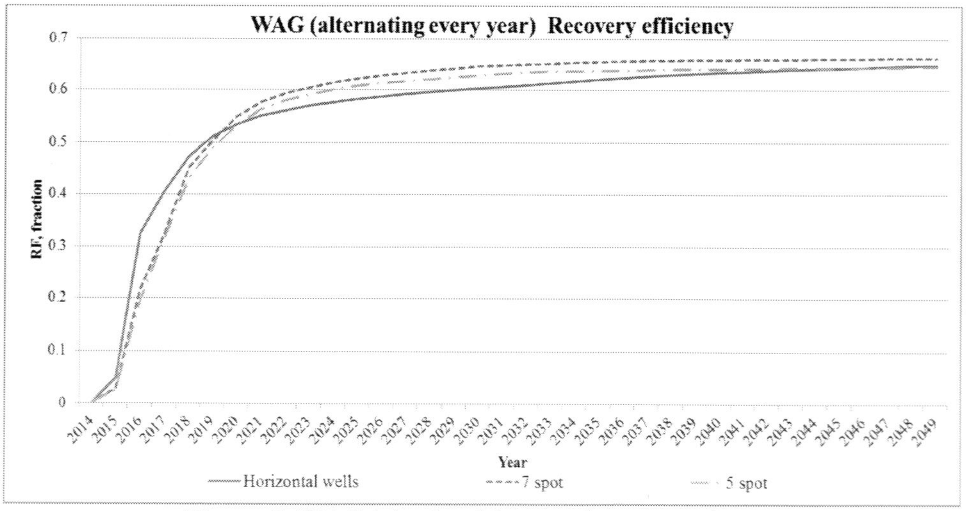

Figure 7. WAG injection (alternated every year) strategy

A trend being similar to the alternation of agents every 6 months can be determined for the WAG injection with alternation performed every year. However, a small decrease in recovery efficiency (less than 1%) is present for all the patterns. The trend persists for WAG injection with alternation to be performed every 2 years, but the decrease in the recovery factor is higher (figure 8) (more than 2% in comparison with alternation occurring every 6 months).

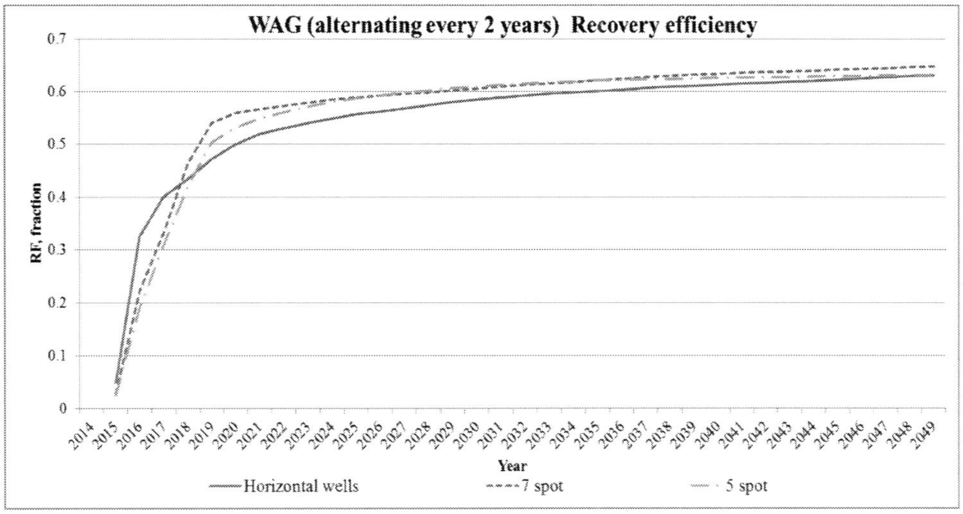

Figure 8. WAG injection (alternated every 2 years) strategy

Simulation and economic data are summarized in Table 6. It can be seen that the highest recovery factor is achieved with a 7-spot pattern and WAG injection with alternation performed every 6 months and equals 67%. The highest NPV and NPVI are achieved with horizontal wells and WAG injection with alternation performed every year and equal 8,461.3 million rubles and 6.7, respectively. The highest IRR is achieved in horizontal wells and miscible gas injection (without restrictions on the quantity of injected gas) and equals 198%.

The pattern with horizontal wells and WAG injection (alternated every year) was recommended for development, as it has the highest NPV (8,461.3 mln rubles) and a high recovery factor (64.8%). The scenario with a 7-spot pattern and WAG injection (alternated every 6 months) has a 2-percent higher recovery factor, but the NPV is smaller by 30%.

Table 6. Development plans summary

Development strategy	Well patters	RF,%	NPV, mln rub	IRR	NPVI
Waterflooding	Horizontal wells	32	1,282.5	0.26	1.3
	7-spot	31.2	-2,521.1	-	-
	5-spot	31	-2,118.1	-	-
Miscible gas injection (reinjection of APG separated from oil)	Horizontal wells	13.5	-512.5	-	-
	7-spot	12.9	-3,729.3	-	-
	5-spot	12.3	-4,040.5	-	-
Gas injection (gas import)	Horizontal wells	53.7	7,851.2	1.9881	6.2
	7-spot	56.5	6,272.4	0.9938	3.2
	5-spot	52.7	5,038.1	0.8141	2.4
WAG (alternated every 6 months)	Horizontal wells	64.6	8,303.4	1.6633	6.5
	7-spot	67	5,952	0.7282	2.6
	5-spot	64.2	5,491	0.6761	2.3
WAG (alternated every year)	Horizontal wells	64.8	8,461.3	1.562	6.7
	7-spot	66.2	6,041.1	0.6689	2.1
	Horizontal wells	64.5	5,533.8	0.6184	1.9

WAG (alternated every 2 years)	7-spot	63.2	7,967.7	1.538	6.3
	5-spot	64.8	6,185.2	0.7078	2.2
	Horizontal wells	63.2	5,458.6	0.6017	1.9

As one can see from Tables 6 and 5, the analytical calculations shows better (but similar to simulation) results in terms of recovery efficiency with the maximum deviation equal to 2% (relative).

The main technical parameters of the recommended development plan are presented in Figures 9-11. Its dynamics shows that there was a slight decrease in reservoir pressure (50 bars) during the early stages and it was subsequently compensated for. The periods of water injection and gas injection are distinguishable in the graphs, and the impact of these periods on the water cut and GOR are distinguishable too. Cumulative oil production equals 4.5 mln tons.

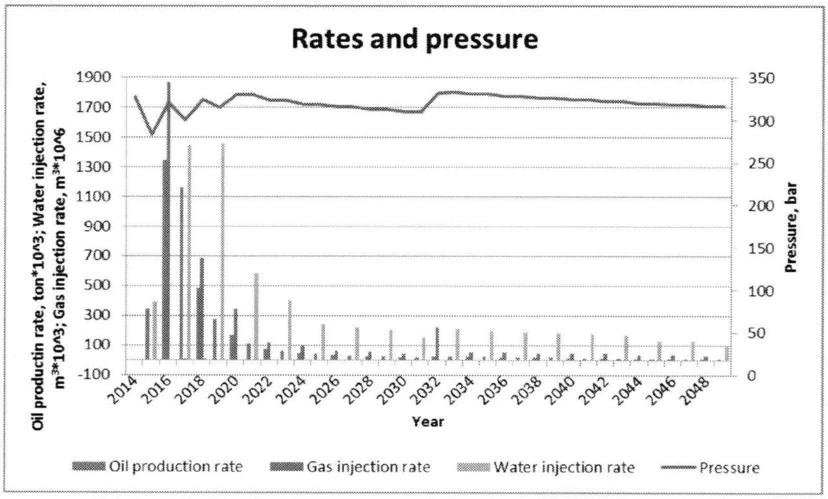

Figure 9. Production rates and average reservoir pressure

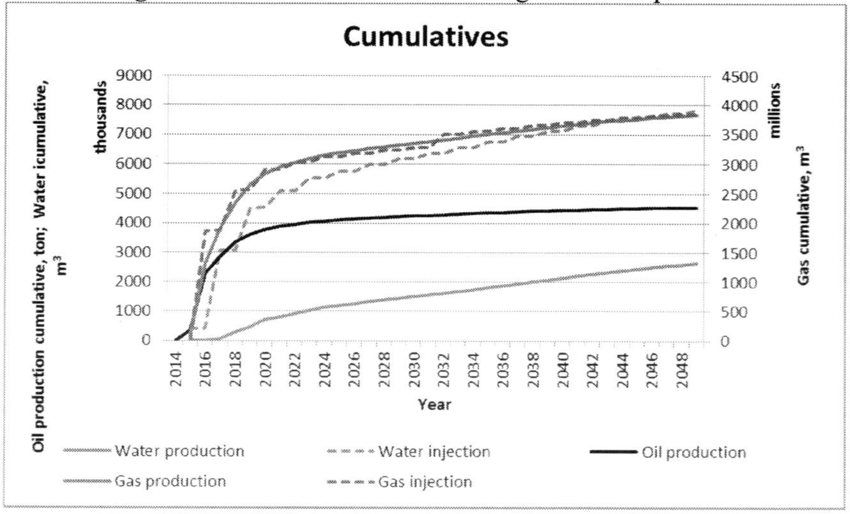

Figure 10. Injection and production cumulative

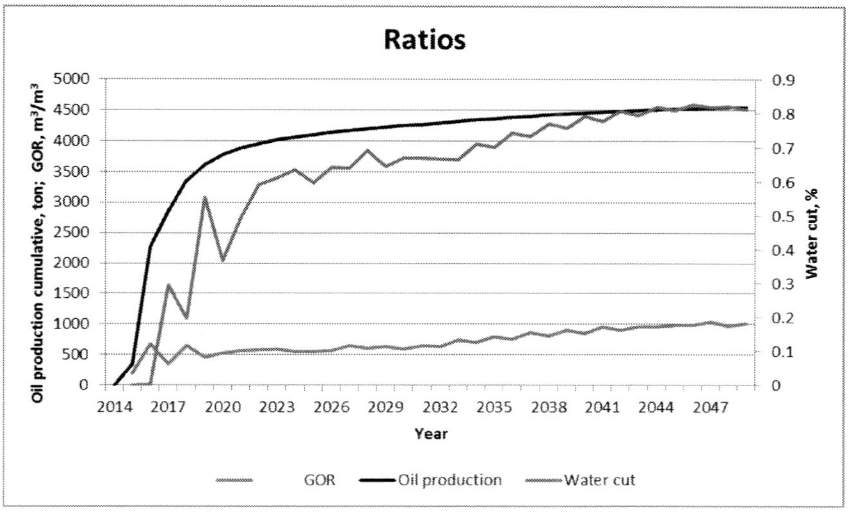

Figure 11. Oil production cumulative, water cut and GOR

7. Economics

Assumptions and input data used for economics calculations:

- Base year: 2014
- Inflation and exchange rate forecast provided by the Ministry of Economic Development and Trade
- Oil price in the domestic market: 12,500 rubles per ton
- Discount factor: 10%
- Unit of production depreciation scheme
- Percentage of oil that can be exported: 30%
- Wells dropping out of production due to the economic limit: water cut being higher than 98%, oil production rate being lower than 1 ton/day, GOR being higher than 1,000 m^3/m^3

The calculation results are presented in figure 12.

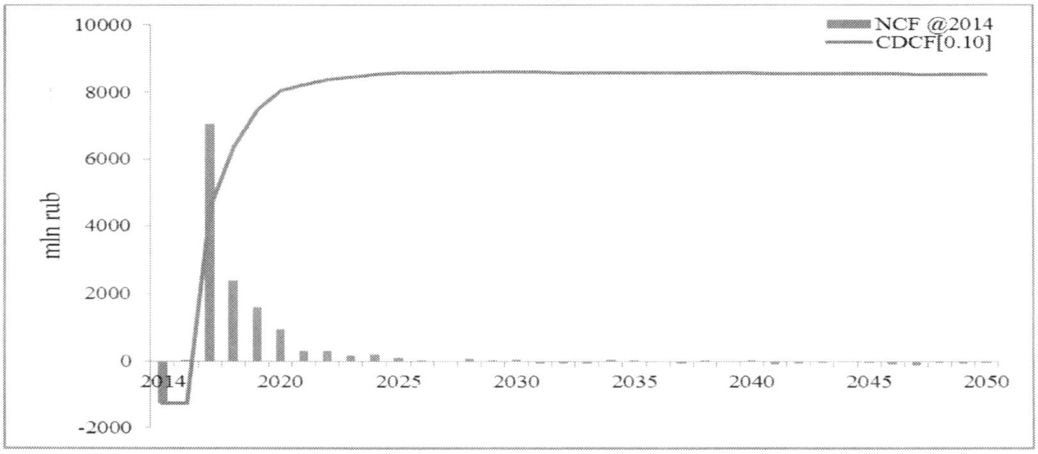

Figure 12. Net cash flow and cumulative discounted cash flow

At the end of development, NPV equals 8,461.3 mln rub, IRR equals 156.3%, NPVI equals 6.74, total CAPEX [@2014] equals 3,001.3 mln rub, total OPEX [@2014] equals 10,166 mln rub, total taxes [@2014] equals 55,182 mln rub.

8. Proxy model

In addition, the computational accuracy and practical applicability of the 2D areal model for Field MTX was assessed. Calculating a 2D model is significantly less time-consuming vs. a full-scale 3D model. The results obtained (Table 7) show that the calculations for the 2D areal model are very inaccurate in comparison with the full-scale 3D model with deviations ranging from 2.9% to 46.2%. This is due to the fact that the 2D areal model does not account for vertical heterogeneity, thus all recovery factors which were obtained with the 2D model have excessively high values. Thus, the 2D areal model cannot be applied to accelerate Field MTX simulation.

Table 7. Comparison of 3D and 2D simulation results

Development strategy	Well patters	RF 3D,%	RF proxy, %	Deviation, %
Waterflooding	Horizontal wells	32	44	37.5
	7-spot	31.2	45.6	46.2
	5-spot	31	44.6	43.9
Miscible gas injection (reinjection of APG separated from oil)	Horizontal wells	13.5	13.9	2.9
	7-spot	12.9	14	8.5
	5-spot	12.3	13.2	7.3
Gas injection (gas import)	Horizontal wells	53.7	69.5	29.4
	7-spot	56.5	71.3	29.2
	5-spot	52.7	75.1	42.5
WAG (alternated every 6 months)	Horizontal wells	64.6	83.4	29.1
	7-spot	67	84.5	26.1
	5-spot	64.2	85.3	32.9
WAG (alternated every year)	Horizontal wells	64.8	83	28.1
	7-spot	66.2	85.5	29.2
	5-spot	64.5	84.6	31.2
WAG (alternated every 2 years)	Horizontal wells	63.2	82.5	30.5
	7-spot	64.8	80	23.5
	5-spot	63.2	85.9	35.9

9. Conclusion

The results of this study show the benefits of miscible displacement and miscible WAG injection over the waterflooding strategy. These methods allow producing more oil from reservoirs than conventional waterflooding. An increase in recovery efficiency can be identified by comparing the waterflooding strategy with miscible gas injection and water alternating miscible gas injection. The highest recovery factor can be achieved with miscible WAG injection (67%), an intermediate result is relevant to the miscible gas injection strategy where the recovery factor equals 56.5% and the worst result was shown by the waterflooding strategy (32%). It was determined that the efficiency of miscible WAG was better than miscible gas displacement. Also, there is an improvement in the economic efficiency of the project with NPV of 1.2 billion rubles for waterflooding up to 8.4 billion rubles for miscible WAG.

The comparison of the analytical and simulation results showed their convergence, moreover, the analytical calculations showed a higher recovery factor vs. simulation by 2%.

A proxy (2D areal) model was also considered. It showed significant differences in the results in comparison with the full -scale 3D model due to the high vertical heterogeneity of Field MTX. Thus, it was established that proxy modeling has no practical applicability in Field MTX simulation.

The sensitivity analysis of miscible WAG displacement to the periods of agent alternation was performed to show that the optimal time of agent injection equals one year (the most cost effective option).

Despite the fact that WAG injection and miscible gas injection are hard to implement and require expensive equipment for pumping gas, the economic calculations show that the additional profit covers the additional costs of capital investments. Thus, an optimal development strategy was chosen. It is the pattern with horizontal wells and miscible WAG injection with agent alteration to be performed every year, because it has the highest NPV (8.4 billion rubles) and a high recovery factor (64.8%).

References

[1] Ali Danesh 2001 PVT and Phase Behavior of Petroleum Reservoir Fluids
[2] Ahmed Tarek H 2007 Equations of State and PVT Analysis: Applications for Improved Reservoir Modeling
[3] Ahmed T 2007 Minimum Miscibility Pressure from EOS PETSOC-2000-001
[4] Srivastava J P et al. 2012 Water-Alternating-Gas (WAG) Injection a Novel EOR Technique for Mature Light Oil Fields – A Laboratory Investigation for GS-5C Sand of Gandhar Field
[5] Van Vark W et al. 2004 Simulation Study of Miscible Gas Injection for Enhanced Oil Recovery in Low Permeable Reservoir SPE – 88717
[6] Latil M 1980 Enhanced Oil Recovery Editions Technip, Paris
[7] Lambeth N and Dawe R A 1988 Boundary and Cross Flow Behavior during Displacement in Nodal Systems SPE No. 16972 *SPE Reservoir Engineering* **3** 1137-42
[8] Yahaya M Al – Wahaibi et al. 2008 Mechanism of Oil Recovery via Multi-Contact Miscible Gas Injection Process within Lensed Systems *SPE* – 110161
[9] Mehran Namani et al. 2013 Advantages of Up-Dip Water-Miscible Gas Injection *SPE* – 164836
[10] Ahmad Aladasani 2010 Recent Developments and Updated Screening Criteria of Enhanced Oil Recovery Techniques *SPE* - 130726
[11] Tarek Ahmed 2007 *Equations of State and PVT Analysis* Gulf Publishing Company, Houston, Texas
[12] Stepanova G S 1991 The estimation of Recovery Efficiency with Different Methods of Gas and WAG Processes *Oil industry* **7**
[13] Maklavani A et al. 2010 New Minimum Miscibility Pressure Correlation for Hydrocarbon Miscible Injections *Brazilian journal of petroleum and gas*

IOP Conf. Series: Earth and Environmental Science **24** (2015) 012046 doi:10.1088/1755-1315/24/1/012046

Geological reasons causing rapid rate decline. A case study of Field T, Russia.

S Kuzmin[1,2] V Rukavishnikov[1] and V Sukovatyi[2]

[1]Tomsk Polytechnic University, Tomsk, Russia
[2]Gazpromneft-Orenburg, Orenburg, Russia

E-mail: rukavishnikovvs@hw.tpu.ru

Abstract. The paper introduces the complex approach to determining possible geological reasons causing rate decline in the wells of Field T, Russia. Therefore, possible geological reasons are sequentially considered and were divided into three main groups: 1) rate decline due to poorer reservoir quality; 2) rate decline due to facies lateral substitution; 3) rate decline due to active fault tectonics. The most appropriate facies models were constructed on the basis of all available data. Besides, in this study, core, well logging, seismic and well test data were integrated for the fullest reservoir characterization. The core from several recently drilled wells was described in detail to determine clue features. Further, seismic data were interpreted: structural interpretation, including faults and attribute analysis, was implemented. Appropriate electrofacies models were chosen as well. At the final stage, all previously-mentioned data were integrated with appropriate facies model construction. However, as it turned out later, the facies model was not the key factor affecting the rapid decline that appeared in some wells. It is suggested that very proximal faults can be a possible explanation. To confirm this suggestion, well test data were additionally used and both analytical and numerical methods were applied to show the consistency of this theory.

1. Introduction

Field T is at the end of exploration and at the beginning of production drilling phases. Several wells (13, 141, 142, 143) have recently been drilled in the western part of the field according to the appraisal drilling program. No special risks were expected in this part of the field. However, the wells experienced a liquid rate decline soon after the production start. Production in well 141 fell from 120 m^3/day to 60 m^3/day within several months (figure 1). In well 142, the rapid rate decline from 150 to 25 m^3/day occurred within just 3 days. The situation in well 143 is very similar to that in well 142. All wells due to the poor reservoir quality were hydraulically fractured before or soon after the production start. The geometry of fractures in the wells under consideration is quite similar and all technological risks are believed to be significantly reduced. Therefore, the true reason for the decline is still unknown and possible geological reasons affecting well production should be analyzed. The correct determination of the reasons is of great applied significance, since the result will have a direct influence on the further drilling program for Field T.

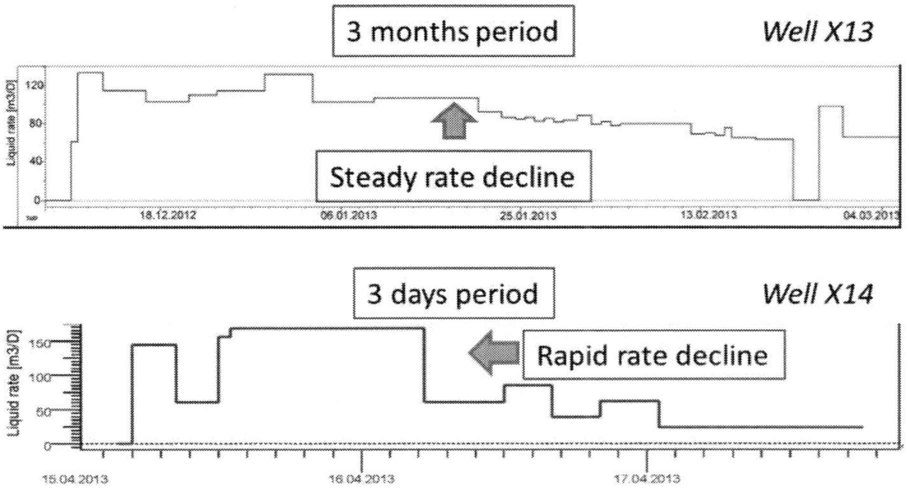

Figure 1. Liquid rate history of wells 141 and 142

2. General information

Field T is located in the western part of Orenburg Region and is known as the largest discovery made in this area in recent years (2005). Thus, a 3D seismic survey was conducted and 18 wells were drilled within the license area (figure 2).

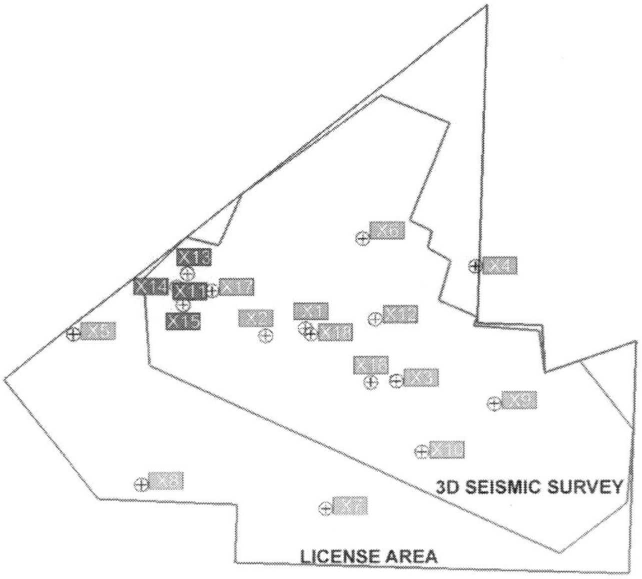

Figure 2. Exploration maturity map

The terrigenous Kolgan formation of the Late Frasnian age is the principal reservoir containing ~85% of the overall STOIIP. There are three main units within the Kolgan formation: 1) upper terrigenous unit; 2) intermediate argillaceous-calcareous unit; and 3) lower terrigenous reservoir unit (figure 3). The lower terrigenous unit consists of a relatively thick (20-30 m) sandstone bed sometimes capped by shale. This sandstone bed is called Dkt-reservoir/unit. It is composed of two principal sandstone types interbedding each other: 1) fine-grained tight gray sandstone with frequent shale laminae without oil shows; 2) massive-to-slightly cross-bedded fine-grained brown sandstone with oil stains. It is the unit for which a facies model was constructed.

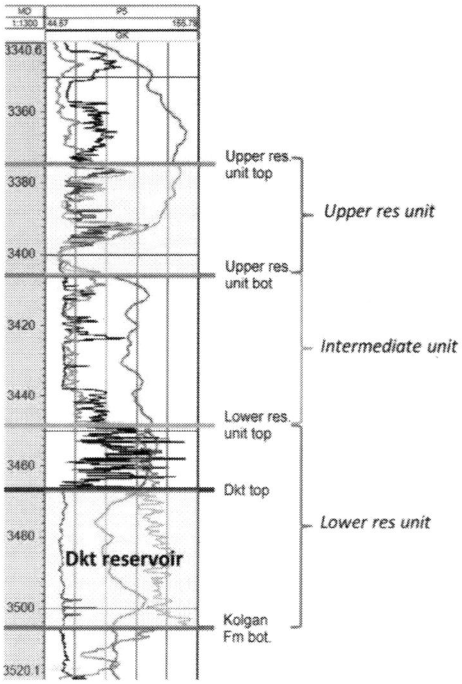

Figure 3. Well 1 section

During core description, several characteristic features were marked and they are the clues to depositional environment determination. In well 12, part of the hummocky cross stratification was recognized to be the indicator of a lower shoreface (figure 4) [9]. In well 8, several indicators such as glauconite, flaser-to-wavy-structure (figure 5) and bidirectional structure were marked out all at once. Glauconite is intrinsic to shallow marine, whereas flaser-to-wavy and bidirectional structures are indicators of tidal activity [6]. During the core analysis, two ichnofacies were recognized: Skolithos (figure 6) and Cruziana (Teichichnus) (figure 7). Bioturbation by Skolithos linearis tubes is typical for the shoreface or tidal flat facies. The latter, Cruziana ichnofacies, in general, are intrinsic to a deeper, sublittoral zone and tidal flat as well [5].

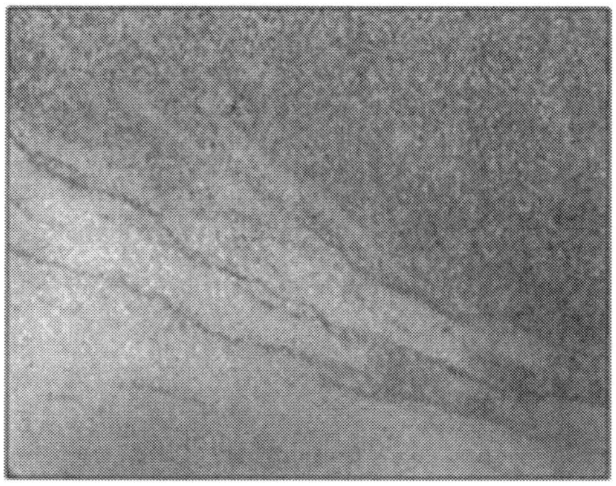

Figure 4. Part of HCS

Figure 5. Flaser-to-wavy structure

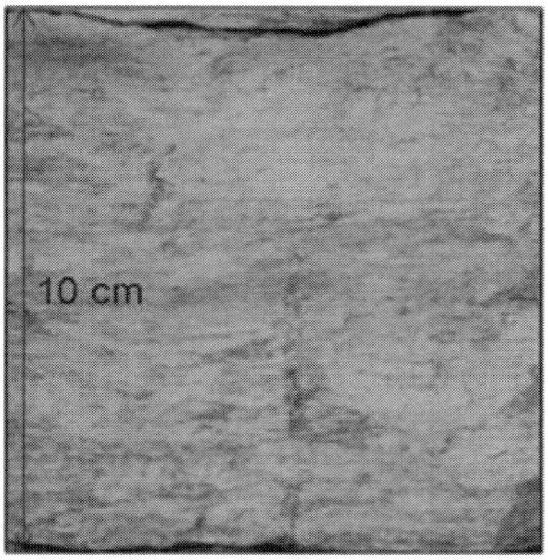

Figure 6. Skolithos ichnofacies

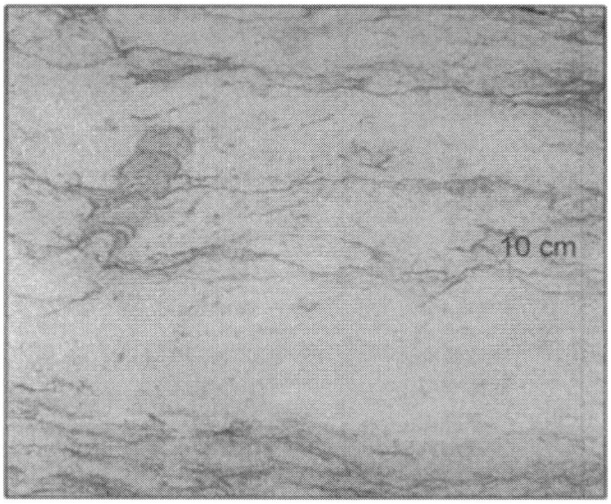

Figure 7. Cruziana ichnofacies

A correlation profile in the SW-NE direction was constructed and it became clear that the Dkt reservoir is laterally extensive and shales out both in the north and in the south. In the north, a pinch-out line is suggested somewhere between wells 14 and 4. Since the Kolgan formation is abnormally radioactive, GR was used to separate it from under/overlying low radioactive carbonate rocks to mark out three distinct units within it. For the same reason, GR logs cannot be used to estimate the volume of shales. Thus, only an SP log was available for Vshale determination, which, however, has a lower resolution and gives less reliable results. At the next stage, well logs were used to estimate porosity in the Dkt-unit. There are three distinct well logging groups allowing porosity estimation: 1) neutron log; 2) density log; 3) acoustic log. It turned out that the acoustic log is matched better with core data than the others. The difference in poroperm properties was considered to be one of the possible reasons that influence the rate decline in the above-mentioned wells. Core plugs for measurements were sampled from wells 1, 2, 4, 11, 14, 222. Permeability for all these wells was estimated using the 'core porosity – core permeability' relationship (R2=0.65). Thus, there is a single poroperm relationship for all wells, and wells with similar porosity profiles have similar permeability profiles as well. This is the case for the 'problem' wells (figure 8). It is obvious that they are quite similar. Therefore, the rate decline in certain wells cannot be explained by the difference in reservoir properties.

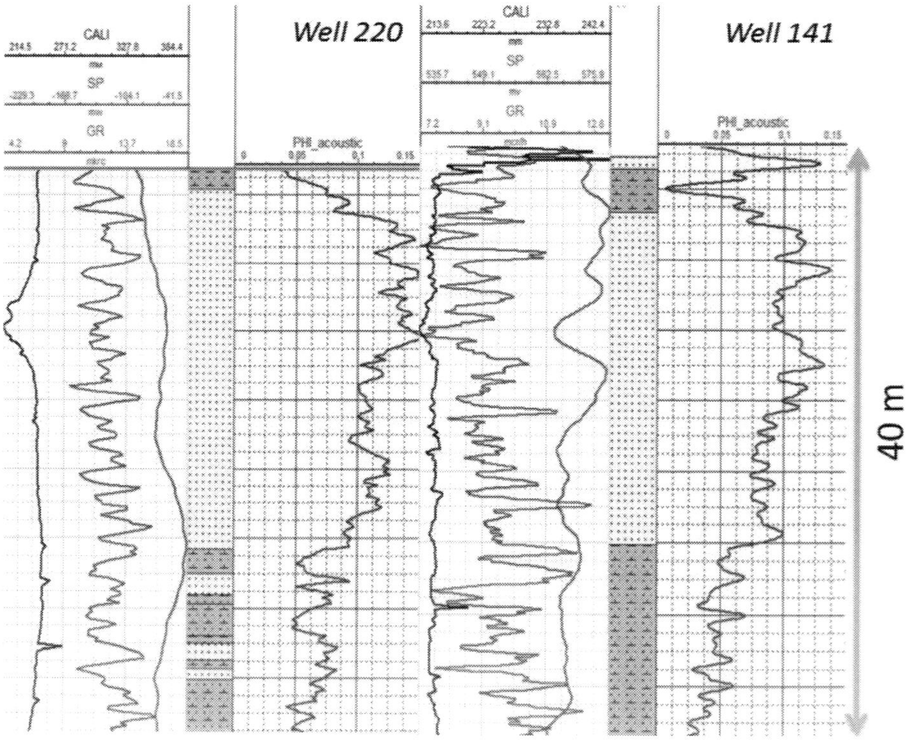

Figure 8. Comparison of porosity profiles in wells 220 and 141

At the next stage, it was necessary to link lithotypes recognized in the core to petrophysical data. Porosity was chosen as a petrophysical property based on which the division into classes could be implemented. Core plugs were taken only from lithotypes 1 and 2, thus, it was meaningful to distinguish core plug data only in two classes (figure 9). The cut-off porosity value is 7.3% with which lithotypes distribution in the wells was implemented and QC'ed on the real core data (figure 10).

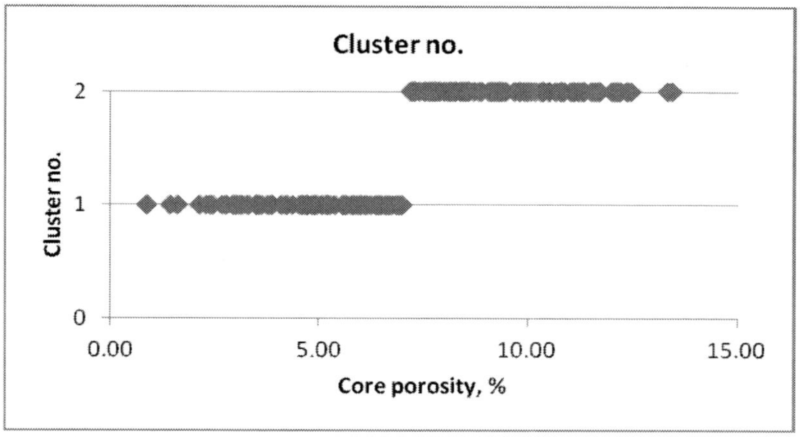

Figure 9. Resulting division of core plug porosity data into two classes

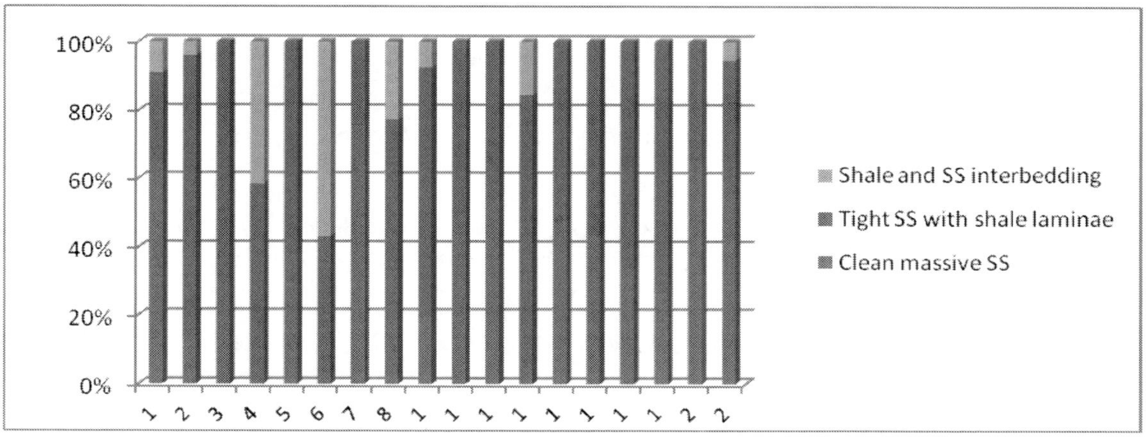

Figure 10. The resulting lithotypes distribution in wells

The electrofacies analysis is the most common sort of facies analysis due to the limited volume of core material. SP logs from 18 wells were divided into three distinct classes: 1) a barrier bar/island and associates with the central part of the suggested sand body [8]; 2) a dissected type curve indicating frequent interbedding of sandstones and shales and may correspond to a transition zone confined to the peripheral parts of the sand body [2]; 3) the 3rd type indicates the total absence of sand bodies within the cross-section and occurs in northern wells 4 and 6. This type of curve is typical for offshore environments. The selected electrofacies curves suggest a prograding barrier bar/island system, so a typical barrier island log curve was matched with a standard sedimentological model and real core observed in wells (figure 11) [10].

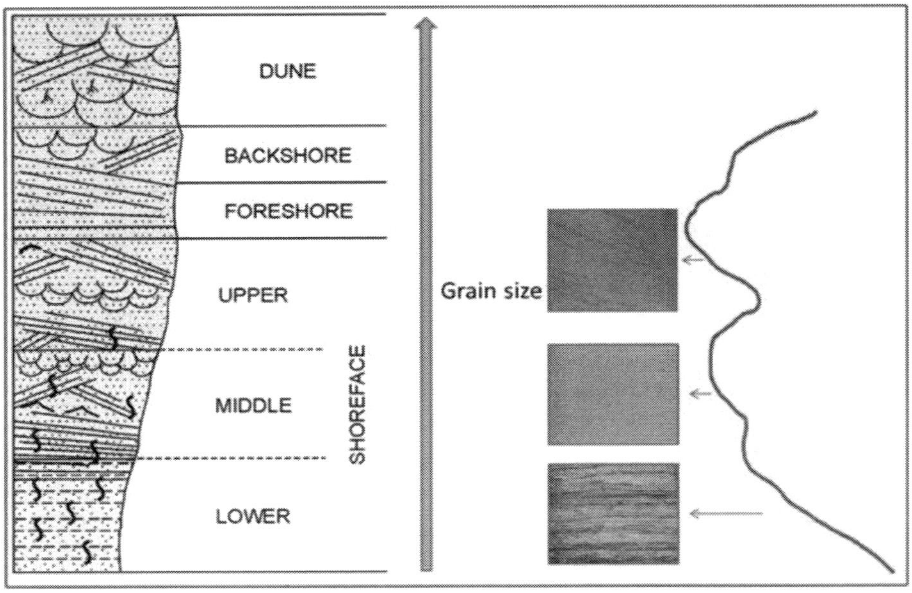

Figure 11. SP log, sedimentary log and core data matching [10]

3. Seismic interpretation and attribute analysis

The area under study has recently been covered with a 3D seismic survey. Since the target Dkt-reservoir is within the lower terrigenous unit, only two reflectors were interpreted: 1) Dk1 – the lower terrigenous unit bottom; 2) Dk2 – the lower terrigenous unit top. It is worth noting that faults were interpreted as well and according to the seismic interpretation, wells 13/141/142/143 are located within an extensively faulted area. In the next step, an attribute analysis was performed. There were two principal problems to solve with this tool: 1) an attempt to recognize sand body boundaries, pinch-out line tracing in the north; 2) calculation of average petrophysical properties (e.g. porosity) using the 'attribute-property' correlation in wells. A set of different volume attributes was calculated (Relative acoustic impedance, instantaneous phase, etc.) and surface attributes using different algorithms (median, most of, etc.) were extracted from 3D attribute cubes. The analysis showed that the 'RMS Amplitude' and 'Relative acoustic impedance' attributes are the most useful ones and at least two different regions are clearly outlined 1) in the north/north-west; and 2) elongated in the west-eastern direction in the southern part of the area. It may be suggested that the boundary between these two regions is the above-mentioned pinch-out line of the Dkt-reservoir that was previously set somewhere between wells 14 and 4 that can now be easily traced. Finally, average porosity was calculated with the obtained porosity logs in wells and distributed within the investigated area with the multiple regression algorithm (figure 12). An average porosity map showed that the upper region is the region with quite low porosities – this is confirmed by dry wells 4 and 6. Thus, this is the area of reservoir absence.

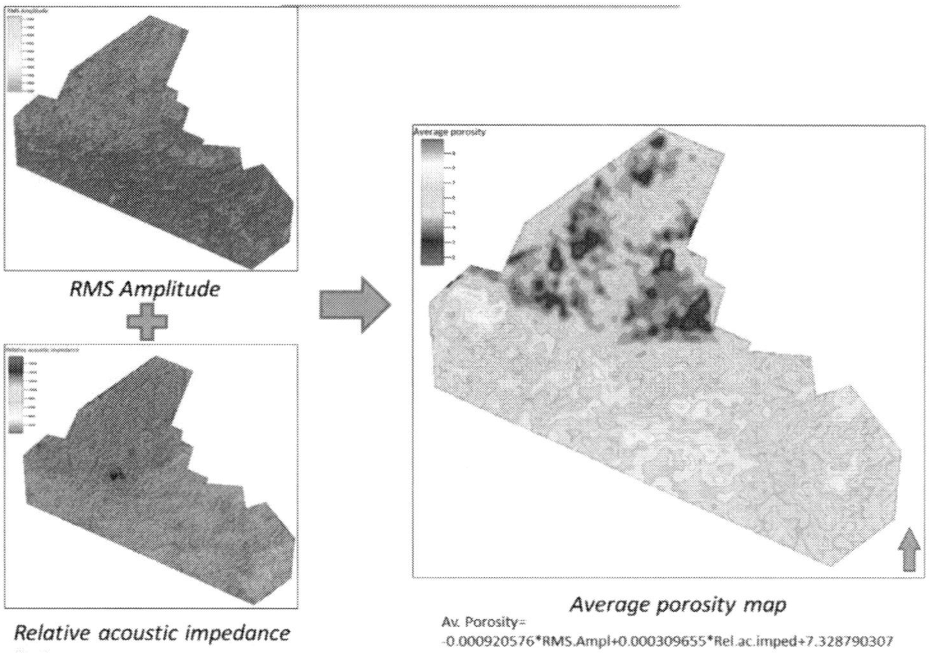

Figure 12. Average porosity map derived from attribute maps

4. Facies model construction

The next logical step after the separate analysis of seismic, well logging and core data was their integration and design of a depositional environment concept being compliant with the whole data set. There is a list of 11 reasons derived from the previously analyzed data and that are in favor of the barrier bar/island depositional environment: 1) elongated shape of the sand body; 2) parallel to the supposed shoreline (regional geology); 3) appropriate size: 15,000x4,000x25 m [3]; 4) well-sorted quartzose sandstone; 5) coarsening upwards sequence; 6) shallow marine ichnofacies; 7) presence of glauconite [1]; 8) flaser-to-lenticular structure (tidal flat); 9) hummocky cross-stratification (HCS); 10) 70-90% of sandstone in the cross-section [4]; 11) appropriate electrofacies model. Thus, with the data available, it is logical to suggest a barrier bar/island depositional environment. Offshore, lower shoreface, upper shoreface/beach and tidal flat facies were recognized. In the lower part of all wells, a regression cycle is obvious, where offshore facies are consistently replaced by shoreface and beach facies. In the upper part, however, a transgression cycle is evident. The final part of depositional environment determination is the spatial facies distribution. Here, attribute and average porosity maps were taken as the basis. By integrating all data available, one can state that the 1st (central) zone of the sand body predominantly consists of middle/upper shoreface and beach facies, the 2nd (peripheral) zone contains lower shoreface/transition zone facies and finally, the 3rd zone is composed of offshore deposits (figure 13). Two 'undefined' zones were marked out. The reason for this is the ambiguousness in tracing boundaries between zones 2 and 3 using 3D raw seismic data and attribute cubes. The obtained facies model is confirmed by oil rates in the wells (figure 14). In this figure, an average porosity map derived earlier from the seismic attribute analysis is used as the background. Three distinct areas are recognized: 1) zone 1 where rates are relatively large, there are no lithotypes 3 or 4 in the wells and the average porosity value is higher than elsewhere (~8-9%). This is the central part of the supposed sand body; 2) transition zone where oil rates are significantly lower, lithotype 2 dominates and a remarkable amount of lithotypes 3 and 4 appears in the wells. This zone is peripheral and rims the central zone of the sand body from the north and the east. Average porosity is ~5-6%; and 3) offshore zone in the north that includes dry wells 4 and 6 and where average porosity is extremely low (2-4%). According to the obtained facies model and corresponding well logs, wells 13, 141, 142 and 143 penetrate the same barrier bar/island deposits with similar poro/perm properties and oil-bearing thicknesses. Moreover, according to the admitted facies model, wells 142 and 143 are located in the more central part of the barrier island and thus, should have less clayey rocks, which is confirmed by well logs interpretation. Despite the fact all wells (13/141/142/143) penetrate very similar deposits, well 141 sustained a steady decline from ~120 m^3/d liquid rate to ~50-60 m^3/d within 3 months after the production start. At the same time, well 142 suffered a rapid liquid rate decline right after 2 days of production: from 150m^3/d to 24 m^3/d. The situation with well 143 is similar to that of well 142. Well 13 is also experiencing a steady rate decline. Thus, it can be concluded that facies distribution is not the key factor leading to the rate decline in wells 13, 141, 142 and 143 and other geological features should be considered.

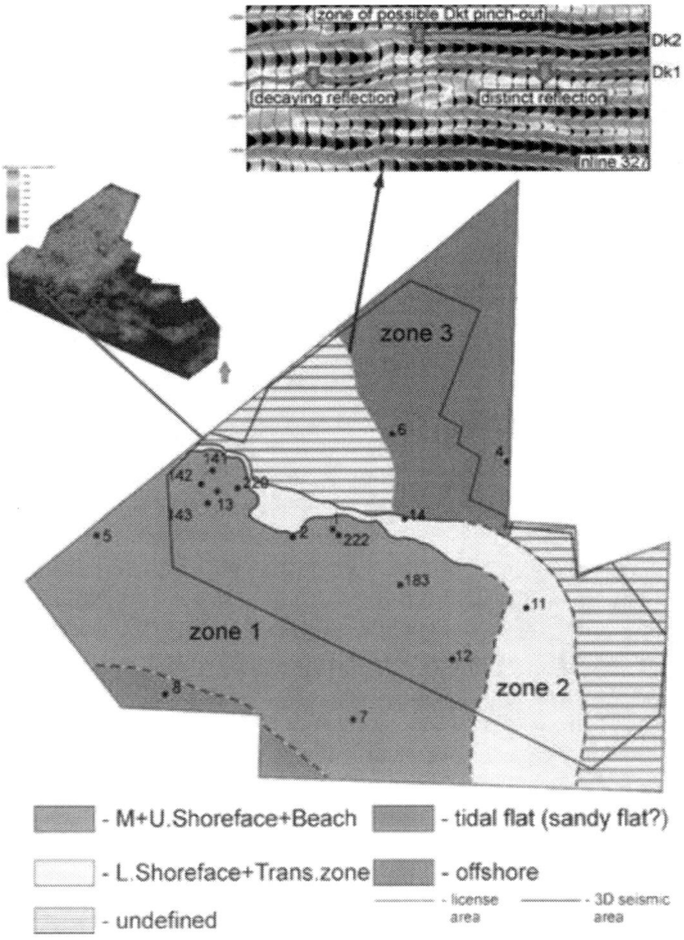

Figure 13. Field T facies model

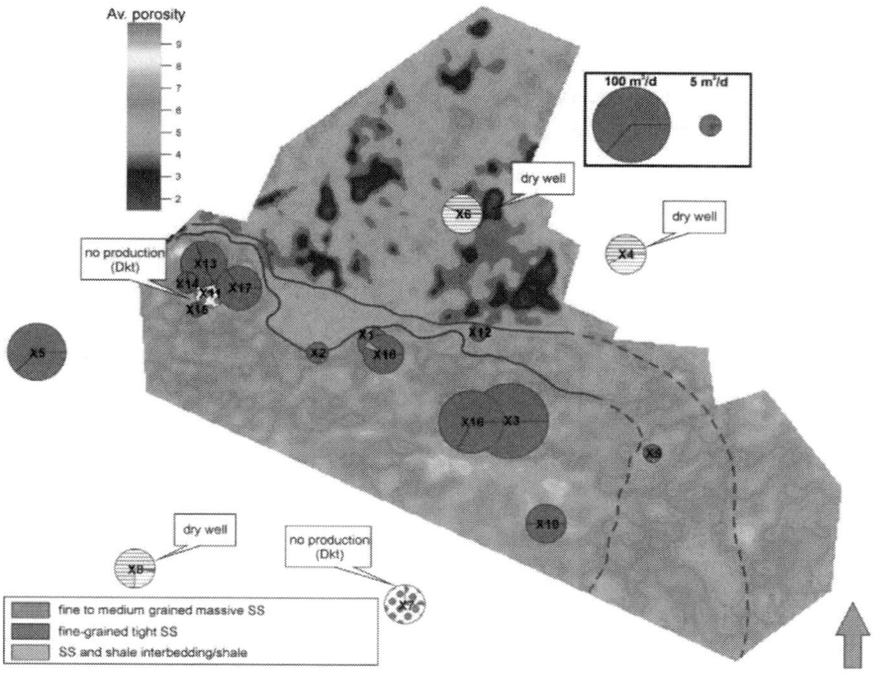

Figure 14. Average porosity map with oil rates, lithotypes distribution in wells and three different zones

5. Fault analysis

Unfortunately, the facies model was unable to fully explain the anomalous behavior of wells 142 and 143. Thus, there is a question: what other geological reasons may cause this effect? As it has already been mentioned, structural seismic interpretation including faults interpretation was carefully conducted in the area of interest. It turned out that wells 13, 141, 142 and 143 are located in an extensively faulted region. One of the characteristic features of the faults is their proximity to the wells (figure 15). The faults drawn with a solid line are confidently recognized with a seismic survey, whereas the dashed-line faults are not so clear and require additional confirmation by other methods. As an additional justifying method, a well test was applied. In the absence of tests with sufficient duration, drawdown (DD) tests from wells 141 and 13 were analyzed. In both tests, the half slope line in the beginning (~100 hours) evidently indicates a fractured well [7]. Then, there is a short-lived derivative plateau (DP) from which average permeability and skin-factor were estimated – approximately 4 mD permeability in both wells and the integral skin-factor being about -5. In well 141, the deviation from DP occurs at the end of the test and was interpreted as a 'double' derivative plateau caused by the suggested single fault in the north of the well. In well 13, approximately in 250 hours from the beginning of the test, the strong influence of boundaries and neighboring working wells becomes apparent because the unit slope on the log-log plot is quite clear (figure 16).

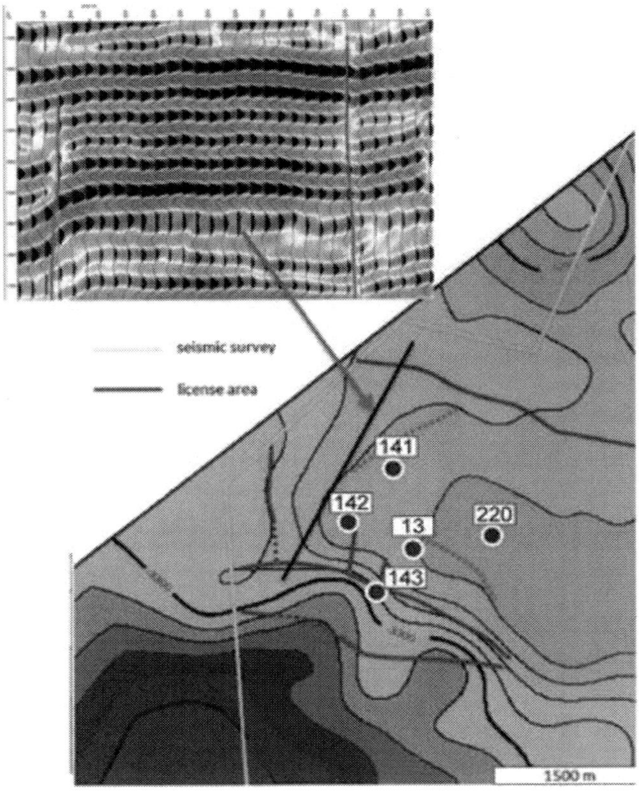

Figure 15. Fragment of the Dk1 structural map with faults

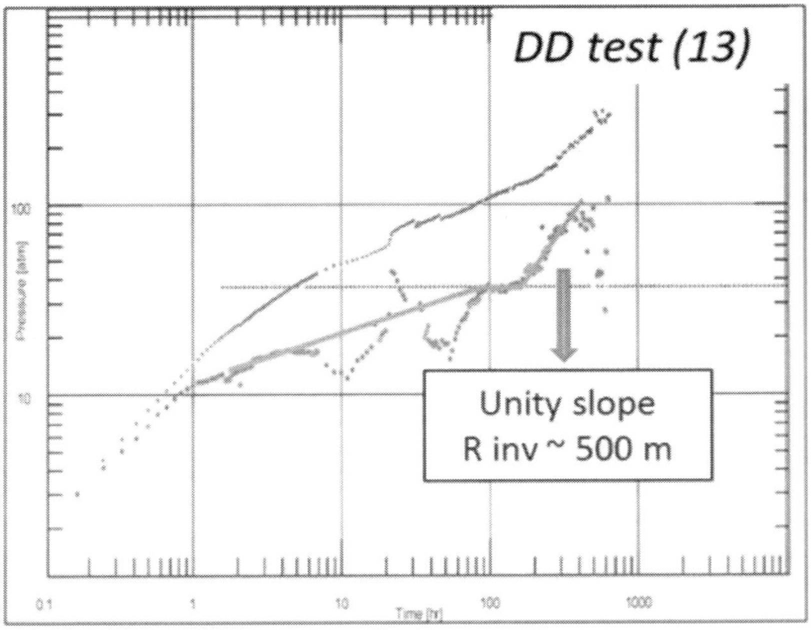

Figure 16. Log-log plot with interpretation (well 13)

6. Numerical model

The previous well test interpretation is analytical and it does not take into account various spatial distributions of faults, their shape, fracture geometry and its orientation, the influence of neighboring working wells, etc. Thus, during the next step, a numerical model was constructed to confirm that the interpreted faults could cause such a drastic rate/pressure decline. Three maps were used as the input data to the numerical model: 1) an average porosity map in reservoir rocks; 2) an effective thickness map; 3) an average permeability map in reservoir rocks (figure 17). Liquid rates, pressure history from neighboring wells, fracture geometry and its orientation were loaded as well. The main purpose of the numerical model construction is the justification of such a rate/pressure decline, demonstration of its possibility in general. The model was QC'ed in several ways: 1) it was matched with wells 141/142 pressure history; 2) Pi control in all wells (the initial pressures in wells calculated numerically should not differ from the real initial pressures) (figures 18, 19). The real initial pressures and those calculated numerically are shown in figure 20. The final match is adequate and the absolute error between the real and calculated initial pressures is not higher than 5 atm. In the numerical model, to match the pressure data, faults were assumed to be sealing.

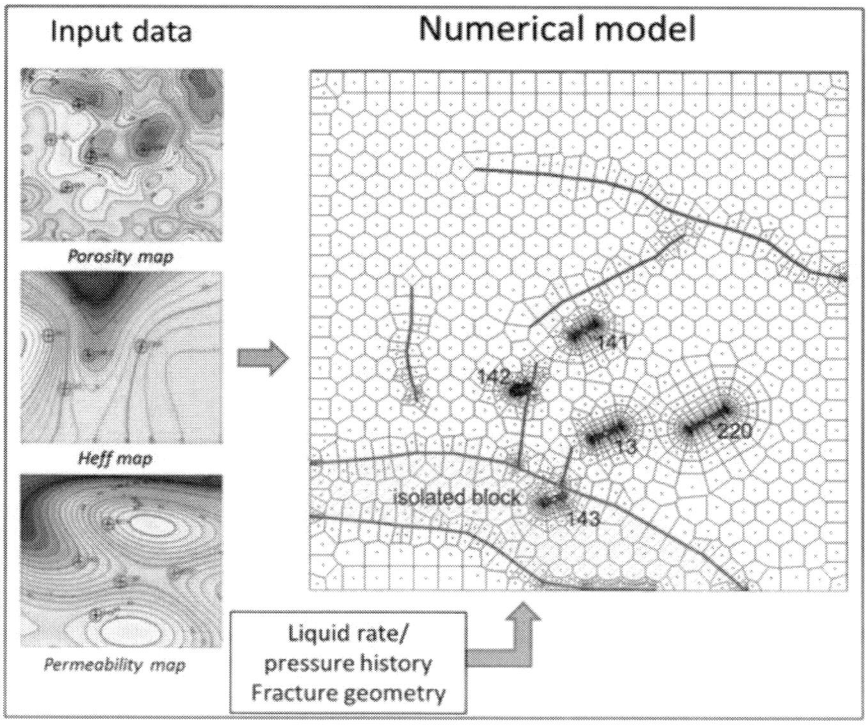

Figure 17. Input data and the resulting numerical model

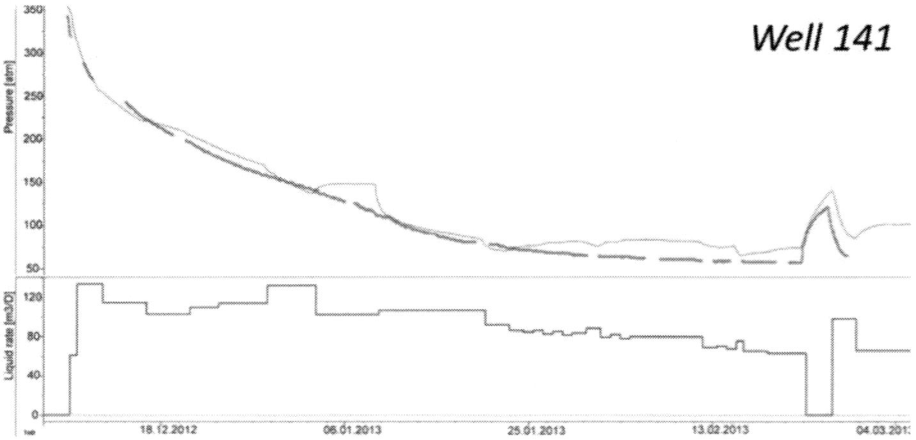

Figure 18. Pressure history matching, well 141

Figure 19. Pressure history matching, well 142

Well no.	Date	Pi [atm]	Pi (model)	Error [atm]
220	06.09.11	387	387	**0**
141	08.12.12	342	346	**4**
142	15.04.13	378	373	**5**
143	20.04.13	370	-	-
13	21.06.13	340	335	**5**

Figure 20. Observed and calculated initial pressures

12. Conclusions

In this paper, a complex approach using all data available to determine possible geological reasons for a liquid rate decline in wells was demonstrated. Possible geological reasons were determined: 1) poorer reservoir quality; 2) facies lateral substitution; 3) fault tectonics. Core, seismic, well logging data were sequentially analyzed and finally integrated and a conceptual barrier bar/island facies model was chosen as the most reasonable one. However, as it turned out later, it was unable to fully explain the problems, such as the rapid liquid rate/pressure decline that appeared in wells 13, 141, 142, 143, and the application of additional data was required. Fault tectonics was considered a possible explanation for such a rapid rate decline. To confirm this suggestion, well test data was additionally used and both analytical and numerical methods were applied to show the consistency of this theory. Therefore, it is important to consider an integrated approach in analyzing the possible geological reasons for problems in the field, since it provides the fullest reservoir characterization and determination of the key factors causing the challenge.

References

[1] Alekseev V P 2003 Lithofacies Analysis: Teaching-Methodical Instruction of Sediments and Self-Learning in Lithology (Ekaterinburg: the Ural Academy of Mining & Geology) 147
[2] Belozerov V B 2011 The Role of Sedimentological Models in Electrofacies Analysis of Terrigenous Deposits *Proc. of Tomsk Polytechnic University* **319** #1.
[3] Chernova O S *Reservoir sedimentology* Tomsk 300
[4] Davies D K and Berg R R 2010 Sedimentary Characteristics of Muddy Barrier-Bar Reservoir and Lagoonal Trap at Bell Creek Field *Montana Geological Society: 20th Annual Conference*
[5] Frey R W and Seilacher A 1980 Uniformity in Marine Invertebrate *Lethaia* **13** 183-207
[6] Galloway W E and Hobday D K 1983 Terrigenous Clastic Depositional Systems. Applications to Petroleum, Coal and Uranium Exploration (Springer-Verlag) 423
[7] Mangazeev P V et al. 2004 Well Test Analysis Tomsk 340
[8] Muromtsev V S 1984 Electrometrical Geology of Sand Bodies in Lithological Oil and Gas Traps (Moscow: Nedra) 259
[9] Reineck H E and Singh I B Depositional Sedimentary Environments 2. M. edition Springer-Verlag Berlin 551
[10] Reinson G E 1984 Barrier-Island and Associated Strand-Plain Systems *Facies Models Geoprint* Canada Reprint Series ed R.G. Walker **1** 119-40

IOP Conf. Series: Earth and Environmental Science **24** (2015) 012047 doi:10.1088/1755-1315/24/1/012047

Lithology evaluation of a Bazhenov formation reservoir using seismic and well log analysis

N Y Lomakina[1], V B Belozerov[2] and N V Bukhanov[3]

[1,2,3]Tomsk Polytechnic University, Tomsk, Russia

E-mail: [1]Lomakinany@hw.tpu.ru

Abstract. The Bazhenov formation is the analog of the Bakken formation; these formations are source rocks and contain tight reservoirs of a tectonic-hydrothermal model. The interpretation of well logs begins with the identification of the Bazhenov marker and tracing a carbonate layer within the Bazhenov formation by GR, neutron and DT logs. The thickness of this layer varies from 1 to 2 m. The next step is seismic interpretation. The creation of 2D seismic forward models is presented in this paper. In these models, the behavior of the amplitude with and without a carbonate layer was studied. These models allow tracing the sign of a carbonate layer (negative amplitude peak). The analysis of 3D seismic cube slices was performed. An amplitude map was created. The distribution of carbonate rock was detected according to this map. Thus, using this approach a thin layer was determined within the Bazhenov formation.

1. Introduction

Attention to the problem of unconventional hydrocarbon resources development has been constantly growing in the past few years. Bazhenov formation development problems are becoming essential. The main goal of this project is the evaluation of the Bazhenov formation reservoir and study of very thin reservoir rocks using wireline logs and seismic interpretation.

This goal may be subdivided into the following objectives:

- Comparison of the Bazhenov formation and other well-known formations;
- Evaluation of organic matter in the Bazhenov formation within Field S;
- Interpretation of logging data and correlation of layers in the wells;
- Analysis of the amplitude and creation of 2D seismic forward models that allows detecting the sign of carbonate distribution;
- Analysis of horizontal and vertical sections of the 3D seismic cube;
- Creation of an amplitude map and analysis of the distribution of reservoir rocks.

2. Comparison of the Bazhenov and Bakken formations

The deposits of the Bazhenov and Bakken formations are oil shale that has the properties of source rocks and characterized by a very high content of organic matter. The main similarity is a tight rock reservoir within source rock. The main difference is the thickness of the pay zone. The Bakken reservoir is 40 m thick, the Bazhenov – 0.5-3 m thick. The reservoir of the Bazhenov formation is confined to a thin bed and restricted in the lateral direction. However, the Bazhenov formation is characterized by higher values of porosity that can reach 10% or more (on average, 8%) and permeability up to 10 mD, but being 1 mD on average. Also, the Bazhenov formation occupies the area of more than 1 million sq. km, while the Bakken formation covers 520 thousand km2. The main feature of the Bazhenov formation is a very different content of rocks (kerogen and mudstone, carbonate and silica components) depending on the location [1, 2].

3. Evaluation of organic matter in the Bazhenov formation within Field S

The dependence of organic matter versus the average gamma ray value was obtained by V.A. Kontorovich in his work [3]. He obtained two dependencies:

- TOC=0.18GK–4.81 (R2=0.9)
- TOC=0.17GK+0.12 (R2=0.94)

The second relationship is applicable for the transition zone between the Bazhenov and Maryanovsky formations (GR has the average values of 20-55). This area covers a significant part of Tomsk Oblast. Field S is located in the transition region, in the Ust-Tym depression. Thus, TOC was determined on the basis of the second relationship (figure 1).

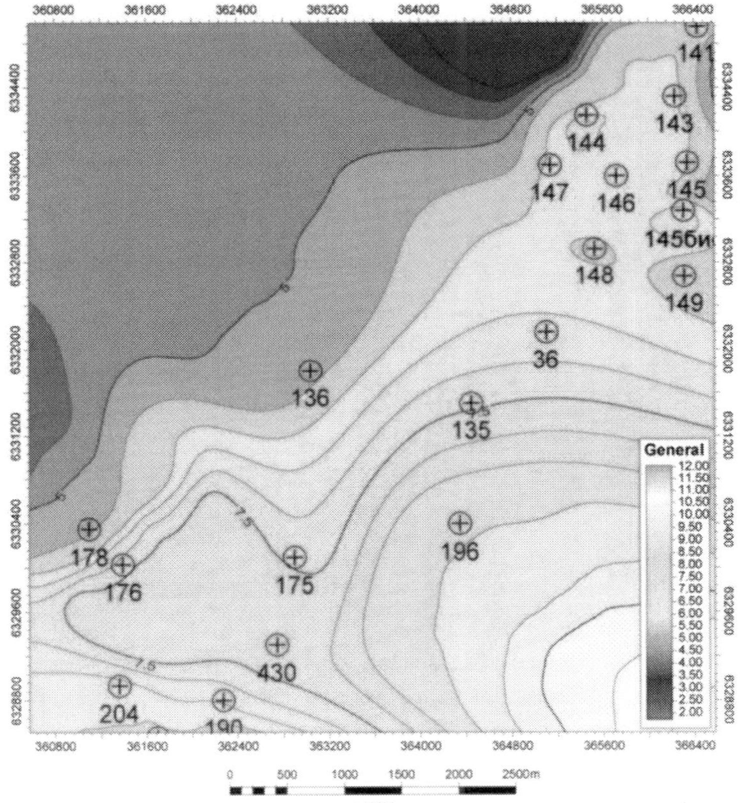

Figure 1. TOC content in the Bazhenov formation at Field S

Three reservoir models are considered below. The model of a fractured reservoir assumes that reservoir-rock was represented by bituminous mudstones ("bazhenite") in which horizontal microcracks (65%), vertical (65.7%) and stylolite cracks comprise up to 11%. This model was described by E M Halimov and V S Melik-Pashayev [4].

The foliated model explains that the reservoir was formed by the fluid auto-fracturing of the formation. This process was explained to be the result of the transformation from organic matter to liquid state by F G Gurari, I I Nesterov in 1977-1985.

The tectonic - hydrothermal model is associated with tectonic and hydrothermal effects on the rocks. This reservoir was characterized by a fracture-cavern-pore texture (M U Zubkov) [5]. Only the third reservoir model was considered with a wireline log analysis and seismic modeling.

IOP Conf. Series: Earth and Environmental Science **24** (2015) 012047 doi:10.1088/1755-1315/24/1/012047

4. Well correlation

Correlation was performed on the basis of log data from eleven wells. Correlation (figure 2) started with the identification of markers in the part of the cross section under study. The Bazhenov formation is a regional seismic marker lying on the Georgiev formation.

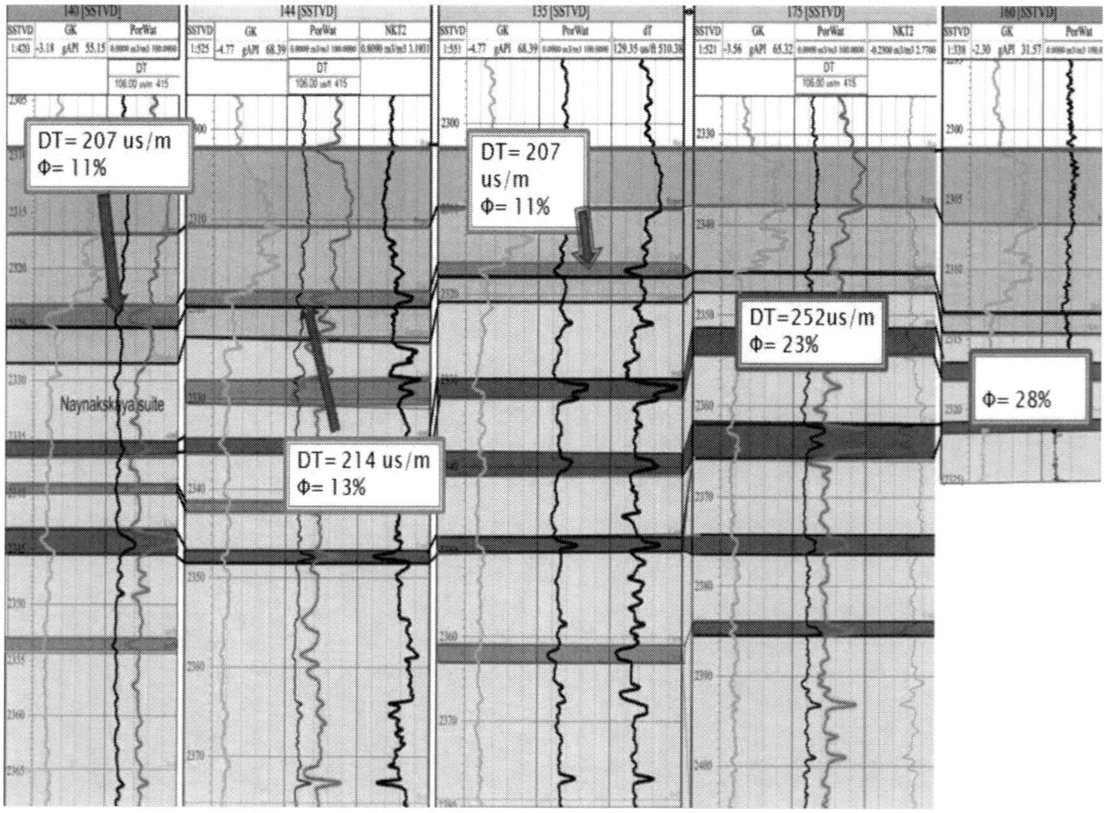

Figure 2. Correlation panel for five wells

This marker was identified with gamma ray logs and a lateral device. The values of gamma ray logging are very high and may reach 30 mR per hour (abnormally high-radioactive bituminous shale). The carbonate layer was indicated in the lower part of the Bazhenov formation with acoustic logs (travel time decreases), and also with neutron porosity logs (porosity decreases too). This carbonate layer was indicated practically in all wells, except two wells (175 and 160). The average thickness of the layer ranges from 1.2 to 2 m. Thus, this layer may be interesting as a potential reservoir within the Bazhenov formation.

The Georgiev formation is also the first-order marker that is represented by black and grey thick mudstones. This marker was determined with induction logs (the minimum value of the induction log). The Naunakskaya formation is subdivided into two sub-formations: coal-overlaying sub-formation and sub-coal sub-formation. The coal layer divides these sub-formations and is 1-1.5 m thick. This layer is laterally continuous at Field S. The Naunakskaya formation is characterized by clastic rocks.

5. Seismic interpretation

One of the basic methods of qualitative interpretation is the seismic amplitudes analysis, while amplitudes are the main parameters of seismic data recorded [6].

IOP Conf. Series: Earth and Environmental Science **24** (2015) 012047 doi:10.1088/1755-1315/24/1/012047

5.1. *Petrophysical analysis of velocity and density of rocks*

The acoustic impedance of rocks is the main factor that affects the reflection coefficient and the amplitude of the wave. Acoustic impedance depends on density and velocity. If (ρ*V) increases, then acoustic impedance increases too. Amplitude increases, if the acoustic impedance of two layers is very different. Thus, if a reflectivity coefficient is high, then the amplitude is also high.

Firstly, the velocity of the Bazhenov formation was determined. As one can see (figure 3), the Bazhenov formation was divided into two parts in accordance with the velocity (low-speed part and high-speed part). The upper part has low velocity (2,800 m/sec), while the lower part has high velocity (3,400 m/sec).

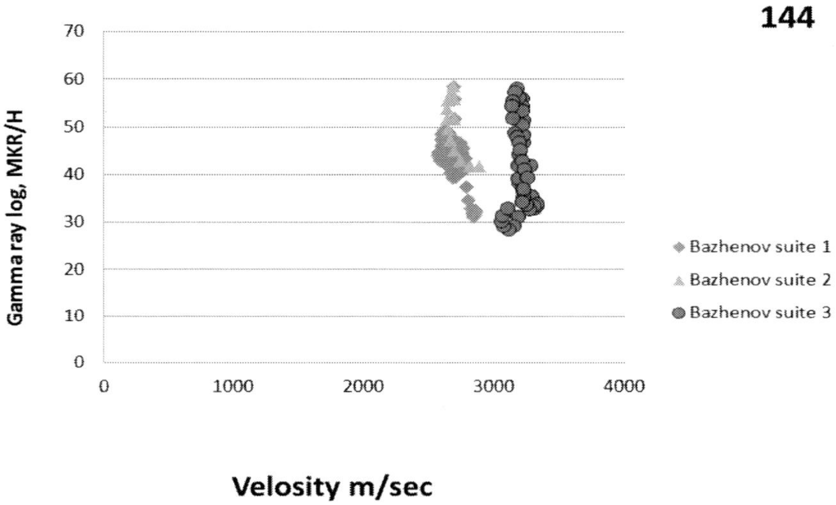

Figure 3. Chart of gamma ray response versus velocity for the Bazhenov formation (well 144)

Secondly, the same plots were created to determine the velocities of each lithology type (figure 4).

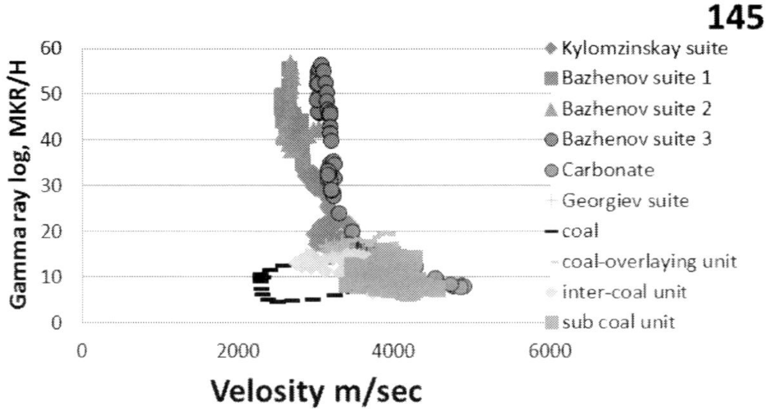

Figure 4. Chart of gamma ray response versus velocity for rocks with different lithology types located under the Bazhenov formation

The velocities determined were used to create 2D seismic forward models. The next figure (figure 5) describes the division of the Bazhenov formation into packs that can be characterized by a fixed velocity value.

IOP Conf. Series: Earth and Environmental Science **24** (2015) 012047 doi:10.1088/1755-1315/24/1/012047

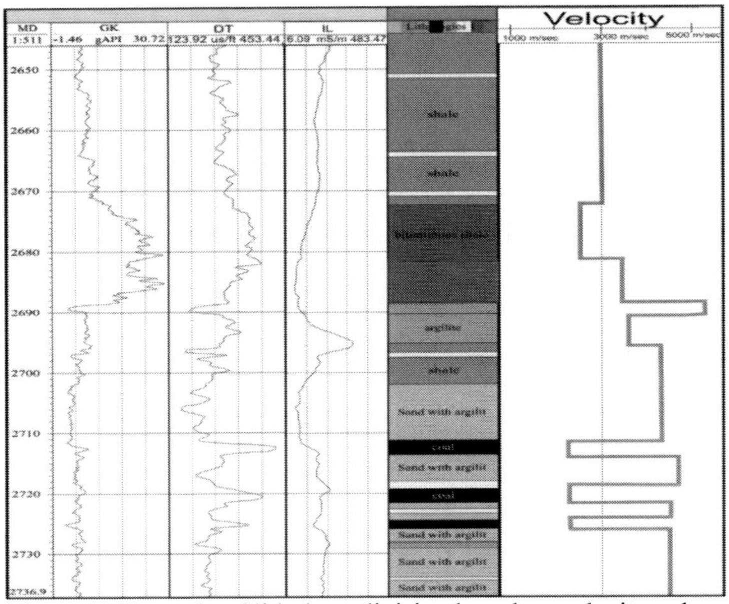

Figure 5. Example of lithology division based on velocity values

2D seismic forward models were created in the following sequence:

- models of acoustic rock properties and layer thicknesses were created in Excel.
- geological study was made for the models in Excel, such as: presence and absence of a carbonate layer, increasing thickness of the carbonate layer, as well as the presence of gas in the coal-overlaying formation (figure 6);

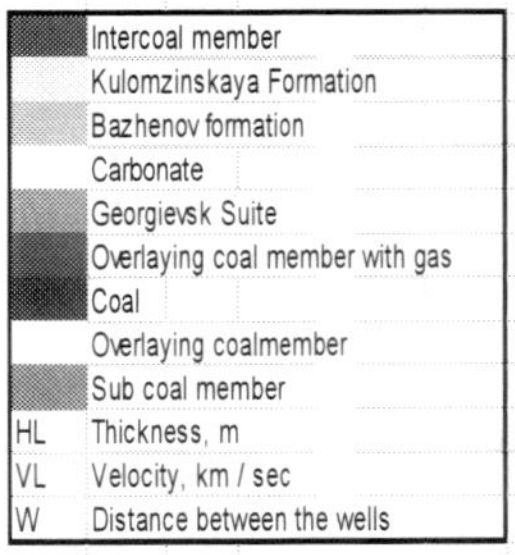

	HL	VL	HR	VL		HL	VL	HR	VL
		W1 =	40.0				W2 =	40.0	
	10.0	3.0	10.0	3.0		10.0	3.0	10.0	3.0
1	6.3	3.0	7.2	3.0	Kulomzin	7.2	3.0	9.3	3.0
2	5.5	2.5	7.7	2.5	Bazhenov	7.7	2.5	5.6	2.5
3	2.4	2.5	2.4	2.5	Bazhenov	2.4	2.5	2.2	2.5
4	1.0	2.5	0.5	2.5	Bazhenov	0.5	2.5	0.4	2.5
5	3.1	3.4	3.5	3.4	Bazhenov	3.5	3.4	5.5	3.4
6	1.5	5.0	1.2	5.0	carbonate	1.2	5.0	1.7	5.0
7	2.6	3.6	3.8	3.6	Georgiev	3.8	3.6	6.2	3.6
8	3.0	4.2	3.0	4.2	coal-overl	3.0	4.2	5.0	4.2
9	0.0	5.0	0.0	5.0	carbonate	0.0	5.0	0.0	5.0
10	4.3	4.2	8.6	4.2	coal-overl	8.6	4.2	9.9	4.2
11	0.5	2.5	2.3	2.5	coal	2.3	2.5	1.4	2.5
12	3.4	4.5	3.0	4.5	intercoal	3.0	4.5	3.0	4.5
13	0.7	5.0	0.0	5.0	carbonate	0.0	5.0	0.0	5.0
14	3.1	4.5	6.6	4.5	intercoal	6.6	4.5	3.7	4.5
15	2.1	2.5	1.1	2.5	coal	1.1	2.5	1.7	2.5
16	7.3	4.2	8.1	4.2	subcoal	8.1	4.2	2.8	4.2
17	0.0	2.5	0.0	2.5	coal	0.0	2.5	1.4	2.5
18	0.0	4.2	0.0	4.2	subcoal	0.0	4.2	2.2	4.2
19	1.1	5.0	2.3	5.0	carbonat	2.3	5.0	0.8	5.0
20	9.5	4.2	13.4	4.2	subcoal	13.4	4.2	10.8	4.2

Figure 6. Part of the model, Legend for all models

The elementary wavelet was matched with the similar characteristics of the real wavelet, knowing that a real seismic section was recorded with the frequency of 19.2 Hz of the wavelet. The wavelet shape and its characteristics are depicted in figure 7 (consideration was performed for 19 Hz and 47 Hz);

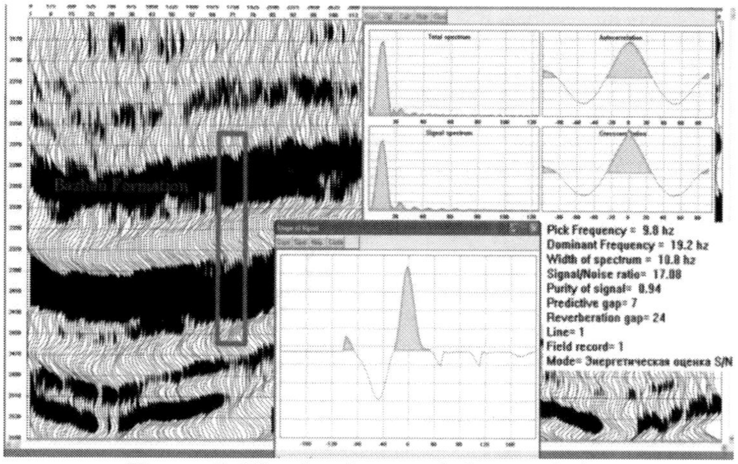

Figure 7. Wavelet shape and characteristics

5.2. *2D seismic forward models with and without a carbonate layer for a seismic impulse at 19Hz*

According to this 2D seismic forward model 1, the sign of carbonate layers was not unambiguously detected (figure 8). Thus, the real seismic cross section cannot be applied to detect a carbonate layer. The amplitude with a carbonate layer is almost the same as the amplitude without a carbonate layer. Thus, in the next 2D seismic forward models, a wavelet with the frequency of 47 Hz was used, because this frequency is applicable to detect exactly the Bazhenov formation and is the sign of a carbonate layer.

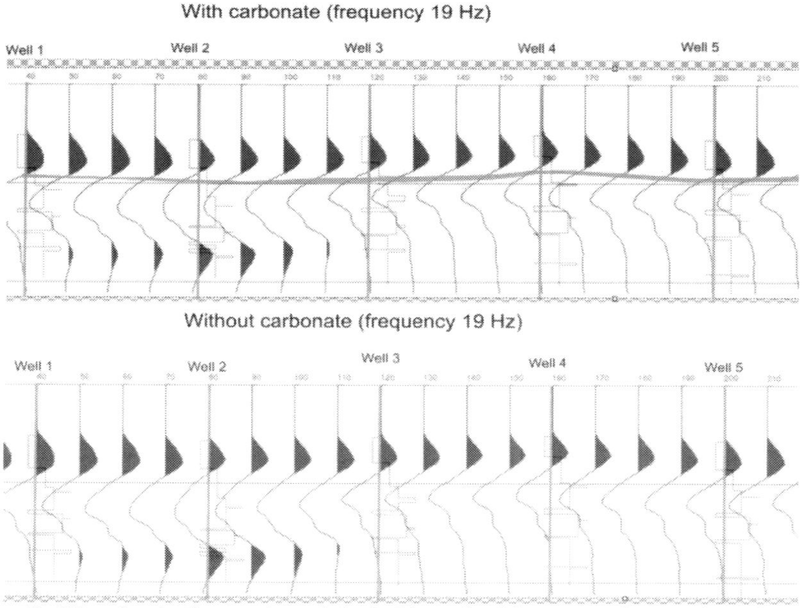

Figure 8. 2D seismic forward models (with and without a carbonate layer) at 19Hz

5.3. *2D seismic forward models at 47 Hz with and without a carbonate layer*

Obtained by one-dimensional simulation (figure 9), the synthetic seismic trace was analyzed for the presence and absence of a carbonate reservoir that was detected with wireline log interpretation. One can see from this figure that the presence of carbonates affects the image, since the negative amplitude increases (is more pronounced) where carbonates occur.

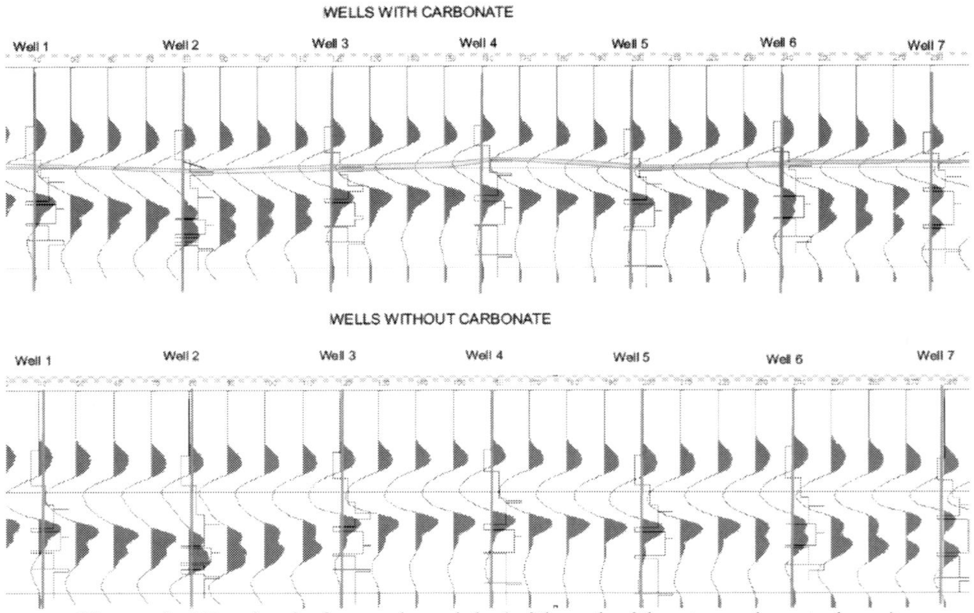

Figure 9. 2D seismic forward models (with and without a carbonate layer)

6. Attribute analysis

The visual analysis of slices in each well over a certain period of time (where a carbonate layer is located) was performed. The information from the slices and that from the 2D seismic forward models was taken as a basis. It is known that the carbonate layer is located immediately in the bottom of the Bazhenov formation, and the seismic surface of the bottom part of the Bazhenov formation was used as initial data (but this surface is slightly lower than the carbonate layer), then a carbonate surface was created. This surface has the same configuration as the bottom part of the Bazhenov formation, however, the surface is located higher where the first largest peak occurs ("carbonate peak") (figure 10).

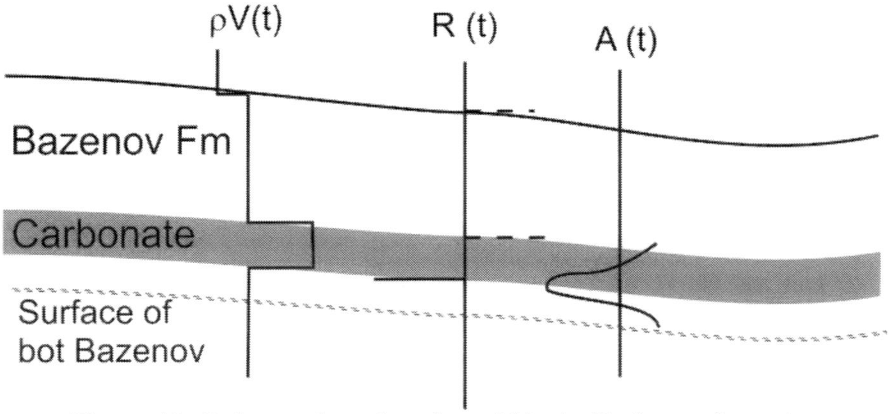

Figure 10. Carbonate layer location within the Bazhenov formation

The next step was to create an amplitude map (figure 11). This attribute retrieves the value of the amplitudes of the existing 3D seismic cube with respect to the carbonated bed.

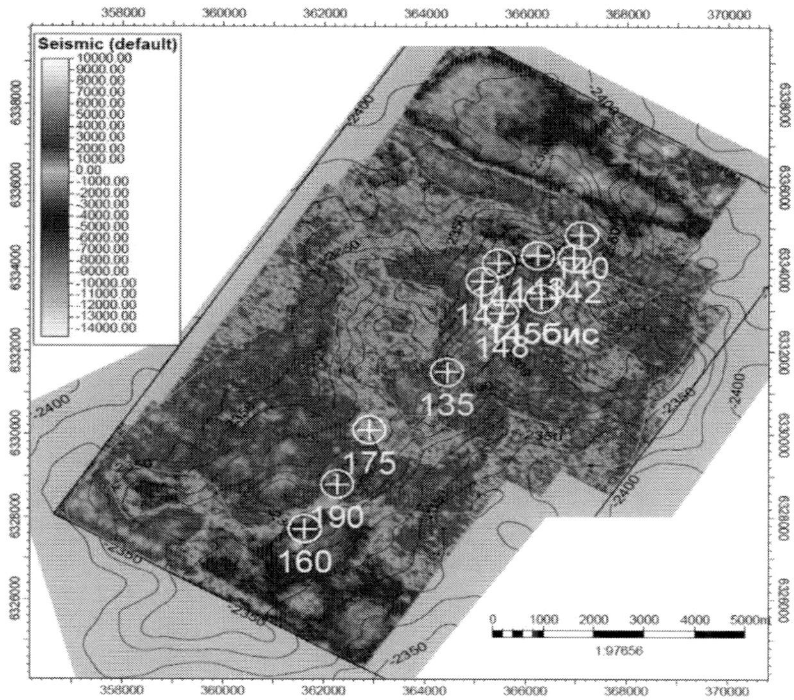

Figure 11. Amplitude map for the carbonated layer

As the result of this map interpretation, the distribution of the carbonate layer was contoured. Thus, it may be concluded that the distribution of the carbonate layer is lateral. Most wells lie exactly on the carbonate layer and only two wells (160 and 175) lie on the transition zone, i.e. grey color. Thus, the carbonate formation wedges out in wells 160 and 175.

7. Generation productivity of the Bazhenov rocks

The average content of organic matter is quite high (around 8%). TOC is one of the main parameters that characterizes the generation potential of rocks, the quality of organic matter, the state of katagenesis and the volume of rocks containing organic matter; all these parameters affect the generation productivity [7].

The following approach in estimating the generation productivity of the Bazhenov formation was applied: the product of the organic hydrocarbon mass with a generation productivity coefficient for the area where the maturity of katagenesis corresponds to the main phase of oil generation. Passive hydrocarbon resources can be assessed with a generation coefficient, as this coefficient can be used to identify specific generation productivity.

The generation productivity map of the Bazhenov formation in Tomsk Oblast presented below (figure 12) was obtained by multiplying the map of organic matter (Corg) in Tomsk Oblast by the generation coefficient (Kgn) and by the organic hydrocarbon mass of the Bazhenov formation (Msp) [2].

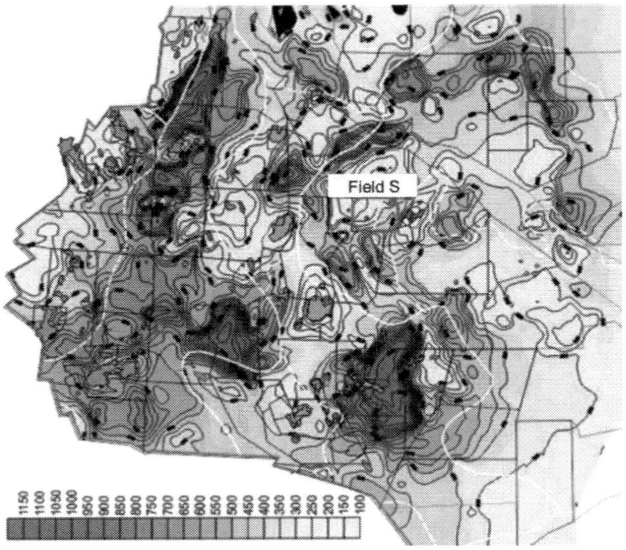

Figure 12. Generation productivity of the Bazhenov formation in Tomsk Oblast

This map shows the amount of hydrocarbons generated from one square area. According to the obtained map, it may be concluded that the rocks of the Bazhenov formation have a high generation potential and it reaches 350-500 kg/m^2 for Field S.

8. Results

The fracture-cavern-pore model of the rocks possibly represents a carbonate layer that was determined through logging interpretation and seismic data. Constructing 2D seismic forward models allowed identifying and tracing the sign of carbonate rocks.

The amplitude map was plotted on the basis of these 2D seismic forward models and the slices were analyzed. This map may be applied to evaluate the spread of the reservoir rocks that are possibly represented by carbonates.

9. Conclusion

1. Unconventional oil reserves are becoming more and more challenging due to the growing demand for oil and the depletion of traditional oil reserves. Thus, special attention should be paid to the Bazhenov formation, which is primarily a source rock and is an unconventional reservoir. The Bazhenov formation is continuous and is represented by bituminous mudstones, interlaying-shale-carbonate-silica and the high content of organic matter. In this project, the fracture-cavern-pore model of the reservoir was considered, which is mainly associated with the layers of secondary transformed radiolarites;
2. Carbonate nature of this layer was confirmed by acoustic, neutron and gamma ray logs (high velocity values (5,000m/s), abnormally low neutron values, and also low gamma ray log values). Silicate rocks do not differ from carbonate rocks as depicted on the mentioned logs, so a photoelectric absorption log is recommended for their more confident separation. The essential difference is that the carbonate rocks have a very high Pe value and the silicates are characterized by the minimum value of Pe;
3. Sign of a carbonate layer can be traced with a 2D seismic forward model;
4. Amplitude Map can be used to analyze the carbonate distribution and create a future development strategy;
5. In addition, this approach based on the visual analysis of the wave pattern can be suggested to determine these thin layers.

References

[1] Vygon G and Rubtsov A 2013 *Nonconventional Oil: Will Bazhen Become Second Bakken?* Moscow School of Management p 69
[2] Braduchan Y V, Gurari F G and Zakharov V A *Bazhenov Horizon of Western Siberia* (Moscow: Nedra Publishing House)
[3] Kontorovich V A 2001 *Generation Potential of the Volga Sediments in the South-Eastern Regions of Western Siberia* Geology of oil and gas
[4] Melik-Pashayev V S, Halimov E M and Seregina V N 1983 *Abnormally High Reservoir Pressures in Oil and Gas Fields* (Moscow: Nedra Publishing House) p 181
[5] Fedorov Y U and Zubkov T A 1989 *Hydrothermal Secondary Reservoirs in Black Shales* Geology of oil and gas
[6] Razin A V, Merkulov V P and Chernov S A 2004 *Modern Methods of Studying Seismic Fields* TPU
[7] Kontorovich A E, Surkov V S and Trofimuk A A 1981 *Oil and Gas Geology of the Siberian Platform* (Moscow: Nedra Publishing House)

.

IOP Conf. Series: Earth and Environmental Science **24** (2015) 012048 doi:10.1088/1755-1315/24/1/012048

Integrated approach to gas accumulation identification in Field M

K Malyshevskaya[1,2], V Rukavishnikov[1] B Belozerov[2] and A Podnebesnikh[2]

[1]Tomsk Polytechnic University, Tomsk, Russia
[2]LLC Gazpromneft Science and Technology Center, Saint Petersburg, Russia

E-mail: rukavishnikovvs@hw.tpu.ru

Abstract. The given paper describes how the integration of different methods, such as core data, well logs, production logging, seismic data and well test analysis, was used to solve the problem of determining gas accumulation boundaries in sediment complex PK1-3 of Field M. This paper is devoted to the block with wells 2, 36, 49, 85, 127, 148 of the field, since it is characterized by high uncertainty, sc. recently drilled wells 1V, 2V and 120 have produced oil, although according to the present-day geological concept they were considered to be gas saturated in the intervals investigated with production logging. Besides, well 127 that was presumably oil saturated has produced gas.

By accounting mismatching production data and the geological concept, the authors have supposed that PK1-3 gas accumulation is characterized by a more complex structure than it was supposed by the predecessors and it is represented by reservoir compartmentalization and high heterogeneity. Therefore, the main goal of the work was to revise the distribution of gas saturated reservoir within the PK1-3 sediment complex. To achieve this goal, the authors have set the following tasks: to revise the geological correlation and gas oil contact; to carry out fault interpretation by means of seismic and well test data; to determine areal facies distribution on the basis of integrated core, perform a log motifs and seismic facies analysis. Thus, the estimation of the gas saturated reservoir portion was implemented in two stages: defining the boundary of gas accumulation in depth on the basis of well logs, production data and fault interpretation; reservoir distribution determination on the basis of the seismic facies analysis within the derived gas accumulation boundary.

1. Introduction

Field M is located in the south-east of the Gydan Peninsula, Tazovskiy Region of the Yamalo-Nenets Autonomous District of the Tyumen province (figure 1). According to the regional tectonics, it is limited to the Srednemessoaykhskiy arch that represents part of the regional megalithic banks system with sublatitudinal extension. This field is located on the eastern local lifted block of the Srednemessoyakskiy arch.

On the basis of its hydrocarbon potential, Field M is related to the Messovskiy petroleum region of the Gydan petroleum area. According to the present-day geological concept, most reserves are concentrated in the PK1-3 sediment complex being part of the Pokurskaya formation (Cenomanian age) overlain by the regional shaly seal of the Kuznetsovskaya formation (Turonian age). The top of the Pokurskaya formation is a regional lithostratigraphic marker and corresponds to regional seismic key reflector G.

IOP Conf. Series: Earth and Environmental Science **24** (2015) 012048 doi:10.1088/1755-1315/24/1/012048

In general, the PK1-3 sediment complex is comprised of grey siltstones and sandstones generated in the alluvial and lacustrine depositional environment and according to the present-day geological concept it is represented by a massive gas and oil reservoir divided into separate blocks by faults . According to [1], the eastern local arch was formed as a structural trap predominantly at the end of the Cretaceous period and at the beginning of the Paleogene period and later it was filled in with hydrocarbons. Most field reserves are concentrated in the block under study within wells 2, 36, 49, 85, 127, 148.

The direct development of the field will start in 2016. Nowadays, 27 exploration wells have been drilled with two water supply wells and two pilot bores 33PL and 13PL among them in Field M. Three horizontal wells have been drilled from pilot wells and the whole area of Field M is covered by 2D and 3D seismic surveys (figure 2).

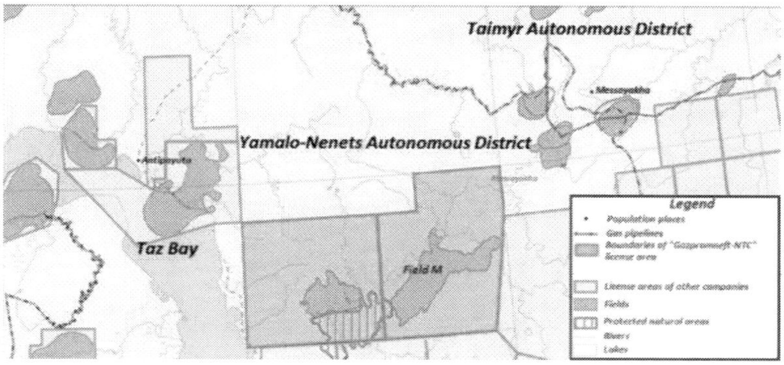

Figure 1. Field location

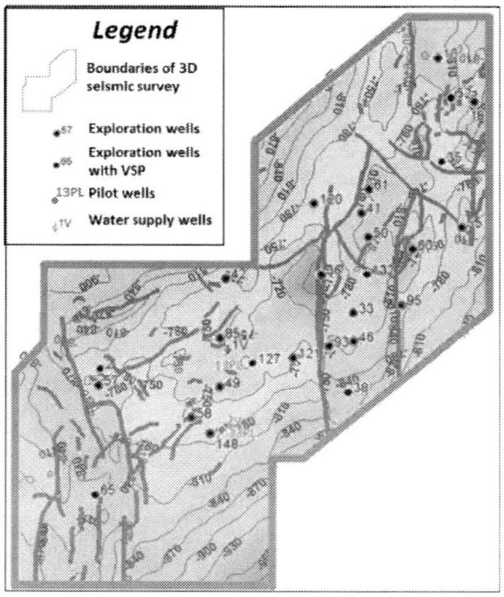

Figure 2. Field M exploration maturity

2. Depositional environment

As for the present-day depositional environment, according to [18], PK1-3 is represented by a subaerial delta plain with three predominant facies: the first one is a floodplain facies comprising fine-grained rocks with rootlets, coal laminae and bioturbation deposited in low energy conditions; the second is floodplain facies deposited in high energy conditions and represented by distinct thin-bedded very fine to fine- grained sandstone with bioturbation and rootlets at the top. These facies are interpreted as the distinct distal parts of sheet flood area. The last facies is distributory channels

IOP Conf. Series: Earth and Environmental Science **24** (2015) 012048 doi:10.1088/1755-1315/24/1/012048

consisting of sheet flood sandstones located near the channels and delta channels themselves. The sheet flood sandstones are represented by sandstones with rare current ripples and vegetation and the delta channels comprise massive to cross laminated fine to medium sandstones. The reservoir facies are delta channels and sheet flood sandstones, floodplain facies are a non-reservoir. The conceptual scheme of the depositional environment is illustrated in figure 3.

It should be noted that from the date when this depositional environment concept was worked out, a significant part of the wells were drilled, therefore, the facies distribution in the recently drilled wells was determined on the basis of matching the derived lithofacies and electrical log motifs (SP, GR) in the wells with and without cores .

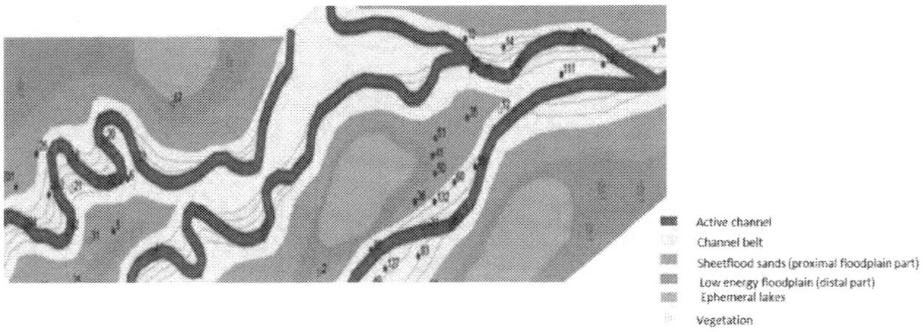

Figure 3 Initial conceptual depositional environment for C interval [16]

3. Geological correlation

Since the recently drilled wells were not used in the present-day geological correlation, well correlation was revised to account data on new wells. According to the well correlation, investigated sediment complex PK1-3 can be divided into three intervals: A, B and C, while interval A can be subdivided into two zones: A1 and A2. Zone A1 includes the interbedding of all three facies and is characterized by shaly floodplain facies at the base. The thickness of A1 varies from 9.8 to 23.5 m and its petrophysical characteristics are the following: effective thickness (Heff) is 8.8 m, porosity (ϕ) is 27.6%, permeability (k) is 72.2 mD and hydrocarbon saturation (Shc) is equal to 59.2%. As for A2, its top is clearly picked as the bottom of the shaly layer in the lower part of A1. A2 is characterized by the predomination of the channel and proximal part of the delta facies, therefore, it has better reservoir properties than A1 and is regarded as a good reservoir. The thickness of this interval lies within the range of 18.8-44.3m and it has the following petrophysical properties: Heff=19.5 m, ϕ=29.7%, k=274.1mD and Shc=58.7%. Zone B is a non-reservoir unit, since it is represented by shaly rocks that can be obtained over the whole investigated area. The genesis of this shale interval (marine or continental) is undefined because of the absence of palynological studies. Perhaps this interval was formed during short-time transgression. The thickness of interval B varies from 1.5 to 7.3 m. Interval C is represented by predominantly thick channel facies and is characterized by the best reservoir properties: Heff=35.9 m, ϕ=33.2%, k=771.6 mD. However, hydrocarbons occur predominantly in interval A (and Shcin interval C=42.3%).

4. Gas oil (water) contact

Gas oil (water) contact was determined on the basis of measurements acquired from resistivity, neutron and density logs and production data. An interval was considered to be gas saturated if it was characterized by different indicators of resistivity tools with different lengths, where the highest value corresponded to the longest tool. Another "fingerprint" of a gas saturated reservoir was small values of density logs and a relatively high value of neutron logs.

Thus, two levels of GOC were determined: the first GOC level is 784 m on average (±4m) in the block limited by wells 85, 49, 58 and the second level is 776 m in the block of well 127. The separate gas accumulation with GOC level being equal to 811 m was defined in interval C of the block of well 127.

Different GOC levels in the above- mentioned blocks were presumed to be caused by reservoir compartmentalization due to faults, therefore, the next stage was the determination of impermeable faults from seismic and well test data.

5. Fault interpretation from seismic and well test data

Direct fault tracking was implemented by analyzing the vertical sections of the origin amplitude cube and horizontal slices of such attributes as "Variance" and "Ant tracking". Besides, the directions of faults were determined with the help of the "Dip angle" attribute derived along the PK1-3 top time surface map (figure 4).

The attributes show that the area under study is characterized by the abundance of tectonic disruptions, and despite, a rather high quantity of faults tracked during (orange color) and before (red color) this work, the full interpretation of all faults in the given area requires a great amount of time. Therefore, fault detection was concentrated on the area next to well 127 and 1V due to the high uncertainty in the geological concept connected with them.

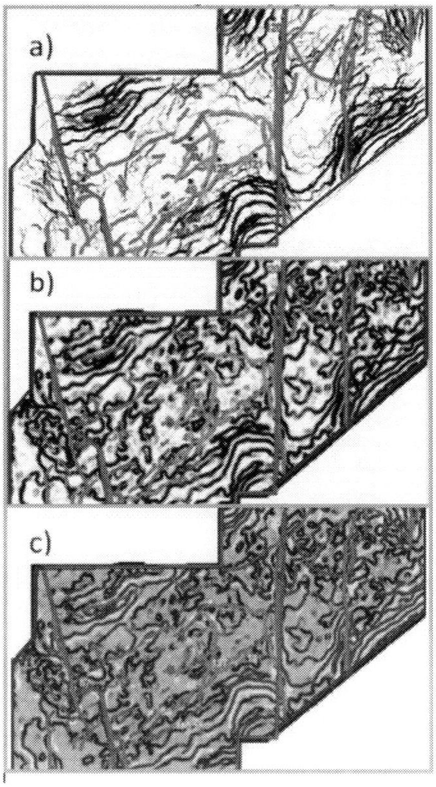

Figure 4. Attributes used in fault interpretation with overlapped fault lines (a) Ant tracking (718 ms); (b) variance horizontal slice (718ms); (c) dip angle extracted from the PK1-3 top time surface overlapped variance horizontal slice (718ms)

In general, all the derived faults (by predecessors and directly in this work) are represented by sub-vertical disruptions, the combination of which generates the system of normal and reverse faults. In seismic sections, faults were determined by the vertical displacement of seismic events in the vertical direction and sharp amplitude alteration on both sides of the fault plane. Most derived faults are characterized by an increasing amplitude displacement magnitude up to the section. The maximum magnitude of amplitude displacement at the level of the PK1-3 top is equal to 100 m. It should be noted that the abundance of tectonic disruptions affects the wave pattern and, consequently, the interpretation results in all.

Fault transmissibility was estimated on the basis of well test data and conducted in horizontal wells 11G, 13G, 15G (pad 1), 31G, 33G (pad 2) and vertical well 120. It should be noted that the well test data analysis in this study was carried out only on the basis of the log-log plots represented in reports describing well test interpretation [15, 21, 22, 23, 24, 29], because raw data was not available.
Since both pads are situated in proximity of the previously extracted faults (figure 5), the foremost well test results of these wells were considered.

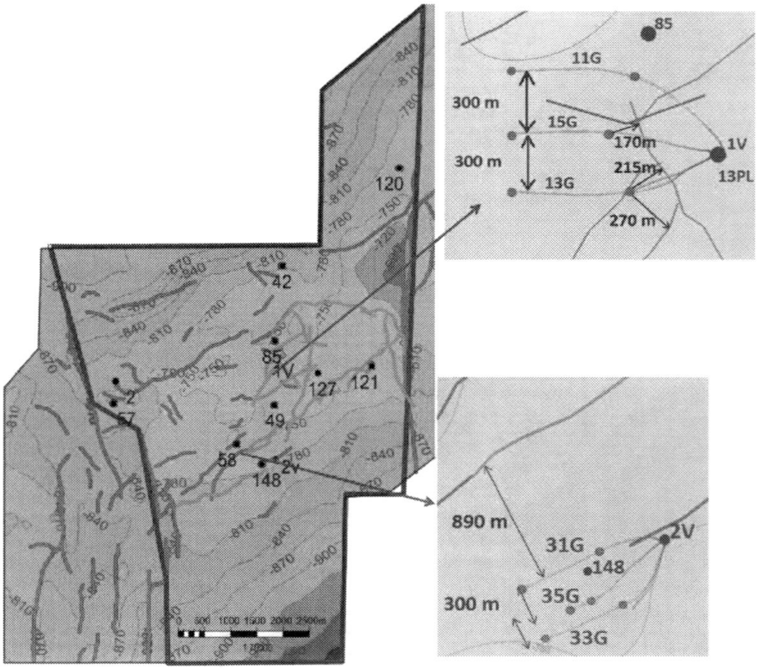

Figure 5. Location of horizontal wells with wells test data used for fault transmissibility analysis

It should be noted that wells in one pad were operating and shut in practically at the same time that resulted in their interference with each other. Thus, the pad 2 drainage area of each well is far smaller than the distance to the fault (from the top of the perforation interval) and more than half of the distance between them (150 m), therefore, the effect from the sealing fault could be masked provided the build-up was long enough to reach LTR. As for the first pad, well 11G has unrepresentative data because of the small drawdown duration and as the remaining wells directly intersect extracted faults Nos. 2, 3 and 4. Analyzing the interference of the wells observed in pressure measurements shows that it is most probable that the presence of faults № 1, 3 and 4 was masked, since the distance from the top of the perforation interval to these faults is less than half of the distance between them (i.e. this fault is permeable at the interval of less than 150 m). As for fault No. 2, it is close to the completion interval top in well 13 G and it is most likely that this fault is permeable, otherwise, it would have been reflected on the log-log plot by a specific ½ slope.

Taking the aforesaid into consideration, it can be concluded that fault No.1 derived next to wells 85 and 1V is permeable and does not act as a barrier to flow, while the fault systems to the north from well 148 is considered to be impermeable because of the absence of well test data to verify this supposition.

6. Integration of results

Gas accumulation boundaries were determined on the basis of well logging data and fault interpretation. According to the interpretation of these data, two separate gas accumulations can be singled out in interval A due to the fault compartmentalization of well 127 and 1V. The first gas accumulation is at the level of 784 m on average (±4 m). Another gas accumulation is limited by the faults tracked around well 127; the depth level of this accumulation is 776 m (figure 6).

A separate gas accumulation was determined in interval C in the block of well 127. The depth level of this accumulation is 811 m and it is isolated from overlain interval A by the impermeable shale of interval B. Besides, the block with oil producing well 1V is also isolated.

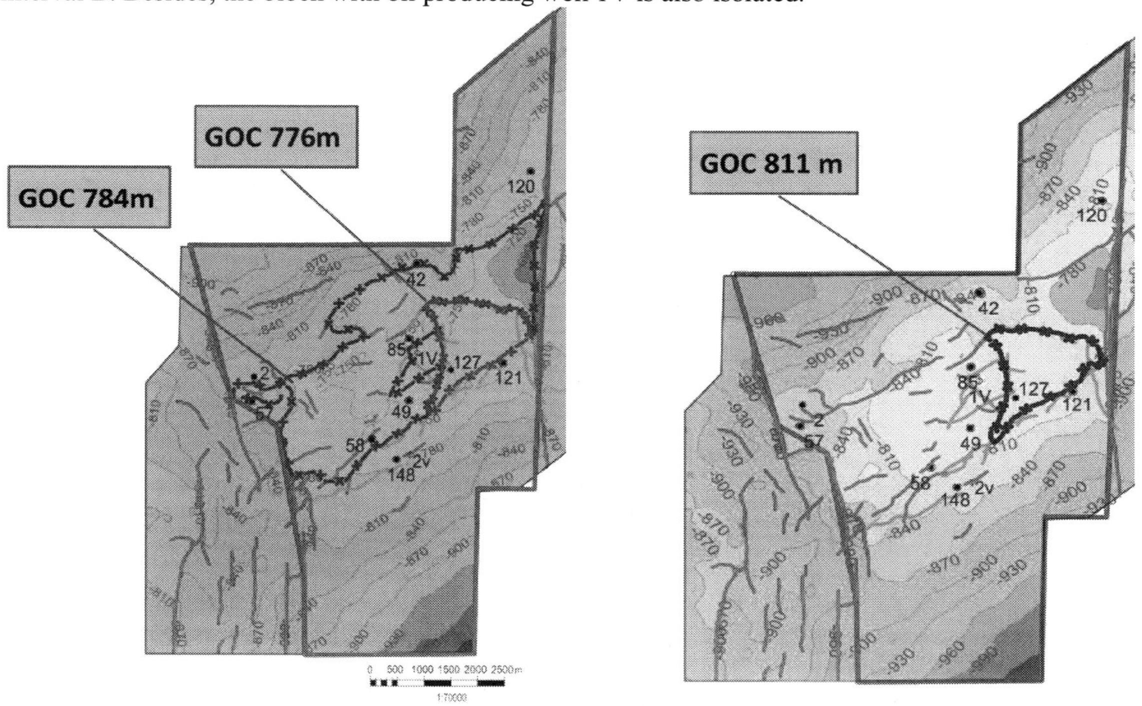

Figure 6. Gas accumulation boundaries in interval A (on the left) and interval C (on the right) determined on the basis of well logs, production data and fault interpretation results

8. Seismic facies analysis

A seismic facies analysis was carried out to determine a gas saturated reservoir within gas accumulation boundaries previously defined by well logs and fault interpretation. As for this study, seismic facies were calculated by classifying the waveform in the original amplitude data, since this technique is supposed to be the most effective one [19, 20].

The possibility to calculate seismic facies within intervals A, B and C was estimated on the basis of the time thickness of a particular interval and correspondence of its depth thickness to the vertical seismic resolution (Table `1). Time thickness was considered to be sufficient for the seismic facies calculation if it covered more than 1.5 cycles of a seismic event (~30ms) for the more accurate estimation of the waveform. Depth thickness was compared with the vertical resolution to evaluate the ability of seismic data to detect facies distribution features.

According to Table 1, the seismic facies calculation is more valid for the whole interval A, since the separate computation of seismic facies within interval A1 is supposed to be inaccurate due to its small time thickness. When it comes to interval B, its time and depth thickness are too small to be reliably reflected on seismic data.

Zone	Time thickness	Depth thickness	Vertical resolution	Seismic facies calculation	
A1	19 ms	22 m	22.4 m	-	more reliable to calculate for A1+A2
A2	29 ms	37.5 m	24.8 m	+	
B	4.7 ms	6.3 m	-	-	
C	36 ms	44.7 m	24.4 m	+	

Table 1. Possibility of calculating seismic facies within intervals A, B, C

9. Seismic facies analysis within interval A

The optimal number of classes (model traces) for calculating seismic facies within this interval was determined on the basis of the correlation function (the degree of similatiry of different model traces i, j as the function of the difference (i-j)), correlation maps and matching core and well logging data). This correlation function had to be represented by a straight line implying a correlation coefficient diminishing with the increase (i-j) and,consequently, accurate extraction of model traces (classes). As for the correlation maps, they had to be represented by the predominant value equal to ±70 %.

The minimum number of classes was set to be 3, since there are three facies as follows from the core data. The maximum number of classes was 15 and further it was chosen for a direct seismic facies calculation (figure 7).

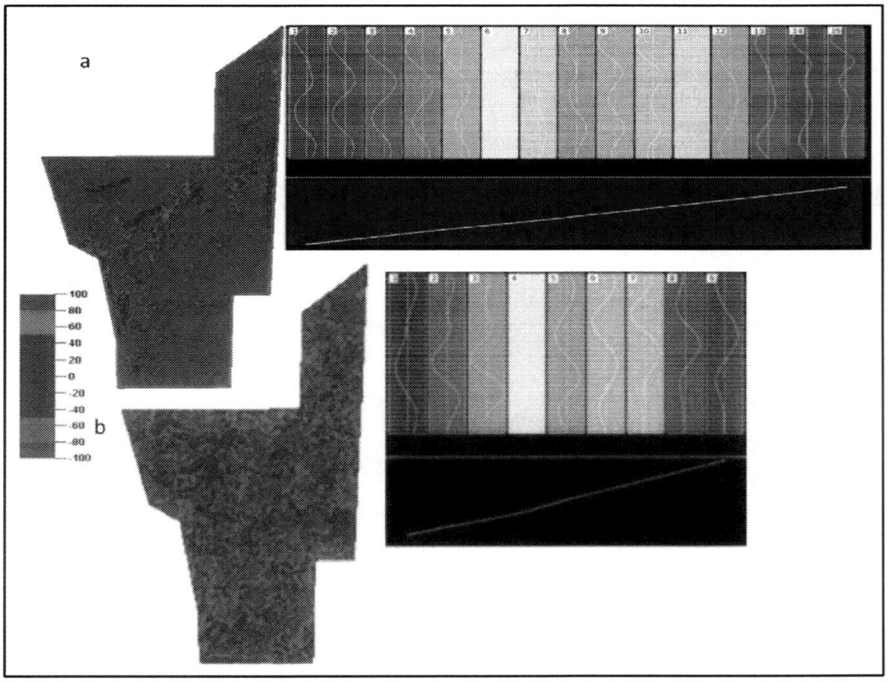

Figure 7. Optimal number of classes for seismic facies calculation within interval A (a) and interval C (b)

The result of the seismic facies analysis is given in figure 8. According to the core data, all wells may be combined into the following groups on the basis of predominant facies:

- Delta channel facies: wells 58, 85, 57
- Proximal part of the delta plain (sheet flood sands): 127, 49, 42
- Alternation of delta and channel facies: 2,148, 2V, 121, 120, 1V

The seismic facies for interval A clearly reveals the anomalies of purple-dark blue colors elongated in the north-east to the south-west direction. These anomalies presumably follow the paleochannel direction (highlighted by white stippled arrows), since they correspond to wells 85 and 58 with the predomination of channel facies. As for the wells characterized by the predomination of the proximal part of the delta plain facies, they are marked blue. Most wells with the alternation of all facies (148, 2V, 1V) do not correspond to one color that is presumably caused by the fact that such a frequent alternation of thin facies is out of seismic vertical resolution. It should be noted that the poor matching of the seismic facies in vicinity of wells 2, 120, 121 (with alternation of facies) and 57 (channel facies) is most likely caused by their location close to the large faults characterized by a collapsed wave pattern around them and, this in turn, affects seismic facies calculation. This supposition can be proved by correlation maps (figure 7) that illustrate a lower correlation coefficient along faults.

In order to make the seismic facies derived comparable with the facies observed in the core data, they were combined in three groups corresponding to the delta channels (1-3 classes of the dark blue color in figure 9), sheet flood sands (4-7 classes) and delta facies (8-15 classes).

The final gas saturated reservoir distribution was determined by extracting only reservoir facies, i.e. delta channels and sheet flood sandstones, within the gas accumulation boundary previously derived with well logs data and fault interpretation.

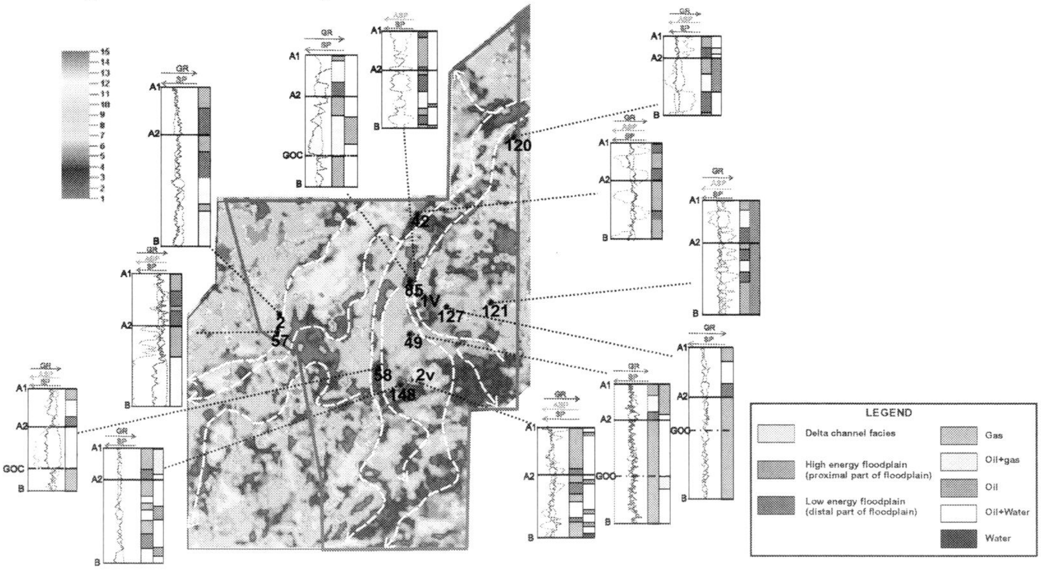

Figure 8. Seismic facies map, electrical log motifs and core facies for interval A

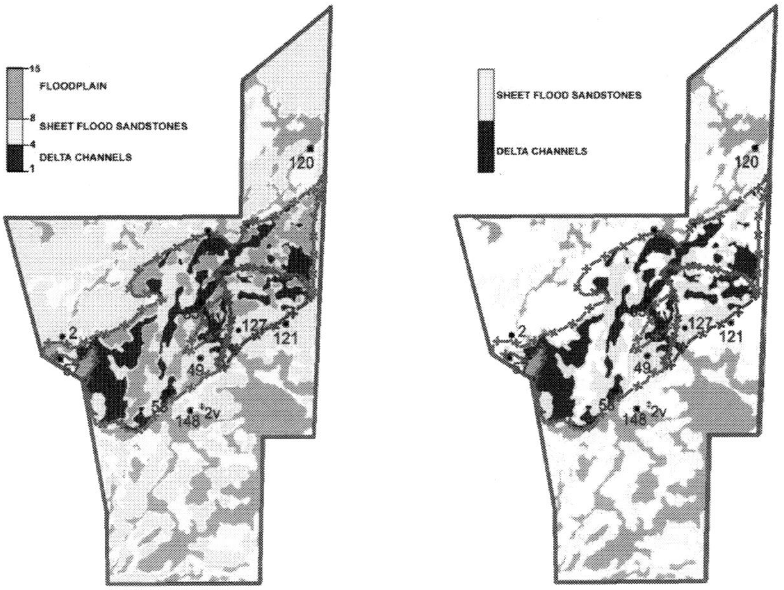

Figure 9. Seismic facies map within interval A with combined classes; reservoir facies within interval A (on the right).

10. Seismic facies analysis within interval C

The minimum number of classes for calculating seismic facies within this interval was set to 3 and the maximum number was set to 15. The optimal number of model traces for computing seismic facies within interval C is 9 (figure 9).

On the basis of the core data, the following well groups in compliance with the predominant facies may be discriminated:

- Wells with the predominance of delta channel facies: 57, 49, 127,121
- Wells with the predominance of sheet flood sandstone facies: 120, 85
- Wells with the equal portion of delta channel facies and sheet flood sandstones facies: 2, 42,1V, 2V, 148

Figure 10 shows a seismic facies map calculated for interval C with the direction of paleochannels highlighted with white stippled arrows. By matching the seismic facies and core data, one can assume that the paleochannels corresponds to the purple-dark blue classes (1, 2) and consequently they were elongated from the north-west to the south-east along the direction of the wells with delta channels and combination of delta channels and sheet flood sandstones facies.

Another notation is contradiction in wells 2, 57, 120 and 121. Wells 2, 57, 121 do not correspond to the purple-dark blue color, unlike the remaining wells with delta channels and delta channels/sheet flood sandstones, while well 120 lies in the purple area, but is characterized by the predominance of sheet flood sandstones. Like in interval A, the aforesaid may be explained by the proximity of these wells to large faults.

The similarly with the interval A gas saturated reservoir distribution was defined by eliminating non-reservoir floodplain facies and putting them on the gas accumulation boundary determined in the previous section (figure 11).

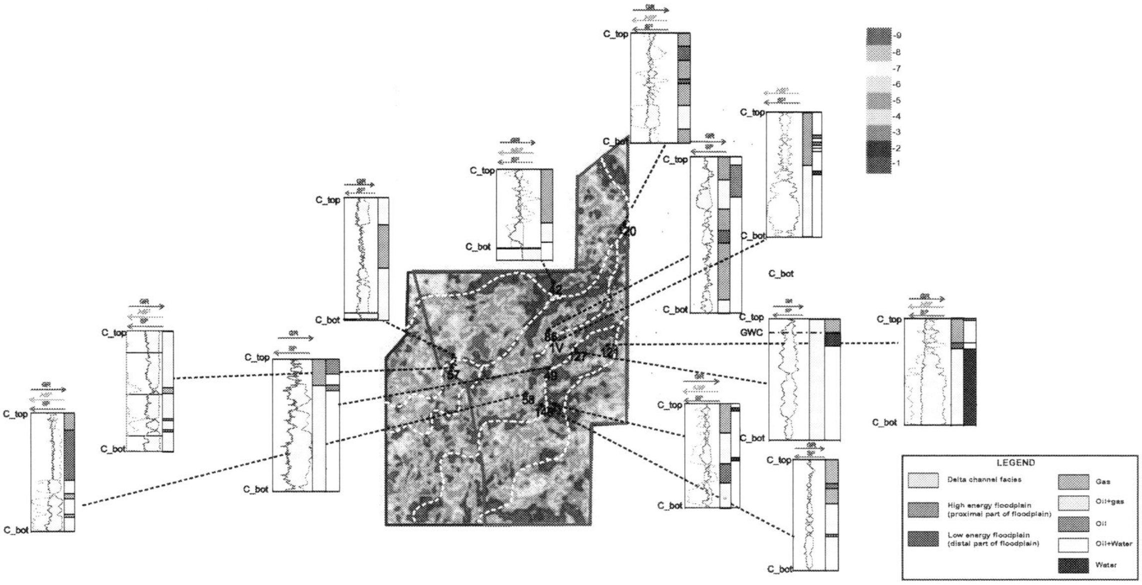

Figure 10. Seismic facies map, electrical log motifs and core facies for interval C

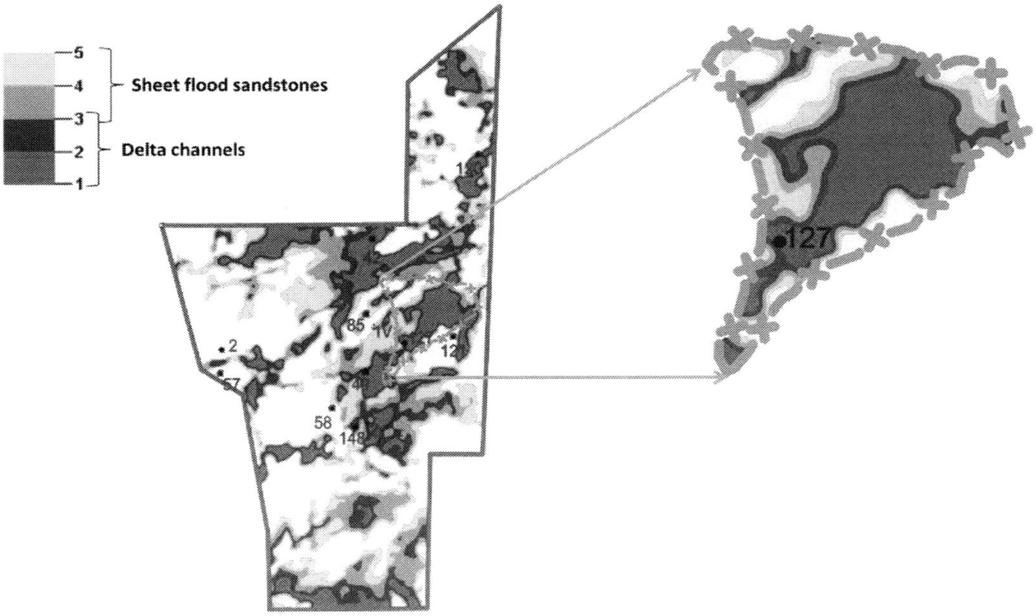

Figure 11. Gas saturated reservoir portion within interval C

11. Comparison with the present-day geological concept

According to the present-day geological concept, PK1-3 is represented by a massive homogeneous accumulation formed as a result of tectonic activity. In compliance with the described study in this paper, the sediment complex under consideration is characterized by high reservoir heterogeneity both in the vertical direction due to shale impermeable barriers to flow and the areal direction due to facies variation.

The present-day concept states a single gas-oil contact within the investigated block, while according to the concept developed in this paper, there are two separate blocks of a gas saturated reservoir in interval A, one of which is at the depth of 784 m and is limited by wells 85, 49, 58 and the second one is a block formed due to faults compartmentalizing, well 127. In addition, a single gas accumulation was determined in interval C of the block with well 127 at the depth of 811 m.

The alteration of the present-day geological concept may lead to changes in the development strategy. A more detailed discussion of this aspect will be given further.

12. Influence on development strategy

Since oil production in Field M is of prime importance, this article concentrates on the influence of geological concept alteration on the development strategy in terms of oil production efficiency. As direct field development has been only recommended currently, the development strategy arises from the geological features of present-day concept and PVT properties of fluids, i.e. a homogeneous reservoir with a single GOC and WOC in the block under study, small net pay thickness, three mobile fluids (gas, oil and water) and highly viscous oil (103.9 cPZ).

A well pattern is given in figure 12 and is represented by a line drive system with horizontal producers and deviated injectors to extract oil from small net pay thickness. As for the position of wells, they are supposed to be drilled within oil saturated thickness for 2 meters along the regional stress direction. Polymer injection is recommended to enhance oil mobility.

In general, the changes in the gas saturated reservoir portion represented by the concept in this paper will lead to the alteration of the oil saturated reservoir portion. The occurrence of different blocks implies their separate development. As for the waterflooding pattern, in case of a highly heterogeneous reservoir, the most common spot patterns used in practice to increase the probability of drilling a reservoir zone are 5, 7 or 9.

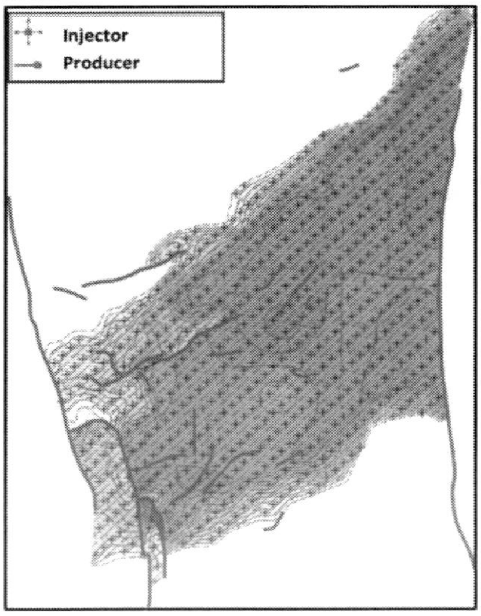

Figure 12. Recommended development strategy [7]

References

[1] Agalakov S E, Baburin A N, Bespalova S N, Bochkarev V S and Korovina I O 2007 *Geological Composition Features and Oil and Gas Bearing Capacity of Srednemessoyakhskiy Arch* Nedra-Consult SibNATS

[2] Ampilov Y P 2008 *From Seismic Interpretation to Modeling and Estimation of Oil and Gas Fields* Spektr, Moscow

[3] Boganik G N and Gurvich Y Y 2006 *Seismic Survey. College Textbook* Tver'
[4] Brown A R 1999 *Interpretation of Three-Dimensional Seismic Data* Published jointly by the American Association of Petroleum Geologists and the Society of Exploration Geophysicists, Tulsa, Oklahoma, USA
[5] Chopra Satinder and Marfurt Kurt J 1985 *Seismic Attributes Mapping of Structure and Stratigraphy*
[6] Desbrandes R 1985 *Encyclopedia of Well Logging* Imprimerie Nouvelle
[7] Fedorov M V, Podnebesnykh A V and Klimov M Y 2013 *Techno-Economic Justification of Obtained Oil Recovery Factor and Condensate Recovery Factor for Field M* Report Gazpromneft NTC Tyumen'
[8] Fedynsky V V, Ivanov L I, Polshkov M K, Puzyrev N N, Rapoport M B, Timoshin Y V and Mihalysev A V 1979 *Application of a Combination of Geophysical Methods of Direct Detection of Oil and Gas Deposits* World Petroleum Cong.
[9] Gogonenkov G N and Timurziev A I 2010 *Strike-Slip Faults in the West Siberian Basin: Implications for Petroleum Exploration and Development* Geology and Geophysics **3**
[10] Heriot Watt University 2005 *Reservoir Geophysics Manual*
[11] Horne R N 2000 *Modern Well Test Analysis* Petroway Inc.
[12] Kasatkin V E 2012 *Hydrocarbon In Place Estimation and Techno-Economic Justification of Obtained Oil Recovery Factor and Condensate Recovery Factor for Field M* Report Gazpromneft NTC Tyumen
[13] Levitskaya T V 2012 *MDT Interpretation Results Acquired in Water Supply Well 1V of M Field* Gazpromneft NTC
[14] Levitskaya T V 2012 *MDT Interpretation Results Acquired in Water Supply Well 2V of M Field* Gazpromneft NTC
[15] Levitskaya T V, Mazhar V A 2013 *Well Test Interpretation Results Acquired in Well 15G of Field M* Gazpromneft NTC
[16] Lopatin A, Latypov E, Baraboshkin E, Petrov E, Aleshina O, Topunova G, Shakirova Y, Shevyakov V and Grobushkina O 2012 *Kinematic and Dynamic Interpretation of 3D Cubes. Structural Interpretation of Horizons and Tectonic Model. Detailed Interpretation for PK1-3 FM. Detailed Interpretation within Cube 4 for MKH8-9, BU6_1-2, BU8* Halliburton Consulting and Project Management report, Moscow
[17] Lopatin A, Latypov E, Gavrilov S, Grobushkina O, Bardina N, Kasiyanov A, Shevyakov V, Kosenkova O and Shakhova E 2012 *Structural Model Building. Building digital 3D Geological Model of Reservoir Within PK1-3 FM (Including all Accumulations), MKH8-9, BU6_1-2, BU8 (Including Accumulations within Cube 4)* Halliburton Consulting and Project Management Report, Moscow
[18] Lopatin A, Latypov E, Tikpenny A, Russel K, Petrov E, Grobushkina O, Shevyakov V, Zavadskaya N Report 2012 "*Lithofacies Model of Sediment Complex PK1-3 of Messoyakhskoe Field*" Halliburton Consulting and Project Management Moscow
[19] Malyarova T N "*Seismic Facies Analysis as Universal Method of Reservoir Composition Interpretation*" Paradigm
[20] Malyarova T N and Ivanova N A 2006 Modern Methods of Seismic Facies Analysis Demonstrated on Case Studies *VIII International Research and Practical Conf. "Geomodel-2006"*
[21] Maslak S G 2013 *Well Test Interpretation Results Acquired in Well 11G of Field M* Gazpromneft NTC
[22] Mazhar V A, Levitskaya T V and Maslak S G 2013 *Well Test Interpretation Results Acquired in Well 13G of Field M* Gazpromneft NTC
[23] Mazhar V A, Levitskaya T V and Maslak S G 2013 *Well Test Interpretation Results Acquired in Well 31G of Field M* Gazpromneft NTC

[24] Mazhar V A and Maslak S G 2013 *Well Test Interpretation Results Acquired in Well 33G of Field M* Gazpromneft NTC
[25] Merkulov V P 2006 *Reservoir Properties Evaluation and Express Analysis of Wireline Logs* Tomsk
[26] Muromtsev V S 1984 *Electrometrical Geology of Sand Bodies in Lithological Traps of Oil and Gas* Moscow: Nedra
[27] Popova O I, Zinatulin M M and Konyaev D N Report 2010 *Building Seismic Models for Reservoir Sediment Complexes of Field M on the Basis of Integrated Analysis of Well and Seismic Surveys by Means of Specific 3D Processing and Interpretation for Geological Exploration Optimization and Subsequent Development* Slafneft-NPC, Tver'
[28] Reading H G et al. 1978 *Sedimentary Environments and Facies* Published by Blackwell Scientific Publications Oxford
[29] Ridel A A 2013 *Well Test Interpretation Results Acquired in Well 120 of Field M* Gazpromneft NTC
[30] Rovnin L I, Nesterov I I, Salmanov F K and Ervye Y G 1975 Laws Governing Oil and Gas Accumulations in Western Siberia *World Petroleum Cong.*
[31] Serra O 1985 *Sedimentary Environments from Wireline Logs* Schlumberger Houston
[32] Shlumberger *Petrel 2011.1* Manual
[33] Sulimov D, Ezerskiy D, Philimonov A, Cyclakov A, and Egorov S 2013 *Report. Interpretation of open Hole Well Logs Acquired in Well 121R of Field M* Schlumberger
[34] Sulimov D, Ezerskiy D, Cyclakov A, Malinin Y and Egorov S Report 2013 *Interpretation of Open-Hole Well Logs Acquired in Well 120 of Field M* Schlumberger

Depositional features of the Middle Jurassic formation of Field N and their influence on optimal drilling schedule

D Mishina[1,2], **V Rukavishnikov**[1], **B Belozerov**[2], **and A Bochkov**[2]

[1]Tomsk Polytechnic University, Tomsk, Russia
[2]LLC Gazpromneft Science & Technology Center, Saint Petersburg, Russia

E-mail: rukavishnikovvs@hw.tpu.ru

Abstract. The Middle Jurassic formation of Field N represented by 4 hydrodynamically connected layers (J5-6, J4, J3 and J2) contains 42% of the field STOIIP. The J2-6 formation is characterized as a gas-oil-condensate massive lithologically and tectonically screened accumulation with a gas cap (J2, J3 layers) and bottom water (J5-6 layer). Oil is predominantly in the J3 and J4 layers. There is a high risk of early gas coning from gas-bearing layers to oil producing wells determined on the basis of production test results, which can significantly decrease the life of the well. To select a more optimal drilling schedule, it is necessary to take the risk of early gas coning into account and determine distinctive features within the gas-saturated zone that can reduce it. The presence of a thick shale barrier between the J2 and J3 layers with thicknesses varying from 0 to 30 m is recognized as the beginning of a transgression cycle, and if the gas cap is only in the J2 layer, this barrier with the thickness of more than 5 m can extensively prevent early gas coning into oil producing wells. The integration of geological information represented by the probability map constructed and petrophysical information represented by the kh map provide the more precise determination of an optimal drilling schedule.

1. Introduction

Field N is the one of the largest gas-oil-condensate fields on the Yamal Peninsula that was discovered in 1964 and has not been developed yet. Field N comprises 19 hydrocarbon-bearing formations within which the J2-6 formation of the Middle Jurassic age contains 42% of the STOIIP and is the target object of this study. This formation is described as a tectonically and lithologically screened united stratified massive accumulation with a gas cap and bottom water and characterized by the high level of compartmentalization due to the presence of discontinuous shale streaks.

Gas coning is the critical issue for oil wells with a gas cap drive, and in the case of early gas invasion, associated gas production increases and consequently the life of this well shortens. Gas invasion can be inhibited by a natural barrier and its distribution depends on a depositional environment. Consequently, the determination of depositional environments of sediments, recognition of typical facies that control such geological features as compartmentalization and the presence of impermeable barrier can help choose the most appropriate areas for drilling where the risk of gas coning is minimal and hence can extend the production life of a field.

This paper implies the following order: providing general information about the field and target object, determining the depositional environment, defining distinctive geological features and their influence on the risk of gas coning with a gas invasion probability map, integrated analysis of the kh map and the probability map with determining the optimal field drilling schedule and evaluating the main uncertainties and recommendations on how to reduce them.

2. General information

Geographically, the field is located in the southeast part of the Yamal Peninsula, Yamal-Nenets Autonomous District, Western Siberia. Field N was discovered in 1964. The geological section of the field is comprised of Pre-Jurassic basement rocks and Jurassic, Cretaceous, Paleogene and Quaternary sedimentary rocks. The sediments of the J2-6 layer are referred to the Middle Jurassic Leontyevskaya and Malyshevskaya formations and were accumulated in a transition coastal marine environment.

The area of the field is located within the Novoportovskoe crestal location that commands a large share of similarly-named second-order structures that are, in turn, part of the higher first-order structure of the South-Yamal megalithic bank. The top of the Middle Jurassic formation has absolute elevations from -1,870 to -2,240 m and is complicated by four crestal locations (figure 1).

There are many multi-directional faults and cracks of various amplitudes within the field area being sufficient for screening the Jurassic formation. The basic fault indicated on the seismic time section from the basement to the Lower Cretaceous sediments separates the area of the field into eastern and western parts.

As for the field hydrocarbon potential, Field N is located in the South-Yamal petroleum bearing region of the Yamal petroleum area in the Western Siberian petroleum province. The J2-6 hydrocarbon reservoir of the Middle Jurassic age is referred to the second hydrocarbon-saturated zone and contains 42% of the STOIIP. In compliance with the fluid content, the J2-6 formation is described as a gas-oil-condensate accumulation with a gas cap and bottom water.

Basically, the Middle Jurassic formation is characterized as a massive lithologically and tectonically screened accumulation consisting of J2, J3, J4 and J5-6 layers being hydrodynamically connected and having common GOC and WOC. Gas is contained in J2 and J3 layers, oil – predominantly in J3 and J4, whereas, J5-6 is a water-bearing layer (figure 2).

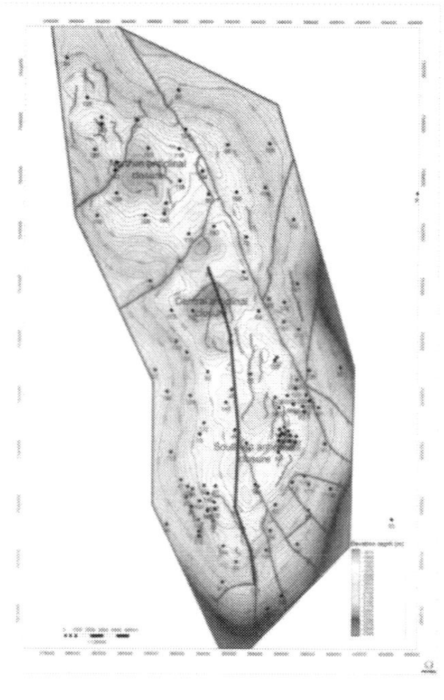

Figure 1. Structural map of J2 top. Field N

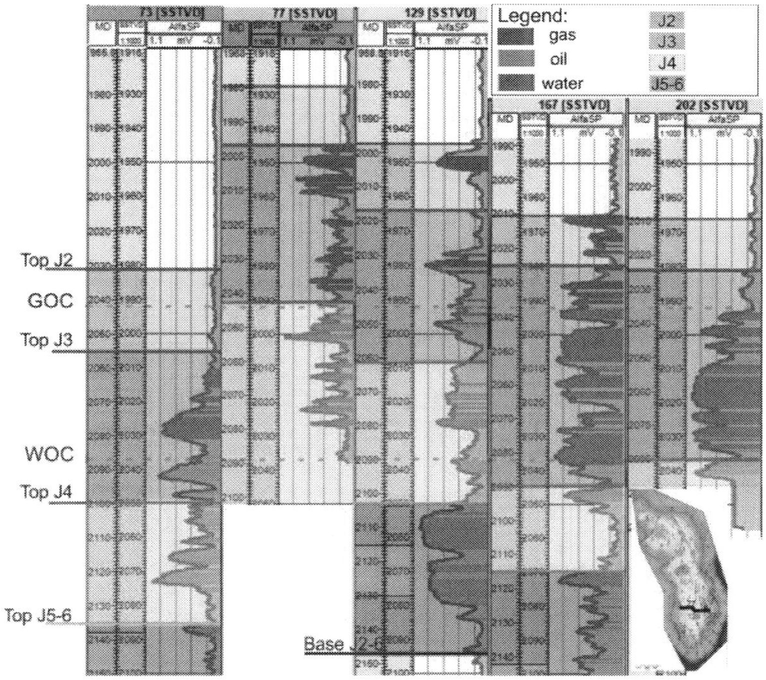

Figure 2. Well section with indicated saturation

3. Deposition environment determination

According to the structural and lithological features distinguished within the core, three main lithotypes can be recognized for the J4 and J3 layers. Typical structures and features are illustrated in figure 3.

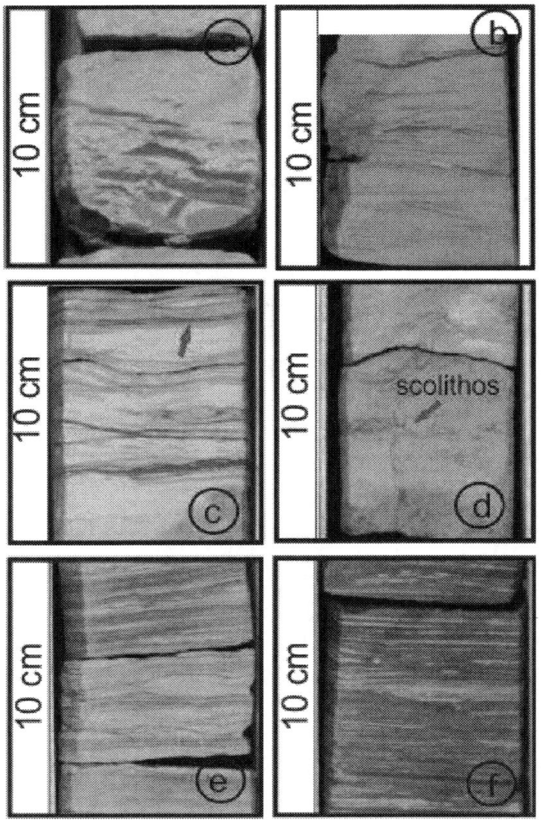

Figure 3. Typical sedimentary structures and features of J4 and J3

Lithotype 1 is represented by medium to fine sandstone with argillaceous and calcite cement. There are mud clasts and clasts of siderite and traces of slump at the bottom of the sediments (figure 3, a) and an upward change in the structure from large cross-bedding to current ripples and herringbone stratification (figure 3, b). Double mud drapes can be recognized and indicate tidal activity (figure 3, c) [3]. There is also the bioturbation of sediments (figure 3, d) at the top and the upward trend towards an increase in the shale content (probably, channel abandonment). Ichnofacies recognized in the core is referred to Scolithos and indicates marine conditions of sedimentation [4]. Also, there are plant debris within the core samples. Lithotype 2 is characterized by the interbedding of sandstone and siltstone with the abundance of a lenticular, wave and flaser structure and mudstone with a planar structure (figure 3, e). Besides, the sediments of Lithotype 2 are highly bioturbated. The change in lithology and typical structures indicate a condition with a frequent variation from a passive environment to the high activity of currents. Lithotype 3 is represented by the thin interbedding of shale, siltstone and sandstone with bioturbation and a planar-bedding structure (figure 3, f) and indicates little or no current activity.

The sediments of the J2 layer can be divided into two types: grey to green marine shale (lithotype 4), thin interbedding of siltstone, sandstone and mudstone highly reworked by the storm activity with high bioturbation and a lenticular and wavy structure and the presence of such ichnofacies as scolithos and palaeophicus that shows a marine environment (lithotype 5). Figure 4 illustrates the basic features of the J2 lithotypes.

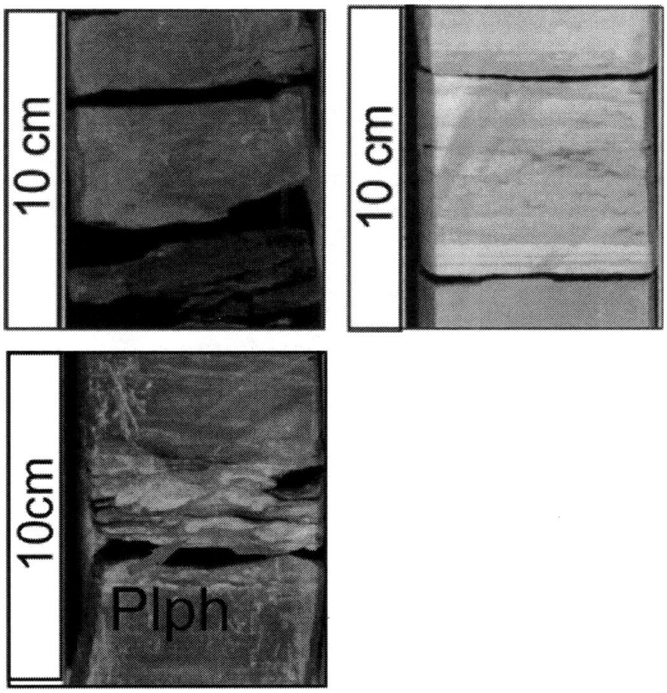

Figure 4. Typical sedimentary structures and features of J2

For the recognition of log motif types, the shape of an SP curve is chosen due to the fact that GR does not illustrate the change in the lithological composition sufficiently, as the shape of resistivity curves is subjected to the influence of hydrocarbons. In compliance with the lithotypes identified within the J4, J3 and J2 layers, four main types of log motifs (electrofacies) based on the shape of the SP curve can be recognized. Type 1 electrofacies is referred to Lithotype 1 and characterized by a predominantly square shape with a sharp horizontal base that gives evidence of erosion in the underlying sediments, a gently or slightly notched vertical curve and gently decreasing top that indicates a change in the size of grains and an increase in shale components in the sediments. Such a shape of the SP curve is typical of channel facies. The standard value of αSP for this electrofacies is 1.0-0.6. Electrofacies 2 is described by the shape of a rectangular triangle with the high level of notch that can be explained by the frequent variation of sandstone and mudstone, the maximum value of αSP is equal to 0.4-0.6. This electrofacies is characterized by the high intensity of compartmentalization due to the frequent changes in conditions from active to passive and typical of Lythotype 2. Electrofacies 3 is characterized by the high level of notch with the maximum value of αSP equal to 0.2-0.3. Such a type of an SP curve shape is typical of Lithotypes 3 and 4. Electrofacies 4 describes the presence of Lithotype 5 within J2 and is characterized by an upward coarsening trend, often has an oblique notch side line. The value of αSP varies from 0.6 to 0.8.[2].

On the basis of the lithotypes distinguished from the core data, log motifs and main vertical trends, the following conclusions about facies and depositional environments can be made.

The sediments in the J4 and J3 layers have similar typical features and are characterized by Lithotypes 1, 2 and 3. There are traces of tidal activity represented by the herringbone structure and mud drapes within Lithotype 1 that can be interpreted as a tidal channel and is characterized by a small thickness, however, with a high level of amalgamation [1]. Away from the areas of strong tidal currents, there are tidal mudflats (Lithotype 3). The succession is predominantly of mud with thin sand sheets present if very high tides or storms were washing the coarser material across the flats. The evidence of vegetation is normally abundant. The same is relevant to bioturbation by the fauna living on the nutrient-rich mudflats [3]. Also, the interdistributary bay facies (Lithotype 2) is widely developed within the area of the field. The dominant position of the lenticular to wavy structure identifies the features of distributary bay sediments. With due account of these features, the sediments of the J4 and J3 layers are referred to as tidal-dominated delta deposits. Tidal processes form sandbodies being parallel to the directions of the tidal currents; they are roughly perpendicular to the regional shorelines [7]. Figure 5 illustrates the modern analog of a tidal-dominated delta.

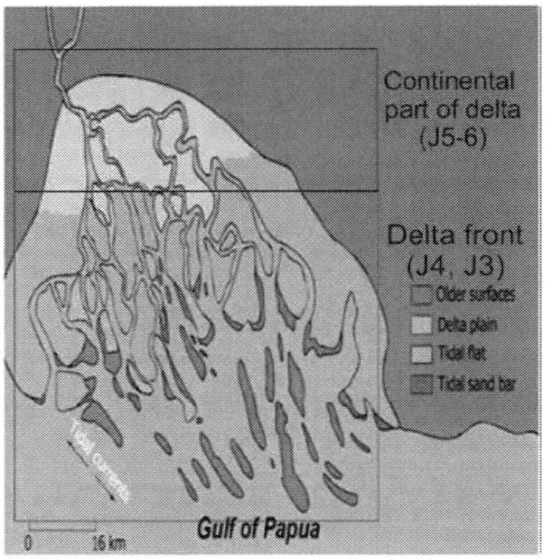

Don Boyd, personal communication, 2007

Figure 5. Tidal-dominated delta in Papua New Guinea (according to Boyd)

There is a shale barrier between J3 and J2 sand bodies that indicates a rise in the sea level. The J2 layer is represented by rare sand lenses within the shale background and is interpreted as tempestites (Lithotype 5) – sands reworked by the storm activity. The abundance of marine bioturbation including phycosiphon and paleophicus ichnofacies indicates that they are probably shallow marine tempestites. These sediments are usually characterized by the presence of hummocky cross stratification as the most distinctive feature, but since this structure is big-sized, it is difficult to recognize it within the core data [4]. These marine sediments indicate the beginning of a transition cycle.

In case of vertical facies distribution, the following features can be distinguished. There is a frequent variation of muddy flat, channel and interdistributary bay facies in the vertical direction. The channel facies has a significant thickness varying from 5 to 30 m due to the vertical accretion and amalgamation of the channels. Thin shale laminae within the channel facies are interpreted as a boundary between the channels, and these facies are usually interchanged by interdistributary bay facies or muddy flat facies that can indicate channel abandonment. The proportional quantity of interdistributary bay and muddy flat facies increases at the top of the J3 layer.

It is necessary to point out that there is shale barrier between the J3 and J2 sediments with a thickness varying from 0 to 30 m. This shale is recognized as marine and can be interpreted as the indicator of transgression beginning. With due account of the vertical trends that are recognized in the correlation panel, it can be concluded that the traces of continental and marine depositional conditions exist simultaneously and the main trend towards a change from a more continental to a more marine environment is distinguished.

Laterally, there are the following trends: within the J4 and J3 layers, channel facies and interdistributary bay facies are predominant. The orientation of the channel facies distribution is from the west-southwest to the east-northeast that corresponds to the main source area location. The J2 layer is characterized by the presence of rare sand lenses within a marine shale background (figure 6).

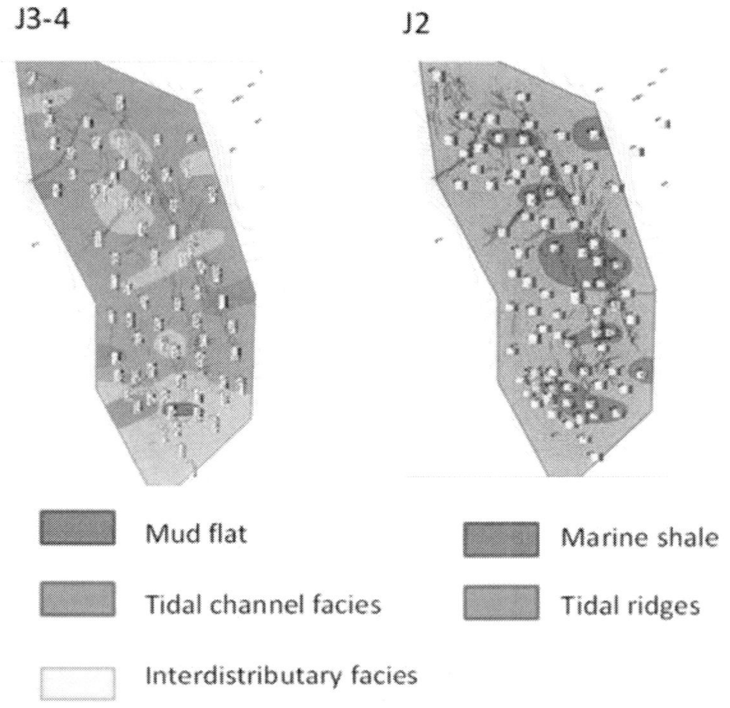

Figure 6. Lateral facies distribution within J4, J3 and J2

4. Determining an optimal drilling schedule

Gas coning is the critical issue for oil wells with a gas cap drive, therefore, the main problem for the cost-effective development of Field N is the high risk of early gas coning and selecting such a drilling schedule that takes it into account and reduces the risks of early gas invasion. Reservoir sand continuity, the presence and distribution of continuous and discontinuous barriers, the level of compartmentalization are referred to a distinctive geological feature that can reduce the risk of gas coning in some way, they are controlled by the depositional environment and facies distribution. Therefore, apart from the STOIIP quantity and reservoir quality, to choose the most optimal drilling schedule, it is necessary to take into account the geological features of the identified facies that can be a possible constraint to gas invasion and reduce the risk of coning.

As it has already been mentioned, three reservoir facies and two non-reservoir facies can be distinguished within the gas-bearing zone including the J3 and J2 layers.

J3 is characterized by the variation of tidal channels and interdistributary bay reservoir facies that can contain hydrocarbons. The non-reservoir facies is rare within J3. There is a tempestite facies represented by rare lenses within the J2 shale background.

From the petrophysical point of view, the facies identified have relatively similar petrophysical properties. There is a difference in the lithological composition between the reservoir facies of the J3 layer controlled by the conditions of a depositional environment. The channel facies is represented by sandstone with the absence or rare thin layers of carbonate sediments or siltstone indicating channel abandonment by the finer-grained material that can be reworked by the next stage channel activity, while the interdistributary facies is characterized by the high level of compartmentalization represented by the interbedding of sandstone, siltstone and mudstone accumulated during the slack periods of the tidal cycle. The big number of mudstone layers in the interdistributary bay facies can prevent or deter a vertical flow between reservoir compartments. Where these layers are discontinuous, they act as baffles for flow, and hence, the probability of early gas coning through this facies is lower vs. in case of the channel facies [6]. The net to gross ratio is the quantitative representation of shale presence. The channel facies is characterized by a range of NTG from 0.46 to 1, while the measured value of NTG for the interdistributary bay facies varies from 0.20 to 0.58.

It is necessary to point out another important geological feature of the Middle Jurassic sediments accumulation: there is an explicit continuous shale barrier between the J2 and J3 layers indicating the beginning of a transgression cycle and separating the tempestites of J2 from the reservoir facies of J3. In the case when gas is contained only in J2, this shale plays the role of a barrier that can prevent gas coning. The thickness of the shale barrier varies from 0 to 30 m within the area of the field. This shale is not a hydrodynamic barrier for J2 and J3 accumulation due to the connectivity between these layers where the thickness of the barrier is small and the shale is discontinuous.

In compliance with the identified geological features that can potentially reduce the risk of early gas invasion, a gas coning probability map can be created. When gas is only in the J2 layer, the shale barrier indicating a transgression cycle can prevent gas invasion from J2 to oil-bearing J3. Figure 7 demonstrates the thickness map of the shale barrier between the J2 and J3 layers. The critical value of the shale thickness above which the risk of early gas coning is minimal depends on such factors as drawdown pressure that can vary during field development. Thus, it is assumed that the critical value is equal to 5 m, this choice will be explained later.

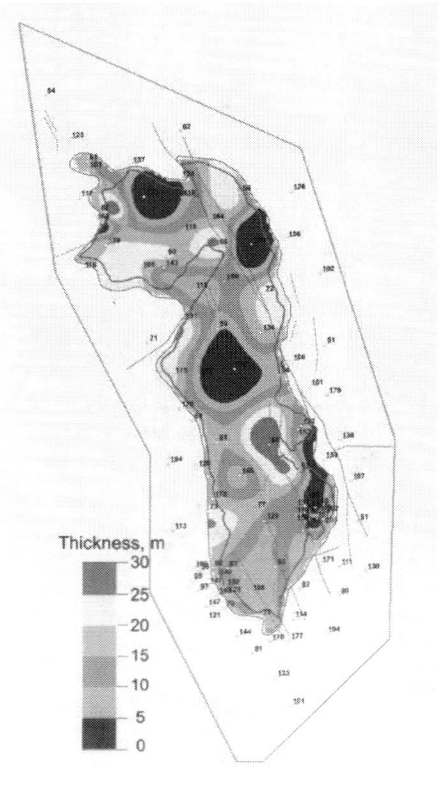

Figure 7. Shale barrier thickness map

When gas is contained in J2 and J3 (the predominant part of the field), the main factor influencing the risk of gas coning is the presence and distribution of discontinuous shale within the gas-saturated zone. To determine a zone with the minimal risk of gas coning, an NTG map above GOC is created (figure 8). Critical values for zonation of the NTG map according to the level of gas coning probability are chosen on the basis of the production test results for appraisal wells. In compliance with the production test results, it can be concluded that when gas is in the interdistributary bay facies characterized by the low value of NTG, gas rates from the perforated oil zone are minimal or absent (the low value of GOR near the gas solubility), while when gas is in the channel facies with high NTG, there are high gas rates and, consequently, the high value of GOR (up to 5,000). The integrated analysis of the production test results and NTG values allows determining the critical values of NTG for each facies. For the interdistributary bay facies, the maximum value at which there is no gas coning is equal to 0.40. For the channel facies, the critical value is equal to 0.53, which means that there is the high probability of gas coning above this value (figure 9). As for well 167 with NTG being 0.40, the thickness of shale is equal to 5 m, and it is enough for preventing early gas coning. Due to the absence of other data that could characterize the critical thickness of the shale barrier in the J2 layer, this value is assumed as a limit for a ranging area in accordance with the thickness of shale barrier. Figure 10 illustrates the ranging of the territory by the level of gas coning probability according to the determined critical value of NTG and the thickness of the shale barrier. Red color illustrates the areas where the risk of gas invasion is lowest, blue color – highest.

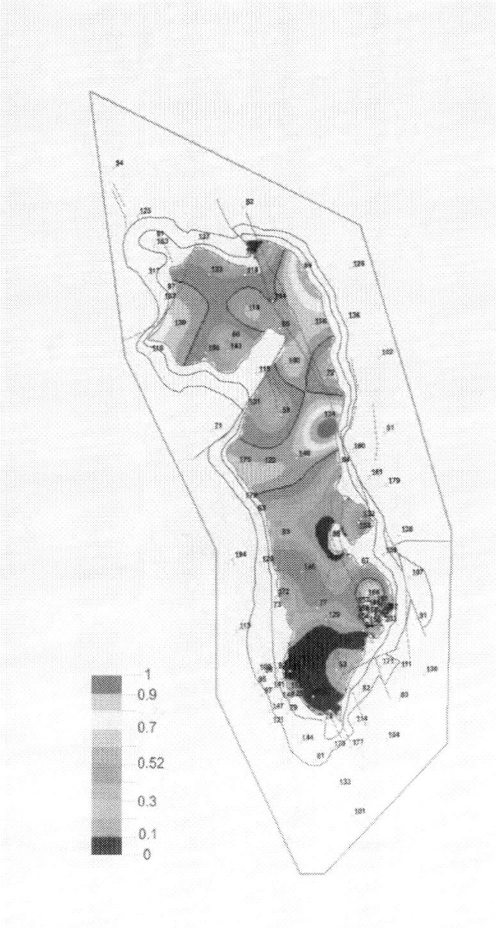

Figure 8. NTG map of J3 gas bearing zone

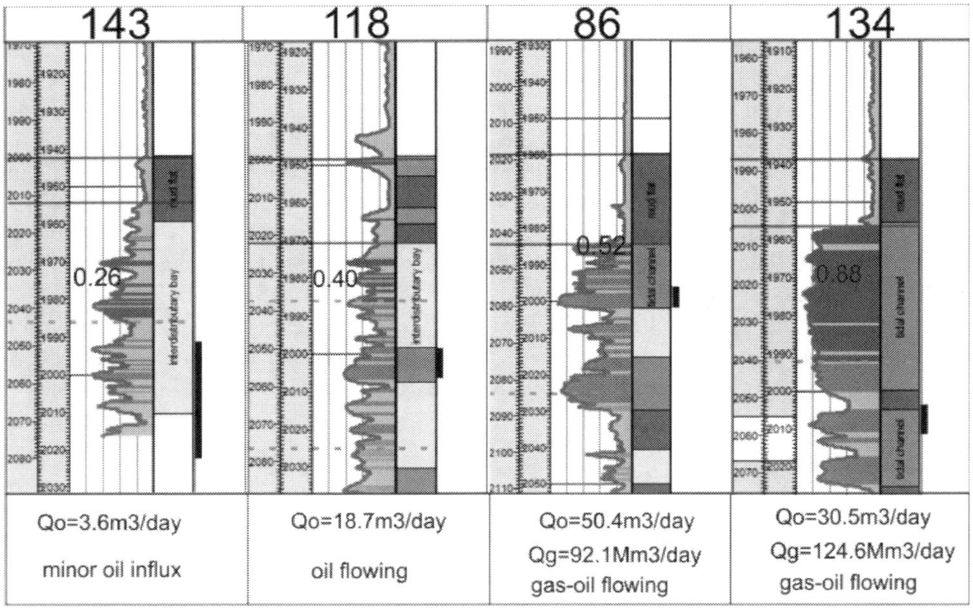

Figure 9. Results of production tests for appraisal wells

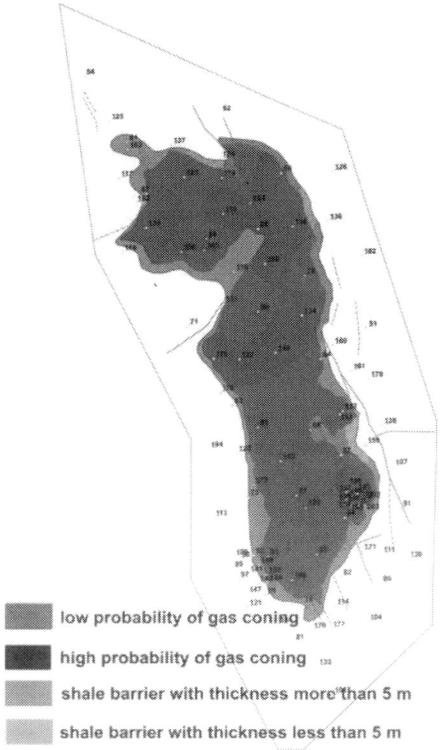

Figure 10. Gas coning probability map

Usually the most optimal drilling schedule is determined on the basis of kh maps that indicate the quantity of STOIIP and reservoir quality simultaneously. When it comes to Field N, one of the basic criteria for selecting a schedule is the quantity of both gas and oil. In compliance with that, a kh map for the oil-bearing and gas-bearing zone was reconstructed. The net gas thickness map was created on the basis of seismic (correlation with a dynamic attribute) and well data. The net oil thickness map was created on the basis of only well data. To determine permeability maps, porosity maps reconstructed with Gaussian simulation based on the variogram analysis results and poropermrelationship from the core data were used. Figure 11 demonstrates the mapping results for the oil-saturated and gas-saturated zones.

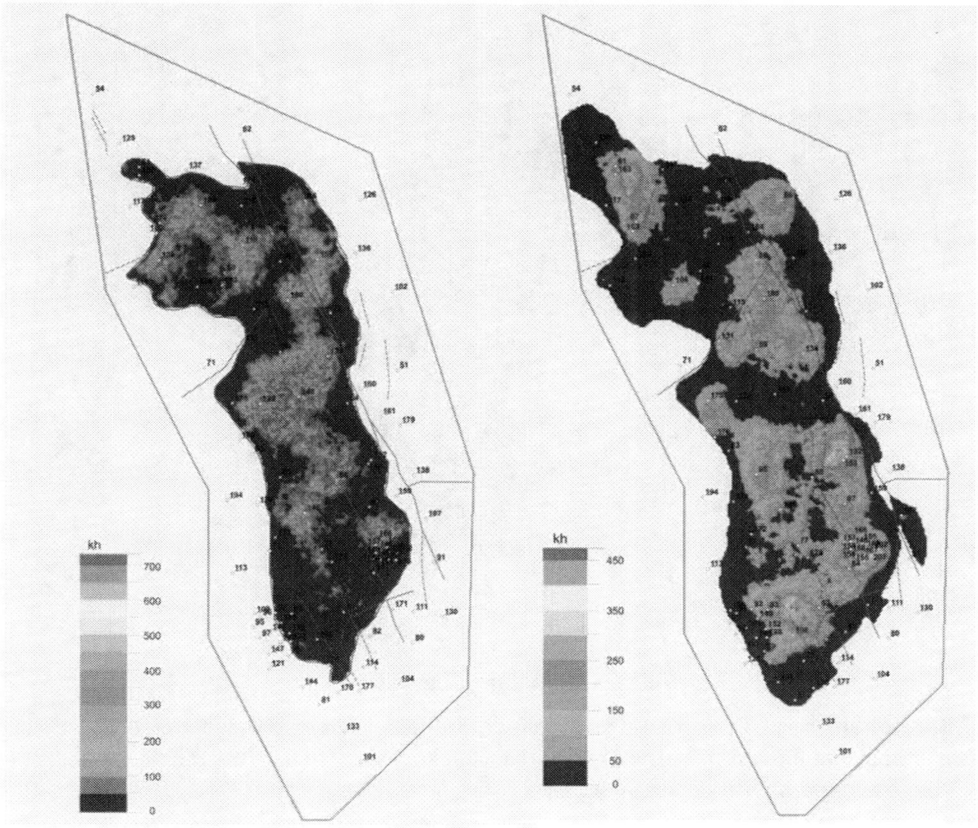

Figure 11. Kh map above GOC (left) and below GOC (right)

The complex analysis of the obtained data, including the kh maps of gas-bearing and oil-bearing zones, and the gas coning probability map, allows determining the most optimal drilling schedule that takes into account the qualitative characteristic of STOIIP indicated by the kh-values, gas cap thickness and the risks of gas invasion into a production well that considerably reduces oil production. As it can be seen from figure 12 that illustrates the results of mapping with an identified promising area for drilling, such a factor as the probability of gas coning plays the role of a controlling factor for taking the final decision. Conditionally, the territory of the field can be divided into three parts: the southern and northern parts being promising for oil production and the central part that is not promising due to the significant thickness of the gas cap and small effective oil thickness. The most promising territory for field development beginning is the southern part and Area 1 where the following factors can be recognized simultaneously: a gas cap is absent or has an insignificant thickness, there is a large anomaly on the kh map below GOC that is the evidence of the high value of STOIIP and improved reservoir quality, the probability of early gas coning is extremely low due to the domination of the interdistributary bay facies. Area 2 is characterized by the presence of a gas cap, but this gas is predominantly in the interdistributary bay facies that implies the low risk of gas coning. The least promising area within the southern part of the field is Area 3, despite the fact that there is sufficient STOIIP and the quantity of gas is the same as for Area 2. Because of the development of the channel facies within this area, and, consequently, the high probability of gas invasion, only the oil reservoir on the edge is promising, as there gas is in the J2 layer that is characterized by the presence of a shale barrier being thick enough to prevent gas invasion.

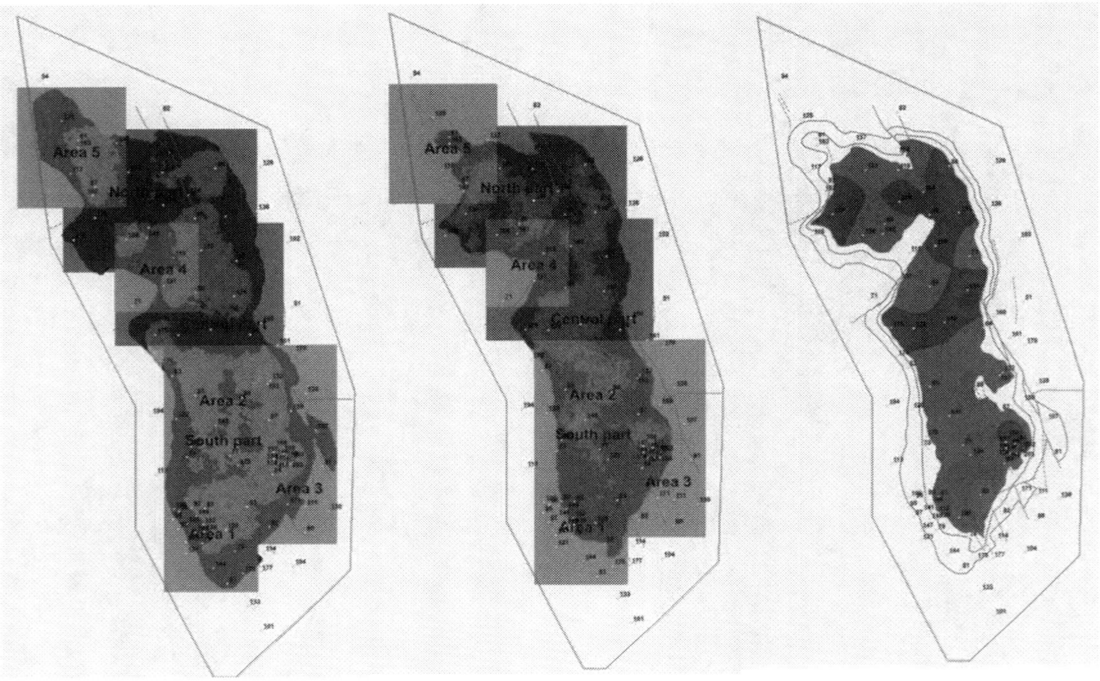

Figure 12. Optimal drilling schedule

Within the northern part of the field, there are two promising areas that were also chosen because of the most optimal combination of the smaller gas thickness, sufficient oil thickness and reservoir quality, and the lowest probability of gas coning.

4. Main uncertainties and recommendations

In compliance with the identified geological features that control the risk of gas coning, the following main uncertainties can be recognized.

Firstly, it is the critical value of NTG that depends on facies determination and production test results. The facies were determined manually and subject to the interpreter's opinion. Besides, there is no explicit discrimination between the values of NTG for a particular facies and the typical ranges are overlapped. The production tests results also imply uncertainty connected with the intervals of perforation. In addition, as it has already been mentioned, the largest value of NTG at which there is no gas production is equal to 0.4, while the lowest value of NTG for the channel facies at which gas invasion occurs is 0.52, so the range of 0.4 to 0.52 comprises values for which there are no production test results and it can indicate the presence or absence of gas coning with equal probability. In this study, the value of critical NTG equal to 0.52 was used as an optimistic scenario, as it is assumed that the channel facies cannot prevent gas invasion sufficiently and the range of NTG typical of this facies is 0.46-1, which does not include the pessimistic critical value of 0.40. Figure 13 illustrates a probability map with the critical value of 0.4. It should be noted that the most promising area (Area 1) does not change, while there is a reduction in the territory within Area 2. The territory of the area within the northern part of the field is also decreasing. As a recommendation for uncertainty reduction, it is necessary to perform production tests for newly drilled wells in order to determine the critical value of NTG more precisely.

The second uncertainty is shale barrier thickness. The change in the critical value by 10% does not have a significant impact and there is only a small area with the high risk of gas coning. Basically, the critical value of thickness depends on certain conditions of field development, including the value of drawdown pressure. With due account for this fact, it is recommended to find a correlation between the critical barrier thickness at which gas invasion is prevented and the drawdown pressure.

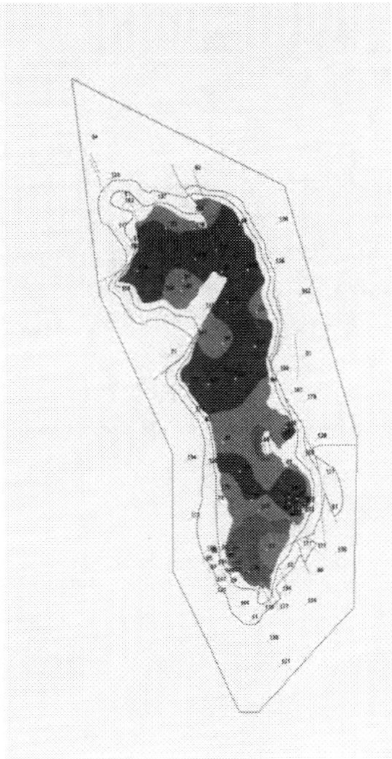

Figure 13. Probability map of gas coning with the critical value of NTG=0.4

4. Conclusions

To determine an optimal drilling schedule, it is necessary to take the risk of early gas coning into account. The classical usage of the kh map for determining an optimal drilling procedure cannot predict the places where the gas coning probability is minimal to the full extent. Thus, the recognition of a depositional environment in order to distinguish geological features that can significantly reduce the risk of gas invasion was carried out. The J4 sediments saturated with predominantly oil and the J3 sediments containing oil and gas are interpreted as tidal-dominated delta sediments and the reservoir is represented by two facies: channel facies and interdistributary bay facies. J2 is represented by the rare sand lenses of tempestites within a marine shale background.

The presence of discontinuous shale in the interdistributary bay facies characterized by the low value of NTG (0.2-0.58) can prevent or inhibit early gas coning. Also, there is a shale barrier between the J3 and J2 layers. The results of the production tests for appraisal wells confirm the conclusion that the interdistributary bay is able to prevent the risk of early gas coning. The critical value of NTG that divides the territory of the field into areas with the high and low probability of gas coning was determined on the basis of the production test results and data on the NTG value for the facies. It varies from 0.4 to 0.52.

To determine the most optimal drilling schedule, the kh maps and the gas coning probability map were used. This integrated analysis allows determining that Area 1 is most appropriate for field development beginning, as it is simultaneously characterized by the high quantity of STOIIP and the best reservoir quality, a small gas cap and the low probability of early gas coning.

An uncertainty analysis was also performed. There are two main uncertainties connected with the critical values of NTG and the shale barrier thickness that are used for probability map reconstruction. This map is very sensitive to the critical value of NTG. There is a sufficient reduction in the territory where the risk of gas coning is minimal based on the probability map with the critical value of NTG being equal to 0.40. It is recommended to perform production tests for the newly drilled wells with the purpose of determining the critical value of NTG more precisely.

References

[1] Coe Angela L 2002 *The Sedimentary Record of Sea-Level Change* Cambridge University Press
[2] Muromtsev V S 1984 *Electrometrical Geology of Sand Bodies in Lithological Traps of Oil and Gas* Nedra 259
[3] Nichols Gary 1999 *Sedimentology and Stratigraphy* Blackwell Publishing
[4] Pemberton S George et al 1992 *Ichnofacies of a Wave-Dominated Shoreline* SEPM
[5] Pemberton S George et al 1992 *Ichnology and Event Stratigraphy: the Use of Trace Fossils in Recognizing Tempestites* SEPM
[6] Tillman R W 1994 Barriers to Flow in Tidal Sandstone Reservoirs, Sun Ranch Field, Wyoming *SPE*
[7] Walker R G and James N P 1992 *Facies Models: Response to Sea Level Change* Geological Association of Canada

IOP Conf. Series: Earth and Environmental Science **24** (2015) 012050 doi:10.1088/1755-1315/24/1/012050

Bazhen Fm matured reservoir evaluation (West Siberia, Russia)

S Parnachev[1], A Skripkin[2], V Baranov[3] and S Zakharov[4]
[1] Associate Professor, Tomsk Polytechnic University, Tomsk, Russia
[2] Senior Researcher, Tomsk Polytechnic University, Tomsk, Russia
[3] Chief of Laboratory of Oil&Gas Geology, Tomsk Polytechnic University, Tomsk, Russia
[4] Chief Geologist, Tomskneft Oil Company, Strezhevoy, Russia

E-mail: parnachevsv@hw.tpu.ru

Abstract. The depletion of the traditional sources of hydrocarbons leads to the situation when the biggest players of the oil and gas production market turn to unconventional reserves. Commercial shale oil and gas production levels in the USA have largely determined world prospects for oil and gas industry development. Russia takes one of the leading place in the world in terms of shale oil resources. The main source rock of the West Siberia, the biggest oil and gas basin in Russia under development, the Bazhen Fm and its stratigraphic and lithologic analogs, is located in the territory of over 1,000,000 square kilometers. Provided it has similar key properties (organic carbon content, porosity, permeability) with the deposits of the Bakken Fm and Green River Fm, USA, it is still extremely poorly described with laboratory methods. We have performed the laboratory analysis of core samples from a well drilled in Bazhen Fm deposits with matured organic matter (T_{max}>435 ^{0}C). It was demonstrated the applicability of the improved steady-state gas flow method to evaluate the permeability of nanopermeable rocks. The role of natural fracturing in forming voids was determided that allows regarding potential Bazhen Fm reservoirs as systems with dual porosity and dual permeability.

1. Introduction

The size of the West Siberian Basin, the largest oil and gas basin developed in Russia, is mainly controlled by the deposits of the Bazhen Fm and the Abalak Fm underlying it (as well as their stratigraphic and lithologic analogs) being the main source rocks of the region. The existing examples of commercial oil development in Bazhen Fm reservoirs – despite the more than 55-year history of studies for its oil-and-gas content (Galyanovskoe, Em-Egovskoe, Maslikhinskoe, Pravdinskoe, Salymskoe, Sredne-Nazymskoe and other fields) – are inevitably accompanied by the shortage of credible laboratory studies for petrophysical properties. The main reason for the latter is the inadaptability of the existing equipment and methods of studying nanopermeable Bazhen deposits. With the prevalent argillaceous-siliceous and argillaceous-carbonate composition, the thickness of the Tithonian-Berriasian (some 150 million years ago) deposits in the Bazhen Fm varies from first meters on the periphery of the Bazhen paleo-sea up to 60 meters at its depocenters. The total organic carbon level in these deposits reaches 25 % [1] with the mean values being about 12 %. The studies of the Bazhen reservoir quality have historically been localized by its productivity areas where oil influxes were associated with abnormally high formation pressures with natural hydraulic fracturing, which also leads to the development of "bazhenit", i.e. foliated shalestones that virtually fall to pieces in hands and do not allow using routine porosity and permeability evaluation methods. The pore volume of such rocks could be determined with the liquid saturation method with specific restrictions, but ony well operation results, such as liquid flow rates or the calculated Kh value, were (and are still) used as a proxy indicator to evaluate permeability.

The target of the research was core samles from the wells drilled in Tomsk Region in an area characterized by the significant thermal heating on the Bazhen Fm deposits established by the value of the T_{max} temperature being somewhat over 435 ^{0}C. In this study, we used about 20 meters of core (approximately 60 % of the thickness of the Bazhen Fm deposits in the area) being 80 meters in diameter from the depth of about 3 kilometers below sea level, which allowed to sample cylindrical core plugs 30 millimeters in diameter parallel to the bedding. The core from this well was mainly composed by silty silicites with clay matter (on average, 11 %) with the prevalence of illite in the latter.

2. Research methodology

The study used standard cylindrical core samples being 30 millimeters in diameter and also crushed samples being 2-5 millimeters in size. The application of crushed core to study shale/tight gas/oil reservoirs allowed significantly reducing the duration of preparatory procedures (extraction) and the study process itself down to practically acceptable values by increasing the specific surface of the sample [2]. Another benefit of the method using crushed core samples is the possibility to measure the permeability and porosity of matrix matter by excluding the impact of the natural fracturing of shaly rocks enhanced by the relief of lithostatic pressure during core sampling and its withdrawal to the surface. The samples (both cylindrical and crushed) were exposed to extraction with a spirits-benzol blend in a centrifugal extractor for 48 hours and subsequent drying at 105 ^{0}C for 16 hours. The samples were saturated with kerosene in two stages, including their vacuumization and resaturation under the pressure of 150 bars for 3+ days (this duration was determined experimentally).

The value of helium permeability in crushed samples was studied in normal conditions with a pressure decay method [3], where the curve of pressure drawdown in an operating chamber was approximated with a standard function the parameters of which were used to calculate permeability. The insignificant change in pressure during the test (on average, 0.5-2.0 % of the absolute value) places higher demands on instrument calibration, absence of leaks and thermal stabilization. The cycle of measurements with one sample is considered correct, if temperature changes during the test do not exceed 0.05 ^{o}C. The resulting experimental pressure drawdown curve is the superposition of pressure drawdown curves for grains with varying sizes and permeability values. The comparison of results provided by different installations can demonstrate a difference in permeability values by several orders. Another obvious drawback of the GRI (Gas Research Instutute, USA) method is measurements in atmospheric conditions.

It is possible to overcome the above listed drawbacks by using the classical method of measuring gas permeability in reservoir thermobaric conditions with geometrized samples. Gas consumption stabilization during the use of standard core samples that are 45 millimeters high takes one week to several months, which makes such measurements technically unfeasible. Therefore we used samples being 30 millimeters in diameter and with a reduced height (5-20 millimeters vs. standard 45 millimeters), which allowed decreasing test duration down to 3-5 days. The pressure set on entry to the coreholder was 100-150 bars higher vs. reservoir pressure, and during the experiment the pressure was kept constant. The units of the filtration installation were thermostated during the experiment (figure 1).

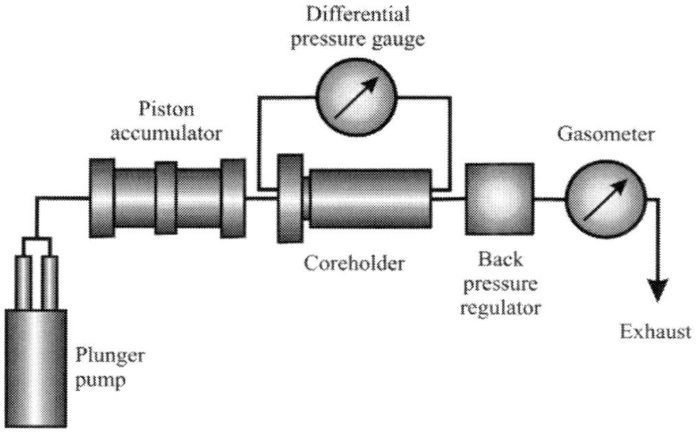

Figure 1. Hydraulic installation scheme for measuring gas permeability in cylindrical samples in reservoir conditions under steady-state flow.

Gas is supplied to the coreholder from a vessel with a dividing piston, outlet pressure is maintained with a back pressure regulator. The volume of gas supplied to the system is registered on the basis of a pump piston position and recorded into a log file. In a steady-state (established on the basis of stabilization of differential pressure gauge indications), the consumption is compared using a gasometer. To determine the flow pattern and possibility to calculate permeability with the Darcy law, we checked the dependence of the steady pressure differential on a sample vs. gas filtration rate under constant effective pressure (figure 2). The test was performed under changes in pressure on entry to the coreholder.

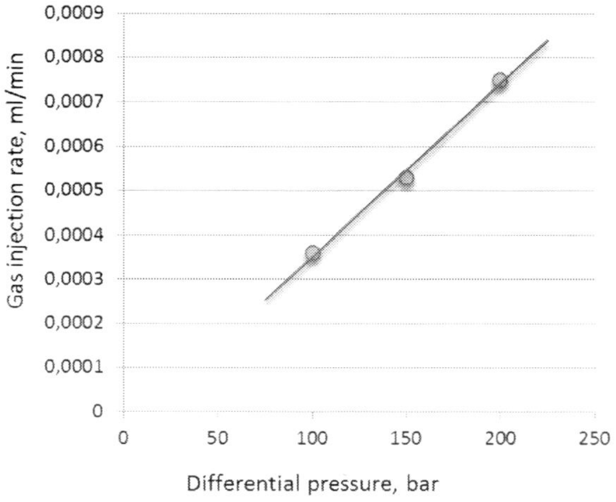

Figure 2. Steady gas consumption dynamics under differential gas pressure changes during tests.

The consumption of gas measured with the gasometer is corrected with due account for the difference in pressures/temperatures and compared with the indicators of the plunger pump. The difference in consumptions allows determining the value of a leak. For a steady-state filtration area, the values are approximated with a straight-line correlation the slope of which characterizes the established consumption through the sample under test (figure 3).

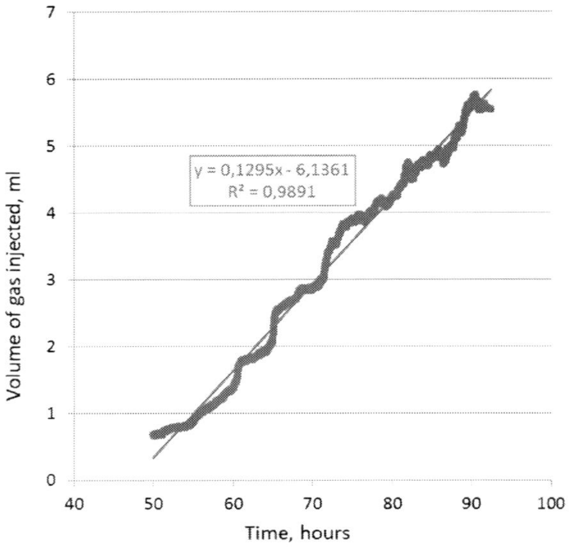

Figure 3. Dynamics of gas volume pumped through the sample.

The porosity of cylindrical samples and crushed samples was determined with the liquid saturation method (kerosene) with sequential weighting.

3. Research findings

The values of gas permeability measured in reservoir conditions in cylindrical samples significantly exceed the values of gas permeability in crushed samples determined with the GRI method (figure 4). At the same time, the GRI permeability corresponding to the permeability of a rock matrix without microfractures demonstrates relative constancy and varies within $0.051 \cdot 10^{-9}$ mkm^2 to $0.208 \cdot 10^{-9}$ mkm. The permeability of cylindrical samples varies within an extremely wide range, which apparently characterizes the properties of the fracture space not collapsing completely even when overburden pressure is exerted [4]. The study of the overwhelming majority of the petrographic thin sections oriented perpendicularly the bedding demonstrated the presence of discontinuous and continuous fractures of different morphology with an aperture making 0.01 to 0.40 mm (figure 5).

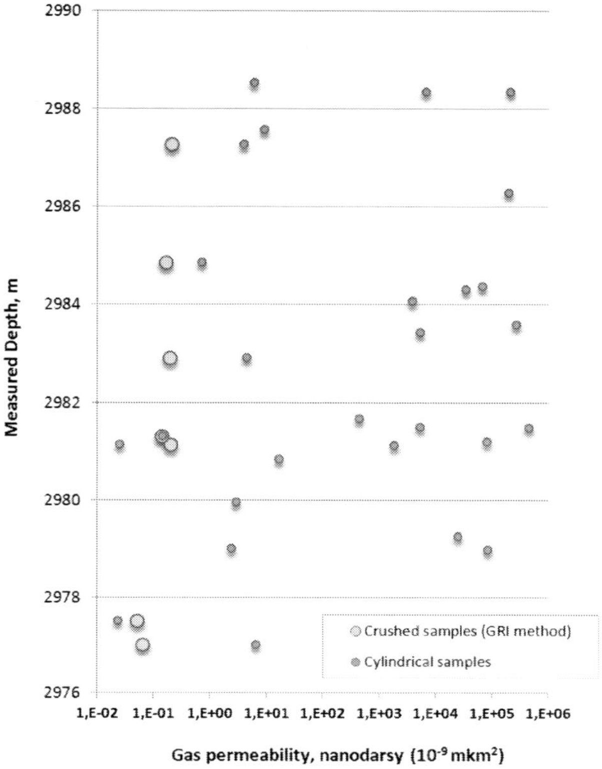

Figure 4. Distribution of gas permeability of core samples measured with different methods.

The values of effective porosity measured in crushed and cylindrical samples demonstrate analogous correlations (Figure 6), which is also regarded as the comparative characteristic of different types (pore and fracture) of voids.

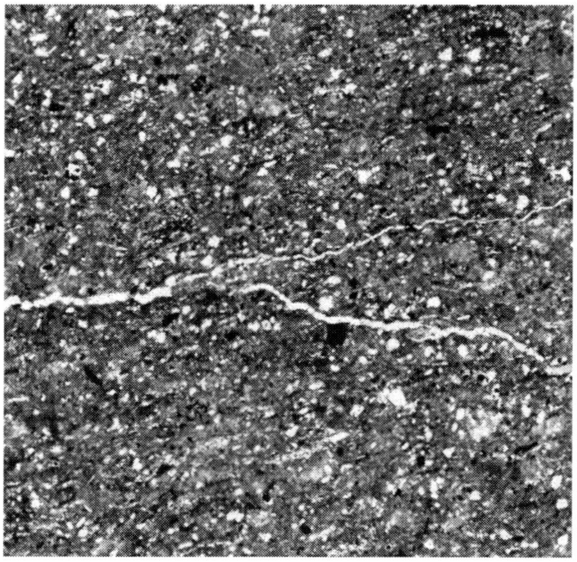

Figure 5. Microphoto of a petrographic thin section (field-of-view width – 1.74 mm). Pyritized silty silicite enriched with organic matter. The aperture of the fractures developed both in parallel to and across the foliation surface equals 0.015 mm.

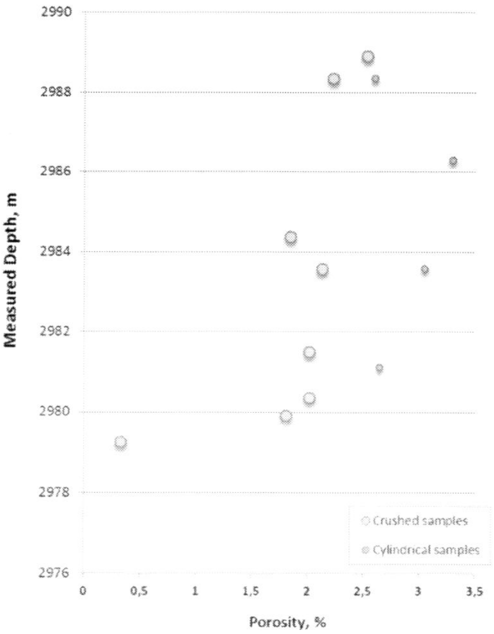

Figure 6. Distribution of kerosene porosity measured in different samples.

4. Conclusion

The present study has revealed the possibility to measure permeability in nanopermeable samples in reservoir conditions under steady-state gas filtration. For the overall section, the permeability of cylindrical samples exceeds matrix permeability by the factor of $10^1 \div 10^6$, which is explained by the presence of fractures that are abundantly diagnosed in petrographic thin sections and visible to unaided eye in core plugs. The reservoir properties (and especially permeability) of the Bazhen Fm deposits is largely determined by the presence and geometry of natural fracturing. The study of this parameter that is apparently an effective pressure function should be performed under different values of overburden pressure with the steady-state gas or liquid filtration method. The research findings confirm observations about the productivity of the fractured and cavernous fractured reservoir at the Bazhen and Abalak Fm that have frequently been mentioned earlier.

References

[1] Dakhnova M V 2009 Geochemical Studies of Core and Oils from Galyanovskaya and Sredne-Nazymskaya Producing Areas in West Siberia VNIGNI Geochemical Center
[2] Luffel D L and Guidry F K New Core Analysis Methods for Measuring Reservoir Rock Properties of Devonian Shales SPE20751
[3] Luffel D L, Hopkins C W and Schettler Jr P D Matrix Permeability Measurements of Gas Productive Shales SPE26633
[4] Handwerger D A, Suarez-Rivera R, Vaughn K I and Keller J F Impoved Petrophysical Core Measurements on Tight Shale Reservoir Using Retort and Crushed Samples SPE147456.

AUTHOR INDEX

Abramova, R. 65, 73, 100, 176
Aliev, F. 81, 89
Ananyeva, L. 95
Antropova, N. 106
Baranov, V. 202, 208, 357
Baranova, A. 81, 106, 122
Barkalova, A. 116
Bashirov, A. 195
Belozerov, B. 330, 343
Belozerov, V. 195, 202, 208, 291, 320
Biletskiy, M. 146
Bochkov, A. 343
Bolsunovskaya, L. 12
Bolsunovskaya, Y. 12
Bondarchuk, I. 17
Boyarko, G. 165
Boyko, I. 38, 43, 49
Brilin, V. 21, 26
Brylin, V. 31
Bubnov, A. 38, 43, 49
Bukhanov, N. 195, 320
Chekanstev, V. 65
Cherdansteva, D. 100
Cherkasova, T. 187
Demidova, E. 215
Dmitriev, A. 202, 208
Dolgaya, T. 116, 127
Evdokimov, A. 195
Ezhova, A. 65, 73
Fadeeva, S. 239
Gavrilov, R. 54
Ginsar, V. 95
Gojkovic, S. 275, 283
Goncharov, I. 229, 239
Gorbenko, M. 81, 89
Isaev, V. 1
Isaev, Y. 81, 89
Ivanova, Y. 152
Kasenov, A. 146
Khorev, V. 38, 43, 49
Kondratyev, A. 251
Konoshonkin, D. 264
Korotchenko, T. 31, 146
Korovin, M. 275, 283
Korovkin, M. 95
Kovalyov, A. 81, 89
Krasnoshchekova, L. 100
Krets, V. 106
Kucherenko, I. 133
Kuntsevich, V. 291
Kuzmin, S. 306
Lobova, G. 1
Lomakina, N. 320
Lukyanov, V. 31, 106, 192
Maksyutin, K. 251
Maksyutina, O. 215
Malyshevskaya, K. 330
Maslovsky, A. 192
Matveenko, I. 152
Mazurov, A. 158
Merkulov, V. 65, 275, 283
Mishina, D. 343
Nedolivko, N. 116, 127
Nevidimova, O. 109
Oblasov, N. 229, 239
Osipova, E. 1
Oskina, Y. 141
Ostankov, A. 73
Panova, E. 73, 176
Parnachev, S. 264, 357
Pasko, O. 122
Perevertailo, T. 116, 127
Podnebesnikh, A. 330
Podobina, V. 60
Pogharnitskaya, O. 182
Pshenichkin, A. 141
Ratov, B. 146
Razva, O. 95
Reshetko, M. 152
Ruban, A. 158
Rudmin, M. 54, 158
Rukavishnikov, V. 251, 306, 330, 343
Ryabchikov, S. 81, 89
Samoilenko, V. 239
Savichev, O. 152
Savinova, O. 187
Shakirzyanov, L. 251
Shenderova, I. 17, 165
Shmurygin, V. 192
Shubina, E. 31
Sizov, A. 165
Skripkin, A. 357
Slivkin, S. 291
Strelnikova, A. 89, 182
Strokova, L. 170
Sukovatyi, V. 306
Ten, T. 176
Terre, D. 1
Timkin, T. 141, 187
Tretyakov, K. 165
Tsibulnikova, M. 182
Ulyanova, O. 21, 26
Yankovich, E. 109
Yankovich, K. 109
Yu, A. 7
Yuxuan, Z. 133
Zakharov, S. 357